Characterisation of Porous Solids VIII
Proceedings of the 8th International Symposium on the Characterisation of Porous Solids

Characterisation of Porous Solids VIII

Proceedings of the 8th International Symposium on the Characterisation of Porous Solids

Edited by

Stefan Kaskel
Institut fur Anorganische Chemie, Dresden, Germany

Philip Llewellyn
CNRS - Universite de Provence, Marseille, France

Francisco Rodriguez-Reinoso
Universidad de Alicante, Alicante , Spain

Nigel A. Seaton
University of Edinburgh, Edinburgh, UK

RSCPublishing

The proceedings of the 8th International Symposium on the Characterisation of Porous Solids held at the University of Edinburgh, UK on 10–13 June 2008.

Special Publication No. 318

ISBN: 978-1-84755-904-3

A catalogue record for this book is available from the British Library

Published by The Royal Society of Chemistry,
Thomas Graham House, Science Park, Milton Road,
Cambridge CB4 0WF, UK

Registered Charity Number 207890

For further information see our web site at www.rsc.org

Preface

The International Symposia on the Characterisation of Porous Solids began in 1987, in Bad Soden in Germany, and a conference in this series is held every three years. This book is the Proceedings of "COPS VIII", the eighth in this series. This is the second COPS in the UK, following COPS IV in Bath in 1996, and the first "north of the border" in Scotland.

I would like to thank my colleagues on the Scientific Committee – Stefan Kaskel, Philip Llewellyn and Francisco Rodríguez-Reinoso – for their support and particularly for their contribution to the scientific programme and to the editing of this volume. I am particularly grateful to Philip, who was Chairman of COPS VII in Aix-en-Provence, for his advice on "the art of COPS-conference-organising". I would also like to thank my colleagues in the Organising Committee – Mark Biggs, Tina Düren and Lev Sarkisov – who contributed greatly to all stages of our project, from the original idea to run the conference in Edinburgh, to helping to make sure things happen as they should at the conference itself.

Finally, my thanks go to Arlene Sievwright, of the Office of Lifelong Learning at the University of Edinburgh, who masterminded the planning and administration of the conference.

On behalf of the conference organisers, and all the participants, I gratefully acknowledge financial support from Micromeritics, Quantachrome, BEL, Rubotherm, Hiden Isochema, the Institution of Chemical Engineers and the Royal Society of Chemistry.

Nigel Seaton
Chairman, COPS VIII Organising Committee

Contents

PORE ACCESSIBILITY IN NANOPOROUS CARBONS: EXPERIMENT, THEORY AND SIMULATION

S. K. Bhatia and T. X. Nguyen

Department of Chemical Engineering, The University of Queensland, QLD 4072, Australia

1 INTRODUCTION

The understanding of pore accessibility in adsorbents is important to adsorbent design for gas mixture separation and storage. It is also a long-standing issue related to experimentally-observed complex adsorption behaviour, such as enhancement of adsorbed quantity with increasing temperature as well as the phenomenon of open loop hysteresis. The accessibility is not simply a result of the difference between the size of the adsorbate molecule and that of pores in the adsorbent, *i.e.* the result of a size exclusion mechanism, but more intricately influenced by the kinetic energy of the adsorbate molecule. The dependence of accessibility on the kinetic energy also leads to its strong temperature dependence.

Despite the considerable effort directed at tackling the issue of accessibility in porous carbons using percolation theory, with the underlying assumption of a size exclusion mechanism, little effort has been made at comprehensive investigation of this phenomenon based on atomistic modelling. The latter enables one to capture the dynamic nature of the pore accessibility, as well as to explain several complex adsorption behaviours in porous carbons, which may provide understanding crucial to optimal adsorbent design for gas mixture separation and storage. Accordingly, in this article we present an overview of our recent atomistic level studies on determination of the temperature dependent accessibility of simple gases (Ar, N_2, CH_4, CO_2) in a Hybrid Reverse Monte Carlo (HRMC) model of saccharose char CS1000a constructed using our proposed approach[1]. Subsequently, we also present the validation of our calculated pore connectivity results against experimental adsorption data.

2 METHOD AND RESULTS

2.1 Determination of Pore Accessibility

In this section, we review two recent methods for determination of pore accessibility in porous carbonaceous materials.

2.1.1 Percolation model. The first method is based on percolation theory modelling[2]. In this model, the accessibility is assumed to be temperature independent, and pore network connectivity is simply characterized by a mean coordination number, Z, which represents the mean number of pores connected at an intersection. For any adsorbate of molecular size d_c the accessible pore volume is then obtained as

$$V_{acc} = \frac{\Omega^a(Z,\Omega)}{\Omega} V_{tot} \int_{d_c}^{\infty} f(H)dH$$

(1)

where V_{tot} is the actual total pore volume, $f(H)$ is the pore volume distribution, Ω is the number fraction of available pores, given by

$$\Omega = \int_{d_c}^{\infty} \frac{f(H)}{H} dH \Big/ \int_{0}^{\infty} \frac{f(H)}{H} dH$$

(2)

and $\Omega^a(Z,\Omega)$ is the fraction of accessible pores. This fraction is given by percolation theory, based on an assumed network model. For a random network of uniform coordination number an expression for the function $\Omega^a(Z,\Omega)$ has been provided by Lopez-Ramon et. al.[2], and utilised by Ismadji and Bhatia[3,4] to study accessibility of ester molecules in carbons. Although the percolation model with the assumption of temperature-independent accessibility does not reflect the experimentally observed kinetic nature of accessibility and its temperature dependence, it may approximate well the behaviour for large molecules such as esters[3]. This is seen in Figure 1 depicting Ideal Adsorbed Solution

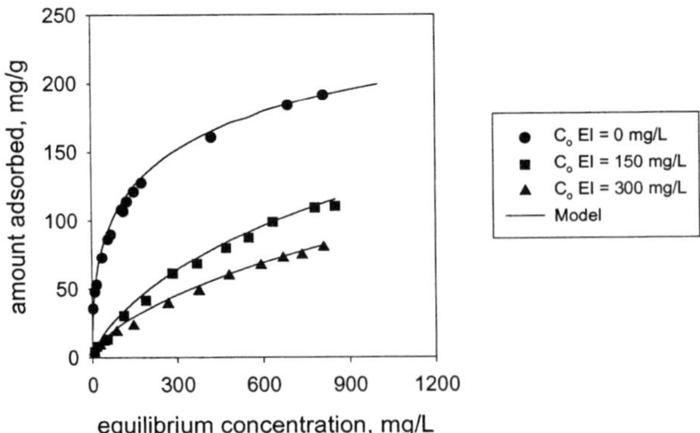

Figure 1 *Prediction of adsorption isotherms of ethyl propionate in the presence of ethyl isovalerate in activated carbon F-400 at 313 K using percolation model with coordination number Z = 2.98.*[4]

Theory based as well as experimental ester mixture isotherms at 313 K on Filtrasorb F-400 activated carbon[4]. The predictions are based on the fit of pure component isotherms with the coordination number, Z, used as a fitting parameter. The subsequent mixture predictions are parameter free, and consider the different zones that are inaccessible to both components or accessible only to the smaller molecule, and the zone accessible to both species. The excellent correspondence with the experimental data provides good support for the approach. On the other hand when the accessibility was assumed to be unity in the pure component isotherm fits, the mixture predictions were less satisfactory[4]

The above success of the percolation theory is due to the fact that in practice observation of the kinetic nature of the accessibility is very much dependent upon the difference between the experimental time scale and that of activated diffusion of adsorbate molecules in the porous solid. The latter is dictated by the ratio of activation energy to kinetic energy, i.e. E_a/k_BT. For large adsorbate molecules, having strong interaction with the adsorbent, the activation energy could be significantly enhanced compared to the maximum kinetic energy in the experimental temperature range. This leads to temperature-independent accessibility, as pores with constricted entries remain inaccessible over a wide temperature range. In contrast, for small adsorbate molecules with weak interactions the time scale of activated diffusion through the pore mouth is very sensitive to temperature, leading to strongly temperature dependent accessibility. Accordingly, atomistic structural modelling of porous carbon is essential for capturing the kinetic feature of the accessibility.

2.1.2 Atomistic model. In this subsection, we briefly present our recently proposed method to determine pore accessibility using an atomistic structural model[1]. In general, our approach is based on the analysis of continuity of a close packed adsorbed phase in the adsorbent, comprising three main steps: (1) Filling all pore spaces of the solid structure with close packed adsorbate using grand canonical Monte Carlo (GCMC) simulation; (2) identification of all adsorbate clusters; (3) determination of pore accessibility based on continuity of each cluster throughout the structure. The approach has been successfully applied to the determination of accessibility of Ar, N_2, CH_4 and CO_2 over a wide range of temperature, with further analysis of its kinetic features using transition state theory (TST)[1]. According to TST, the mean crossing time between cages A and B is given as

$$\tau_{A \to B} = \frac{1}{k_{A \to B}} \tag{3}$$

where $k_{A \to B}$ is a rate constant, given as[1]

$$k_{A \to B} = \kappa \sqrt{\frac{k_B T}{2\pi m}} \frac{\int_{DS} e^{-\beta \phi_{sf}(\mathbf{r})} d^2\mathbf{r}}{\int_{V_{cageA}} e^{-\beta \phi_{sf}(\mathbf{r})} d^3\mathbf{r}} \tag{2}$$

where the integral in the numerator is taken over the dividing surface of maximum potential energy, between cages A and B. Here κ is a transmission coefficient shown to be nearly unity[1], k_B is the Boltzmann constant, T is temperature and m is the mass of the particle. ϕ_{sf} is the interaction potential between the adsorbate particle i at position $\mathbf{r}$ and all solid atoms of the adsorbent phase, given as

$$\phi_{sf}(\mathbf{r}) = \sum_{j=1} u(|\mathbf{r}_j - \mathbf{r}|) \tag{3}$$

where u is the solid-fluid pair potential, taken to be the 12-6 Lennard-Jones potential. As an example Figure 2a depicts the temperature dependence of the crossing time of Ar and N_2 between cages A and B in the atomistic model of saccharose char, shown in Figure 2b.

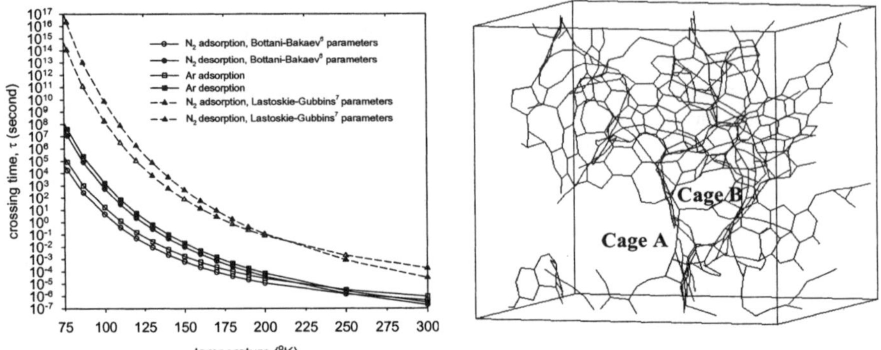

Figure 2 (*a*) *Variation of the crossing time of Ar and N_2 between cages A and B with temperature, estimated using transition state theory[1]; and (b) HRMC constructed model of saccharose char CS1000a illustrating detected open cage A and closed cage B.*

2.2 Impact of Pore Accessibility

2.2.1 Adsorption Equilibrium. Pore accessibility has a major impact on adsorption equilibrium. In particular, from Figure 2a, it can be predicted that Ar and N_2 may have a pore accessibility problem at low temperatures (<100 K) due to extremely long crossing times (τ > 100 seconds), but not at higher temperatures (>150 K) where the crossing time is much shorter (< 10^{-2} second). The former prevents the attainment of equilibrium, while the latter enables the adsorption system to instantly reach equilibrium at experimental time scales. This is consistent with our recent experimental observation[5] of an increase in adsorbed amount of argon with increase in temperature in coals, depicted in Figure 3. The increase in adsorption capacity with temperature, evident in Figure 3, is due to the increase in kinetic energy, and the resulting higher accessibility, with increase in temperature.

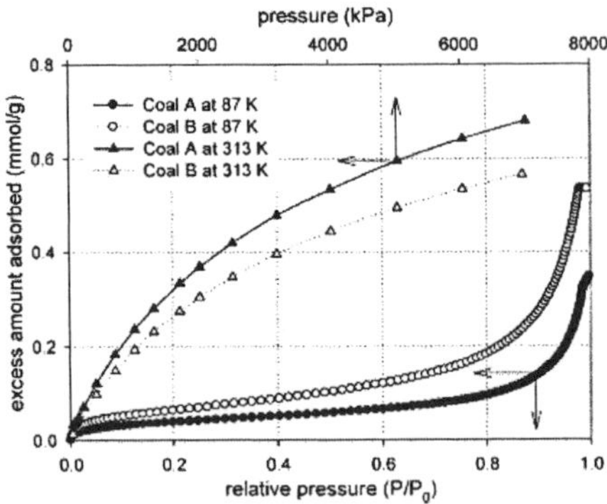

Figure 3 *Adsorption isotherms of Ar at 87 K and 313 K in coals[5].*

Figure 2a shows that the estimated crossing time, or adsorption time, of N_2 at 77 K from cage A to B is extremely large (about 5 hrs to ~ 3.7 million years), for both choices of LJ parameters for N_2[7,8],while that of argon at 87 K is of the order of minutes. This suggests an accessibility problem for N_2 at 77 K, but not for Ar 87 K. This is supported by the results shown in Figure 4, where experimental adsorption data for CO_2 in BPL carbon at temperatures of 273 K and 323 K[8] (triangles and squares respectively) are correctly predicted by the use of structural parameters derived from experimental Ar adsorption data at 87 K (solid and dashed lines) using our finite wall thickness based density functional theory model[9], but significantly underpredicted by the use of parameters derived from experimental N_2 adsorption data at 77 K (dotted and dashed dotted lines

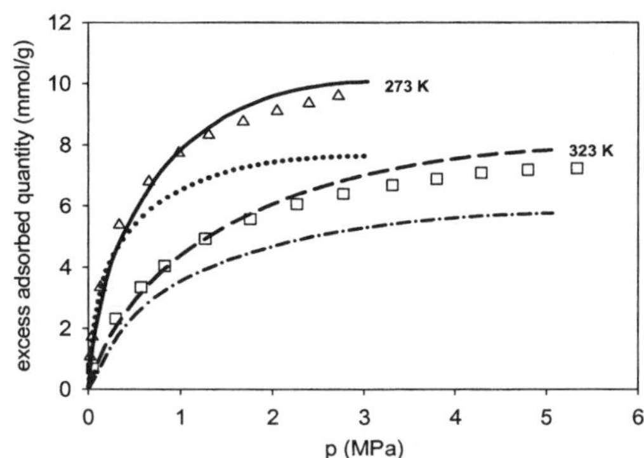

Figure 4 *Comparison between theoretical predictions and corresponding experimental adsorption data[8] for CO_2 in BPL carbon at 273 K and 323 K.*

2.2.2 Open Loop Hysteresis. For long, hysteresis in porous materials has been of much interest to adsorption scientists. In general, experimentally observed hysteresis phenomena can be grouped into two classes. The first class of hysteresis, or closed loop hysteresis, involves adsorbate condensation in mesoporous materials[10], while the other class of hysteresis, or open loop hysteresis, is normally observed in ultra-microporous carbons such as coals and molecular sieve carbons. In particular, closed loop hysteresis shows irreversible adsorption only in the vicinity of the condensation region, but is reversible outside this region. Such irreversibility is related to the high activation energy of adsorbate droplet growth[10] resulting from the confinement. In contrast, open loop hysteresis shows the occurrence of irreversible adsorption down to the lowest experimental pressure. From Figure 2a, it can be seen that the crossing time of N_2 and Ar at low temperatures (77K<T<100 K) from cage A to cage B, or adsorption time ($4s<\tau_{ads}<5hrs$), is within the experimental adsorption time scale, while their crossing time from cage B to cage A, or desorption time ($588s<\tau_{des}<134$ days), is far beyond the experimental time scale. This suggests that the open loop hysteresis is related to the high activation energy arising from constriction of the pore mouth. For further clarification, it is evident from Table 1[7] that the crossing time of carbon dioxide at 273 K between cages A and B is very short, leading its instantaneous equilibrium under practical adsorption experimental conditions, i.e. no open loop hysteresis is expected to be observed. This prediction is consistent with experimental observation of adsorption hysteresis of N_2 at 77 K in Takeda 3 Å, but not for CO_2 at 273 K in the same carbon, as shown in Figures 5(a, b).[11]

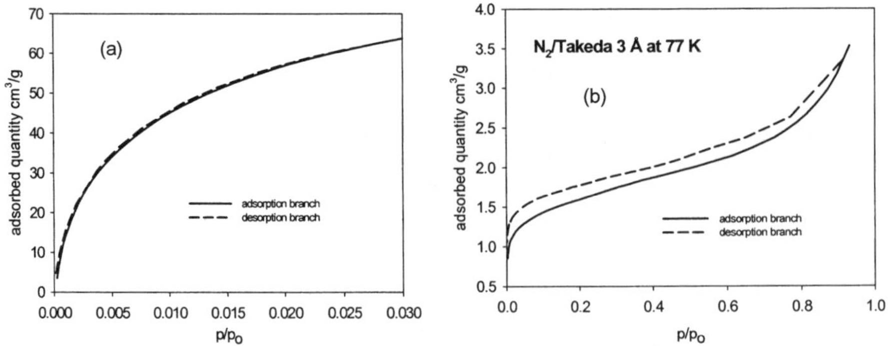

Figure 5 *Experimental adsorption isotherms of CO_2 at 273 K and N_2 at 77 K in Takeda 3Å* carbon molecular sieve.

2.2.3 Gaseous Mixture Selectivity. An intriguing feature of carbonaceous adsorbents is their excellent capability for mixture separation. For instance, molecular sieve carbon (MSC) has long been industrially utilised as an adsorbent in the pressure swing adsorption (PSA) technique due to its extremely high gas selectivity. However, for a typical example of a CO_2/CH_4 mixture under supercritical condition theoretical adsorption modelling of the mixture using a slit-pore model showed at best only small value of selectivity (<4.5) regardless of pore size[12], while extremely high selectivity of this mixture in molecular sieve carbons is experimentally shown.[13] The latter is very consistent with extremely high crossing time of CH_4 at ambient temperature in comparison with that of CO_2,[11] as shown in Table 1. Consequently, it is clearly evident that extremely high adsorption selectivity in disordered porous carbons cannot be achieved based on difference in adsorption affinity of individual species in the mixture, unless the size of the pore mouth is reduced to the molecular dimension at which activated diffusion occurs.

Table 1. Calculated results of the crossing time of CO_2 and CH_4 between cages A and B, for the HRMC based structural model of saccharose char, using the TST approach.

CO₂			CH₄		
Temperature (K)	$\tau_{ads(A\text{->}B)}$ (second)	$\tau_{des(B\text{->}A)}$ (second)	Temperature (K)	$\tau_{ads(A\text{->}B)}$ (second)	$\tau_{des(B\text{->}A)}$ (second)
273	5.92×10^{-5}	1.80×10^{-4}	200	4.46×10^{7}	5.34×10^{7}
283	3.80×10^{-5}	9.54×10^{-5}	253	8.21×10^{3}	3.88×10^{3}
293	2.53×10^{-5}	5.29×10^{-5}	263	2.48×10^{3}	9.90×10^{2}
303	1.75×10^{-5}	3.05×10^{-5}	273	8.28×10^{2}	2.79×10^{2}
313	1.24×10^{-5}	1.82×10^{-5}	283	3.01×10^{2}	8.63×10^{1}
323	9.08×10^{-6}	1.12×10^{-5}	323	1.05×10^{1}	1.63×10^{0}

2.2.4   Gasification. High surface area and pore volume is essentially created in the activation stage of preparation of carbonaceous porous materials, whereby a small amount of carbon atoms is removed by partial oxidation. In this activation stage, it is interesting to observe that majority of the surface area of activated carbon is initially created with removal of small carbon fraction (< 5%),[14] as shown in Figure 6. Accordingly, the opening of new pores must occur in this initial stage of the activation process. However, if these new pores are formed from etching of carbon planes, such small carbon fraction removed cannot account for the dramatically increased surface area unless a significant increase in pore accessibility of the existing pore structure occurs as a result of the opening of pore mouths. The latter mechanism would appear reasonable considering that the pore mouth normally contains highly reactive defects. Consequently, the pore mouth will be enlarged as the reactive defects are removed, leading to significant increase in pore accessibility. This is further supported by the results in Figure 7, showing a steep increase in helium density of coke during gasification as a consequence of increase in helium accessibility early in the gasification process.[15]

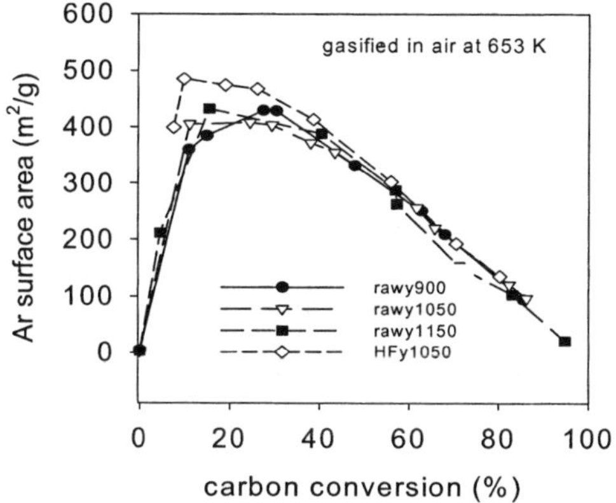

Figure 6 *Variation of surface area (per unit initial mass) with conversion, for air oxidation of Yarrabee coal char heat treated at various temperatures.*[14]

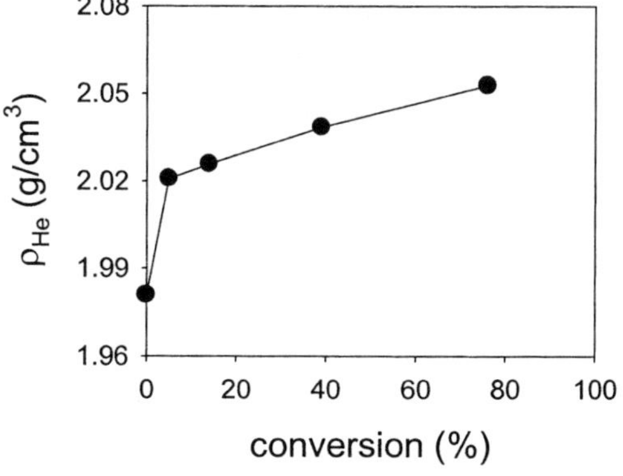

Figure 7 *Variation of helium density with conversion, for coke calcined at 1225 °C.*[15]

3 CONCLUSION

We have presented here an overview of our studies on pore accessibility in nanoporous carbons, covering both the semi-empirical approach (percolation model) and the atomic level approach using transition state theory. Furthermore, the impact of pore accessibility on adsorption in porous carbons is also presented. In particular, our review concentrates on explanation of the anomalous adsorption behaviour of adsorbate fluids in porous carbons such as open loop hysteresis, increase in adsorbed quantity with increasing temperature, extremely high selectivity, and significantly increased surface area encountered at the early stage of the process of activation of carbons. Such atomistic modelling provides a realistic picture of the adsorption process in porous carbons, which will ultimately facilitate the design of carbons with optimal structure for gas mixture separation and storage.

Acknowledgement

This work has been supported by a grant from the Australian Research Council, under the Discovery Scheme.

References

1 T.X. Nguyen and S.K. Bhatia, *J. Phys. Chem. C*, 2007, **111**, 2212.
2 M.V. López-Rámon, J. Jagiello, T.J. Bandosz, and N.A. Seaton, *Langmuir*, 1997, **13**, 4435.
3 S. Ismadji and S.K. Bhatia, *Langmuir*, 2000, **16**, 9303.
4 S. Ismadji and S.K. Bhatia, *AIChE J.*, 2003, **49**, 65.
5 J.S. Bae and S.K. Bhatia, *Energy&Fuels*, 2006, **20**, 2599.
6 E.J. Bottani and V.A. Bakaev, *Langmuir*, 1994, **10**, 1550.
7 C.M. Lastoskie, K.E. Gubbins, N. Quirke, J. Phys. Chem., 1993, **97**, 4786.
8 S. Himeno, T. Komatsu, S. Fujita, *J. Chem. Eng. Data*, 2005, **50**, 369.
9 T.X. Nguyen and S.K. Bhatia, *Langmuir*, 2004, **20**, 3532.
10 R. Valiullin, S. Naumov, P. Galvosas, J. Kärger, H.-J. Woo, F. Porcheron, and P. A. Monson, *Nature*, 2006, **443**, 965.
11 T.X. Nguyen and S.K. Bhatia, *Langmuir*, 2008, **24**, 146.
12 D. Nicholson and K.E. Gubbins, *J. Chem. Phys.*, 1996, **104**, 8126.
13 D. Lozano-Castello, J. Alcaniz-Monge, D. Cazorla-Amorós, A. Linares-Solano, W. Zhu, F. Kapteijn, J. A. Moulijn, 2005, **43**, 1643.
14 B. Feng and S.K. Bhatia, *Carbon*, 2003, **41**, 507.
15 K. Tran, S.K. Bhatia, and A. Tomsett, *Ind. Eng. Chem. Res.*, 2007, **46**, 3265.

CHARACTERIZATION MEASUREMENTS OF COMMON REFERENCE NANOPOROUS MATERIALS BY GAS ADSORPTION (ROUND ROBIN TESTS)

J. Silvestre-Albero[a], A. Sepúlveda-Escribano[a], F. Rodríguez-Reinoso[a,*], V. Kouvelos[b], G. Pilatos[b], N. K. Kanellopoulos[b], M. Krutyeva[c], F. Grinberg[c], J. Kaerger[c], A. I. Spjelkavik[d], M. Stöcker[d], A. Ferreira[e], S. Brouwer[e], F. Kapteijn[e], J. Weitkamp[f], S. D. Sklari[g], V. T. Zaspalis[g], D. J. Jones[h], L. C. de Menorval[h], M. Lindheimer[h], P. Caffarelli[i], E. Borsella[i], A. A. G. Tomlinson[i], M. J. G. Linders[j], J. L. Tempelman[j], E. A. Bal[j]

[a] *Laboratorio de Materiales Avanzados, Departamento de Química Inorgánica, Universidad de Alicante, Ap. 99, E-03080 Alicante, Spain; email: reinoso@ua.es*
[b] *Institute of Physical Chemistry, NCRS "Demokritos", Agia Paraskevi, 15310 Athens, Greece*
[c] *Department of Interface Physics, University of Leipzig, Linnéstr. 5, D-04103 Leipzig, Germany*
[d] *SINTEF Materials and Chemistry, P.O. Box 124, Blindern, N-0314 Oslo, Norway*
[e] *The Pore-Catalysis Engineering, DelftChemTech, Delft University of Technology, Julianalaan 136, 2628 BL Delft, The Netherlands*
[f] *Institute of Chemical Technology, University of Stuttgart, D-70550 Stuttgart, Germany*
[g] *Laboratory of Inorganic Materials, Chemical Process Engineering Research Institute, Center for Research and Technology Hellas, 57001 Thermi-Thessaloniki, Greece*
[h] *Institut Charles Gerhardt, Laboratoire des Agrégats, Interfaces et Matériaux pour l'Energie, UMR CNRS 5253, Université Montpellier II, 2 Place Eugène Bataillon, 34095 Montpellier Cedex 5, France*
[i] *Institute for the Study of Nanostructured Materials-CNR, Via Salaria km 29.3, 00016 Monterotondo Stazione (RM), Italy*
[j] *TNO Defence, Security and Safety, Lange Kleiweg 137, 2288 GJ Rijswijk, The Netherlands*

1 INTRODUCTION

The correct characterization of the textural characteristics of porous solids is of paramount importance in order to understand their behaviour in a certain application.[1] Traditionally, textural characterization has been performed using adsorption of probe molecules (Ar, N_2, CO_2, etc.); among them, N_2 adsorption at low temperature (77 K) is the most widely applied.[2] N_2 adsorption covers the relative pressure (P/P_0) range from 10^{-6} to 1 and provides information about the whole microporosity (up to 2 nm). Unfortunately, the low adsorption temperature becomes detrimental for samples exhibiting kinetic restrictions due to the limited accessibility of the N_2 molecule to the narrow microporosity. This limitation can be overcome using CO_2 adsorption at a higher temperature (273 K).[3, 4] CO_2 adsorption covers a shorter relative pressure range (from 10^{-6} to 0.03) and it is very helpful to complement N_2 in the characterization of the narrow microporosity (up to 0.7 nm).

As described above, the correct characterization of the porous structure is a major drawback in research fields related to the preparation, characterization and application of porous materials.[5, 6] However, the reliability of the experimental results in adsorption measurements is highly affected by several factors. There are errors associated with the characteristics of the experimental equipment (pressure-transducer precision), the calibration of the equipment, the experience of the operator (e.g. time left to ensure that the equilibrium has been reached), the correct application of the mathematical models (eg. Brunauer-Emmett-Teller (BET), Dubinin-Radushkevich (DR), etc.) and so on.

With this in mind, this manuscript reports the results of a Round Robin adsorption test performed under the auspicious of the European Network of Excellence INSIDEPORES (In-SItu study and DEvelopment of processes involving nano-PORous Solids).[7] The Round Robin is aimed to compare the textural properties of 4 reference samples (Takeda 4A, Takeda 5A, Spherical Carbon and silicalite H-ZSM5 Si/Al=400) obtained from the N_2 adsorption isotherms at 77 K and the CO_2 adsorption isotherms at 273 K (adsorption measurements on H-ZSM5 are not included due to the limited space available). This comparison will allow to identify and to emphasize possible weaknesses in the different steps involved in the characterization process. The Round Robin tests involved 10 groups from different countries all of them core members of the INSIDEPORES network of Excellence.

2 METHODS

N_2 adsorption isotherms at 77 K and CO_2 adsorption isotherms at 273 K have been performed in automated equipments following a series of protocols both for the pre-treatment of the sample and for the application of the different mathematical models (application of the BET method to the N_2 adsorption isotherm in order to calculate the "apparent" surface area and application of the DR equation to the N_2 and CO_2 adsorption isotherms in order to calculate the volume of micropores (V_o) and the volume of narrow micropores (V_n), respectively). Before the adsorption experiments, the protocol established a heat treatment in vacuum (outgassing) at 523 K during 4 h in order to remove the adsorbed impurities.

A critical point in the assessment of the textural properties is the analysis of the experimental data. In order to unify the criteria, a protocol was established for the calculation of the specific or "apparent" surface area, the micropore and mesopore volume and the volume of narrow micropores. A summary of the protocol is described below.

N_2 and CO_2 adsorption data

The N_2 adsorption isotherms at 77 K and the CO_2 adsorption isotherms at 273 K provide the volume adsorbed in STP conditions as a function of the relative pressure. The treatment of these data can be used to obtain the apparent BET surface area, the volume of micropores and mesopores, using the N_2 adsorption data, and the volume of narrow micropores, using the CO_2 adsorption data.

Specific or "apparent" surface area (S_{BET})

The surface area can be obtained from the BET equation:

$$\frac{\left(\dfrac{P}{P_0}\right)}{n\left(1-\dfrac{P}{P_0}\right)} = \frac{1}{n_m C} + \frac{C-1}{n_m C}\frac{P}{P_0} \qquad\qquad \textit{equation 1}$$

Plotting $(P/P_0)/[n(1-P/P_0)]$ vs. P/P_0 gives a straight line whose intercept and slope allow to calculate the monolayer capacity (n_m) and the constant C.[8]
The surface area can be calculated from the monolayer capacity (n_m) using the following equation:

$$S\left(\frac{m^2}{g}\right) = n_m \cdot 6.023\ 10^{23} \cdot 0.162 \cdot 10^{-18} \qquad\qquad \textit{equation 2}$$

In the same way as the DR equation, not all the experimental points are meaningful to calculate the BET surface area.

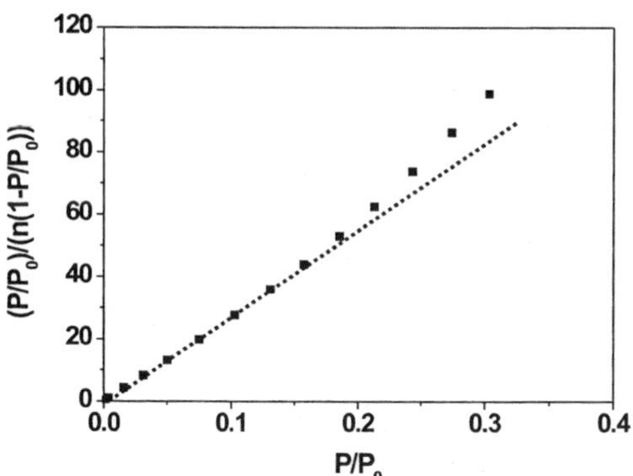

Figure 1 *Typical BET plot obtained from the N_2 adsorption data at 77 K on carbon molecular sieve Takeda 4A.*

The right points are those giving the best fit, usually between $0 \le P/P_0 \le 0.10$. However, it is important to remember that the intercept must be always positive. In order to certify that the selected points are correct, an easy calculation can be done.

$$\bullet \quad C \rightarrow \left(\frac{P}{P_0}\right)_m = \frac{1}{\sqrt{C}+1}$$

$$\bullet \quad n_m \xrightarrow{\ isotherm\ } \left(\frac{P}{P_0}\right)_m$$

The selected values are correct if the difference between the two values of $(P/P_0)_m$, obtained in these two ways, is smaller than 10%.

Volume of micropores (V_0) and mesopores (V_{meso})

The volume of micropores can be obtained using the Dubinin-Radushkevich (DR) equation:

$$\log V_{ads\ liq} = \log V_0 - D \log^2 \left(\frac{P_0}{P} \right)$$

equation 3

where $V_{ads\ liq}$ is the volume of liquid adsorbed, calculated from the adsorbed volume (STP conditions) using a density for nitrogen of 0.808 g/cm^3; V_0 is the volume of micropores and D is a system parameter.

The plotting of log ($V_{ads\ liq}$) vs. $\log^2(P_0/P)$ usually gives rise to a plot with three well-defined regions (Figure 2).

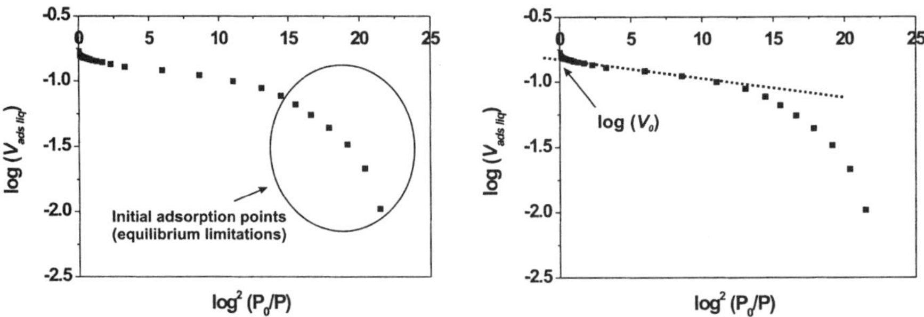

Figure 2 *Typical Dubinin-Raduskevich plot obtained from the N$_2$ adsorption data at 77 K on carbon molecular sieve Takeda 4A.*

The three regions characteristic of the DR plot correspond to the adsorption on mesopores ($\log^2(P_0/P)$ below ~ 2), adsorption on micropores ($\log^2(P_0/P)$ between ~ 2-15) and the region corresponding to the initial adsorption points ($\log^2(P_0/P)$ above ~15) where equilibrium limitations are present. Considering only the micropore region, experimental results can be fitted to a straight line the extrapolation of which allows the calculation of the micropore volume (V_0).

The volume of mesopores can be obtained from the total amount adsorbed at high relative pressures ($V_{ads\ liq}$ at P/P_0 ~ 0.95 and/or 0.99) using the following equation:

$$V_{meso} = V_{ads\ liq\ (\approx 0.95-0.99)} - V_{micropores}$$

equation 4

Volume of narrow micropores (V_n)

The volume of narrow micropores is calculated by applying the DR equation (equation 3) to the CO_2 adsorption data (adsorbed volume vs. relative pressure).[3,4] The DR plot gives a straight line the intersection of which allows to calculate the volume of micropores (Figure 3). In this case, the density of CO_2 is taken as 1.023 g/cm^3 and the saturation pressure at 273 K is taken as 3.49 MPa.

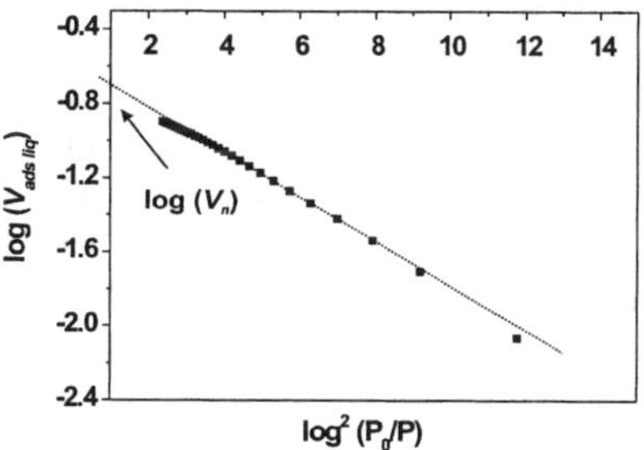

Figure 3 *Typical DR plot obtained from the CO_2 adsorption isotherm at 273 K on carbon molecular sieve Takeda 4A.*

Commonly, the DR plot from the CO_2 adsorption data does not exhibit equilibrium problems, thus allowing considering all the experimental points. However, equilibrium limitations can appear on samples with very narrow microporosity.[9] In this case, the useful points will be those that can be fitted to a straight line (normally those at medium relative pressures).

3 RESULTS AND DISCUSSION

As an example of the Round Robin experiments, N_2 adsorption isotherms at 77 K corresponding to the carbon molecular sieves Takeda 4A and Takeda 5A are shown in Figure 4. For both samples, the adsorption isotherms are of type I according to the IUPAC classification, characteristics of microporous carbon materials.[10]

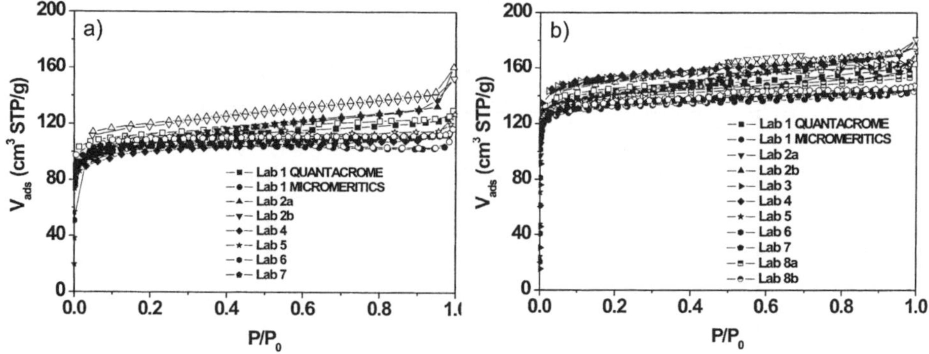

Figure 4 *N_2 adsorption (closed symbols) - desorption (open symbols) isotherms at 77 K for samples a) Takeda 4A and b) Takeda 5A.*

The textural characterization of the reference materials Takeda 4A and Takeda 5A highly depends on the laboratory where the measurements are performed. Additionally, even within the same laboratory the adsorption measurements are highly affected by the adsorption equipment used (compare the isotherms obtained in Lab 1 using equipment from Quantacrome and from Micromeritics). In principle, these differences in the adsorption results could be attributed to several factors such as the operator, the precision of equipment (the pressure-transducer), etc. However, it is noteworthy to mention that the divergence in the adsorption results is more evident on sample Takeda 4A than on sample Takeda 5A and also more important than on Spherical Carbon (not shown).

Spherical Carbon is an activated carbon with wider pores and a broad pore size distribution (PSD) while samples Takeda 4A and Takeda 5A are carbon molecular sieves (CMS) with very narrow pores and a narrow PSD. In this sense, immersion calorimetry studies using liquids with different molecular dimensions confirm that spherical carbon exhibits a broad PSD, with mainly all the porosity being accessible to a molecule such as 2,2-dimethylbutane with 0.56 nm. However, on carbon molecular sieve Takeda 5A even a molecule such as cyclohexane with only 0.48 nm has kinetics restrictions to access the microporosity, this effect being more drastic for Takeda 4A (cyclohexane is not able to access the porosity on Takeda 4A). Considering the information obtained from the calorimetric measurements, the larger divergence observed in the adsorption measurements for carbon molecular sieves could be attributed to the presence of kinetics restrictions due to the narrow pore entrance. In this sense, previous studies on these carbon molecular sieves have shown that the "equilibrium time", i.e. the time elapsed in the automated equipment between consecutive pressure readings to ensure that the equilibrium has been reached, is a critical parameter when the samples exhibit a pore diameter below 0.4-0.3 nm (Takeda 4A and Takeda 3A).[9]

The divergence in the adsorption measurements is also reflected in the apparent surface area and in the micropore and mesopore volume obtained following the protocol described above. In this sense, Table 1 and Table 2 shows the different values of S_{BET}, V_0 (N_2), V_{meso} (N_2) and V_n (CO_2) obtained in the different laboratories for samples Takeda 4A and Takeda 5A.

Table 1. *Apparent BET surface area, micropore volume (V_0) and mesopore volume (V_{meso}) obtained from the N_2 adsorption isotherms at 77 K on sample Takeda 4A. The volume of narrow micropores (V_n) obtained from the CO_2 adsorption isotherms at 273 K is also included.*

Laboratory	S_{BET} (m²/g)	V_0 (N_2) (cm³/g)	V_n (CO_2) (cm³/g)	V_{meso} (cm³/g)
Lab 1 (Q)	402	0.16		0.03
Lab 1 (M)	422	0.16		0.00
Lab 2	380	0.14		0.11
Lab 4	313			
Lab 5	386	0.16	0.27	0.02
Lab 6	418	0.15		
Lab 7	399	0.17	0.22	

Table 2. *Apparent BET surface area, micropore volume (V_0) and mesopore volume (V_{meso}) obtained from the N_2 adsorption isotherms at 77 K on sample Takeda 5A. The volume of narrow micropores (V_n) obtained from the CO_2 adsorption isotherms at 273 K is also included.*

Laboratory	S_{BET} (m²/g)	V_0 (N_2) (cm³/g)	V_n (CO_2) (cm³/g)	V_{meso} (cm³/g)
Lab 1 (Q)	573	0.22		0.02
Lab 1 (M)	527	0.20		0.01
Lab 2	545	0.20		0.07
Lab 3	589	0.20		
Lab 4	476			
Lab 5	547	0.22	0.27	0.02
Lab 6	524	0.19		
Lab 7	516	0.21	0.23	
Lab 9	550	0.21	0.25	

Even with a well-defined protocol for the calculation of the different textural parameters, the divergence in the adsorption measurements is also transferred to the textural parameter. In accordance with previous observations, a fundamental parameter such as the BET surface area exhibits a larger deviation on Takeda 4A (S_{BET} : 389 m²/g) with a standard deviation of 37 and a coefficient of variance of 9%, followed by Takeda 5A (S_{BET} : 539 m²/g) with a standard deviation of 33 and a coefficient of variance of 6% and, finally, the Spherical Carbon (S_{BET} : 1234 m²/g) with a standard deviation of 68 and a coefficient of variance of only 5.5%. The similar coefficient of variance ($\sim$ 6%) obtained for the last two samples clearly reflects that even when using samples where kinetic restrictions are not expected and even when using a well-established protocol, textural parameters calculated on different laboratories exhibit an inherent error commonly associated with the equipment and the operator. However, the larger coefficient of variance associated with Takeda 4A confirms that on certain samples there can be additional errors associated with the porous structure of the material.

In summary, the large deviation on the textural parameters obtained for carbon molecular sieve Takeda 4A points out that on samples with narrow pores (below 0.3-0.4 nm), adsorption of N_2 at 77 K and CO_2 at 273 K may be kinetically restricted. Additionally, these effects must be more important for the N_2 molecule due to the low temperature of the adsorption measurement (77 K). Thus, even when using a well-defined protocol for the pre-treatment of the sample, the experimental procedure during the measurement and the treatment of the experimental data, care must be taken by the operator to ensure that the adsorption isotherms are performed under true equilibrium conditions. Factors such as the equilibration time, i.e. time elapsed between consecutive pressure readings, the number of pressure readings and the deviation allowed within these pressure readings must be carefully defined in the automated equipments to ensure that equilibrium has been reached if the presence of narrow microporosity is suspected. In this sense, a conclusive proof to assess the presence of true equilibrium conditions would be to check the effect on the adsorption isotherms either of: i) an increase in the stringency of the pressure readings criteria, e.g. an increase in the "equilibrium time", ii) the effect of an increase in the adsorption temperature and/or iii) to check the amount uptake in the desorption branch.[9] In the absence of kinetic restrictions: i) the amount uptake must be independent of the pressure reading criteria used over the whole relative pressure range, ii) at a higher adsorption temperature the adsorption capacity must decrease over the whole relative pressure range, as corresponds to an exothermic process and iii) the desorption branch of the isotherm must always close with the adsorption branch once the hysteresis loop is closed (it must never remain above or below).

Acknowledgement

The authors acknowledge the support from the European Network of Excellence INSIDEPORES (NMP3-CT2004-500895).

References

1 H. Marsh and F. Rodriguez-Reinoso, *Activated Carbon*, Elsevier, London, **2006**.
2 K. S. W. Sing, *Advances in Colloid and Interface Science*, 1998, **76-77**, 3.
3 J. Garrido, A. Linares-Solano, J. M. Martin-Martinez, M. Molina-Sabio, F. Rodriguez-Reinoso and R. Torregrosa, *Langmuir*, 1987, **3**, 76.
4 F. Rodriguez-Reinoso, J. Garrido, J. M. Martin-Martinez, M. Molina-Sabio and R. Torregrosa, *Carbon*, 1989, **27**, 23.
5 F. Rodriguez-Reinoso, *Carbon*, 1998, **36**, 159.
6 H. Marsh and F. Rodríguez-Reinoso, *Activated Carbon*, Elsevier, London, **2006**.
7 European Network of Excellence INSIDEPORES, website: www.pores.gr.
8 S. J. Gregg and K. S. W. Sing, *Adsorption, surface area and porosity*, 2 ed., Academic Press, London, **1982**.
9 R. V. R. A. Rios, J. Silvestre-Albero, A. Sepúlveda-Escribano, M. Molina-Sabio and F. Rodríguez-Reinoso, *J. Phys. Chem. C*, 2007, **111**, 3803.
10 IUPAC Recommendations, *Pure Appl. Chem.*, 1985, **57**, 603.

COMPETITIVE NAPHTHALENE ADSORPTION ON ACTIVATED CARBONS. EFFECT OF POROSITY AND HYDROPHOBICITY

B. Cabal[1], CO Ania[1], PAM Mourão[2], MML Ribeiro Carrott[2], PJM Carrott[2], JB Parra[1], JJ Pis[1]

[1] Instituto Nacional del Carbón, CSIC, Apartado 73, 33080 Oviedo, Spain
[2] Centro de Química de Évora and Departamento de Química, Universidade de Évora, Colegio L. A. Verney, 7000-671 Évora, Portugal

1 INTRODUCTION

Liquid phase adsorption is more complex than gas adsorption due to the interactions that arise inside the multi-component liquid solution.[1,2] Besides the expected/desired adsorbate-adsorbent interactions, the solvent also plays an important role in establishing the degree of retention. In other words, adsorption capacity can be dominated by the selectivity and thus the preferential adsorption of the targeted probe may be significantly suppressed by competitive retention of a second component.[1-7] This is particularly remarkable in diluted solutions (i.e, low surface coverage) where the solvent concentration is very high, being noteworthy the solvent-adsorbent interactions. Thus, a 3-way approach covering adsorbate-adsorbent, adsorbate-solvent and the likely competitive adsorbent-solvent interactions must be considered.

The goal of this work was to investigate the mechanism of competitive adsorption of naphthalene on activated carbons from diluted solutions. The polarity of the adsorption media was varied, by selecting solvents ranging from high to low dielectric constants (ethanol, cyclohexane and heptane). By altering the hydrophobicity of the adsorption media, the preferential adsorption of the organic probe is changed; therefore the selectivity and efficiency of naphthalene adsorption may be tuned by optimising the features of the adsorbent upon the adsorption media.

2 METHODS

2.1 Characterisation of the adsorbents

The carbons chosen for this study have been described in detail elsewhere.[8] A coal based activated carbon -B- supplied by Agrovin (physical activation) was used as starting material. The sample was oxidised using nitric acid (BN60) and ammonium persulfate (BS). Details of the oxidation procedure have been described elsewhere.[9] Textural characterisation was carried out by measuring the N_2 (Micromeritics ASAP 2010M) adsorption isotherms at -196 °C. Before the experiments, the samples were outgassed under vacuum at 120 °C overnight. The N_2 isotherms were used to calculate the specific surface area, S_{DFT}, total pore volume, V_{TOTAL}, and micro-, mesopore volumes, which were evaluated by means of the DFT model. Further characterisation was performed by ultimate analysis.

The heptane, cyclohexane and ethanol vapour adsorption isotherms at 25 °C were determined gravimetrically in an apparatus equipped with a CI Electronics MK2 vacuum microbalance and Edwards Barocel 600 (0–1000 torr) capacitance manometers. Blank measurements indicated that buoyancy corrections were negligible. Initially all samples were outgassed at 120 °C for 8 h. The saturation pressure and density of each adsorptive at the working temperature were calculated from the data and equations given in the literature.[10] The physical properties of the organic vapours used are shown in Table 1.

Table 1 *Physical properties of the selected organic vapours[10] and naphthalene solubility[11]*

	d_{min} [nm]	Molar volume (cm³/g)	Dielectric constant (ε)	Kow	Density (g/cm³)	Naphthalene Solubility (mole fraction x_i)
Heptane	0.43	146.6	1.9	31623	0.680	0.130
Cyclohexane	0.48	108.1	2	2200	0.774	0.149
Ethanol	0.45	58.4	24	0.48	0.784	0.040

2.2 Adsorption from solution

Details of the experimental procedure for evaluating naphthalene adsorption isotherms have been reported elsewhere.[8] Briefly, naphthalene equilibrium adsorption isotherms were obtained from batch experiments at 30 °C using cyclohexane, heptane and ethanol as solvents.

3 RESULTS AND DISCUSSION

Adsorption of naphthalene on activated carbons from different media –including water and hydrocarbon solutions- has been reported in a previous study.[7, 8] The effects of the carbon polarity and solvent hydrophobicity were discussed there. Briefly, the results showed that when adsorption takes place from an aqueous solution, the preferential adsorption of non-polar naphthalene takes place on hydrophobic carbons; simultaneous water anchoring to the hydrophilic centres on polar carbon surfaces leads to reduced naphthalene adsorption upon oxidation treatments of the carbons. In contrast, naphthalene uptake is favoured in adsorbents of increasing polarity when adsorption takes place from organic media (i.e., heptane). This finding was explained in terms of competitive solvent adsorption and solvent-adsorbate affinity (i.e., solvation effects), disregarding the effect of the accessibility of the solvents to the porous structure of the carbons.

The purpose of this paper was to extend the previous study, incorporating a third approach: the effect of the accessibility of the solvent to the porous structure of the carbons on the competitive naphthalene adsorption on activated carbons. Taking into account the differences in the molecular dimensions of the selected organic solvents (Table 1), it seems reasonable to think that the accessibility of the adsorbate/solvent system to the inner pore network (and therefore to the active sites for adsorption) may be limited by the ratio pore width:diameter of the solution constituents (naphthalene and the corresponding solvents). Thus, a key factor resides on a reliable characterisation of the porosity of the activated carbons. For this reason, we have carried out adsorption of the selected organic vapours to complement the characterisation of the porous features obtained from gas adsorption.

3.1 Textural and chemical properties of the studied carbons

The main textural parameters of the series of modified carbons obtained from gas adsorption are shown in Table 2. A detailed analysis about the effect of the oxidation treatment on the textural properties of the activated carbons has been previously reported. Here we merely report the main parameters for data interpretation.[8] Briefly, oxidation with ammonium persulfate (sample BS) slightly modifies the porosity of the raw carbon, as seen by a small decrease in the pore volume and surface area. In contrast, oxidation in 60 % nitric acid brought about a partial collapse of the porous network, particularly in the micropore range. In addition, wet oxidation incorporated a large amount of oxygen groups mainly of acidic nature, as corroborated by the oxygen content and the values of point of zero charge, varying from 9 pH units for sample B and between 2-3 units for the oxidised counterparts.

Table 2 *Main chemical and textural parameters of the studied activated carbons obtained by N_2 adsorption isotherms at -196 °C*

Oxygen [wt. %]		S_{DFT} $[m^2g^{-1}]$	V_{TOTAL} $[cm^3g^{-1}]$	$V_{MICROPORES}$ $[cm^3g^{-1}]$	$V_{MESOPORES}$ $[cm^3g^{-1}]$
1.9	**B**	1095	0.646	0.337	0.134
11.6	**BS**	1000	0.559	0.326	0.094
17.2	**BN60**	798	0.447	0.242	0.091

The changes in the surface chemistry and the porosity of the carbons due to oxidation will play an outstanding role in the performance towards the adsorption of naphthalene from solution. While the textural features control the accessibility of the adsorbate-solvent systems, the polarity of the carbons (i.e., hydrophobic character) governs the appearance of competitive effects that could lessen the retention of the targeted probe. The accessibility to the porosity may be restricted by the pore sizes of the carbons and the diameters of the probes (adsorbate and solvents). Molecules such as heptane and ethanol, having small molecular diameters (Table 1) should be able to enter the narrowest pores of the adsorbents, whereas some small pores could be excluded for larger hydrocarbons. In order to investigate the effect of the accessibility, we have measured the adsorption isotherms of the organic vapours that have been employed as solvents in the removal of naphthalene from solutions.

3. 2   Adsorption of organic vapours

For each sample, the adsorption isotherms of ethanol, cyclohexane and heptane at 25 °C are shown in Figure 1. The adsorbed amounts in this figure are expressed in terms of equivalent liquid volumes, obtained by assuming the values of density compiled in Table 1, to facilitate the comparison among the various molecules.

The adsorption isotherms of all the organic vapours are type I, regardless of the nature of the activated carbon, characterised by a rapid and high adsorption at very low relative pressures; the saturation plateau is not parallel to the relative pressure axis, indicating that the microporous adsorbents contain a large portion of mesopores. These results are in good agreement with previous characterisation based on nitrogen adsorption data (Table 2).

It should be noted that the adsorption of heptane and cyclohexane is enhanced by the elimination of oxygen on the carbon surface.[12] Both isotherms are parallel in the whole range of relative pressures, regardless of the hydrophobic/hydrophilic nature of the adsorbent. In contrast, the ethanol isotherm of carbon B shows two types of behaviour depending on the degree of coverage. This observation enables two different regimes to be distinguished: i) the effect of the surface chemistry modification and ii) of the pore volume.

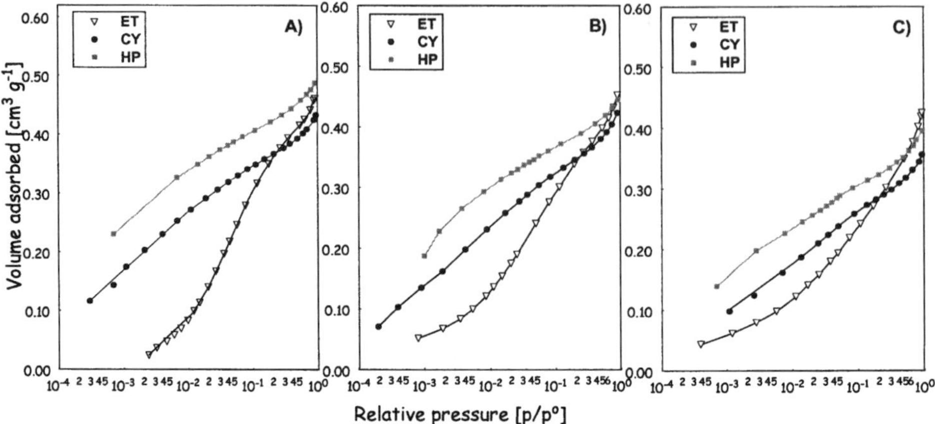

Figure 1 *Adsorption isotherms of ethanol (ET), cyclohexane (CY) and heptane (HP) at 25 ºC on A) the as received carbon, B) and C), oxidised BS and BN60 samples, respectively.*

The saturation region of the isotherms is more sensitive to the porosity of the adsorbents (accessibility). For a given carbon, the maximum uptake at the *plateau* (i.e., relative pressures close to unit) follows the general trend: heptane > ethanol > cyclohexane, with the exception of the most oxidised carbon BN60. The trend is in good agreement with the molecular diameters computed for these probes (Table 1). This result indicates the presence of size exclusion effects - heptane, having the smallest diameter, presents the largest uptake, which confirms that it is able to enter to a larger fraction of the porosity of the adsorbents, whereas some small pores (narrow micropores) seem to be excluded for the larger cyclohexane. This finding was further corroborated by the analysis of the vapour isotherms by means of the DR. The derived values of the micropore volume (Wo) and adsorption energies (E) are compiled in Table 3.

For this vapour (ca. cyclohexane), small decreases in pore volumes are observed; $W_{o\,CY}$ is ca. 10 % lower that the corresponding values for heptane in all the carbons. Taking into account that the micropore size distribution 'seen' by a particular adsorptive for a given carbon is a function of that adsorptive's size,[14-16] this behaviour can be interpreted in terms of molecular sieving, whereby the effective size of cyclohexane appears to be sufficiently large for that molecule to be physically excluded from certain pores which are yet accessible to smaller probes (heptane and ethanol).

The adsorption energy decreases from heptane to cyclohexane to ethanol, in agreement with the isotherms. For heptane and cyclohexane, the E value decreases with the hydrophilic character of the carbons, whereas for ethanol the opposite trend is observed, indicating an increasing specific contribution to the adsorption energy as the surface is oxidised.

In contrast, the general trend followed at low coverage is heptane > cyclohexane > ethanol. This is attributed to the stronger dispersive interactions of heptane and cyclohexane with the hydrophobic carbon pore walls, as compared to ethanol (shorter alkyl chain). The differences between the amount of ethanol and non-polar hydrocarbons adsorbed at low coverage decrease with the polarity of the adsorbent, confirming that the low relative pressure region is more sensitive to the surface chemistry of the carbons.

The ethanol isotherms of the three carbons are similar in the high pressure -low adsorption potential- regions, although clearly showing the effects of the surface chemistry in the low pressure (high potential) domains. Indeed, ethanol isotherms cross-over at relative pressures between 0.02 and 0.04. The amount adsorbed increased after oxidation (seen in the upward displacement of the initial region of the isotherms in BS and BN60), as a result of the formation of additional centres for specific adsorption via hydrogen bonding. The opposite trend is observed for the adsorption of the hydrocarbons (heptane and cyclohexane), indicating in this case the higher affinity of hydrophobic carbon surfaces towards non-polar hydrocarbons.

Table 3 *Characteristics of the activated carbons derived from the DR equation applied to the vapour isotherms*

	Wo_{HP}	E_{HP}	Wo_{CY}	E_{CY}	Wo_{ET}	E_{ET}
	$[cm^3 g^{-1}]$	$[kJ\ mol^{-1}]$	$[cm^3 g^{-1}]$	$[kJ\ mol^{-1}]$	$[cm^3 g^{-1}]$	$[kJ\ mol^{-1}]$
B	0.426	24.13	0.384	19.02	0.453	9.04
BS	0.394	21.84	0.362	17.47	0.384	10.95
BN60	0.322	20.73	0.292	16.12	0.268	12.94

In the high pressure range, ethanol adsorption isotherms of non-polar B, and hydrophilic BS carbons are very similar. This fact confirms that the modification of the surface chemistry brought about by a mild oxidation treatment (accounting for the higher ethanol adsorption at low relative pressures), is accompanied by a slight change in the pore texture (Table 2), explaining the similar adsorbed amounts at relatives pressures close to unity. On the other hand, the fall in the porous features with drastic oxidation explains the lower uptake of BN60 carbon at saturation, whereas at low relative pressures it superimposes with BS isotherms, as a consequence of the specific interactions.

In summary, the adsorption isotherms of the organic vapours revealed restrictions in the accessibility of the solvents to the inner porous network of the studied carbons. The adsorption of the larger cyclohexane molecule is restricted for the small pores, this not being the case for heptane or ethanol. Such limitations should be taken into consideration when investigating the overall performance towards adsorption capacity from the liquid phase. On the other hand, ethanol adsorption is sensitive to surface chemistry at low relative pressures, and it is more affected by texture at high relative pressures. This behaviour has also been observed for water vapour, methanol and ethanol in non-porous carbon blacks.[13-15]

3.3   Naphthalene retention from the liquid phase

Compared to water, naphthalene uptake is largely suppressed when adsorption occurs from organic solutions.[7] This is expected given the higher solubility of naphthalene in organic solutions (i.e., solvent affinity) and the simultaneous (competitive) adsorption of the solvent in dilute solutions. However, based on the results inferred from the adsorption of vapours, the effect of the accessibility of the solvents to the porous structure of the carbons should not be disregarded. Therefore, it is expected that competitive solvent uptake would depend not only on the affinity of the carbon surface for the solvent, but also on size exclusion effects. Additionally, given the naphthalene structure -flat molecule composed of two fused aromatic rings-, it seems reasonable that it would be preferentially adsorbed in a prostrate position (aromatic ring parallel to the pore walls, maximising dispersive interactions). Hence, we may assume that access of naphthalene to the carbon porosity is not restricted.

The experimental equilibrium adsorption isotherms for naphthalene retention are shown in Figure 2. It is interesting to note that for the most hydrophobic carbon -B-, the amount of naphthalene adsorbed does not correlate with the polarity of the solvent. For this sample, naphthalene adsorption from ethanol practically overlays the isotherm in cyclohexane and outperforms the uptake from heptane solution. Considering the hydrophilic nature of ethanol ($K_{ow} \sim 0.48$), this fact indicates that competitive adsorption of ethanol occurs at a low extent in carbons of intrinsic hydrophobic nature. However, one would expect a lower uptake from cyclohexane solution, due to competitive retention of the solvent and to the high solubility of naphthalene in this solvent (Table 1). The results obtained suggest that size exclusion effects may be occurring for this probe. If the solvent cannot access the porosity of the carbon, then the competitive effect is hindered and therefore the retention of naphthalene increases, overcoming the uptake from the alcoholic phase.

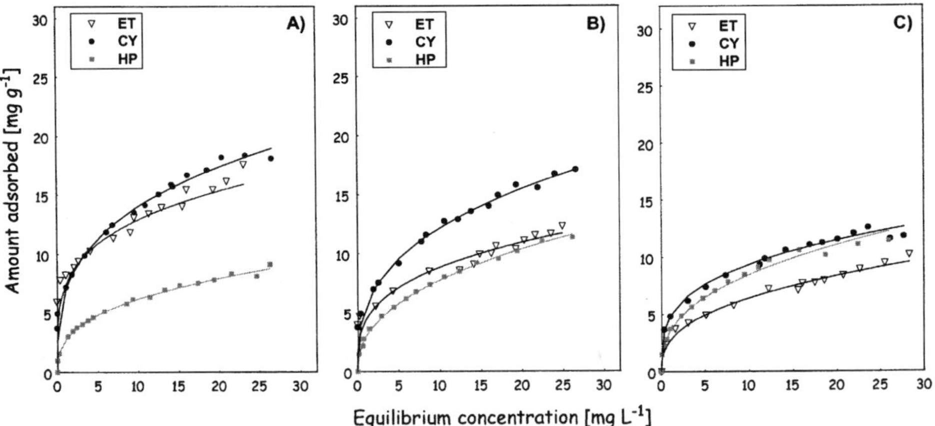

Figure 2 *Naphthalene equilibrium adsorption isotherms from ethanol (Et), cyclohexane (CY) and heptane (HP) solutions on the as-received carbon (A), and its oxidised BS (B) and BN60 (C) counterparts.*

In fact, if the amount adsorbed is normalised *vs* the solubility of naphthalene in the different solvents (Figure 3) the effect of the restricted accessibility of cyclohexane is underlined; the normalised isotherms do not follow the trend due to solvent affinity, and the uptake from cyclohexane is larger than expected.

A different trend was obtained for the oxidised samples, whose corresponding isotherms from alcoholic solution superimpose those in heptane. The behaviour is analogous to that found for adsorption from aqueous solutions, namely, the ability of the carbons to adsorb naphthalene from ethanol solutions strongly decreased with the polarity of the carbons. The retention is hindered by the formation of specific interactions between ethanol molecules and the surface functionalities. Such H-bonding probably occurs at the pore entrances -where most surface functionalities are created-,[17] leading to the formation of alcoholic clusters that impede the access and adsorption of naphthalene molecules in the inner porous structure.

Comparatively, naphthalene adsorption capacities in cyclohexane solutions are larger than in heptane solutions. This behaviour is an indirect consequence of their different molecular dimensions and hydrophobicity. Bearing in mind that cyclohexane present restricted access to the porosity of the carbons, it is clear that the retention of the adsorbate is favoured over the solvent. Analogously, a straightforward consequence of the higher accessibility of heptane to the porous structure is a strong competition for the adsorption sites. Heptane is more hydrophobic than cyclohexane (K_{ow} 31623 and 2200, respectively), thus it is expected to compete more effectively with naphthalene for the adsorption sites on the carbon. This is in good agreement with expectations, and similar results have been reported for other aromatic adsorbates.[6]

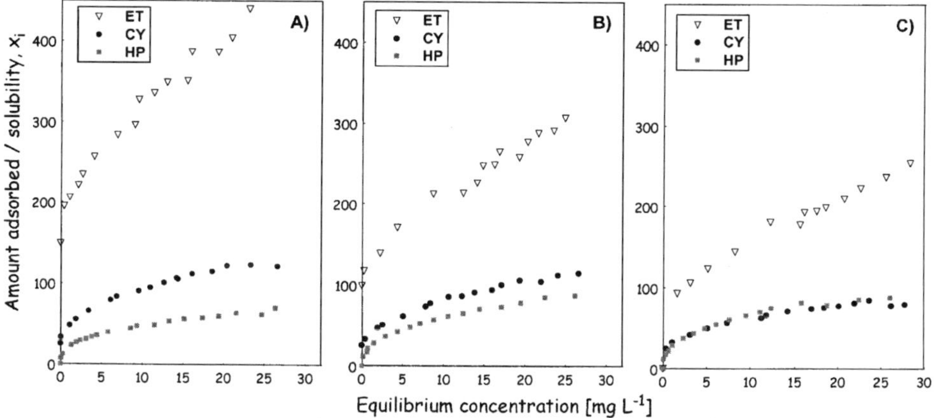

Figure 3 *Normalised naphthalene equilibrium adsorption isotherms vs the solubility of naphthalene in the studied solvents (mole fraction, x_i): ethanol (ET), cyclohexane (CY) and heptane (HP) on the as-received carbon (A), and its oxidised BS (B) and BN60 (C) counterparts.*

Moreover, the differences in the adsorption isotherms from cyclohexane and heptane gradually become smaller with increasing the carbon polarity. Due to the lower competitive effect of cyclohexane, the naphthalene retention experiences a notorious decrease as the polarity of the adsorbent increases. On the other hand, the decrease in the competitive heptane retention leads to a slightly higher naphthalene adsorption on the oxidised carbons in comparison with the original B carbon. Only in the case of the most oxidised carbon, BN60, both isotherms practically superimpose, indicating that for this sample the competitive retention of heptane is compensated (minimised) by the polarity of the carbon.

Summarising, although accessibility to the porosity of carbon adsorbents plays an important role in all adsorption processes, it is demonstrated that the affinity of the adsorbate-solvent and adsorbent-solvent systems govern the adsorption extent, due to undesirable but parallel competitive reactions.

4 CONCLUSIONS

Liquid phase adsorption using solvent molecules of different sizes and hydrophobicity can be an effective means of characterising the accessibility and mechanism of competitive adsorption. The results show that the surface polarity of the carbons can be modulated to favour the retention of a non-polar adsorbate (i.e., naphthalene), by controlling the competitive adsorption of the solvent. Oxidation of an activated carbon might result in a better adsorbent even of non-polar compounds, when the competitive adsorption of the solvent becomes important. However, one must also take into account the dimensions of the molecules, and therefore the accessibility to the porosity of the adsorbents. These factors would compensate the slight loss in pore structure and gain in polarity of oxidised carbons, when adsorption takes place from organic media.

Acknowledges
This work was partially supported by the Spanish MICINN (project CTM2008-01956). C.O.A. thanks the Spanish MEC for a Ramon y Cajal Research Contract.

5 REFERENCES

1. L.R. Radovic, C. Moreno-Castilla, J. Rivera Utrilla, in Carbon materials as adsorbents in aqueous solutions, ed. L.R. Radovic, Chemistry and Physics of Carbon, vol. 27, Dekker, New York 2000, p. 227.
2. P.J.M. Carrott, P.A.M. Mourao. M.M.L. Ribeiro Carrott, E.M. Gonçalves, *Langmuir,* 2005, **21**, 11863.
3. A.W. Marczewski, A. Derylo-Marczewska, M. Jaroniec, *Chemica Scripta,* 1988, **28**, 173.
4. A. Derylo-Marczewska, A.W. Marczewski, *Polish J. Chem.*, 1997, **71**, 618.
5. J. Pires, N.L. Pinto, A.P. Carvalho, M.B. de Carvalho, *Adsorpt.*, 2003, **9**, 303.
6. F. Ahnert, H.A. Arafat, N.E. Pinto, *Adsorp.*, 2003, **9**, 311.
7. C.O. Ania, B. Cabal, J.B. Parra, A. Arenillas, B. Arias, J.J. Pis, *Adsorpt.,* 2008, **14**, 343.
8. C.O. Ania, B. Cabal, J.B. Parra, J.J. Pis, *Wat. Res.*, 2007, **41**, 333.
9. C.O. Ania, J.B. Parra, J.J. Pis, *Adsorpt. Sci. Technol.*, 2004, **22**, 337.
10. R.C. Reid, J.M. Prausnitz, B.E. Poling, The Properties of Gases and Liquids, McGraw-Hill, New York, 1986.
11. IUPAC-NIST Solubility Database http://srdata.nist.gov/solubility/index.aspx
12. R.C. Bansal, T.L. Dhami, *Carbon,* 1980, **18**, 137.
13. F. Cosnier, A. Celzard, G. Furdin, D. Begin, J.F. Mareche, O. Barres, *Carbon*, 2005, **43**, 2554.
14. R. H. Bradley, B. Rand, Fuel 1993, 72, 389.
15. A. Andreu, H.F. Stoeckli, R.H. Bradley, *Carbon* 2007, 45, 1854.
16. F. Rodrıguez-Reinoso, M. Molina-Sabio, Adv. Colloid and Interf. Sci. *1998*, **76-77**, 271.
17. J.B. Donnet, *Carbon*, 1982, **20**, 267.

STUDY OF HYDROGEN SULPHIDE ADSORPTION ON MIL-47(V) AND MIL-53(AL, CR, FE) METAL-ORGANIC FRAMEWORKS BY ISOTHERMS MEASUREMENTS AND *IN-SITU* IR EXPERIMENTS

L. Hamon,[1] A. Vimont,[2] C. Serre,[3] T. Devic,[3] A. Ghoufi,[4] G. Maurin,[4] T. Loiseau,[3] F. Millange,[3] M. Daturi,[2] G. Férey,[3] and G. De Weireld[1]

[1]Laboratoire de Thermodynamique et Physique mathématique, Faculté Polytechnique de Mons, boulevard Dolez 31, 7000, Mons, Belgium
[2]Laboratoire Catalyse et Spectrochimie, UMR 6506, CNRS-ENSICAEN-Université de Caen, 6 boulevard du Maréchal Juin, 14050, Caen Cedex, France
[3]Institut Lavoisier, UMR CNRS 8180, Université de Versailles Saint-Quentin-en-Yvelines, 45 avenue des Etats-Unis, 78035, Versailles Cedex, France
[4]Institut Charles Gerhardt, UMR CNRS 5253, UM2, ENSCM, Université de Montpellier II, place Eugène Bataillon CC 003, 34095, Montpellier Cedex 5, France

1 INTRODUCTION

Natural gas and biogas are now of main interest due to the energetic transition from quasi-total petroleum products consumption to reasonable clean and more energetic products consumption.[1,2] Moreover, this transition is coupled with a global environmental interest to reduce pollutant emissions. Natural gas and more specifically biogas contain natural sulphur compounds such as hydrogen sulphide and many added odorous compounds such as mercaptans. The combustion of these species produces SOx compounds responsible of acid rains and industrial equipments corrosion. Therefore, in order to reduce and remove sulphur compounds from gases, gas industry developed absorption processes with amine treatments.[3] However these processes do not allow to obtain a sulphur-free gas: ca. 50 ppm mol. residual concentration sulphur compounds is usually observed after the treatment. Adsorption processes have thus been developed to reach the requirement issued from European directives (10 mg m^{-3} of sulphur compounds in the reference gas for automotive applications, 293.2 K, 101.3 kPa).[4] In the aim to separate sulphur compounds from gas and to purify gas, pressure swing adsorption (PSA) or temperature swing adsorption (TSA) processes designs need pure adsorption thermodynamic data on well-known or new adsorbents.

The most important required specification for such gas separation or purification technique is a high selectivity for sulphur compounds in order to remove them from the matrix gas. Moreover, adsorbent materials have to present a strong chemical resistance to acid and corrosive sulphur gases. The adsorbent material regeneration is also a main point for the process design.[5,6] Thus, adsorbent material choice focuses on resistant solids such as new metal-organic frameworks (MOFs) built-up from inorganic sub-units linked by organic moieties such as poly-carboxylates or -phosphonates.[7,8] These microporous solids present excellent adsorption properties.[9-17]

In this paper, we present hydrogen sulphide adsorption isotherms measurements recorded on two isostructural MOFs exhibiting one-dimensional pores. One of these solid (MIL-47(V), MIL: Matériaux Institut Lavoisier) presents a rigid structure whereas the second one (MIL-53(Cr, Al, Fe)) has got a flexible structure.

2 EXPERIMENTAL SECTION

2.1 MOFs synthesis and characterization

2.1.1 MIL-47 and -53 structure and main properties.
The two different types of MOFs used in this study present a three-dimensional framework built from inorganic chains of corner-sharing [MO_6] octahedra (M = Cr^{III}, V^{IV}) connected by terephthalate moieties. This structure presented in figure 1 defines a one-dimensional pore system according to the classification by Kitagawa *et al.*[18-20] In these solids, formulated $V^{IV}O\{O_2C\text{-}C_6H_4\text{-}CO_2\}$ (later denoted MIL-47(V)) and $M^{III}(OH)\{O_2C\text{-}C_6H_4\text{-}CO_2\}$ (M = Al, Cr, Fe, MIL-53(M)), the connection between the [MO_6] octahedra is ensured either by oxo or hydroxo bridge, depending on the oxidation state of the metal. Whereas the first solid (MIL-47(V)) presents a rigid structure and a porous volume of 0.51 cm^3g^{-1},[14-21] the second one (MIL-53(Cr, Al, Fe)) was shown to exhibit a large breathing effect under particular polar atmospheres with porous volumes of 0.48 and 0.49 cm^3g^{-1} for MIL-53(Al) and MIL-53(Cr) respectively.[14,21-26]

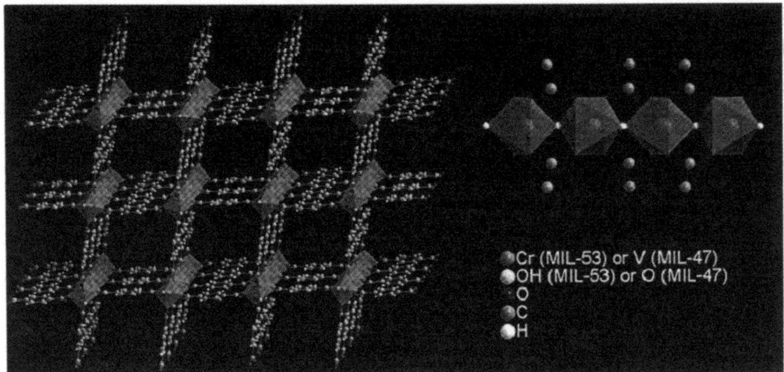

Figure 1   *Structure of MIL-47(V) and MIL-53(Cr)*

2.1.2 MIL-47(V) synthesis.
MIL-47(V) is synthesized under hydrothermal conditions during four days at 473 K from a mixture of vanadium(III) chloride, terephtalic acid, and deionized water using the published procedure.[21] Activation was performed by (i) exchange of the free terephthalic acid present in the ores by DMF, followed by (ii) calcination in air at 250°C for 48 hours.
2.1.3 MIL-53(Cr, Al, Fe) synthesis.
MIL-53 are also synthesized by hydrothermal way during six days at 493 K after mixing metal nitrate, terephtalic acid, hydrofluoric acid, and desionized water in molar ratio of 1:1.5:1:280.[14,22,25] A purple powder is obtained and then purified by eliminating remaining terephatlic acid using powder dispersion in dimethylformamide (DMF) for *ca.* 10 min. The excess free terephtalic acid is then removed by solvothermal exchange with DMF in steel autoclave for 12 hours at 423 K. After cooling, the solid obtained is calcinated at 523 K for 12 hours in order to remove DMF molecules.

2.2   Isotherm adsorption measurements

2.2.1 Gravimetric apparatus.
Adsorption isotherm measurements are carried out using a magnetic suspension balance developed by Lösch *et al.*[27,28] and now commercialized by Rubotherm, Germany. The analytic balance (resolution: 10 µg; thermal deviation: 1.5 ppm K^{-1} between 283 K and 303 K; time deviation: 1.5 ppm year^{-1}) is separated from the adsorption chamber in which stand pressure, temperature, and corrosive operating conditions. The magnetic suspension system couples an electromagnet linked to the balance with a permanent magnet linked with a crucible containing the adsorbent. This system avoids condensation of subcritical gases[29,30] and data acquisition is automated.[31,32] Adsorption isotherm measurements can be performed from 273 K up to 373 K for pressure from secondary vacuum up to 10 MPa. The adsorbent is exposed to gas at constant temperature and mass variation is measured until thermodynamic equilibrium is reached. The adsorbed mass determination needs a correction of the measured mass m_{meas} due to the buoyancy effect of the gas phase on the adsorbent and crucible volumes according to:

$$m_{exc} = m_{meas} + \rho_{gas} \cdot V_{ads} \tag{1}$$

The mass obtained m_{exc} is commonly called excess adsorbed mass. The gas phase density ρ_{gas} is determined using an appropriate equation of state (EOS); adsorbent and crucible volumes V_{ads} are determined by a direct helium buoyancy effect measurement assuming that helium is not adsorbed.[33] Helium density is evaluated using a modified Benedict-Webb-Rubin EOS.[34] Two pressure transmitters are used: for low pressure measurement, a thermoregulated (318 K) MKS Baratron piezzo-resistive transmitter (627 B) allows to perform measurement up to 133.3 kPa with a 2.7 Par accuracy; an Endress-Hauser pressure transmitter (Cerabar S PMP 635) is used for high pressure determination up to 10 MPa: this second pressure transmitter presents adjustable ranges from 0-0.5 MPa up to 0-10 MPa with an accuracy of 0.1% of the full range value. Temperature is measured using a Pt100 probe (A class) with a 0.15 K accuracy; the probe is located nearby the crucible containing the adsorbent. All the apparatus except the analytical balance is located into an incubator (Binder, KB400) able to keep constant temperature from 263 K up to 373 K. Measurement error is calculated according to the Guide to the expression of Uncertainty in Measurement (International Organization for Standardization, Geneva, 1995). For these measurements of hydrogen sulphide adsorption, maximum uncertainty is estimated to 5% of the excess adsorbed measured quantities with an interval of confidence of 95%.

2.2.2 Hydrogen sulphide adsorption measurements.
Prior to adsorption measurements, MIL samples are outgassed *in-situ* under secondary vacuum (less than 10^{-3} mbar) at 393 K for MIL-53 and at 473 K for MIL-47(V) using a heating-belt surrounding the adsorption chamber (280 W); samples are considered to be out gassed when mass variations are less than 20 µg for *ca.* 15 min, *i.e.* outgassing lasts *ca.* 8 hours. During adsorption measurements, thermodynamic equilibrium is considered to be reached when mass variations are less than 80 µg for 15 min. Figure 2 shows adsorption isotherms for pressures up to 1.8 MPa; adsorbed presented quantities are corrected from the gas phase buoyancy effect using a Helmholtz type EOS for gas phase density computation.[35] A zoom for low pressure under 200 kPa is presented in figure 3.

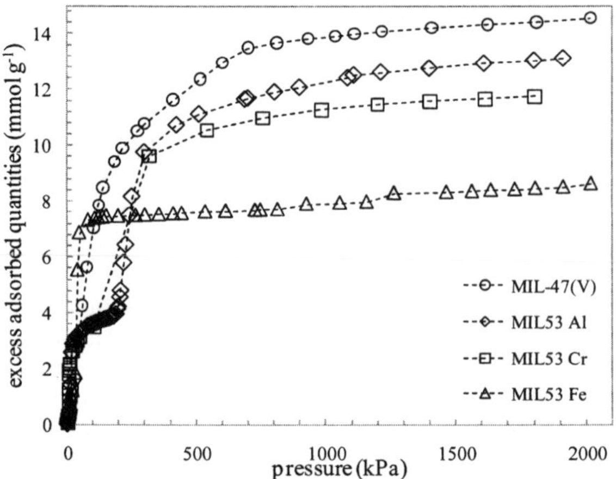

Figure 2   *Excess adsorbed quantities of hydrogen sulphide as a function of pressure on MIL-47(V) and MIL-53(Al, Cr, Fe) at 303 K*

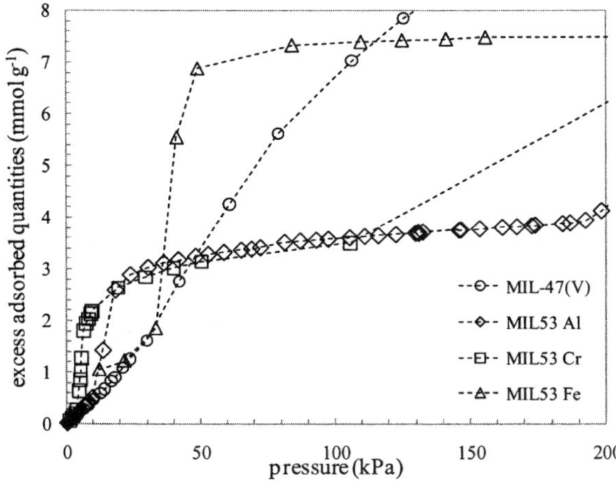

Figure 3   *Zoom on the excess adsorbed quantities of hydrogen sulphide as a function of pressure on MIL-47(V) and MIL-53(Al, Cr, Fe) at 303 K for pressures up to 200 kPa*

The rigid MIL-47(V) presents a I-type monotonic isotherm with a initial Henry constant of 47.0 mmol g^{-1} MPa^{-1} and a maximum adsorption capacity of 14.6 mmol g^{-1} whereas flexible MIL-53 isotherms show two steps: after a quasi-Henry zone for very low pressures, a first step begins at 9.0 kPa, 4.5 kPa and 9.0 kPa for the Al, Cr and Fe solids respectively. The second step begins at 210 kPa, 118 kPa and 33 kPa respectively, with maximum adsorption capacities of 11.7, 13.1 and 8.5 mmol g^{-1}. As previously observed with polar molecules like water or carbon dioxide, these steps can be attributed to a breathing effect of solids: MIL-53s could exist in an open structure at very low pressures, closed for intermediate pressures, and reopen for high hydrogen sulphide pressures. Infrared spectroscopic measurements under hydrogen sulphide atmosphere can confirm this hypothesis. It is important to note that after the adsorption of hydrogen sulphide, MIL-

53(Fe) is destroyed: the formation of a black powder likely an iron sulphur and a deposit of white crystallized powder likely terephtalic acid is observed in the crucible containing the adsorbent and at the surface of the adsorption chamber. That is the reason why IR spectroscopic measurements are not performed on this sample. IR spectroscopic study focuses on the MIL-47(V) and the MIL-53(Cr). MIL-53(Al) has not been studied because of very similar effects to MIL-53(Cr) solid.

2.3.   IR measurements

MILs samples are deposited on a slice of silicon after dispersion in pure-grade ethanol; the slice is then placed in a quartz cell equipped with KBr windows. The cell is connected to a vacuum line for the outgassing and for the adsorbate gas inlet. IR spectra are recorded using a Nicolet Nexus spectrometer equipped with an extended KBr beam splitting device and a mercury cadmium telluride cryodetector. IR spectra are recorded in the range 500-5600 cm^{-1} with a resolution of 4 cm^{-1}. Samples are cooled at 220 K with an octane-carbo ice mixture because the *in-situ* system is limited to a pressure of one atmosphere; this technique allows to qualitatively observe the second step presented by adsorption isotherm measurements on IR spectra. Figures 4 and 5 present IR spectra for MIL-53(Cr) and MIL-47(V) solids.

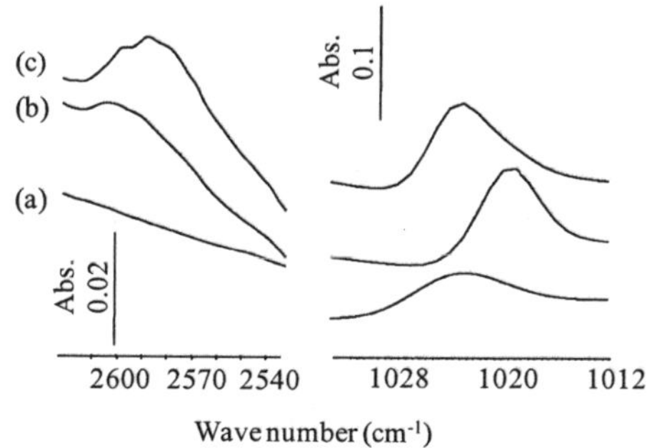

Figure 4   *IR spectra of MIL-53(Cr) at 220 K: (a) vacuum, (b) 1.33 kPa Torr of H₂S, (c) 6.93 kPa of H₂S*

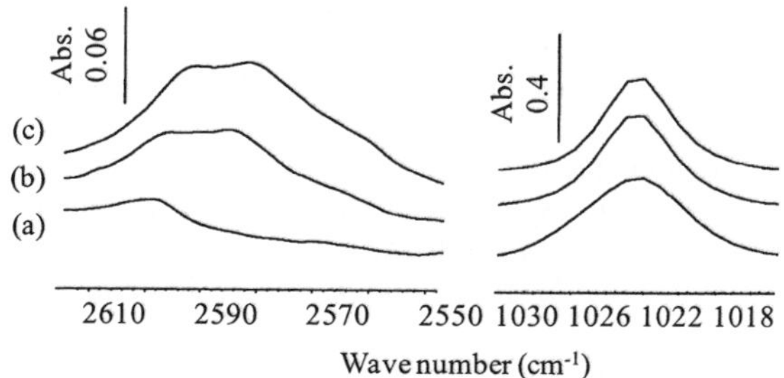

Figure 5   *IR spectra of MIL-47(V at 220 K: (a) vacuum, (b) 1.33 kPa of H₂S, (c) 6.53 kPa of H₂S*

The spectra of both samples display an absorption band in the 1015-1025 cm^{-1} range assigned to the ν18a ring mode of terephtalate entities.[36] This band in the spectrum of activated MIL-53(Cr) is observed at 1022 cm^{-1}, and correspond to the open-form.[25] When hydrogen sulphide pressure increases, this band downward shifts to 1018 cm^{-1} . This red shift is associated to the closing of the structure, *i.e.* it corresponds to the first step observed on the adsorption isotherm. After the second uptake its position comes back to 1022 cm^{-1}, showing the structure reopening for high hydrogen sulphide loading. For MIL-47(V), this ν18a band is not sensitive the hydrogen sulphide uptake: the structure does not breathe and the rigidity of the MIL-47 framework could explain the monotonic isotherm measured.

3 CONCLUSION

In this paper, we have shown that MIL-47(V) MOF exhibits a rigid structure under hydrogen sulphide atmosphere whereas the MIL-53(Cr) solid presents a breathing effect under the same polar conditions. Moreover the metallic cluster presenting OH-functions is a preferential adsorption site. MIL-47 and MIL-53 MOFs hydrogen sulphide and methane adsorption capacities are similar to commercial activated carbons,[37] but these MOFs are regenerable, no chemisorption appears during hydrogen sulphide adsorption. Due to differences between hydrogen sulphide and methane adsorption capacities, MIL-47 and MIL-53 could be good candidates for hydrogen sulphide removing. Thus these first pure compound adsorption measurements have now to be extended to mixture adsorption to characterize the microporous solids under separation and purification conditions using *e.g.* methane as matrix gas to simulate natural gas or biogas desulfuration.

Acknowledgements
Belgian authors acknowledge the European Program InterReg III and the financial support of the Environment and Natural Resources of the Walloon Region (DGRNE, Belgium). French authors acknowledge the ANR ("NoMAC" project) and STREPS ("DeSANNS" project) programs.

References

1   K. S. Lackner and J. D. Sachs, *Brookings Papers on Economic Activity*, 2005, **2**, 215.
2   The BP Statistical Review of World Energy 2007, available at http://www.bp.com.
3   A. Rojey, B. Durand, C. Jaffret, S. Jullian, and M. Valais, *Le gaz naturel, production, traitement, transport*, Technip-Publications de l'Institut Français du Pétrole, Paris, 1994.
4   European Commission Directive 2001/27/EC of 10 April 2001.
5   D. M. Ruthven, *Principles of adsorption and adsorption processes*, Wiley-Interscience, New-York, 1984, pp 336-342.
6   R. T. Yang, *Gas separation by adsorption processes*, Imperial College Press, London, 1997, pp 9-26.
7   G. Férey, C. Mellot-Draznieks, C. Serre and F. Millange, *Acc. Chem. Res.*, 2005, **38**, 217.
8   O. M. Yaghi, M. O'Keeffe, N.W. Ockwing, H. K. Chae, M. Eddaouddi, and J. Kim, *Nature*, 2003, **423**, 705.
9   Férey, G., *Chem. Soc. Rev.* **2008**, 37, 191.

10 G. Férey, M. Latroche, C. Serre, F. Millange, T. Loiseau, and A. Percheron-Guegan, *Chem. Comm.* 2003, **24**, 2976.

11 M. Latroche, S. Surblé, C. Serre, C. Mellot-Draznieks, P. L. Llewellyn, J.-H. Lee, J.-S. Chang, S. H. Jhung, and G. Férey, *Angew. Chem., Int. Ed.*, 2006, **45**, 8227.

12 A. G. Wong-Foy, A. J. Matzger, and O. M. Yaghi, *J. Am. Chem. Soc.*, 2006, **128**, 3494.

13 X. Lin, J. Jia, X. Zhao, K. M. Thomas, A. J. Blake, G. S. Walker, N. R. Champness, P. Hubberstey, and M. Schröder, *Angew. Chem., Int. Ed.*, 2006, **45**, 1.

14 S. Bourrelly, P. L. Llewellyn, C. Serre, F. Millange, T. Loiseau, and G. Férey, *J. Am. Chem. Soc.*, 2005, **127**, 13519.

15 A. R. Millward and O. M. Yaghi, *J. Am. Chem. Soc.*, 2005, **127**, 17998.

16 M. Eddaoudi, J. Kim, N. Rosi, D. Vodak, J. Wachter, M. O'Keeffe, and O. M. Yaghi, *Science*, 2002, **295**, 469.

17 K. Seki and W. Mori, *J. Phys. Chem. B*, 2002, **106**, 1380.

18 S. Kitagawa, R. Kitaura, and S.-I. Noro, *Angew. Chem. Int. Ed.*, 2004, **43**, 2334.

19 A. J. Fletcher, K. M. Thomas, and M. J. Rosseinsky, *J. Solid State Chem.*, 2005, **178**, 2491.

20 G. Férey, *Chem. Soc. Rev.*, 2008, **37**, 191.

21 K. Barthelet, J. Marrot, D. Riou, and G. Férey, *Angew. Chem. Int. Ed.*, 2002, **41**, 281.

22 F. Millange, C. Serre, and G. Férey, *Chem. Commun.*, 2002, 822.

23 C. Serre, F. Millange, C. Thouvenot, M. Noguès, G. Marsolier, D. Louër, and G. Férey, *J. Am. Chem. Soc.*, 2002, **124**, 13519.

24 T. Loiseau, C. Serre, C. Huguenard, G. Fink, F. Taulelle, M. Henry, T. Bataille, and G. Férey, *Chem. Eur. J.*, 2004, **10**, 1373.

25 C. Serre, S. Bourrelly, A. Vimont, N. A. Ramsahye, G. Maurin, P. L. Llewellyn, M. Daturi, Y. Filinchuk, O. Leynaud, P. Barnes, and G. Férey, *Adv. Mater.*, 2007, **19**, 2246.

26 A. Vimont, A. Travert, P. Bazin, J.-C. Lavalley, M. Daturi, C. Serre, G. Férey, S. Bourrelly, and P. L. Llewellyn, *Chem. Commun.*, 2007, **31**, 3291.

27 H. W. Lösch, R .Kleinrahm, and W. Wagner, *Neue Magnetschwebewaagen für gravimetrische Messungen in der Verfahrenstechnik*, "Verfahrenstechnik und Chemieingenieurwesen", VDI-Gesellschaft Verfahrenstechnik und Chemieingenieurwesen (GVC), VDI-Verlag, Düsseldorf, 1994, pp 117-137.

28 F. Dreisbach, R. Staudt, M. Tomalla, and J. U. Keller, *Fundamentals of Adsorption, Proceedings of the 5th International Conference on Fundamentals of Adsorption*, 1996, 259.

29 F. Dreisbach, R. Seif, and H. W. Lösch, *J. Therm. Anal. Calorim.*, 2003, **71**, 73.

30 Y. Belmabkhout, M. Frère, and G. De Weireld, *Meas. Sci. Technol.*, 2004, **15**, 848.

31 G. De Weireld, M. Frère, and R. Jadot, *Meas. Sci. Technol.*, 1999, **10**, 117.

32 M. Frère, G. De Weireld, and R. Jadot, *Fundamentals of Adsorption, Proceedings of the 6th International Conference on Fundamentals of Adsorption*, 1998, 279.

33 F. Rouquerol, J. Rouquerol, and K. Sing, *Adsorption by powders and porous solids*, Academic Press, London, 1999.

34 R. D. McCarty and V. D. Arp, *Adv. Cryo. Eng.*, 1990, **35**, 1465.

35 E. W. Lemmon and R. Span, *J. Chem. Eng. Data*, 2006, **51**, 785.

36 J. F. Arenas and J. I. Marcos, *Spectrochim. Acta.*, 1979, **35A**, 355.

37 L. Hamon, M. Frère, and G. De Weireld, *Adsorption*, 2008, in press.

H$_2$ ADSORPTION IN PRISTINE AND LI-DOPED CARBON REPLICAS OF FAU AND EMT ZEOLITES

T. Roussel[1-2], C. Bichara[1], R. J.-M. Pellenq[1*], K. E. Gubbins[2]

[1] CINAM, CNRS, Campus de Luminy, Case 913, 13288, Marseille Cedex 09, France
[2] Department of Biomolecular and Chemical Engineering, North Carolina State University, Raleigh, North Carolina, 27695-7905,USA
* to whom correspondence should be sent

1 INTRODUCTION

Carbon materials with a large porous volume are widely used in environmental and energy applications. One of the key points for the performances of these materials is their large amount of micropores. The synthesis of conventional carbons by carbonization of an organic precursor and activation by physical or chemical means are one of the routes to generate carbon materials with large microporous domains [1]. Nevertheless, the tailoring of the porosity in such materials is difficult and the pore size distribution of the final material is often wide. Several methods have been proposed to overcome this limitation[2]. One of the most promising techniques is the negative templating of organized porous materials. This synthesis method proceeds through two main steps: the infiltration of a carbon precursor in the porosity of the template and the removing of the template by acid etching. During the last decade, this synthesis technique has been applied with organized mesoporous silicas such as MCM-48 or SBA-15[3,4,5,6] involving essentially mesoporous organized carbon (CMK). Their amount of microporosity is nevertheless too small for storage application at room temperature[7]. Among the possible templates, zeolites can be used to obtain ordered carbon materials with very high microporous volumes[8,9,10,11,12]. The intrinsic ordered microporosity of zeolites leads to well organized carbon replicas that are mainly microporous, and can be an interesting media for hydrogen storage. In this paper, we present atomistic simulations of the synthesis of these new forms of porous carbon based on carbon adsorption in the porosity of the hexagonal (EMT) and cubic (FAU-Y) siliceous faujasite zeolite frameworks[13,14,15]. We first analyze the resulting carbon structures, and then present their H$_2$ adsorption performances both in the pure and lithium doped version of these new porous carbon materials.

2 METHOD

2.1 Grand Canonical Monte-Carlo simulation technique (GCMC):

We performed standard Grand Canonical Monte Carlo (GCMC) simulations on a periodic box containing a unit cell of cubic faujasite zeolite –FAU– (~ 24.85 Å), and the smallest orthorhombic unit cell of the hexagonal faujasite –EMT– (corresponding to two hexagonal unit cells in <x,y> directions: a_x = 17.386 Å , a_y =30.114 Å and a_z = 28.346 Å). Starting with a C_2 dimer, we gradually raised the chemical potential and recorded the average number of adsorbed carbon atoms. This allowed us to explore the accessible configurations of space of the system. Note that the chemical potential of the carbon atoms is referred to that of a fictitious monoatomic ideal gas. This explains the large values of the chemical potential at which the adsorption takes place, between –7.5 and –6.8 eV/atom at 3000K. We use high temperatures far above the experimental conditions to insure a thorough exploration of the accessible states. A minimum of 5.10^5 Monte Carlo macro-steps was performed. The porosity calculated for these faujasite zeolites is about 25%. We thus expect the carbon replica itself to have a porosity of 75% which is very large.

2.2 Carbon/Zeolite, Carbon/Carbon and Lithium/Carbon Interaction Potentials:

The adsorbate-zeolite interactions (carbon with Si, and O atoms of the siliceous form of faujasite frameworks) are assumed to remain weak, *i.e.* in the physisorption energy range. We have used a PN-TrAZ potential function as originally reported for adsorption of rare gases and nitrogen in silicalite-1[16,17]. The PN-TrAZ potential function is based on the usual partition of the adsorption intermolecular energy restricted to two body terms only. The interaction energy of a (neutral) carbon atom at position i is calculated by summing over all atomic sites in the zeolitic matrix[13]. Carbon-carbon interactions are based on the AIREBO potential for carbon materials (Adaptative Intermolecular REactive Bond Order potential)[18]. Concerning doped structures, we obtained the locii of Li ions by energy minimization using the GULP program[19], assuming their *ab initio* charges (+0.7 e⁻) and the composition (LiC_6) giving the electrical charge of carbon atoms.

2.3 Simulation of hydrogen adsorption in doped/undoped carbon replicas:

We also used the GCMC technique to simulate the adsorption of molecular hydrogen in a single unit cell of cubic and hexagonal faujasite replicas called respectively C-FAU and C-EMT (C-FAU-LiC_6 or C-EMT-LiC_6 when Li-doped) using Lennard-Jones based potential functions with parameters given in Table 1. Note that the H_2-H_2 parameters were fitted to reproduce the (ρ,P) equation of state above the critical point[20]. Note that quantum effects at 77 K are not important on the adsorbed amount (less than 1%[21]. The H_2-C_{sp^2} and H_2-H_2 values were taken from reference[22]. The parameterization of an effective H_2-{C-FAU/EMT-LiC_6} potential is based on *ab initio* calculations as discussed below.

3   RESULTS

3.1 Structures Analysis:

Pure carbon replicas of the cubic faujasite Y (C-FAU) and the hexagonal carbon replica of EMT (C-EMT) are represented in the Figures 1(a) and 1(b) below. C-FAU can be seen as a set of tetrahedrically interconnected single wall nanotubes. Its unit cell contains 629 atoms. The carbon replica C-FAU shows a hexagonal ordered distribution of spherical interconnected pores in agreement with experimental TEM observations[23]. On the other hand, C-EMT structure can be considered as a pillared bundle of single wall undulated nanotubes, hexagonally interconnected. It is denser than the cubic phase since its unit cell contains 744 atoms for a similar volume. The densities for the C-FAU and C-EMT are 0.81 and 0.96 g/cc respectively (see Table 1).

	σ (Å)	ε / k (K)
$H_2 - H_2$	2,96	34,2
H_2-C	3,18	30,945
H_2-Li	2.66	24.76
$\alpha(H_2) = 1.33\ Å^3$		$q_{Li} = 0.7\bar{e}$

Table 1 *Potential parameters for GCMC simulation of H_2 adsorption*

We have determined carbon replica Pore Size Distributions (PSD) using a Monte Carlo integrations related to those developed for previous stereological studies of nitrogen in disordered porous silica by Gelb and Gubbins[24]. The PSD felt by confined hydrogen molecules in C-FAU shows a very sharp and centered peak at 11-12 Å (Figure 2), *i.e.* larger than previously evaluated with nitrogen parameters[25]. The PSD of C-EMT structure shows smaller interconnected pores of which a peak at 6 Å corresponding to the inner tubes parallel to the (001) direction. Both structures represent new allotropic phases of carbon; these are the negative of zeolite voids and consist in cages delimited by tubular walls made of carbon atoms in their sp^2 hybridization state. The extreme curvature is stabilized by the presence of five and seven member rings. The proportions of six member rings are on the order of 60 %, and 40% are shared by 5, 7 and some 8 member rings. It is noticeable that these structures do not have dangling bonds, or sp^3 carbon atoms. The cohesive energy of both structures is around -7.2 eV/atom. Thus one can say that these carbon phases represent two degenerative states with possible phase coexistence. It has been shown that cubic and hexagonal faujasites can also coexist within the same grain[26]. Interestingly, this new class of synthesizable carbon materials can be considered as minimal surfaces[27] like the well-known but hypothetical carbon schwarzites[28], among which is D-schwarzite (C_{168}). Vanderbilt and Tersoff calculated that this "theoretical" porous material has an extremely high bulk modulus, B_o, of around 0.11 TPa[29] close to that of non porous silicon. The bulk modulus for our carbon replicas of faujasite zeolites is extremely large (see Table 2). By fitting the cohesive (total) energy of

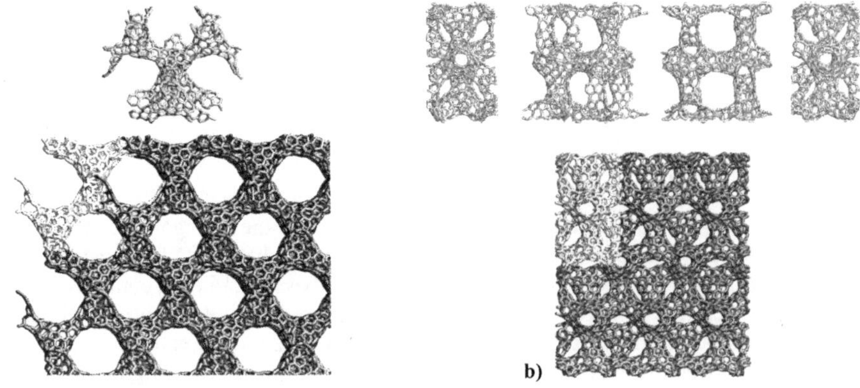

Figure 1 *Ideal pure carbon replicas of faujasite (left : C-FAU ; right : C-EMT; top : the simulation boxes used and bottom : duplicated 16 times.*

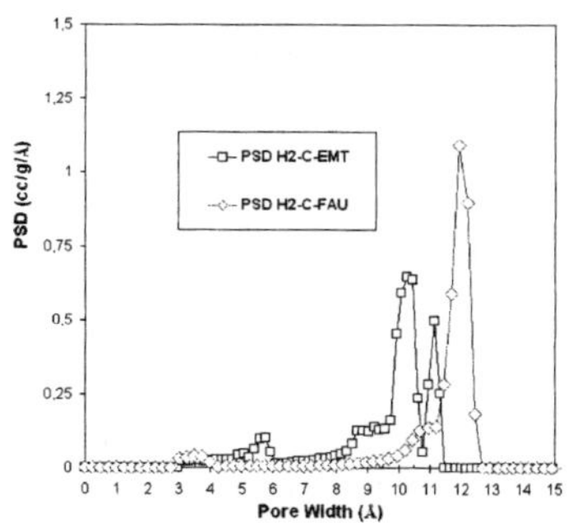

	B_0 (TPa)	ρ (g/cm³)
C-FAU	0.11	0.81
C-EMT	0.10	0.96
Diamond	0.53 - 0.46 [30]	3.5
Quartz	0.06	2.2
C$_{(168)}$	0.11	1.2

Figure 2 *Pore Size Distribution (PSD) viewed by hydrogen gas of the C-FAU (diamond) and C-EMT (square)[24].*

Table 2 *Bulk modulus of carbon replicas, diamond, quartz and schwarzite D (C$_{(168)}$) and their densities.*

our carbon replicas *versus* unit cell volume variation curve around the equilibrium point (following a compression/dilatation path) with the Birsh-Murnaghan equation of state[31], we obtained roughly the same values for the bulk modulus. It is very interesting to note that our carbon replicas of faujasites have a bulk modulus much larger than that of many non-porous minerals such as quartz with a rather low density (less than 1g/cc). It has been shown that bundles of SWNTs exhibit high adsorption energy site for hydrogen confined in their interstitial sites[32]. However, these materials deteriorate at high pressures due to their weak tube-tube cohesive energy compared to the strain imposed by a pressure larger than 40 bar[33]. As a consequence, interstitial sites responsible for the confinement of hydrogen are not preserved. By contrast zeolite carbon replicas can withstand large pressures and thus seem very promising for high pressure hydrogen storage devices.

3.2 Hydrogen storage properties of pure carbon replicas:

We evaluated the gravimetric (wt. %) and volumetric (kg/m³) hydrogen capacities of C-EMT and C-FAU carbon replicas at 77K and 298 K (Figure 3). At low adsorption, the effect of the confinement is clear: due to the H$_2$-matrix interactions, one can store at a same given pressure a larger density of hydrogen in the porous material than in the bulk; the H$_2$

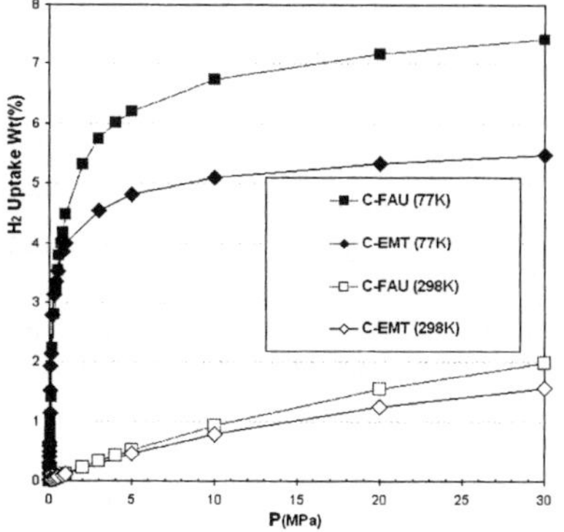
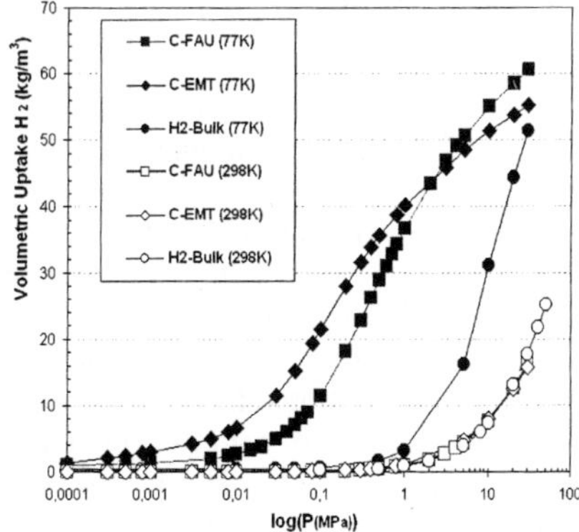

Figure 3 *Gravimetric (wt. %) and volumetric (kg/m³) hydrogen storage capacities.*

uptake is enhanced in the porous carbon phase in agreement with experimental data for different microporous carbon replicas[34]. By contrast at room temperature, it is interesting to see that both the bulk and adsorbed densities are almost identical at all pressures. This is an important result since it demonstrates that even in pores with optimized pore geometry, such as our carbon replicas of faujasite zeolite, hydrogen cannot be stored at room temperature better than in a conventional gas cylinder. The reason for this situation is the intrinsically low interaction of hydrogen molecules with a carbonaceous interface. It is noticeable that the H_2 capacities at low temperature (77 K) show that the amount of void space is sufficiently large to reach the DOE target. Thus C-EMT H_2 storage capacities are better compared to C-FAU until 2 bars due to having smaller pore sizes and a larger isosteric heat of adsorption. Industrial application of H_2 storage at cryogenic temperature (77 K) using carbon replicas of faujasite zeolite could be envisaged since we obtain a density larger than 50 kg.m^{-3} at 50 bars that has to be compared with the H_2 liquid density: 70 kg.m^{-3} at 20 K.

3.3 Hydrogen storage properties of alkaline-doped carbon replicas:

A possible route to achieve an enhanced H_2 physisorption energy was identified in alkaline-doped carbon materials[35,36,37]. Locii of Li ions were obtained by energy minimization using the GULP program[19] assuming their *ab initio* charge and the composition (LiC_6); the charge of carbon atoms being homogenously distributed over all atomic sites. One can see that Li ions stick on the pore wall leaving free the pore core for H_2 adsorption (Figure 4). Using *ab initio* DFT-B3LYP calculations, Maresca *et al* were able to demonstrate that doping the carbon skeleton using alkaline elements such as lithium leads to a strong physisorption of hydrogen in carbonaceous substrates[35]. The underlying mechanism for such a super-physisorption is the electron transfer from alkaline elements to the carbon skeleton atoms. It allows the creation of a polarization (*or induction*) term in the hydrogen-(doped) matrix Hamiltonian that is attractive and proportional to the square of the local electric field in the pore voids due to the doped-matrix electric charges. For the sake of simplicity and transferability in the case of the doped system, we retain the LJ form and parameters as they are for the H_2-C interactions in the case of the un-doped matrix and keep the hydrogen molecular polarizability as a disposable parameter in order to fit *ab initio* calculations. The H_2-Li-doped carbon matrix potential function (in atomic units) is:

$$u_i = -4\sum_{Li,C} \varepsilon_{H_2-[Li,C]} \left[\left(\frac{\sigma_{H_2-[Li,C]}}{r_{H_2-[Li,C]}} \right)^6 - \left(\frac{\sigma_{H_2-[Li,C]}}{r_{H_2-[Li,C]}} \right)^{12} \right] - \frac{1}{2}\alpha_{H_2}\vec{E}^2 \qquad (1)$$

The molecular hydrogen polarizability is then obtained by adjusting such a potential function on *ab initio* plane-wave DFT-GGA points (ABINIT code) in the case of a (periodic) interface made of two graphene sheets confining a single hydrogen molecule and Li species (LiC_{12}, see figure 4a). All hydrogen adsorption potential parameters are given in Table 2. Note that the *ab initio* Li partial charge is 0.7 ē and that the adsorption energy is -15 kJ/mol. Further calculations for the same system in the LDA approximation give an adsorption energy of -24 kJ/mol. These two values are lower and upper bounds of the DFT-B3LYP calculations that give a H_2 adsorption energy of -21 kJ/mol for a LiC_6 doped molecular system[35]. After lithium intercalation, we evaluate the micropore volume losses that are approximately of 0.71 cc/g for the C-FAU$_{(LiC6)}$ instead of 0.78 cc/g for the pristine, and 0.41 cc/g (0.48 cc/g) for the C-EMT$_{(LiC6)}$. Gravimetric (wt. %) and volumetric (kg/m^3) hydrogen storage capacities at room temperature of C-EMT-LiC_6 and C-FAU-LiC_6 are shown in Figure 5. We have assumed that the templates and the nanocomposites (C-LiC_6)

are rigid in these adsorption simulations. Our study reveals that the H_2 uptakes at room temperature can be considerably improved leading to ~ 4 Wt % at 350 bars, close to the 2007 DOE target.

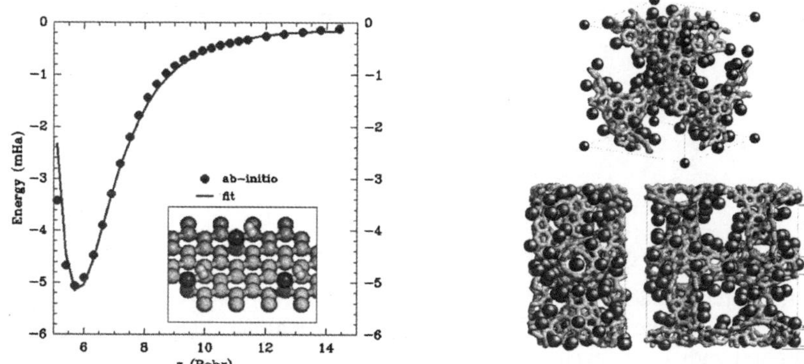

Figure 4 *(left) Fit of ab initio data points using equation (1). Inset: ABINIT (0K) relaxed configuration of two doped (LiC₁₂) graphene sheets in the presence of a single H₂ molecule (z is the H₂ mass centre perpendicular to one of the surface planes taken as reference; for clarity only one grapheme plane is presented). (right) GULP doped relaxed structures C-FAU-(Li-C₆) and C-EMT-(Li-C₆).*

2007. The main point is that the overall evaluation of the interest of any storage materials for automotive applications requires estimating confined hydrogen density and comparing it to that of a H_2 gas cylinder. These data are usually not given or not accessible experimentally since this requires measuring the pore volume. We show that in terms of density too, our alkaline-doped carbon replicas can reach the 2007 DOE target (37 kg.m^{-3} at 350 bars). This is the key issue to envisage this gas as reversible energy storage in the field of transportation. In the light of the paper of Bhatia and Myers[38], from the analysis of the isosteric heat of adsorption curves (see Figure 6), it is clear that neither our pure carbon replicas nor any pure carbon structures can reach the DOE target (4.5 Wt% and 36 kg.m^{-3}).

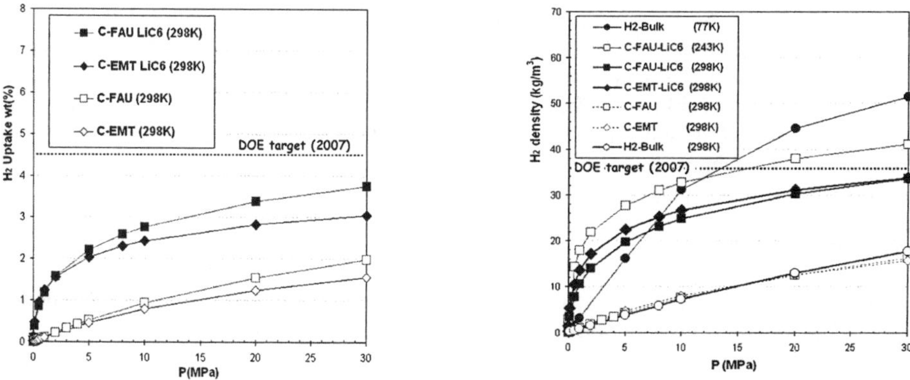

Figure 5 *Gravimetric (wt. %) and volumetric (kg/m³) room temperature (298K) hydrogen adsorption isotherms of hexagonal C-EMT (diamond) and cubic C-FAU (square) carbon replicas of Faujasites and their doped LiC₆ structures (C-FAU-LiC₆ and C-EMT-LiC₆). H₂-bulk data at 77 K are the filled circles, hollow circles at room temperature.*

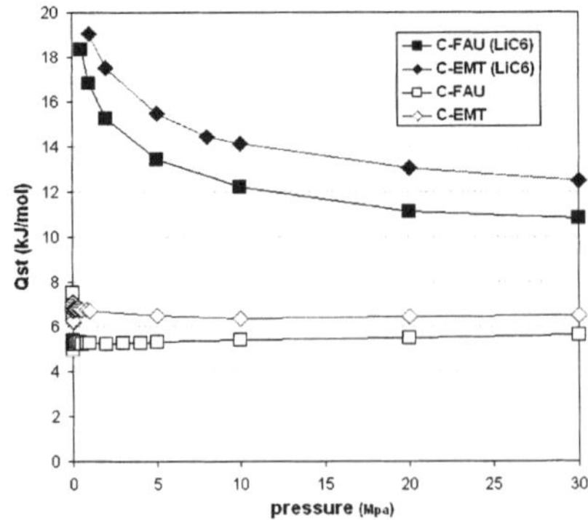

Figure 6 *Isoteric heat of adsorption as a function of pressure.*

However, isosteric heat of adsorption in doped carbon replica exceeds 15 kJ.mol^{-1} at 50 bars. If we consider the filling–exhaustion pressure at 300 bar, the mean integral value of the adsorption enthalpy is even of 16.61 kJ/mol for C-EMT(LiC$_6$), 13.56 kJ/mol for the C-FAU(LiC$_6$). To summarize, doped carbon replicas of faujasite zeolites are a possible route toward an efficient hydrogen storage device working at room temperature under reasonable pressure, acknowledging their ordered structure and high mechanical stability. By comparison Li-doped Metal-Organic Frameworks (Li-MOF)[39] although promising for hydrogen storage, cannot stand large pressure strains.

4 CONCLUSION

Hydrogen storage is the key issue for this gas to be viable as an energy vector in the field of transportation. Porous carbons are possible candidates. In this work, we have considered well-controlled microporous carbon nanostructures, namely carbon replicas of zeolites. We have numerically synthesized these carbon nanostructures at the atomic scale from the replication of the cubic and hexagonal faujasite (FAU and EMT) using a nanocasting procedure based on the Grand Canonical Monte-Carlo simulation technique. We have obtained new forms of crystalline carbon made of tetrahedrally or hexagonally interconnected nanotubes. The pore size networks for these materials give optimized hydrogen molecular storage capacities. However, we demonstrate that these new carbon forms are not interesting for efficient H$_2$ storage at room temperature when compared to the performance of a classical tank. We showed that by doping with an alkaline element such as lithium, one could store the same quantities at 350 bars as a classical tank at 700 bars, very close to the DOE targets set in 2007. This result demonstrates that doping is a possible route to achieve interesting performances for on-board storage systems based on carbon microporous materials.

Acknowledgments

We thank the French Ministry for Higher Education and the National Science Foundation (grant CTS-0626031) for support of this work.

References

1   J. Pikunic, C. Clinard, N. Cohaut, K. E. Gubbins, J.-M. Guet, R. J.-M. Pellenq I. Rannou, J.-N. Rouzaud, *Langmuir*, 2003, **19**, 8565.

2   T. Kyotani, *Carbon*, 2000, **38**, 269.
3   R. Ryoo, S. H. Joo, S. Jun, *J. Phys. Chem. B*, 1999, **103**, 7743.
4   R. Ryoo R, S. H. Joo, M. Kruk, M. Jaroniec, *Adv. Mat.*, 2001, **13**, 677.
5   C. Vix-Guterl, S. Boulard, J. Parmentier, J. Werckmann, J. Patarin, *Chem. Lett.*, 2002, **31**, 1062.
6   C. Vix-Guterl, S. Saadallah, L. Vidal, M. Reda, J. Parmentier, J. Patarin, *J. Mat. Chem.*, 2003, **13**, 2535.
7   T. Roussel, R. J.-M. Pellenq, M ; Bienfait, C. Vix-Guterl, R. Gadiou, M. Johnson, *Langmuir*, 2006, **22**, 4614.
8   Z. X. Ma, T. Kyotani, A. Tomita, *Carbon*, 2002, **40**, 2367.
9   N. Wang , Z. K. Tang, G. D. Li, J.S. Li, *Nature,* 408, **50**, 2000.
10   T. Kyotani T., Z. Ma, A. Tomita, *Carbon*, 2003, **41**, 1451.
11   Z. Ma, Kyotani T., Tomita, A., *Carbon*, 2002, **40**, 2367.
12   K. Matsuoka, Y. Yamagishi, T. Yamazaki, N. Setoyama, A. Tomita, T. Kyotani, *Carbon*, 2005, **43**, 876.
13   FOM. Gaslain, J. Parmentier, VP. Valtchev, J. Patarin, *Chem. Comm.*, 2006, **9**, 991.
14   T. Roussel, R. J.-M. Pellenq, C. Bichara, *Phys. Rev. B,* 2007, **76**, 235418.
15   T. Roussel, A. Didion A., R. J.-M. Pellenq, R. Gadiou, C. Bichara, C. Vix-Guterl, *J. Phys. Chem. C,* 2007, **111**, 15863.
16   R. J.-M. Pellenq, D. Nicholson, *J. Phys. Chem.*, 1994, **98**, 13339.
17   R. J.-M. Pellenq, D. Nicholson D., *Langmuir*, 1995, **11**, 1626.
18   D. W. Brenner, O. A. Shenderova, J. A. Harrison, S. J. Stuart, *J. Phys. Cond. Mat.*, 2002, **14**, 783.
19   J. D. Gale, A. L. Rohl, *Mol. Sim.*, 2003, **29**, 291.
20   A. Wang, J. K. Johnson, *Fluid Phase Equilibria*, 1997, **132**, 93.
21   F. Darkrim, D. Levesque, *J. Chem. Phys.*, 1998, **109**, 4981.
22   W. A. Steele, in *The Interaction of Gases with Solid Surfaces*, Pergamon Press: Oxford, 1974
23   M. Inagaki, K. Kaneko, T. Nishizawa, *Carbon*, 2004, **42**, 1401.
24   L. D. Gelb, K. E. Gubbins, *Langmuir,* 1998, **14**, 2097.
25   T. Roussel, R. J.-M. Pellenq, J. Jagiello, M. Thommes, C. Bichara, *Mol. Sim.,* 2006, **32**, 551.
26   O. Terasaki, T. Ohsuna, *Topics in Catalysis*, 2003, **24**, 1.
27   H. A. Schwartz, in *Gesammelte Mathematische Abhandlungen,* Springer, Berlin 1980.
28   H. Terrones, M. Terrones M., *New J. Phys.*, 2003, **5**, 126.
29   D. Vanderbilt, J. Tersoff J, *Phys. Rev. Lett.*, 1992, **68**, 4, 511.
30   M. Hebbache, *Solid State Com.*, 1999, **110**, 559.
31   F. D. Murnaghan, *PNAS,* 1944, **30**, 244 ; F. Birch, *Phys. Rev.,* 1947, **71**, 809.
32   M. Bienfait,P. Zeppenfeld, N. Dupont-Pavlovsky, M. Muris, M. R. Johnson, T. Wilson, M. DePies, O. E. Vilches, *Phys. Rev. B*, 2004, **70**, 035410.
33   Y. Ye, C. C. Ahn, C. Witham, B. Fultz, J. Kiu, A. G. Rinzler, D Colbert, K. A. Smith, R. E. Smalley, *Appl. Phys. Lett.*, 1999, **74**, 2307.
34   Z. Yang, Y. Xia, X. Sun, R. Mokaya, *J. Phys. Chem.B*, 2006, **110**, 18424.
35   O. Maresca, R. J.-M. Pellenq, F. Marinelli, J. Conard, *J. Chem. Phys.*, 2004, **121**, 24.
36   I. Cabria, M. J. López, J. A. Alonso, *J. Chem. Phys.*, 2005, **123**, 204721.
37   Y. Zhang, L. G. Scanlon, M. A. Rottmayer, P. B. Balbuena, *J. Phys. Chem. B*, 2006, , **110**, 22532.
38   S. K. Bhatia S. K., A. L. Myers, *Langmuir*, 2006, **22**, 1688.
39   S. S. Han, W. A. Goddard, *J. of Am. Soc.*, 2007, **129**, 8422.

COMPARATIVE ANALYSIS OF LOW-TEMPERATURE NITROGEN AND ARGON ADSORPTION ON NONPOROUS AND MESOPOROUS SILICEOUS MATERIALS

E.A. Ustinov[†‡]

[†]Ioffe Physical Technical Institute of RAS, 26 Polytechnicheskaya, St Petersburg, 194021, Russia
[‡]Research & Production Company "Provita", 3-7, 24th Linia, St Petersburg 199106, Russia

1 INTRODUCTION

An effective method of characterization of porous materials is based on analysis of nitrogen adsorption isotherms at 77 K using nonlocal density functional theory (NLDFT). The appearance of highly ordered mesoporous siliceous materials such as MCM-41 and SBA-15 opened up an opportunity for verification of a number of theories. The test of different models against experimental data proved to be some disappointing. Thus, molecular approaches (both NLDFT and numerical simulation) are unable to completely reproduce the adsorption/desorption reversibility. The condensation/evaporation pressure dependence on the pore diameter defies precise description. The same difficulty is encountered in attempts to quantitatively describe the low-pressure region of the N_2 and Ar adsorption isotherms. Recent extension of NLDFT to subcritical N_2 and Ar adsorption on amorphous solids[1-3] substantially improved the description of the isotherms at low pressures because the modified NLDFT versions accounted for the surface heterogeneity inherent to amorphous materials. Nevertheless, some uncertainty still remains on whether capillary condensation or evaporation corresponds to the equilibrium phase transition. We have shown[4-6] that the equilibrium transition pressure in cylindrical pores of MCM-41 samples predicted by NLDFT is surprisingly close to the condensation pressure determined experimentally for pores ranging from 3 to 6-7 nm. Some evidence that the condensation pressure in cylindrical pores occurs as equilibrium transition is obtained by Morishige et al.,[7] although this result is still not completely understood. Recently we found[8] that the adsorption potential in MCM-41 pores is nearly insensitive to the surface curvature in the region of monolayer coverage where adsorption seems to be governed by highly short-ranged forces.[6] Besides, it was shown that in the case of silica surface the potential decays as the inverse forth power of the distance.[2] Such a behaviour suggests that in the low-pressure region N_2 and Ar adsorption on amorphous solids obeys the localized adsorption yielding formation of a peculiar surface layer, which behaves as a powerful secondary source of the potential field. This viewpoint is additionally supported by the fact that the 2D layer exerting the potential field seems to be slightly shifted away from the surface.[6] For this reason, the current work is aimed at developing a new approach based on a combination of the localized adsorption in the monolayer coverage and standard NLDFT in the region of multilayer coverage.

2 MODEL

The physical model is based on the following assumptions. (1) The heterogeneous surface of an amorphous solid is presented as a number of active sites distributed over the adsorption energy, which adsorb gas molecules according to the well-known Langmuir scheme of localized adsorption. (2) The adsorbed molecules adjacent to the surface form a 2D layer located at a distance δ from the surface. This layer is an additional source of adsorption potential. (3) Multilayer non-localized adsorption occurs parallel to the monolayer coverage and can be modelled by means of conventional NLDFT.

Mathematically the task can be reduced to minimization of a thermodynamic functional. In the case of one-dimensional task the thermodynamic functional Ω for open system can be written as follows:

$$\Omega[\rho_S(\alpha),\rho(z)] = \int\rho_S\{kT[\ln\theta + (1/\theta-1)\ln(1-\theta) - g] + q(\alpha) + V(\delta) - \mu\}d\alpha' \qquad (1)$$
$$+ \int\rho\{kT[\ln(\Lambda^3\rho) - 1] + f_{ex}(\bar\rho) + u(\rho) + V(z') + V_S(z'-\delta) - \mu\}dz'$$

Here $\rho_S(\alpha)$ is the surface density of molecules localized in the surface layer on sites, having internal energy $q(\alpha)$, with α being the fraction of the surface where the sites are located; $\theta(\alpha)$ $(= \rho_S/\rho_m)$ is a fraction of sites type α filled by the gas molecules; ρ_m is the capacity of the surface layer; $\rho(z)$, $\bar\rho(z)$ are the local and smoothed densities at a distance z from the surface, respectively; $V(z)$, $V_S(z-\delta)$ are the potential exerted by the solid surface and the peculiar surface layer at a distance z from the surface, respectively; $f_{ex}[\bar\rho(z)]$ is the molar excess Helmholtz free energy defined from the Carnahan–Starling equation of state[9] for the hard sphere fluid; $u[\rho(z)]$ is the intermolecular (attractive) potential calculated in the framework of mean field approximation using the Weeks–Chandler–Andersen (WCA)[10] perturbation scheme; μ is the chemical potential; g is the entropy term of the Langmuir equation assumed to be independent of temperature T and the surface fraction α; k is the Boltzmann constant, and Λ is the de Broglie wavelength.

The smoothed density $\bar\rho(z)$ is defined as a weighted average according to Tarazona theory.[11] Note that in the current model the weighted average accounts for the contribution of molecules residing on the surface layer. The potential exerted by the adsorbed surface layer $V_S(z-\delta)$ is proportional to the surface layer density $\rho_m \int\theta(\alpha')d\alpha'$. The external potential $V(z)$ is assumed to be exerted by the surface oxygen atoms. Therefore, this potential is proportional to the group $\rho^{(s)}\varepsilon_{sf}\sigma_{sf}^2$, where $\rho^{(s)}$ is the surface atoms density, ε_{sf}, σ_{sf} are the solid–fluid potential and collision diameter, respectively.

Minimization of the thermodynamic functional at a specified value of the chemical potential allows one to determine two functions. The first one is the surface density distribution over the surface, i.e., the $\theta(\alpha, \mu)$ function. The second function is the density profile along the distance z from the surface and refers to multilayer adsorption.

The dependence of a group $(q/kT-g)$ on the fraction α can be determined by fitting theoretical results against experimental adsorption isotherm. This is equivalent to evaluating of active sites distribution over the surface. If the second integral in the RHS of eq 1 is omitted, the minimization procedure leads to the superposition of Langmuir equations for the surface layer. Generally, the multilayer adsorption enhances the surface layer adsorption and vice versa. The least square fitting procedure also allows us to determine the capacity of the surface layer and the solid–fluid pair potential.

3 RESULTS AND DISCUSSION

In this section we consider results of application of the model to N_2 at 77.3 K and Ar at 87.3 K adsorption on non-porous silica LiChrospher Si-1000[12,13] and in mesopores of MCM-41 series.[13,14]

3.1 Adsorption on Non-Porous Reference Solid

For the above mentioned systems the function $q(\alpha)/kT{-}g$, parameters δ, ρ_m, and the group $\rho^{(s)}\varepsilon_{sf}\sigma_{sf}^2$ were determined by the least squares fitting. An important problem is determination of the surface area of the sample because the results of analysis are extremely sensitive to the choice of its value. The Brunauer–Emmett–Teller (BET) method often gives unreliable surface estimation and we attempted to circumvent this issue without using the BET. It has been done by comparison of N_2 adsorption isotherms on MCM-41 samples with that on non-porous silica. We briefly consider this technique in section 2.2.

Figure 1 presents N_2 and Ar adsorption isotherms on LiChrospher Si-1000 (a) and the dependence of $(q/kT{-}g)$ on the variable α (b).

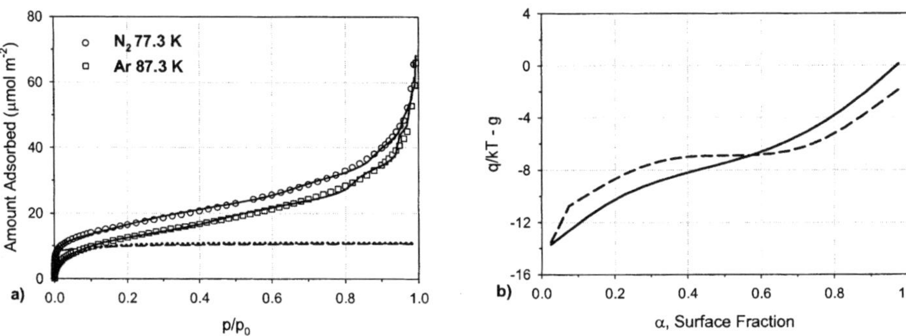

Figure 1   *(a) Nitrogen and argon adsorption isotherms on non-porous silica. Solid lines are plotted with the model. The dashed line (N_2) and dotted line (Ar) show the adsorption in the surface layer. (b) Active site energy distribution over the surface. The solid and dashed lines are for N_2 and Ar, respectively.*

Figure 1a also shows a contribution of the surface layer to the total amount adsorbed. Interestingly, the capacity of the surface layer determined with nitrogen (10.99 $\mu mol/m^2$) is nearly the same as that determined with argon (10.64 $\mu mol/m^2$) in spite of significant difference in the liquid density. This means that molecules of nitrogen and argon occupy the same sites on the surface. The surface density of the surface layer proved to be very close to that for nitrogen in the BET theory equal to 10.25 $\mu mol/m^2$. It gives rise to suppose that the current approach can be applied for estimation and comparison of the surface area of various materials without resort to BET method. Dependences depicted in Figure 2b just reflect strong energetic heterogeneity of the silica surface.

The distance δ between the silica surface and the 2D adsorbed surface layer proved to be one-third of the collision diameter. That is because of the surface roughness, so that adsorbed molecules reside in troughs of the surface. The determined value of the group $\rho^{(s)}\sigma_{sf}^2\varepsilon_{sf}/k$ is 60.08 K for nitrogen and 42.56 K for argon.

The density distribution of nitrogen adsorbed on the silica surface at various relative pressures is presented in Figure 2.

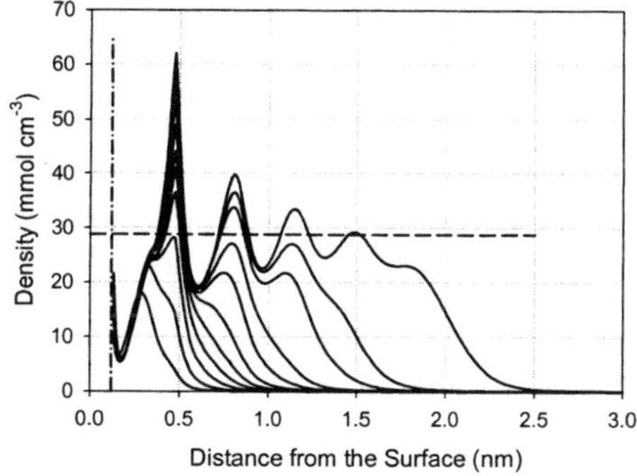

Figure 2　*Nitrogen density profile at the silica surface. Solid lines denote density distributions at relative pressures p/p$_0$ (from left to right): 0.1, 0.2, 0.3, 0.4, 0.5, 0.6, 0.7, 0.8, 0.9, 0.95, 0.99. Dashed line is the liquid density. Dash-dotted line shows the location of the adsorbed surface layer.*

One can see from the figure that the adsorbed nitrogen has a layered structure near the surface, with the distance between neighboring density peaks being close to the N_2 collision diameter as expected.

3.2　Application of the Combined Model to Adsorption in Cylindrical Pores

Analysis of adsorption in cylindrical pores has been carried out using minimization of a thermodynamic potential analogous to eq 1 rewritten for cylindrical geometry. We used experimental data on N_2 at 77.3 K and Ar at 87.3 K adsorption on a series of the same MCM-41 samples[13,14] having diameters from 3 to 6 nm.

3.2.1　Estimation of Surface Area of MCM-41 Samples and Reference Silica

We relied on experimental evidence that t-plots are very close to straight lines in the low-pressure region up to nearly capillary condensation. It does not mean, however, that there is no any potential enhancement due to the surface curvature. Analysis of the theoretical adsorption isotherm on cylindrical surface against that on the flat surface has confirmed the linearity of the t-plot with high accuracy even for pore diameter of 2 nm. To account for the enhancement of the amount adsorbed on the cylindrical surface per unit surface area compared to that on the flat surface we introduced a coefficient $\xi(D)$. Thus, for the pore diameter of 2, 3, 4, 5, and 6 nm the coefficient is 1.099, 1.054, 1.037, 1.029, and 1.023, respectively. As is seen, the effect of the surface curvature is not significant for mesopores, which has already been shown earlier.[8] Hence, once the reference surface area is known, one can determine the surface area of the porous sample using the low-pressure t-plot and the coefficient ξ, and vice versa. On the other hand, the pore wall surface of MCM-41 sample S_p can be determined using XRD data and the pore volume V_p as follows:

$$S_p = 2\sqrt{3}d_{100}^{-1}\left[V_p\left(V_p + 1/\rho_0^{(s)}\right)\right]^{1/2} \tag{2}$$

Here d_{100} is the interplanar spacing; $\rho_0^{(s)}$ is the density inside the solid (2.2 g/cm^3). The above equation accounts for the hexagonal pore geometry, which leads to 5 % larger surface area compared to the equivalent cylindrical pore, having the same volume. The diameter of the equivalent cylindrical pore is given by[15]

$$D = Cd_{100}\left[\rho_0^{(s)}V_p /(1+\rho_0^{(s)}V_p)\right]^{1/2} \qquad (3)$$

where $C = (8/(3^{1/2}\pi))^{1/2}$. An important point is assessment of the pore volume V_p, which is usually determined using the liquid density ρ_L. It is known, however, that the adsorbed phase density at the saturation pressure is substantially larger than ρ_L due to the compression and layered structure of the adsorbate, especially in small mesopores. Therefore instead of ρ_L we use the density $\rho(1,D)$ calculated with the model at $p/p_0 = 1$. Thus, for pore diameter 2, 3, 4, 5, and 6 nm the calculated nitrogen density $\rho(1,D)$ is 38.03, 35.53, 34.04, 33.10, and 32.44 mmol/cm^3, while the liquid N_2 density is 28.84 mmol/cm^3. This is a significant difference, which results in overestimation of the pore diameter for small pores determined with the classical approach. The unexpectedly large average density of the adsorbed gas in the completely filled pore is because of significant contribution of the surface layer into the amount adsorbed, which sharply increases with decrease of the pore diameter. To summarize this issue, we used an iterative scheme to determine the reference surface area. Each of iterations included four steps: (1) determination of D and S_p with eqs 2, 3; (2) calculation the reference surface area using the t-plot and the coefficient $\xi(D)$ for each pore diameter, followed by averaging; (3) recalculation of all molecular parameters by fitting the reference adsorption isotherm at a new surface area; (4) correction the function $\xi(D)$ and $\rho(1,D)$. After several iterations the obtained reference surface area for LiChrospher Si-1000 was 19.73 m^3/g. Some structural parameters for the MCM-41 samples together with the $\xi(D)$, $\rho(1,D)$, and the reference surface area S_{ref} determined with the t-plot are listed in Table 1.

Table 1   *Structural parameters of MCM-41 samples*

| MCM-41 | d_{100} | D | S_p | V_p | $\xi(D)$ | $\rho(1,D)$ | S_{ref} |
	(nm)	(nm)	(m^2 g^{-1})	(cm^3 g^{-1})		(mmol cm^{-3})	(m^2 g^{-1})
(3.1)	3.44	3.03	701.4	0.376	1.054	35.53	19.47
(3.9)	4.17	3.88	708.2	0.639	1.039	34.21	19.73
(4.2)	4.51	4.24	676.7	0.667	1.035	33.81	19.62
(4.6)	5.04	4.53	512.9	0.541	1.032	33.53	19.80
(5.1)	5.37	5.07	579.6	0.685	1.028	33.06	19.74
(5.5)	5.52	5.47	702.6	0.898	1.026	32.78	19.95
(6.0)	5.89	5.91	700.6	0.969	1.024	32.51	19.79

One can see from the Table that the reference surface area determined from various MCM-41 samples is centered at 19.73 m^3/g with a small dispersion.

3.2.2 Modeling of N_2 and Ar Adsorption in Cylindrical Pores

Once determined the reference surface area and all molecular parameters, one can predict N_2 and Ar adsorption isotherms in cylindrical pores. Figure 3 presents some selected isotherms in MCM-41 silica samples, having pore diameter from 3.1 to 6.0 nm. The amount adsorbed in Figure 3 is expressed in terms of density for the sake of generalization of experimental data. The solid lines are calculated for the condition of global minimum of the thermodynamic functional. Dashed lines correspond to metastable state of the adsorbed gas, the vertical section of which refers to the gas-like spinodal pressure.

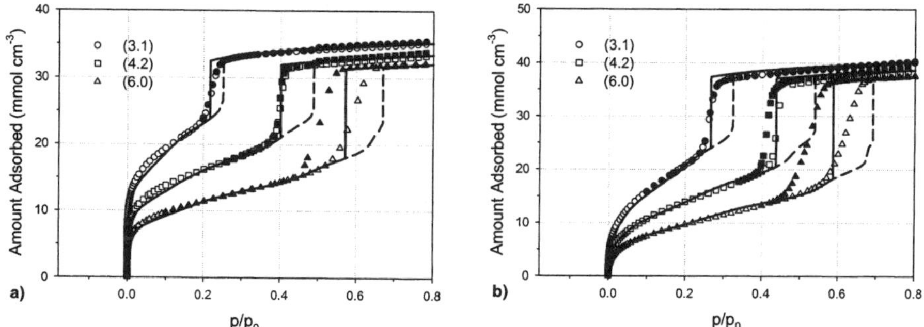

Figure 3　*Nitrogen (a) and argon (b) adsorption isotherms on three selected MCM-41 samples designated as (3.1), (4.2), and (6.0). Open symbols, adsorption; filled symbols, desorption. Solid and dashed lines are calculated with the new model and correspond to desorption and adsorption branches, respectively.*

As is seen from the figure, the new approach fits experimental adsorption isotherms with much success, even though we did not involve the pore size distribution (PSD) function. A good fit extends to relative pressures as low as 10^{-5}, although the predicted loading is slightly underestimated. The reason is that calculations are fulfilled for cylindrical pores, while the actual pore shape is hexagonal. This means that actual pore wall surface area is 5 % larger than that of the equivalent cylindrical pores, which explains the deviation.

Another consequence derived from the figure is that there is no need to resort to the external surface to account for the increase of the amount adsorbed in the upper part of the isotherm. The latter occurs due to the adsorbed phase compression, rather than adsorption on the open surface. The NLDFT excellently describes this feature. The use of the external surface decreases the primary pore volume. On the other hand, this volume increases due to the neglect of the adsorbed fluid compressibility. These two shortcomings of the classical technique partly compensate each other, leading to the pore diameter, which not drastically differs from that determined with the current sophisticated method.

It should be emphasized that the present technique does not involve any free parameters to be adjusted. Nevertheless, one can see from Figure 3 that the condensation step observed experimentally is very close to the evaporation step predicted theoretically. This rather puzzling feature is persistently reproduced by different versions of NLDFT[5,6] and even continual approaches.[6,15] An additional illustration of this phenomenon is presented in Figure 4. Figure 4a shows that in the case of nitrogen adsorption the experimental pore filling pressure is close to that of the equilibrium transition. In the region of reversibility up to the diameter of 4.4 nm both filling and evaporation should occur as equilibrium phase transition. The fact that predicted equilibrium transition pressures (the solid line) in this region coincide with experimental data confirms the correctness of the above conclusion. At the same time, there are some deviations at larger pores, especially in the case of argon. This suggests that if the pore is large enough, the condensation end evaporation should occur at the spinodal and the equilibrium transition pressure, respectively, in complete agreement with classical representations. In pores having diameter less than at least 7 nm there is a cause, which promotes earlier condensation and delayed evaporation.

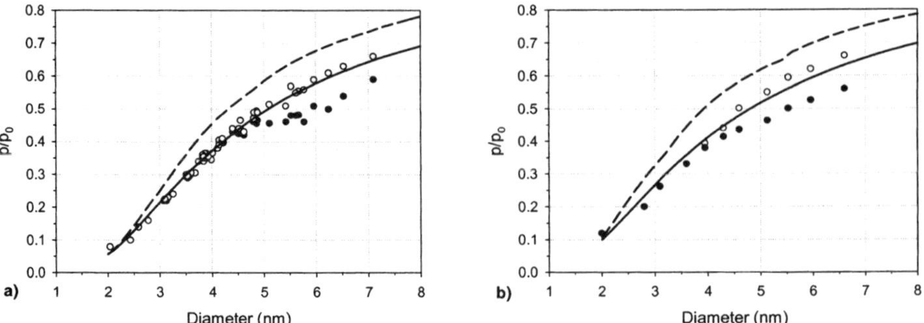

Figure 4 *Condensation/evaporation pressure versus the pore diameter for N_2 at 77.3 K (a) and Ar at 87.3 K (b). Open symbols, condensation; filled symbols, evaporation. Solid lines are prediction for the equilibrium transition. Dashed lines are for the spinodal condensation,*

This mechanism seems to be weaker in the case of argon adsorption. Note that it hardly could be ascribed to constrictions because they would increase nearly proportionally to the pore diameter. Anyway, for pore sizes within approximately 10 nm the PSD analysis can be relied on the N_2 adsorption branch of the isotherm using a kernel obtained for global minimum of the thermodynamic functional.

3.2.3 Application for Pore Size Distribution Analysis of MCM-41 Silica
Pore size distribution function was determined using the Tikhonov regularization procedure.[17] Figure 5 presents PSDs of the MCM-41 silica samples obtained from the adsorption branch of isotherms with nitrogen and argon.

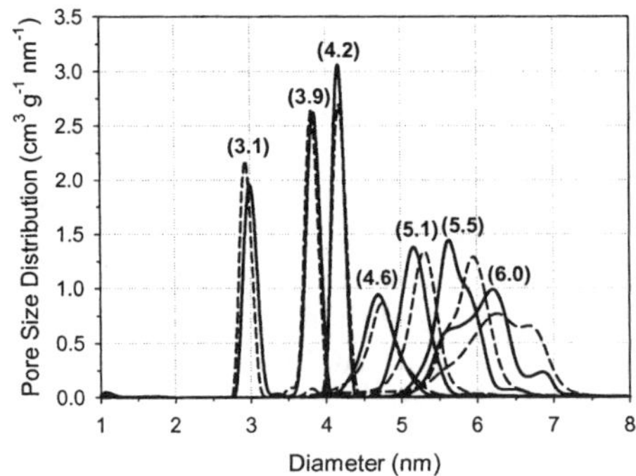

Figure 5 *Pore size distribution of MCM-41 silica obtained from N_2 (solid lines) and Ar (dashed lines) adsorption isotherms.*

In the case of argon the pore diameter is slightly overestimated for larger pores. This is expected because the experimental condensation pressure of argon exceeds the pressure predicted for the equilibrium transition (see Figure 4b). This means that better results of the PSD analysis can be obtained from N_2 adsorption isotherms.

We already presented PSDs of the series of MCM-41 silica obtained with NLDFT[5] and a generalized thermodynamic approach.[16] In those cases the PSD function always indicated the presence of artificial micropores centered at around 1.5 nm. Contrary to that, the present approach does not show any micropores. This is a consequence of the precise description of the low-pressure range of experimental adsorption isotherms with the current approach.

4 CONCLUSION

A new model is developed based on a combination of localized adsorption in the peculiar 2D surface layer and the conventional nonlocal density functional theory for non-localized 3D adsorption. The new concept has been applied to N_2 adsorption at 77.3 K and Ar adsorption at 87.3 K on non-porous and mesoporous siliceous materials. The same number of surface sites on non-porous silica determined with nitrogen and argon confirms the feasibility and consistency of the theory. A comparative analysis of nitrogen adsorption on non-porous silica and MCM-41 samples shows unequivocally that the capillary condensation occurs as the equilibrium phase transition up to the pore diameter as large as 6-7 nm. In the case of argon a mechanism provoking the condensation in the metastable pressure region is weaker than that for nitrogen, so the reliable PSD analysis should be based on N_2 adsorption isotherms.

Acknowledgements

Support from the Russian Foundation for Basic Research is gratefully acknowledged (project 06-03-32268-a).

References

1   E.A. Ustinov, D.D. Do and M. Jaroniec, *Appl. Surf. Sci.*, 2005, **252**, 548.
2   E.A. Ustinov, D.D. Do and M. Jaroniec, *Langmuir*, 2006, **22**, 6238.
3   P.I. Ravikovitch and A.V. Neimark, *Langmuir*, 2006, **22**, 11171.
4   E.A. Ustinov, D.D. Do and M. Jaroniec, *Appl. Surf. Sci.*, 2005, **252**, 1014.
5   E.A. Ustinov, D.D. Do and M. Jaroniec, *J. Phys. Chem., B*, 2005. **109**, 1947.
6   E.A. Ustinov, *Langmuir*, 2008, **13**, 6668.
7   K. Morishige and Y. Nakamura, *Langmuir*, 2004, **20**, 4503.
8   E.A. Ustinov and D.D. Do, *Journal of Colloid and Interface Science*, 2006, **297**, 480.
9   N.F. Carnahan and K.E Starling, *J. Chem. Phys.*, 1969, **51**, 635.
10 J.D. Weeks, D. Chandler and H.C. Andersen, *J. Chem. Phys.*, 1971, **54**, 5237.
11 P. Tarazona, *Phys. Rev. A.*, 1985, **31**, 2672.
12 M. Jaroniec, M. Kruk and J.P. Olivier, *Langmuir*, 1999, **15**, 5410.
13 M. Kruk and M. Jaroniec, *Chem. Mater.*, 2000, **12**, 222.
14 M. Kruk, M. Jaroniec and A. Sayari, *Langmuir*, **13**, 1997, 6267.
15 M. Kruk, M. Jaroniec and A. Sayari, *J. Phys. Chem. B*, 1997, **101**, 583.
16 E.A. Ustinov and D.D. Do, *Colloids and Surfaces A: Physicochem. Eng. Aspects*, 2006, **272**, 68.
17 A.N. Tikhonov and V.Y. Arsenin, *Solutions of ill-posed problems*, Willey, New York, 1977.

CHARACTERIZATION OF MICROPOROUS CARBONS: FROM MATHEMATICAL MODELING TO ATOMISTIC CONSTRUCTION

T. X. Nguyen[1], N. Cohaut[2], J.-S. Bae[1], and S. K. Bhatia[1]

[1] Department of Chemical Engineering, The University of Queensland, QLD 4072, Australia
[2] Centre de Recherche sur la Matière Divisée, UMR 6619 1b rue de la férollerie, 45071 Orléans cedex 2, France

1 INTRODUCTION

Microporous carbon is one of most important adsorbents for gaseous mixture separation and energy storage due to its extremely high surface area, as well as ease in fabrication and modification. The microstructure of the carbon is normally semi-amorphous with short and medium range order induced by complexity of its pore structure. Indeed, local structural images of carbon taken by transmission electron microscopy (TEM) normally show fragmented carbon sheets with various degrees of curvature intercalated by voids or pore spaces. As a result, the slit-pore model, whereby the actual microstructure of the carbon is approximated as an ensemble of disconnected graphitic slit-pores, has been widely used to characterize the microstructure of the carbon. Despite such crude approximation, the slit-pore model with its latest improved version[1], the Finite Wall Thickness (FWT) model, provides correct prediction of adsorption isotherms in carbons, free of pore connectivity problems over a wide range of temperatures and pressures.[2] Unfortunately, the pore connectivity in microporous carbons such as molecular sieve carbons (MSC) and coals is too important to be disregarded due to its serious impact on capability of the adsorbent in gaseous separation and storage.[3] Furthermore, the slit-pore model is unable to correctly predict isosteric heat of adsorption at low coverage[4] as well as to directly determine transport coefficient. This underscores the need for a realistic characterization method.

The realistic characterization of microporous carbons using both Reverse Monte Carlo (RMC)[5-8] and Hybrid Reverse Monte Carlo (HRMC)[9] techniques is now well established.[10-12] The HRMC technique equipped with a short-range carbon potential is currently preferable. Pikunic et. al. sufficiently obtained the atomistic structural model of rather high density saccharose char (ρ_c > 1.5 g/cm^3) which provides correct prediction of argon adsorption at 77 K.[13] However, it is not clear if use of the current HRMC procedure is adequate and even feasible to capture pore structure of low density carbon such as activated carbon fibre ACF15 (ρ_c ~0.9g/cm^3) if initial configuration for the HRMC construction is arbitrarily assigned. This is due to the fact that carbon atoms lying between two opposite pore walls are far beyond the cut-off of the used short-range potentials[14-15], while the experimental pair distribution function is only an averaged structural parameter. Therefore, in this paper we present our newly proposed HRMC construction method for microporous carbons, whereby the resultant pore structure (pore size and pore wall thickness distributions) of the porous carbon obtained by interpretation of argon adsorption at 87 K, based on our FWT model, is utilised to prepare the initial configuration for HRMC construction simulation. Such incorporation of the approximated pore structure into the initial configuration may drive the system to converge more readily to the correct pore structure. The proposed approach was applied to construct the microstructure of the activated carbon fibre ACF15, investigated in our previous work.[16] The HRMC constructed configuration of the ACF15 carbon is validated against experimental adsorption isotherms of simple gases (CO_2, CH_4) over a wide range of temperatures and pressures. Furthermore, comparison between adsorption isotherms of CO_2 and CH_4 in the ACF15 carbon predicted by grand canonical Monte Carlo (GCMC) simulations using the HRMC constructed carbon model and the FWT model is also discussed.

2 METHOD AND RESULTS

2.1 HRMC Atomistic Construction Method for the Microstructure of ACF15 Activated Carbon

This newly proposed HRMC atomistic construction method has been described in detail in our recent paper.[17] Therefore, in the current paper we outline here only the main points of this approach as follows:

2.1.1 Preparation of Initial Configuration. As mentioned above, the structural parameters comprising of pore size distribution (PSD) and pore wall thickness distribution (PWTD) of the ACF15 carbon obtained by interpretation of Ar adsorption at 87 K, taken from our previous work[16], are used to build the initial configuration for the HRMC simulation. From the PSD, a set of specific surface areas $\{s_i\}$ representing corresponding regions in vicinity of most prominent peaks, as shown in Figure 1, can be estimated. The difference in pore width between two consecutive peaks is approximately equal to the interlayer spacing distance (= 3.35 Å) in graphite.

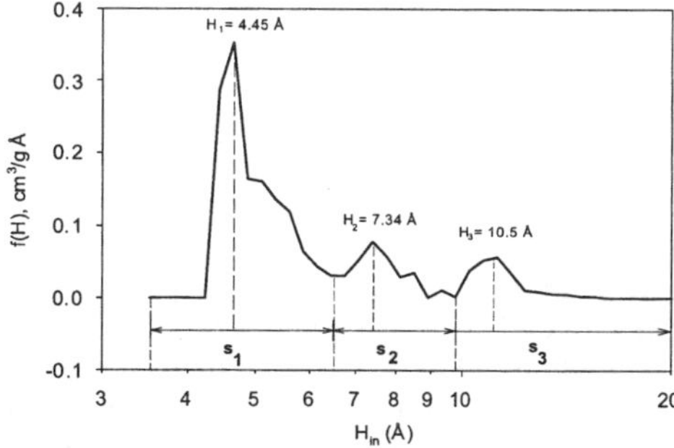

Figure 1 *PSD of the ACF15 carbon derived from Ar adsorption at 87K using FWT model. H_i denotes specific domains, s_i, in the vicinity of most prominent peaks, mentioned in the text.*

Thus, from $\{s_i\}$ and PWTD $\{p_i\}$ we are now able to construct an initial graphitic slit-like pore structure. Subsequently, if needed, the density of this slit-like pore structure is further reduced to that of the actual ACF15 carbon by random removal of a similar number of carbon atoms from each graphitic carbon sheet. Figure 2 shows the resultant slit-like pore configuration, used as the initial configuration for the HRMC simulation described below.

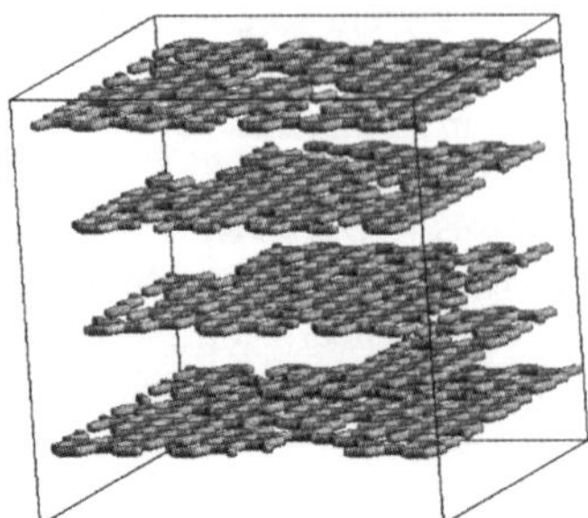

Figure 2   A graphitic slit-like pore configuration used as the initial configuration for the HRMC simulation as discussed in the text.

2.1.2   HRMC Simulation Details. The HRMC construction procedure, as described in detail elsewhere[17], was used. In brief, the HRMC construction simulation procedure is carried out using an annealing process[18], utilizing multiple canonical ensembles which differ only in temperature. The temperature, T, and the standard deviation of experimental pair distribution function[17], σ, the HRMC simulated system are gradually reduced following

$$T = T_o a^{I_{anneal}}$$ (1)

$$\sigma = \sigma_o a^{I_{anneal}/2}$$ (2)

where T_o and σ_o are an initial temperature and adjusting structural parameter for equilibration stage of the system, and taken to be 1000 K and 0.06, which is within the common range of experimental conditions. The parameter a is an annealing rate and given a value of 0.95. I_{anneal} represents the annealing step used in the HRMC simulation and takes the values 1, 2 … N_{anneal}. The total number of annealing steps, N_{anneal}, is determined from the final temperature of the HRMC simulation. In this work, the final temperature has the value of 100 K.

In summary, the ACF15 carbon is modelled as a periodic porous material with dimensions of its unit cell given as 2.95 nm x 2.98 nm x 3.02 nm. Subsequently, 1166 carbon atoms are placed in the unit cell such that simulated bulk carbon density is equal to that of the ACF15 carbon ($\rho_c = 0.87605$ g/cm^3). For the HRMC construction the simulation system is equilibrated at 1000 K for 500,000 Monte Carlo (MC) steps, and followed by further 40x10^6 MC steps at production stage during which the simulation temperature and adjusting structural parameter is gradually reduced to 100 K as described in eqs.1 and 2.

2.2 HRMC Constructed Model of the Activated Carbon Fibre ACF15

Figure 3a illustrates simultaneously a monotonic decrease in the total energy, E, and the structural deviation[17], χ^2, of the simulated pair distribution function (PDF) from the experimental PDF, with the number of MC steps, with both levelling off in the final stage of the annealing process, indicating convergence of the HRMC simulated system. Figure 3b shows an excellent agreement between the converged PDF and the corresponding experimental one.

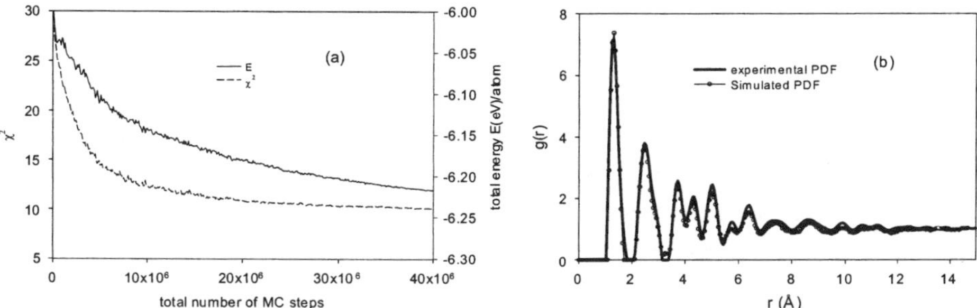

Figure 3 *(a) Variation of total energy, E, and structural deviation, χ^2, with Monte Carlo (MC) steps; (b)Comparison between the converged PDF and experimental one.*

Figure 4a depicts the converged configuration of the ACF15 carbon which contains curly carbon sheets, which are energetically more favorable than defect containing flat carbon sheets created in the initial configuration, shown in Figure 1. The formation of these curved carbon sheets is consistent with curly black fringes, commonly observed in TEM images of activated carbon fibres such as TEM image of ACF7[16] given in Figure 4b

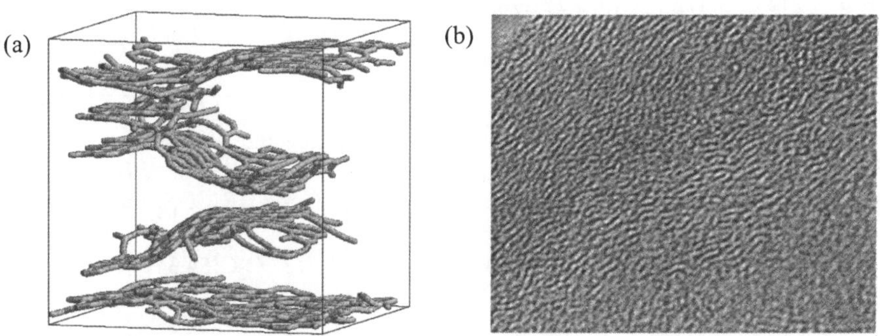

Figure 4 *(a) HRMC converged configuration of the ACF15 carbon and (b) TEM image of the activated carbon fiber ACF7.[16]*

2.3 Validation of the HRMC Constructed Carbon Model

In this section, we briefly present validation of the HRMC constructed carbon configuration of the ACF15 against data of experimental adsorption, as well as of the differential heat of adsorption over a wide range of temperatures and pressures. Further details of the validation are given elsewhere.[17] In order to perform the validation, the Lennard Jonnes (LJ) well depth of carbon-fluid interaction, ε_{sf} for the case of HRMC constructed carbon configuration, is enhanced by a factor of 1.134 in comparison with corresponding value of Steele, $\varepsilon_{sf}^{(1)} = 58K$.[19] Such enhancement, which accounts for positive curvature of carbon sheets[20] in the HRMC constructed configuration, provides best fit of grand canonical Monte Carlo (GCMC) simulated argon adsorption at 87 K to the corresponding experimental one in the actual ACF15 carbon, measured in our laboratory.[17] Thus, the LJ carbon-fluid parameters (ε_{sf}, σ_{cf}) for the HRMC constructed carbon model are given as

$$\varepsilon_{cf} = 1.134\varepsilon_{cf}^{(1)} = 1.134(\varepsilon_{cc}\varepsilon_{ff})^{1/2} \tag{3}$$

$$\sigma_{cf} = 0.5(\sigma_{cc} + \sigma_{ff}) \tag{4}$$

where $\varepsilon_{cc}/k_B = 28$ K and $\sigma_{cc} = 3.4$ Å are the values of Steele for graphitic surface. The LJ fluid-fluid interaction parameters ($\varepsilon_{ff}, \sigma_{ff}$) were obtained by matching between GCMC bulk simulated and corresponding experimental data. All LJ interaction parameters of CO_2 and CH_4 used for GCMC simulations are given in Table 1. It is noted that typo of σ_{cf} for one center model of CO_2 in our recent work[17] is corrected and given in Table 1.

Table 1   (LJ) parameters used with NLDFT calculations and GCMC simulations for the investigated gases used in this work

NLDFT					
Gas	σ_{ff}(Å)	ε_{ff}/k_B (Å)	σ_{cf}(Å)	ε_{sf}/k_B (Å)	Source
CH_4	3.6177	146.91	3.509	64.14	2
CO_2	3.472	221.98	3.436	78.84	2
GCMC					
CH_4	3.751	148.00	3.576	73.00	17
CO_2_1LJ	3.720	236.10	3.56	92.20	17
$CO2\_3LJ$	C:2.824	C:28.68	C:3.112	C-C:32.14	17
	O:3.026	O:82.00	O:3.213	O-C:54.34	
l_{oo}(Å)	2.324				
q(e)	C:+0.6640				
	O:-0.332				

2.3.1 Experimental Adsorption Isotherms. Figure 5 shows excellent agreement between the GCMC simulated CO_2 adsorption isotherm in the HRMC constructed configuration of the ACF15 carbon at 273 K at sub-atmospheric pressure, using a three center molecular model of CO_2, and the corresponding experimental adsorption data, obtained in our laboratory.

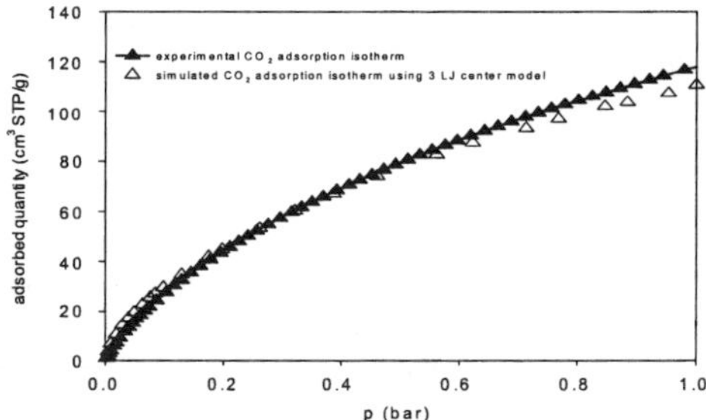

Figure 5 *Comparison between the GCMC simulated adsorption isotherm of CO$_2$ at 273 K in the HRMC converged configuration, and corresponding experimental data.*

Figures 6 (a), (b) show excellent agreement between the simulated high pressure adsorption isotherms of CO$_2$ and CH$_4$ in the HRMC constructed configuration at supercritical conditions and corresponding experimental data. The simulated adsorbed quantity is estimated[17] with consideration of helium pore volume calibration, and deformation of the adsorbent under high pressure adsorption, as

$$m_{ex}^f = m_a^f - \frac{m_a^{He}}{\rho_a^{He}}\rho_b^f - \Delta V_{max}\rho_b^f \qquad (5)$$

where m_a^f and m_a^{He} are the GCMC simulated absolute adsorbed quantities of adsorbing gas and helium. The GCMC simulated CO$_2$ isotherms were obtained using a three center molecular model of CO$_2$. ρ_b^f and ρ_b^{He} are the bulk densities of adsorbing gas and helium respectively, and ΔV_{max} is the maximum excess adsorbent volume under the experimental adsorption condition. ΔV_{max} is estimated to be -0.14 cm^3/g, indicating compression of ACF15 carbon under high pressure adsorption.

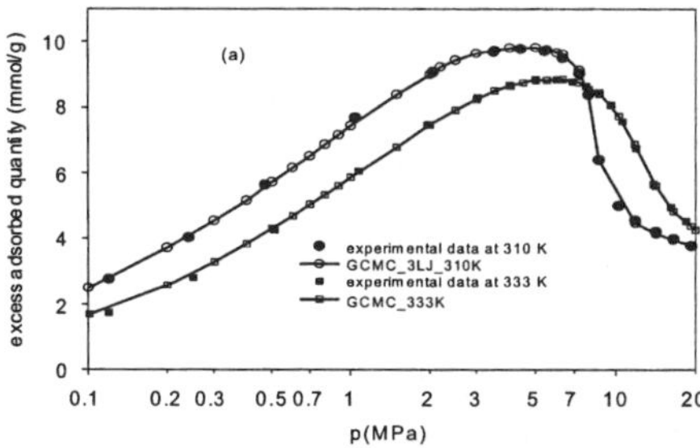

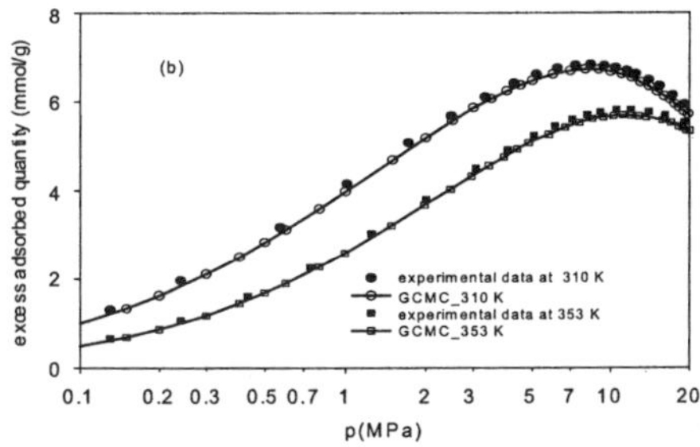

Figure 6 *Comparison between the GCMC simulated adsorption isotherms in the HRMC constructed configuration of the ACF15 carbon using eq.5 and corresponding experimental data for ACF15 carbon. (a) CO$_2$ and (b) CH$_4$.*

2.3.2 Comparison with the FWT model. Here we make comparison between the HRMC constructed carbon model and the FWT model in terms of prediction of adsorption isotherms in the ACF15 carbon. For consistent comparison with the FWT model for the case of CO$_2$, one center model was used to obtain GCMC simulated adsorption isotherms of this gas using the HRMC constructed carbon model. LJ interaction parameters of CH$_4$ and CO$_2$ used for GCMC simulations and Nonlocal Density Functional Theory (NLDFT) calculations are given in Table 1. CO_2 adsorption isotherms were predicted by the FWT model using characterization results (pore size and pore wall thickness distributions) obtained from interpretation of experimental Ar adsorption isotherm at 87 K using the FWT model.[16]

Figures 7(a), (b) show that the FWT model and HRMC constructed carbon model using the one center model of CO_2 overpredict experimental adsorption data of this gas at subcritical and supercritical conditions due to the use of inaccurate molecular model of this gas. Furthermore, it is interesting that the FWT model and HRMC constructed carbon, using the one center model, provide very similar adsorption isotherm at 273 K but not at 310 K, especially in the high pressure range (> 0.2 MPa). This suggests difference in accessible pore volume between Ar and CO_2 at high pressure due to the difference in their molecular geometry.[21]

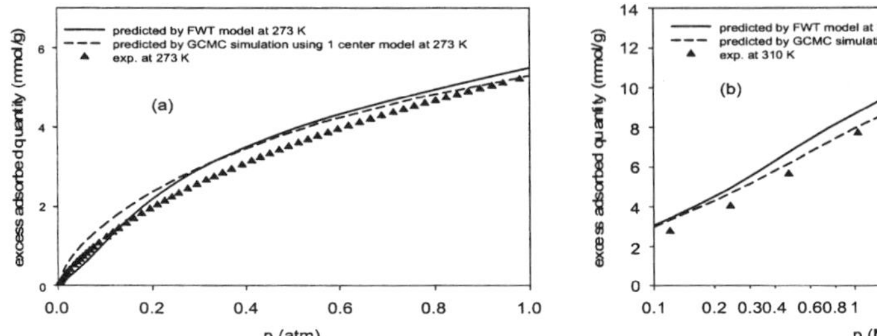

Figure 7   *Comparison between CO_2 adsorption isotherms in ACF15 predicted by the FWT model and GCMC simulation using the HRMC constructed model, and corresponding experimental data. (a) at 273 K and (b)310 K.*

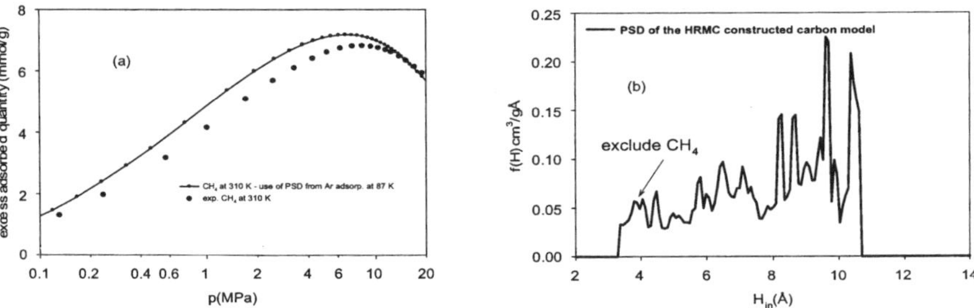

Figure 8   *(a) Comparison between CH_4 adsorption isotherm in the ACF15 carbon at 310 K predicted by the FWT model and corresponding experimental data. (b) Pore size distribution of the HRMC constructed model of the ACF15 carbon was determined by spherical approximation method.*

Figures 8 (a) shows the prediction of high pressure adsorption of CH_4 by the FWT model. From this Figure, it can be seen that the FWT model using the characterization results (pore size and pore wall thickness distributions), extracted from experimental adsorption isotherm of a smaller molecule such as Ar, slightly overpredicts the experimental adsorption isotherm of larger molecule such as CH_4. This was not observed for the case of the HRMC constructed carbon model, as presented in Figure 6b. Further examination of the pore size distribution (PSD) of the HRMC constructed carbon model probed by the spherical approximation method, as presented in Figure 8 b, shows this PSD contains a significant fraction of small pores whose dimension (<3.8 Å) does not accommodate the CH_4 molecule, although CH_4 may transport through these constricted pores within the experimental time scale at supercritical condition.

3 CONCLUSIONS

We have presented here a novel HRMC construction method for low density porous carbons. The use of the graphitic slit-like pore model as initial configuration enables much lower initial temperature (1000 K) to be used for the annealing process than that with the use of an arbitrary carbon model (normally 5000 K at which carbon is liquid form). Accordingly, our proposed method is able to speed up the convergence of HRMC run several times. The proposed method was specifically applied to construct the microstructure of the activated carbon fibre ACF15. Subsequently, the successful verification of the HRMC constructed carbon configuration against experimental adsorption data of the simple gases (CO_2, CH_4) over a wide range of temperatures and pressures was also shown in this work.

Acknowledgements

This research has been supported by a grant from the Australian Research Council, under the Discovery scheme.

References

1   T.X. Nguyen and S.K. Bhatia, *Langmuir*, 2004, **20**, 3532.
2   T.X. Nguyen, S.K. Bhatia, and D. Nicholson, *Langmuir*, 2005, **21**, 3187.
3   T.X. Nguyen and S.K. Bhatia, *Langmuir*, 2008, **24**, 146.
4   Y. He and N.A. Seaton, *Langmuir*, 2005, **21**, 8297.
5   R.L. McGreevy and L. Putsai, *Mol. Sim.*, 1988, **1**, 359.
6   K.T. Thomson and K.E. Gubbins, *Langmuir*, 2000, **16**, 5761.
7   J. Pikunic, C. Clinard, N. Cohaut, K.E. Gubbins, J.-M. Guet, R.J.-M. Pellenq, I. Rannou, and J.-N. Rouzaud, *Langmuir*, 2003, **19**, 8565.
8   S.K. Jain, J.P. Pikunic, R.J.-M. Pellenq, K.E. Gubbins, *Adsorption*, 2005, **11**, 355.
9   G. Opletal, T.C. Petersen, D.G. McCulloch, I.K. Snook, I. Yarovsky, *Mol. Sim.*, 2002, **28**, 927.
10 T.C. Petersen, I.K. Snook, I. Yarovsky, D.G. McCulloch, B. O'Malley, *J. Phys. Chem.*, 2007, **111**, 802.
11 T.X. Nguyen, S.K. Bhatia, S.K. Jain, and K.E. Gubbins, *Mol. Sim.*, 2006, **32**, 567.
12 S.K. Jain, R J.-M. Pellenq, J.P. Pikunic, and K.E. Gubbins, *Langmuir*, 2006, **22**, 9942.
13 J. Pikunic, P. Llewellyn, R.J.-M. Pellenq, K.E. Gubbins, *Langmuir*, 2005, **21**, 4431.
14 D.W. Brenner, *Phys. Rev. B*, 1990, **42**, 9458.
15 N.A. Marks, *J. Phys.: Condens. Matter*, 2002, **14**, 2901.
16 T.X. Nguyen and S.K. Bhatia, *Carbon*, 2005, **43**, 775.
17 T.X. Nguyen, N. Cohaut, J-S. Bae, and S.K. Bhatia, *Langmuir*, **2008**, 24,7912.
18 S. Kilpatrick, *Science*, 1983, **220**, 671.
19 W.A. Steele, *J. Phys. Chem.*, 1978, **82**, 817.
20 J.B. Klauda, J. Jiang, and S.I. Sandler, *J. Phys. Chem.*, 2004, **108**, 9842.
21 S.K. Bhatia, K. Tran, T.X. Nguyen, and D. Nicholson, *Langmuir*, 2004, **20**, 9612.

HYBRID REVERSE MONTE CARLO SIMULATIONS OF MICROPOROUS CARBONS

J. C. Palmer[1], S. J. Jain[1], K. E. Gubbins[1], N. Cohaut[2], J. E. Fischer[3], R. K. Dash[4], Y. Gogotsi[4]

[1]Department of Chemical and Biomolecular Engineering, North Carolina State University, Raleigh, NC 27695-7905, U.S.A.
[2]Centre de Recherche sur la Matière Divisée, CNRS-Orléans University 1B, rue de la Férollerie 45071, Orléans Cedex 02, France
[3]Department of Materials Science and Engineering, University of Pennsylvania, 3231 Walnut Street, Philadelphia, Pennsylvania 19104-6272, U.S.A.
[4]Department of Materials Science and Engineering, Drexel University, 3141 Chestnut Street, Philadelphia, Pennsylvania 19104, U.S.A.

1 INTRODUCTION

Disordered microporous carbons are amorphous materials synthesized from organic precursors such as nut shells, woods, or charcoals by high temperature pyrolysis and activation, or from inorganic precursors, such as metal carbides, by thermo-chemical treatment[1,2]. As a result of the low cost of production and very high surface area of these materials, they have found widespread use in many industrial applications such as gas and liquid purification, catalytic processes, gas storage, batteries, fuel cells and electrochemical capacitors[1]. Due to their lack of periodic structure, little is known about the detailed structure of these materials and how their bonding and disordered structure affects the nano-scale phenomena (e.g. adsorption, diffusion, chemical reaction) that take place within them. This poses a significant problem in developing porous carbons that have carbon skeletons and pore structures that are optimized for specific applications.

In order to develop an understanding of the link between the structures of microporous carbons and their physical properties, Jain et al. adapted a reconstructive reverse Monte Carlo (RMC) simulation technique known as Hybrid Reverse Monte Carlo (HRMC), previously developed by Opletal et al.[3] for amorphous carbons, to microporous carbons[4]. The HRMC method generates realistic molecular models of microporous carbon structures from experimental scattering data. Unlike its predecessor, constrained reverse Monte Carlo (CRMC)[5], which uses nearest neighbor and bond angle constraints to control local carbon bonding structure, the HRMC method uses a many-body potential in conjunction with the reverse Monte Carlo routine to simultaneously minimize the error in fit to experimental diffraction data and the total energy of the system. The use of the many-body potential ensures that the resulting models capture the correct chemistry, minimizing the presence of thermodynamically unstable structures such as three and four member carbon rings. Jain et al. used the HRMC method to generate model structures for three saccharose derived carbons (CS400, CS1000, and CS1000a). Through structural analysis, Jain et al. demonstrated that the resulting models were not only consistent with X-ray diffraction data but also have heterogeneous carbon ring structures that are energetically stable. Using grand canonical Monte Carlo simulations, they showed that the models obtained from HRMC are capable of predicting adsorptive properties for simple gases such as argon that are in excellent agreement with experimental measurements.

The purpose of this work is to demonstrate the range of applicability of the HRMC method in developing realistic molecular models for a wide variety of microporous carbons. We use the HRMC technique to generate models for two types of microporous carbons that have shown promise in numerous applications, particularly in gas storage. We present a detailed structural analysis of models for MeadWestvaco's chemically-activated wood-derived ultra-microporous carbon (UMC, density 0.69 g/cc) and a titanium carbide derived carbon synthesized by chlorination of TiC at 800°C. (TiC-CDC, theoretical apparent density 0.98 g/cc)[6]. The model structures are further characterized by examining the adsorption characteristics of a Lennard-Jones model for argon at 77K using grand canonical Monte Carlo simulations. We explain how the HRMC method is able to capture subtle differences in the microstructures of the various carbon samples and comment on how these differences give rise to distinct thermodynamic properties among the different carbons.

2 SIMULATION DETAILS

The primary aim of reverse Monte Carlo methods is to develop atomistic models of solids or liquids that have structural features that match experimental structural measurements of real systems. As initially proposed by McGreevy and Pusztai[7], the structural property most commonly used in RMC methods is the structure factor. The structure factor, $S(q)$, is a measurement obtained from X-ray or neutron diffraction experiments that describes the scattering characteristics of a material. The RMC procedure minimizes the cost function

$$\chi^2 = \sum_{i=1}^{N} [S_{sim}(q_i) - S_{exp}(q_i)]^2 \tag{1}$$

where S_{sim} and S_{exp} are the simulated and experimental structure factors, respectively, and N is the number of experimental data points. Starting with an initial atomistic configuration, S_{sim} is first calculated. Next, the atomic configuration is altered by randomly displacing a single atom or cluster of atoms. The new configuration is accepted using the Metropolis acceptance probability

$$P_{acc} = \min\left[1, \exp\left(-\frac{1}{T_\chi}(\chi_{new}^2 - \chi_{old}^2)\right)\right] \tag{2}$$

where T_χ is a temperature-like parameter.

For systems where only two-body interactions are important in determining the structure, the RMC procedure will produce a unique structure. However, for bonded systems, in which many-body forces dictate the microstructure, uniqueness of the resulting structure is not guaranteed[8]. Thus, the HRMC method uses an additional energetic constraint to account for these many body-forces and uses the acceptance probability

$$P_{acc} = \min\left[1, \exp\left(-\frac{1}{T_\chi}\left((\chi_{new}^2 - \chi_{old}^2) + \frac{1}{w}(U_{new} - U_{old})\right)\right)\right] \tag{3}$$

where U_{new} and U_{old} are the energies of the new and old configuration and w is a weighting factor.

In our simulations, we use the reactive empirical bond order (REBO) potential[9] to calculate the energy of the system

$$U = \sum_i \sum_{j(>i)} \left[V^R(r_{ij}) - b_{ij}V^A(r_{ij})\right] \tag{4}$$

The total energy is calculated from repulsive and attractive potential contributions, $V^R(r_{ij})$ and $V^A(r_{ij})$, respectively. A many-body term, b_{ij}, that accounts for the strength of covalent bonding between atoms i and j is used to weight the relative strengths of the attractive and repulsive terms. This many-body weighting factor is affected by variables in the local environment of atoms i and j, such as coordination numbers, conjugation effects, and bond angles. A more detailed discussion on the calculation of b_{ij} can be found in Ref [9].

In addition to the use of a many-body reactive potential, we use the radial distribution function (RDF), $g(r_{ij})$, instead of the structure factor, as the target function in our HRMC simulations. The radial distribution function is a measure of the probability of finding two atoms with a separation distance, r_{ij}. The use of the RDF is equivalent to using the structure factor. However, the RDF has been chosen because it is computationally less expensive and has a more intuitive physical interpretation.

3 RESULTS AND DISCUSSION

3.1 Microstructure Characteristics of UMC and TiC-CDC

The experimental radial distribution functions for UMC and TiC-CDC samples were calculated from experimental structure factors data using the Monte Carlo (MCGR) method as first proposed by Soper[10]. We then used these RDFs in the HRMC method to develop models for both UMC and TiC-CDC carbons. Cubic simulation cells with periodic boundary conditions and cell edge lengths of 30 and 35 Ångströms were used for UMC and TiC-CDC, respectively. The density used in the simulation was 0.69 g/cc for UMC, as measured by Hg porosimetry. The density used for TiC-CDC was 1.0 g/cc, which is the close to the theoretical ideal density of 0.98 g/cc. The theoretical ideal density calculation for the TiC-CDC assumes that during the chemical synthesis process, the exterior dimensions of the precursor crystallite remain unchanged.

The carbon-carbon RDFs calculated from experiments were compared with the simulated RDFs for both carbon samples, and are shown in Figure 1. The model RDFs were in excellent agreement with experiment over the entire range of the measured RDFs. The first peaks of both the simulated and experimental RDFs were centered at 1.41 Å. The nearest neighbor coordination number, $n(1)$, was calculated from

$$n(1) = \int_{r_1}^{r_2} 4\pi r^2 \rho_c g(r) dr \tag{5}$$

where ρ_c is the carbon density and r_2 and r_1 are the upper and lower bounds of the first peak in the RDF. For our calculations, values of 1.2 Å and 1.6 Å were chosen for r_2 and r_1, respectively. Integrating over the first peak of the RDF, the nearest neighbor coordination number was found to be approximately 2.75 for UMC and 2.6 for TiC-CDC. While the carbon-carbon bonding distance of 1.41 Å was consistent with that of sp^2-hybridized carbon, the coordination number deviated significantly from the value of 3 for graphite.

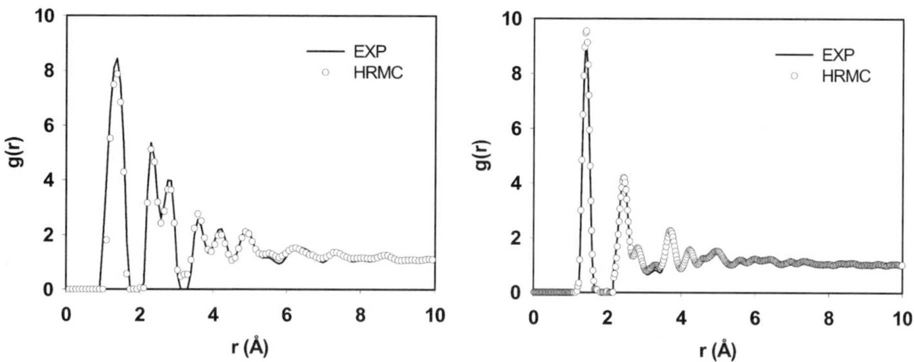

Figure 1: *Experimental and HRMC model radial distribution functions for UMC (left) and TiC-CDC (right).*

To characterize the medium range order of our structures, the ring size distributions were calculated for both of our model structures using the shortest path criteria of Franzblau[11]. The results for both models are summarized in Table 1. Both distributions peaked around 6 member rings, which is the ring size found in graphite. A high presence of five and seven member rings as well as non-planar six member rings indicated a significant deviation from graphitic structures and was in agreement with the interpretation of the RDFs. Moreover, there was an absence of high energy structures such as three and four member rings that are commonplace in model microporous carbon structures generated with the RMC and CRMC methods.

The bond angle distributions shown in Figure 2 for the model carbon samples also confirmed the presence of 6 member rings. Both distributions were centered at 120°, which is the bond angle for hexagonal ring structures and had similar variances and smaller peaks near 180°. The presence of linear carbon chains in CDC has been suspected[12], but no experimental evidence has ever been provided. The peak near 180° was substantially larger for the UMC model than for the TiC-CDC model. This indicated that the UMC model contained a higher fraction of chain-like carbon structures. These chain-like structures were clearly seen in the snapshot of the UMC model, while the TiC-CDC appeared to have more ring structures (Figure 3). The bond angle distributions for the CS400 (density 1.2 g/cc) and CS1000 (density 1.5 g/cc) models developed by Jain et al. using the HRMC technique also had peaks at 180°, while the bond angle distributions for the models of these two carbons generated using the CRMC method did not[4, 5]. The presence of these structures was not aptly captured using the RMC methods because of the absence of appropriate energetic constraints.

Table 1: *Ring statistics for UMC and TiC-CDC*

Ring Size	% of Rings	
	UMC	TiC-CDC
3	0.0	0.0
4	0.0	0.0
5	4.7	9.5
6	36.8	33.7
7	24.9	24.2
8	17.1	17.6
9	16.6	15.0

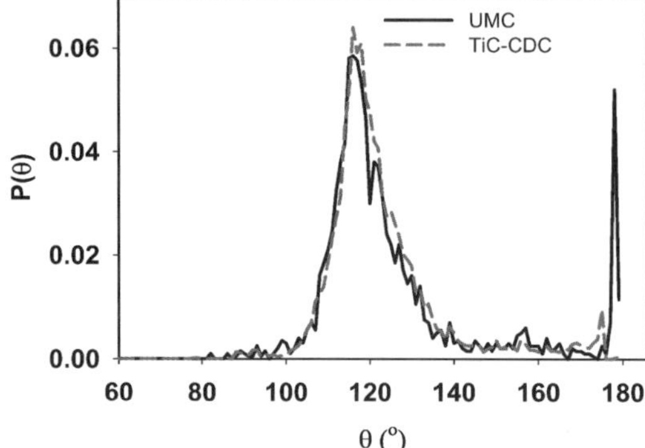

Figure 2: *Bond angle distributions for UMC and TiC-CDC.*

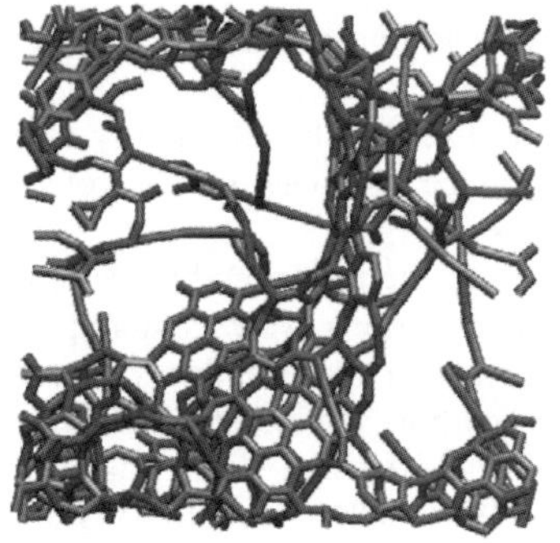
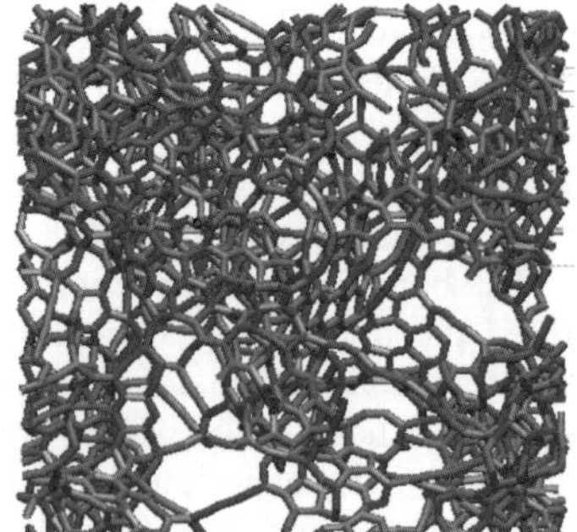

Figure 3: *Simulation snapshots of the HRMC models for UMC (left) and TiC-CDC (right)*

3.2 Porous and Adsorptive Characteristics of UMC and TiC-CDC

To further characterize our model structures, we examined the geometric pore size distribution (PDS), porosity, and adsorptive properties. We calculated the PSD for UMC and TiC-CDC using the method of Gelb and Gubbins[13]. The calculated PSDs for both carbons are shown in Figure 4. While the shape of the PSD distributions were similar for the two carbons, the UMC structure had a larger fraction of larger pores, with an average pore size of 8.1 Å as compared to 6.8 Å for the TiC-CDC model. The porosity and volume occupied by solid matter was calculated as described by Pikunic et al[5]. The UMC model also has a significantly larger porosity than the TiC-CDC model (Table 2). Since the porosity is density dependent, we calculated the volume occupied by solid matter, v^s, which is the fraction of volume occupied by carbon atoms divided by the density. The volume occupied by solid matter is greater for UMC than TiC-CDC. This means that if UMC and TiC-CDC had an equal density, UMC would have a smaller accessible pore volume for adsorption.

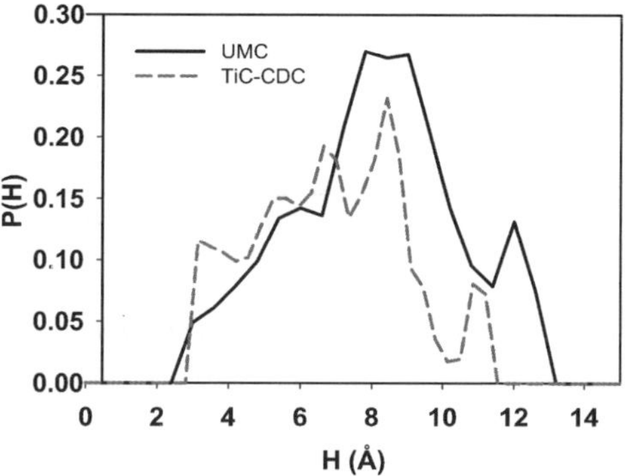

Figure 4: *Pore size distributions for UMC and TiC-CDC*

The adsorptive properties of the two carbon models were investigated using grand canonical Monte Carlo (GCMC) simulations of argon adsorption at 77K. The carbon and argon atoms were modeled as Lennard-Jones spheres, with diameters of $\sigma_C = 3.36\,\text{Å}$ and $\sigma_{Ar} = 3.41\,\text{Å}$ and interaction parameters of $\varepsilon_C / k_B = 28.0K$ and $\varepsilon_{Ar} / k_B = 120.0K$ for carbon and argon, respectively. The argon-carbon parameters were calculated using the Lorentz-Berthlot combining rules.

Table 2: *Calculated Porosity and Volume occupied by solid matter for UMC and TiC-CDC*

	UMC	TiC-CDC
Porosity	0.37	0.22
$v^s(cc/g)$	0.90	0.78

The simulated adsorption isotherms for both microporous carbons are show in Figure 5. The isotherms are examples of type I isotherms, which are experimentally observed for microporous adsorbents. The total amount of argon adsorbed at high pressure is twice as high for UMC as compared with TiC-CDC. This is a result of UMC's higher porosity, yielding a greater available volume for argon adsorption. At lower pressures, TiC-CDC adsorbs more argon due to its higher density and smaller pore size. These two factors both increase the effect of confinement of the argon gas, causing the argon adsorbate molecules to feel the presence of a greater number of carbon atoms when confined within the microporous structures.

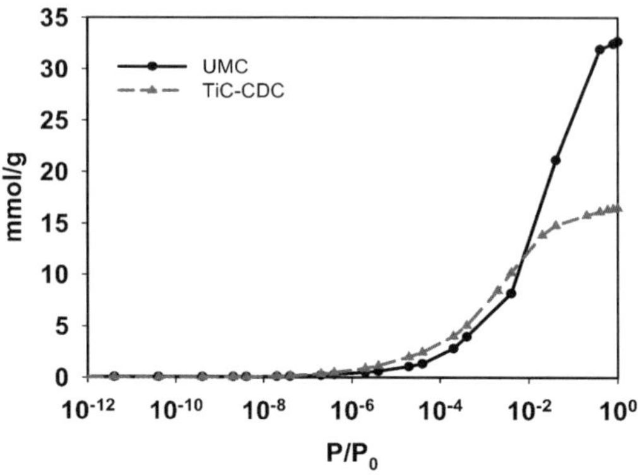

Figure 5: *Simulated adsorptions isotherms for argon at 77K*

4 CONCLUSIONS

We have developed atomistic carbon models for two microporous carbons, UMC and TiC-CDC, using the HRMC method. The RDFs for both carbon models reproduced the ones obtained from experimental neutron and X-ray scattering measurements with a high degree of accuracy. The ring size distribution showed an absence of high energy structures such as 3 and 4 member rings, which are unphysical artifacts seen in microporous carbon models produced with other reverse Monte Carlo methods. Thus, the HRMC method is effective in producing model structures that are both energetically stable and consistent with experimental structural measurements. The bond angles and lengths, ring size distributions, and nearest neighbor coordination numbers indicated that while our structures had microstructures that were on average similar to that of graphite, the microstructures were heterogeneous and had local chemistries that were very different. The large variance and irregular shape of the pore size distributions indicated that the model structures had a wide range of pore sizes with complicated pore geometries. The adsorption isotherm calculated using grand canonical Monte Carlo showed typical type I behavior expected for microporous solids. The trends in the isotherms were consistent with the subtle differences in the pore size distribution of our two models, capturing the expected low and high pressure behavior for the two models. However, the amount of argon adsorbed in TiC-CDC model deviates significantly from the experimentally measured isotherm even in the low pressure range[14]. This deviation is possible a consequence of using Lennard-Jones parameters for the argon-carbon interaction derived for flat graphite surface instead of curved surfaces with heterogeneous ring structures[15].

The two carbon structures presented in this work have much lower densities and are synthesized from different precursors than the saccharose based carbons investigated by Jain et al.[4]. While our models reproduced diffraction data with the same high degree of accuracy and showed similar trends in the local chemistry, they had much broader pore size distributions and larger average pore sizes. These results demonstrate that the HRMC method is capable of producing reasonable geometric models for a wide variety of microporous carbons and is able to capture subtle differences in the structural heterogeneity. Moreover, the impact of these subtle differences can be seen on the adsorptive properties. Thus, the HRMC method provides model structures that can be used to understand the impact of the microstructure of microporous carbon on a wide variety of phenomena that can be studied using molecular simulation techniques.

Acknowledgements

We thank the National Science Foundation (grant. No. 526879) for supporting this research. Work at Drexel and Penn was supported by the US Department of Energy (grant No. DE-FC36-04GO14280). The neutron structure factor measurements of TiC-CDC at Argonne National Laboratory's Intense Pulsed Neutron Source were supported by DOE (Contract W-31-109-ENG-38). We gratefully acknowledge the contributions of C. Benmore, J. Sewienie, J. P. Singer and T. Yildirim to various aspects of the CDC experiments.

References

1.   R. Bansal and M. Goyal, *Activated Carbon Adsorption*, CRC Press, Boca Raton, 2005.
2.   Y. E. Gogotsi (Ed.), *Carbon Nanomaterials*, CRC Press, Boca Raton, 2006.
3.   G. Opletal, T. Petersen, B. Malley, I. Snook, D. McCulloch, N. A. Marks and I. Yarovsky, *Mol. Sim.*, 2002, **28**, 927-938.
4.   S. Jain, R. Pellenq, J. Pikunic and K. E. Gubbins, *Langmuir*, 2006, **22**, 9942.
5.   J. Pikunic, C. Clinard, N. Cohaut, K. E. Gubbins, J. M. Guet, R. Pellenq, I. Rannou and J. Rouzaud, *Langmuir*, 2003, **19**, 8565.
6.   R. Dash, J. Chmiola, G. Yushin, Y. Gogotsi, G. Laudisio, J. Singer, J. Fischer and S. Kucheyev, *Carbon*, 2006, **44**, 2489.
7.   R. McGreevy and L. Pusztai, *Mol. Sim.*, 1988, **1**, 359.
8.   R. Henderson, *Phys. Lett. A*, 1974, **49**, 197.
9.   D. Brenner, *Phys. Rev. B*, 1990, **42**, 9458.
10.   A. Soper, in *Inst. Phys. Conf. Ser. 81*, Institute of Physics Publishing, Bristol, 1990.
11.   D. Franzblau, *Phys. Rev. B*, 1991, **44**, 4925.
12.   A. Kravchik, J. Kukushkina, V. Sokolov and G. Tereshchenko, *Carbon*, 2006, **44**, 3263.
13.   L. Gelb and K. E. Gubbins, *Langmuir*, 1999, **15**, 305.
14.   R. Dash, Ph.D. Thesis, 2006. Philadelphia: Drexel University
15.   J. Klauda, J. Jiang and S. Sandler, *J. Phys. Chem. B*, 2004, **108**, 9842.

CAN CHEMOMETRICS PROVIDE USEFUL DATA ABOUT POROUS SOLIDS CHARACTERIZATION?

Christothea Attipa, Rebecca Kokkinofta and Charis R. Theocharis*

Porous Solids Group, Department of Chemistry, University of Cyprus, P.O. Box 20537, 1678, Nicosia. Cyprus

1. INTRODUCTION

We have been interested in the synthesis and study of porous ceria,[1-3] including ceria doped with known amounts of heteroatoms, because of the importance of these solids in several catalytic applications.[4-6] Oxides such as CeO_2, V_2O_5 and CuO are used mainly pollution.[7-9] control applications. The combination of those three oxides can lead to a desirable catalytic system with interesting physicochemical attributes.

The solids used in the present work were the mixed ternary oxides $V_xCu_xCe_{1-2x}O_{2-y}$ ($0.05 \leq x \leq 0.45$), $V_xCu_{0.4-x}Ce_{0.6}O_{2-y}$ ($0.05 \leq x \leq 0.4$), $V_xFe_xCe_{1-2x}O_{2-y}$ ($0.05 \leq x \leq 0.45$) and $V_xFe_{0.4-x}Ce_{0.6}O_{2-y}$ ($0.05 \leq x \leq 0.4$) and the binary mixed oxides $V_xCe_{1-x}O_{2-y}$ ($0 \leq x \leq 0.9$), $Fe_xCe_{1-x}O_{2-y}$ ($0 \leq x \leq 0.9$) and $Cu_xCe_{1-x}O_{2-y}$ ($0 \leq x \leq 0.9$).

Chemometrics is the application of mathematical (statistical) methods to the solution of chemical problems of all types, where large quantities of data are produced. It is a relatively recent addition to the arsenal of chemical analysis and is applicable in those cases where many (three or more) parameters are measured for each sample, making comparisons and the extraction of useful inferences difficult. This is the so-called "Multivariate Analysis", which has first found application in the social sciences (e.g. econometrics), but its applicability to chemistry was recognized in the 1970's and increasingly so in the past decade.

The chemometric methods can be broadly classified in two main groups:[10, 11] (a) Date Structure Analysis (DA), Principal Component Analysis (PCA) and cluster analysis (CA), and (b) Classification of unknown samples and regularized Discriminant Analysis (CART, KNN, DA), including Canonical (CDA) and Linear Canonical Analysis (LDA). In the very recent past, Neural Networks were also applied to chemometric problems.[10] In the present study, we made use of PCA, Cluster Analysis, and CART analysis. The uses of Chemometrics lie in Experimental Design, Signal Processing, Pattern Recognition, Calibration, and the study of Evolutionary Signals. The work presented here can be construed as an application of pattern recognition.

Table 1 *Example of PCA Analysis*

Sample	Variable 1	Variable 2	Variable 3	Variable 4	Variable 5	Variable 6
A	A1	A2	A3	A4	A5	A6
B	B1	B2	B3	B4	B5	B6
C	C1	C2	C3	C4	C5	C6
F	F1	F2	F3	F4	F5	F6
E	E1	E2	E3	E4	E5	E6
F	F1	F2	F3	F4	F5	F6

Principal Component Analysis (PCA) is the most widely used technique. Quite often, PCA is applied before proceeding to other techniques, such as (hierarchical) cluster analysis, or different types of Discriminant Analyses. For each sample we measure several variables or components. We then find Principal Components (PC1, PC2, etc) which are linear combinations of the original variables describing each sample. For the example in Table 1, we have 6 samples for each of which we have measured 6 parameters. Therefore, the first three PC's for Samples A and B, would be as follows.

For Sample A:

$$PC1 = a_{11}A1 + a_{12}A2 + a_{13}A3 + ... + a_{1n}An$$
$$PC2 = a_{21}A1 + a_{22}A2 + a_{23}A3 + ... + a_{2n}An$$
$$PC3 = a_{31}A1 + a_{32}A2 + a_{33}A3 + ... + a_{3n}An$$

....

For Sample B:

$$PC1 = a_{11}B1 + a_{12}B2 + a_{13}B3 + ... + a_{1n}Bn$$
$$PC2 = a_{21}B1 + a_{22}B2 + a_{23}B3 + ... + a_{2n}Bn$$
$$PC3 = a_{31}B1 + a_{32}B2 + a_{33}B3 + ... + a_{3n}Bn$$

....

For *n* variables, *n* PC's are obtained for each sample. However, if the variables are correlated, as they would be where the variables are points on an adsorption isotherm, then we need far fewer PC than variables, usually only the first three, to fully describe the variation in the sample set. The Parameters a_{nm} in the equations above are called "eigenfactors". The eigenfactors are chosen in such a way that variables are no longer correlated. By plotting graphs of PC1 vs PC2 (usually), or PC1 vs PC3, we obtain plots that group similar samples together. Although we can pre-assign samples (objects) to a specific group, the mathematical method makes no such assumptions. We can therefore check the correctness or otherwise of our assignments.

Cluster Analysis is a method of dividing a group of objects into classes (clusters) so that similar objects are in the same class. Groups are not known prior to analysis, thus no assumptions are made about the distribution. The analysis searches for objects which are close together in the variable space. The distance d between two points in n-dimensional space is taken as the euclidean distance *d*.

$$d = \sqrt{(x_1 - y_1)^2 + (x_2 - y_2)^2 + ... + (x_n - y_n)^2}$$

We plot the objects (observations) in a dendrogram, in which the vertical axis is the similarity s_{ij} defined from the distance d_{ij} between two points i and j as follows:

$$s_{ij} = 100(1 - \frac{d_{ij}}{d_{max}})$$

where d_{max} is the maximum distance between any two point, ie the biggest difference.

2. EXPERIMENTAL

The mixed cerium oxide samples were prepared by co-precipitation method from aqueous solutions, using 1M ammonia solution as base. The metal precursors used were cerium (IV) ammonium nitrate, ammonium metavanadate(V) and copper(II) nitrate or iron(III) nitrate. The concentrations of all the solutions used were 0.01M throughout. The appropriate solutions were mixed in the appropriate rations, to achieve the desired results. The precipitate was left to stand in the mother liquor for 24h, centrifuged and dried at 373K for 24h. Nitrogen adsorption isotherms were obtained at T=77K using a Micromeritics ASAP 2010 apparatus, with a starting pressure of 0,15Pa.

Pattern recognition analyses were performed by means of the statistical software package SCAN (Software for Chemometric Analysis).

3. RESULTS AND DISCUSSION

The 66 samples studied by the chemometric method, are presented in Table 2. The variables used in the analysis for each sample were the adsorption branches of their nitrogen isotherms. For each isotherm, 30 measurements of amount of nitrogen adsorbed were obtained at predetermined values of partial pressure. Therefore, the equivalent Table to that shown for the example in Table 1, would be 30 columns wide by 66 rows deep. At the start of the analysis, each sample was assigned to one of eight groups as shown in Table 2. The basis for assigning a group, was the chemical identity of the heteroatoms in the sample. Hierarchical Cluster Analysis was employed first. Clusters were defined using an agglomerative algorithm.[10] Figure 1 shows the dendrogram that displays the amalgamation of clusters in the form of a binary tree. The tree can be cut at any similarity level. At a similarity level of 99%, one cluster was found. The cluster was composed of 31 oxide samples (Table 3). The cluster was formed by nearly all of the samples containing V and Fe as heteroatoms. Thus, the V-Fe-Ce ternary solids can easily be classified independently from the V:Fe heteroatoms ratio. It was further observed that all the samples that were classified in the V-Fe-Ce group have hysteresis loops with sizes in the range 0.21-0.27 units of partial pressure, as measured by the difference between the starting and finishing value of partial pressure for the hysteresis loop, in contradiction with the rest of the samples which have hysteresis loops covering a wider range of partial pressure values.

Table 2 *Identification of Samples Used in the Chemometric Analysis*

[1]No	Sample	ΔP	No	Sample	ΔP	No	Sample	ΔP
1	$V_{0.05}Cu_{0.05}Ce_{0.9}O_2$	0.146	3	$V_{0.04}Fe_{0.04}Ce_{0.92}O_2$	0.252	5	$V_{0.9}Ce_{0.1}O_2$	0.399
1	$V_{0.1}Cu_{0.1}Ce_{0.8}O_2$	0.149	3	$V_{0.05}Fe_{0.05}Ce_{0.9}O_2$	0.25	6	$Cu_{0.1}Ce_{0.9}O_2$	0.269
1	$V_{0.15}Cu_{0.15}Ce_{0.7}O_2$	0.421	3	$V_{0.1}Fe_{0.1}Ce_{0.8}O_2$	0.248	6	$Cu_{0.2}Ce_{0.8}O_2$	0.2
1	$V_{0.2}Cu_{0.2}Ce_{0.6}O_2$	0.251	3	$V_{0.15}Fe_{0.15}Ce_{0.7}O_2$	0.254	6	$Cu_{0.3}Ce_{0.7}O_2$	0.212
1	$V_{0.25}Cu_{0.25}Ce_{0.5}O_2$	0.249	3	$V_{0.2}Fe_{0.2}Ce_{0.6}O_2$	0.248	6	$Cu_{0.4}Ce_{0.6}O_2$	0.148
1	$V_{0.3}Cu_{0.3}Ce_{0.4}O_2$	0.299	3	$V_{0.25}Fe_{0.25}Ce_{0.5}O_2$	0.312	6	$Cu_{0.5}Ce_{0.5}O_2$	0.213
1	$V_{0.35}Cu_{0.35}Ce_{0.3}O_2$	0.415	3	$V_{0.35}Fe_{0.35}Ce_{0.3}O_2$	0.252	6	$Cu_{0.6}Ce_{0.4}O_2$	0.2
1	$V_{0.4}Cu_{0.4}Ce_{0.2}O_2$	0.521	3	$V_{0.4}Fe_{0.4}Ce_{0.2}O_2$	0.251	6	$Cu_{0.7}Ce_{0.3}O_2$	0.213
1	$V_{0.45}Cu_{0.45}Ce_{0.1}O_2$	0.503	3	$V_{0.45}Fe_{0.45}Ce_{0.1}O_2$	0.25	6	$Cu_{0.8}Ce_{0.2}O_2$	0.219
2	$V_{0.05}Cu_{0.35}Ce_{0.6}O_2$	0.145	4	$V_{0.05}Fe_{0.35}Ce_{0.6}O_2$	0.27	6	$Cu_{0.9}Ce_{0.1}O_2$	0.162
2	$V_{0.1}Cu_{0.3}Ce_{0.6}O_2$	0.199	4	$V_{0.1}Fe_{0.3}Ce_{0.6}O_2$	0.3	7	$Fe_{0.1}Ce_{0.9}O_2$	0.195
2	$V_{0.15}Cu_{0.25}Ce_{0.6}O_2$	0.255	4	$V_{0.15}Fe_{0.25}Ce_{0.6}O_2$	0.249	7	$Fe_{0.2}Ce_{0.8}O_2$	0.171
2	$V_{0.25}Cu_{0.15}Ce_{0.6}O_2$	0.3	4	$V_{0.25}Fe_{0.15}Ce_{0.6}O_2$	0.25	7	$Fe_{0.3}Ce_{0.7}O_2$	0.171
2	$V_{0.3}Cu_{0.1}Ce_{0.6}O_2$	0.269	4	$V_{0.3}Fe_{0.1}Ce_{0.6}O_2$	0.42	7	$Fe_{0.4}Ce_{0.6}O_2$	0.314
2	$V_{0.35}Cu_{0.05}Ce_{0.6}O_2$	0.3	4	$V_{0.35}Fe_{0.05}Ce_{0.6}O_2$	0.35	7	$Fe_{0.5}Ce_{0.5}O_2$	0.269
3	$V_{0.001}Fe_{0.001}Ce_{0.998}O_2$	0.212	5	$V_{0.1}Ce_{0.9}O_2$	0.283	7	$Fe_{0.6}Ce_{0.4}O_2$	0.484
3	$V_{0.002}Fe_{0.002}Ce_{0.996}O_2$	0.149	5	$V_{0.2}Ce_{0.8}O_2$	0.32	7	$Fe_{0.7}Ce_{0.3}O_2$	0.59
3	$V_{0.003}Fe_{0.003}Ce_{0.994}O_2$	0.147	5	$V_{0.3}Ce_{0.7}O_2$	0.25	7	$Fe_{0.8}Ce_{0.2}O_2$	0.489
3	$V_{0.004}Fe_{0.004}Ce_{0.992}O_2$	0.15	5	$V_{0.4}Ce_{0.6}O_2$	0.298	7	$Fe_{0.9}Ce_{0.1}O_2$	0.332
3	$V_{0.005}Fe_{0.005}Ce_{0.99}O_2$	0.255	5	$V_{0.5}Ce_{0.5}O_2$	0.287	8	CeO_2	0.248
3	$V_{0.01}Fe_{0.01}Ce_{0.98}O_2$	0.25	5	$V_{0.6}Ce_{0.4}O_2$	0.389			
3	$V_{0.02}Fe_{0.02}Ce_{0.96}O_2$	0.249	5	$V_{0.7}Ce_{0.3}O_2$	0.249			
3	$V_{0.03}Fe_{0.03}Ce_{0.94}O_2$	0.212	5	$V_{0.8}Ce_{0.2}O_2$	0.298			

[1]: number of the group where the metal oxide belongs

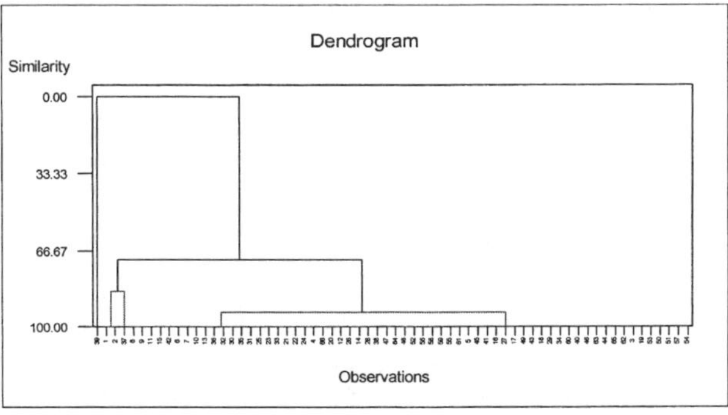

Figure 1 *Dendrogram of cluster analysis of 66 oxides*

Table 3 *Proposed group of samples that resulted from the HCA for resemblance bigger than 99%.*

	Group
Samples	4, 5, 12, 14, 16, 20, 21, 22, 23, 24, 25, 26, 27, 28, 30, 31, 32, 33, 35, 36, 41, 45, 47, 48, 52, 55, 56, 58, 59, 61, 66

PCA was performed using the program SCAN.[12] The singular value decomposition (SVD) algorithm calculates all components together, and so the computer space and time it requires both increase with an increase in the number of variables. Each component is orthogonal and is a linear combination of the original variables.[10] Table 4 reports the cumulative percentage of the total variance provided by the first five principal components obtained from the whole data set. The 99.7% of the total variance is explained by the first two components. In Figure 2 we shown the scores for the first two principal components (PC1 and PC2). From the scatter plot it is possible to discern that the ternary cerium oxides form a cluster (labeled in Figure 2 as 1) separately from the binary cerium oxides (labeled in Figure 2 as 2). The assignment of each sample to a cluster is shown in Table 5. It should be noted that the assignment by the method of a given sample to a cluster (or class of sample) is independent of the initial assignment given by the user. It is noteworthy that samples can be distinguished by the number of heteroatoms they contain, whether these are two or three, but not the nature of the guests. PCA analysis of each cluster separately was not able to separate the cluster into smaller ones, according to the chemical nature of the guests. Furthermore, the concentration of the heteroatoms did not appear to be more discriminant either, than the number of different species present.

Finally, samples 1 and 2 appear to behave differently from all the other samples, and can be assigned to a cluster of their own (3 in Figure 2). It appears that the factor that distinguishes these two samples from all the other ones was that they exhibited hysteresis loops with the smallest range in partial pressure. Specifically for the two samples, the difference between the finishing and starting partial pressure for the hysteresis loop was between 0.146 and 0.149. units of partial pressure.

Table 4 *Principal Component Analysis: Cumulative Proportion of Total Variation (5), Calculated from Correlation by SVD*

component	1	2	3	4	5
cumulative	98.4	99.7	99.8	99.9	99.9

Table 5 *Proposed group that resulted from the PCA*

	Group 1	Group 2	Group 3
Samples	4, 6, 7, 8, 9, 10, 11, 12, 13, 14, 15, 19, 20, 21, 22, 23, 24, 25, 26, 28, 30, 31, 32, 33, 34, 35, 36, 38, 42, 66	3, 5, 16, 17, 18, 27, 29, 40, 41, 43, 44, 45, 46, 47, 48, 49 , 50, 51, 52, 53, 54, 55, 56, 57, 58, 59, 60, 61, 62, 63, 64, 65	1, 2

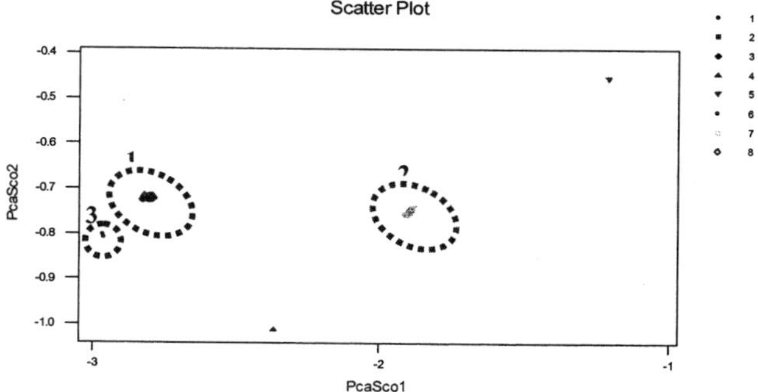

Figure 2 *PCA Distribution of the distillates in the plane defined by the first two principal components*

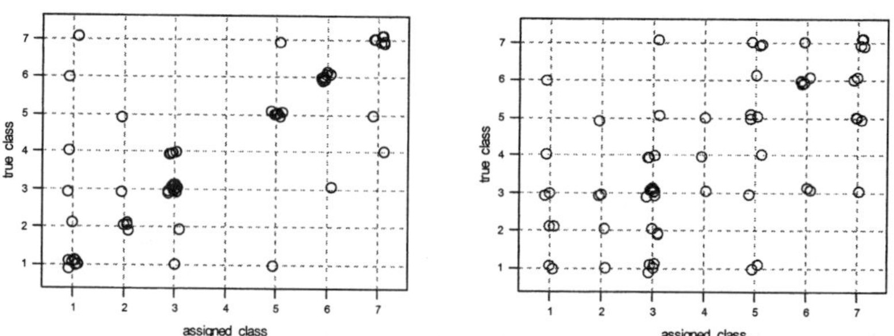

Figure 3 *Class assignment plot for 66 samples*

Figure 3 shows the CART analysis for the samples under investigation. These Figures compare the classification of samples into various classes by Cluster Analysis with the classes assigned by the user, referred to as the model. The first diagram describes the behavior of all samples in terms of the proposed model, and in the second diagram the classification of all samples on the basis of the existing model. From the first diagram, it can be seen that the classification of the samples according to the model had a fault level of 27.6%, that is to say 18 of the 66 samples were unable to obey the model. In detail, one sample from the first group was classified in the third group and one sample in the fifth group. From the second group, one sample was classified in the first group and a second one in the third group. From the third group three samples were classified in the first group, in the second group and in the sixth group, respectively. From the fourth group one sample was classified in the first group, four samples in the third group and one sample in the seventh group. From the fifth group one sample was classified in the second group and one in the seventh group. As for the sixth group two samples were classified in the seventh group, one sample in the fifth group, and one further sample was classified in the first group. Finally a sample assigned to the seventh group was classified in the fifth group and another sample in the first group. In the second diagram the classification of samples having as a base the existing model, in other words the one proposed by the user and presented in Table 2, had an error level of 63%, that is to say 41 of the 66 samples were classified erroneously.

These results appear disappointing but in reality they reflect the fact that a better classification would be one consisting of two groups only, one made up of the current groups 1, 2, 3, and 4, and   second group consisting of groups 5, 6, and 7. This result is in complete agreement with picture obtained from the PCA analysis.

From the discussion above, it can be concluded that a statistical analysis of the adsorption data, in this case of the nitrogen adsorption isotherms can lead to useful inferences which would be otherwise very difficult to obtain. For the series of ceria samples with one or two doping heteroatoms under investigation, the number of different doping species had a more profound effect on the adsorptive behaviour of the samples, than either the nature, or the concentration of the doping species. The analysis presented here can be thought of as a way to compare the shapes of the adsorption isotherms, and classify the samples according to those. The size of the hysteresis loop, as described by the difference between the starting and finishing values of partial pressure was also a characteristic capable of distinguishing samples.

We acknowledge the financial assistance of the Cyprus Foundation for the Promotion of Research (ΙΠΕ) and the University of Cyprus for financial support.

References

1. G. Kyriacou, I. Paschalidis and C.R. Theocharis, *Fundamentals of Adsorption (Editors: K.Kaneko, H. Kanoh and Y. Hanzawa)*, 2002, **7**, 201.
2. C. R. Theocharis, G. Kyriacou and M. Christophidou, *Adsorption*, 2005, **11**, 763.
3. C. Attipa and C. R. Theocharis, *"Studies in Surface Science and Catalysis"*, *Elsevier Science Publishers*, 2006, **160**, 615.
4. T. Bunluesin, R. Gorte and G. Graham, *Applied Catalysis B Environmental*, 1998, **15**, 107.
5. S. H. Overbury and D. R. Mullins, *Journal of Catalysis*, 1999, **186**, 296.
6. W. J. Shen and Y. Matsumura, *Physics Chemistry*, 2000, **2**, 1519.
7. J. Matta, D. Coursot, E. Abi-Aad and A. Aboukais, *Chemistry Materials*, 2002, **14**, 4118.
8. R. Cousin, S. Capelle, E. A.-. Aad, D. Coursot and A. Aboukais, *Chemistry Materials*, 2001, **13**, 3862.
9. E. A.-. Aad, J. Matta, D. Coursot and A. Aboukais, *Journal of Materials Science*, 2006, **41**, 1827.
10. R. I. Kokkinofta and C.R. Theocharis, *Journal of Agricultural and Food Chemistry* 2005, **53**, 5067.
11. D. Ballabio, R. Kokkinofta, R. Todeschini and C.R. Theocharis, *Chemometrics and Intelligent Laboratory Systems*, 2007, **87**, 52.
12. R. Kokkinofta, Authenticity of the Cypriot traditional spirit Zivania, PhD Thesis, University of Cyprus, 2003.

ORIGIN OF THE VARIATION OF ADSORPTION AND DIFFUSION PROPERTIES OF ZEOLITE 5A

K. Barthelet, M. Vidalie, C. Bouvry, D. Marti, A.A. Quoineaud

IFP-Lyon, BP3, 69390 Vernaison, FRANCE

1 INTRODUCTION

Type LTA molecular sieves are part of numerous adsorption processes in the field of refining, petrochemicals or natural gas treatment. The calcium form, *i.e.* zeolite 5A, is one of the most extensively used zeolites because of its ability to adsorb linear alkanes inside the microporosity. Thanks to this property, zeolite 5A is efficient in the separation of linear C_{10}-C_{14} alkanes from a kerosene cut, which are precursors of Linear Alkyl Benzene (LAB). Adsorption properties of zeolite 5A are known to be strongly correlated to the nature and amount of cations in the structure.[1] Moreover, zeolite A shows a low stability when it is calcinated under steam water. Especially, these kind of treatment induce a significant decrease of the molecule's diffusion coefficient, i.e. an apparition of resistance to the mass transfer into the zeolite network. To explain this phenomenon, some authors suggest the formation of diffusional surface barriers consequently to a partial destruction of the structure on the surface of crystallites, evidenced by the detection of aluminum species on extra-framework positions detected by X-ray diffraction[2] and ^{27}Al MAS NMR.[3,4] In other studies, variations of diffusion properties are attributed to cationic rearrangement.

To try to reach a better understanding of the origin of the observed variation of adsorbent performances (capacities and mass transfer), we study the influence of cation size and distribution as well as extra-framework species contain on the adsorption and desorption behaviors for several adsorbents towards n-paraffins. Considered solids have been prepared by ionic exchange from the same zeolite 4A but with various divalent cations and at various exchange levels. Some of them have even been submitted to a thermal treatment in order to study cation mobility and its impact on adsorption properties. By trying to correlate adsorption performances with adsorbent structural characteristics, we will propose an explanation for the variation of adsorption properties and even give some indications to optimize the adsorbent.

2 METHOD AND RESULTS

2.1 Materials

The starting material was a commercially available zeolite A in sodium form ($Na_{12}Al_{12}Si_{12}O_{48}.nH_2O$ or 4A) manufactured by Ceca.
From this solid, zeolites 5A with different compensating divalent cation (Mg, Ca, Sr and Ba) at several contents (between 40 and 96%) were prepared by ionic exchange. The applied cationic exchange procedure is a classical one. Cationic solutions at different concentrations are prepared by dissolution of salts ($MgCl_2,2\ H_2O$, $CaCl_2, 6\ H_2O$, $SrCl_2, 2\ H_2O$ or $BaCl_2, 2H_2O$) in water. Concentrations were chosen on the basis of the M/Na exchange isotherms published in the literature for each cation.[5] Then, zeolite A is added to this solution which is maintained under magnetic stirring at 90°C during 4 hours. When the exchange was completed, samples were dried at 125°C overnight. Recovered solids are referred to Mg-5A, Ca-5A, Sr-5A and Ba-5A and their exchange level (determined by atomic emission spectroscopy with inductively coupled plasma (ICP-AES)) indicated into brackets. They are characterized as prepared and also after thermal treatment. Before adsorption experiments, they will be partially dehydrated at 300°C.
According to the literature, zeolite A has three types of cationic sites (Figure 1). Site I is located near the centre of the six-membered oxygen ring corresponding to the window between the sodalite cage and the supercage. Depending on the cation size, it could be displaced into the supercage ; it will be then referred to site Ib. Site II lies close to the centre of the eight-membered oxygen ring that constitutes the window between two supercages. Site III is positioned into the supercage, near the centre of the four-membered ring of one face of D4R unit connecting two sodalite cages. So, cations on site II will partially block the aperture of the supercage's window and thus probably hinder the access to the supercage so that diffusionnal limitations could appear. However, depending on the position of site Ib, i.e. its distance from the center of the six-ring window which increase with cation size, cations on this site will also create steric hindrance. Moreover, accessibility and availability of porosity will also probably vary with cation nature and more particularly with their size (hydration sphere will have to be taken into account) and polarisability.
It is usually admitted that among the twelve sodium cation of NaA, eight of them occupy site I (complete occupancy of this type of site), three of them are located on site II and the last one lies on a site III. From theoretical[6, 7] and X-ray[8] studies, it appears that site selectivity is in favor of site I for Li^+, Na^+, Ca^{2+} and Sr^{2+} and in favor of site II for K^+, Cs^+ and Ba^{2+} which is probably a consequence of their big size. So, for M^{2+} such Mg^{2+}, Ca^{2+} and Sr^{2+}, the first site to be exchanged is the site III (less energetically stable) followed by sites II that become entirely empty when more than 2/3 of Na^+ are exchanged. The only doubt that subsists is for the sixth divalent cation that is sometimes said to prefer site II (Table 1).

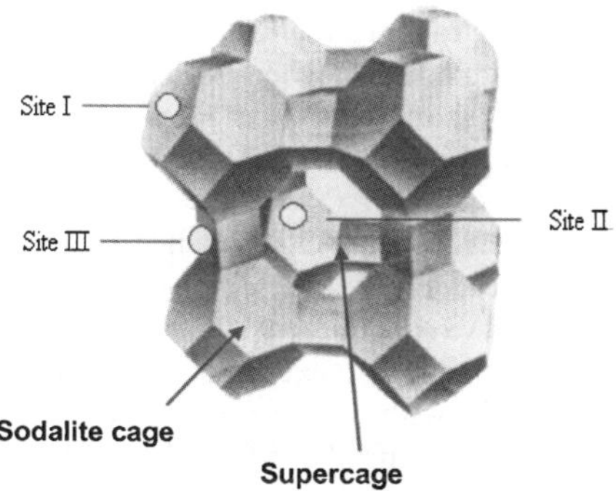

Figure 1 *LTA structure with site's localisation*

Table 1 *Prediction of cationic distribution in function of exchange level*

Exchange level (%)	Cations	Site occupancy		
		Site I	Site II	Site III
0	12 Na$^+$	8	3	1
33	8 Na$^+$ / 2 M^{2+}	6 Na$^+$ + 2 M^{2+}	2	0
50	6 Na$^+$ / 3 M^{2+}	5 Na$^+$ + 3 M^{2+}	1	0
67	4 Na$^+$ / 4 M^{2+}	4 Na$^+$ + 4 M^{2+}	0	0
100	6 M^{2+}	6 or 5	0 or 1	0

Another phenomenon also occur when some divalent cations are introduced that is the formation of some extra-framework aluminium species.[3] Although it remains an hypothesis, several authors proposed as formula $Al(OH)_4^-$. Indeed, some specific experiment proved that Al is surrounded by protons[9] and thin peak observed by ^{27}Al NMR imposes a symmetrical specie. In reality, this also suggests that they are all in the same environment that is comprehensive only if they are located in the centre of the sodalite cage. If it is really the case, they should have no influence neither on the diffusional nor on the adsorption properties of the zeolite A.

2.2 Characterizations

2.2.1 NMR Spectroscopic measurements. ^{27}Al MAS NMR is applied to characterize aluminium environment and to verify the exchange level. By ^{23}Na MAS NMR, exchange level is once more verified and sodium positions are characterized.

Previous to ^{27}Al NMR measurements, solids are saturated in water. On the contrary, prior to the ^{23}Na NMR investigations, the powder materials were heated at 403 K in order to eliminate the physisorbed water. Indeed, previous tests on saturated solids showed that water confer a so good mobility to the cations that we just achieve in an average position, i.e. that the different sites could not be distinguished anymore. Then, the samples were filled into the 4mm MAS NMR rotors under dry nitrogen using a glovebox and tightly sealed with Kel-F rotor caps. NMR experiments were performed on a Avance 400 MHz Bruker spectrometer in a 4 mm MQMAS probehead at 14 kHz spinning speed.

^{27}Al MAS spectra were obtained using a $\pi/16$ flip angle and a 35 kHz radio-frequency field. Under these conditions, verified to be selective, ^{27}Al MAS spectra were quantitative.[10] [27]Al MQMAS spectra were realized using the z-filter MQMAS sequence[11] synchronized on the rotor spinning speed.[12] The NMR parameters of each aluminum sites, chemical shift and quadrupolar parameters, were obtained by decomposition of the both [27]Al MQMAS and MAS spectra using the Massiot's dmfit program.[13] [27]Al chemical shifts were referenced to $Al(NO_3)_3$, 1N.

^{23}Na NMR experiments were performed at a resonance frequency of 105.8 MHz with a sample spinning rate between ca. 12-15 kHz using a 4mm probe optimized for [23]Na MQMAS experiments. [23]Na MAS experiments were carried out under selective and quantitative conditions using a single pulse excitation corresponding to a $\pi/12$ flip angle at a 35kHz radiofrequency field and with a 0.5s relaxation delay. Tests of ^{23}Na MQMAS NMR experiments were performed using the z-filter MQMAS sequence synchronized on the rotor spinning rate proposed by Frydman et al. and Amoureux et al.[11, 14-15] The spinning speed of the sample was 15 kHz. All ^{23}Na NMR chemical shifts are referenced to the solid NaCl. Two methods were applied to obtain NMR parameters of the different sodium contributions. The first one was based on a direct estimation of the isotropic chemical shift (δ_{iso}) and the quadrupolar coupling constant (Qcc) from the [23]Na MQMAS spectra. The isotropic chemical shift was estimated as the center of gravity of the signal obtained from the chemical shift projections in the isotropic and the MAS dimensions of the MQMAS spectrum. The quadrupolar coupling constant was estimated from the composite quadrupolar parameter P_Q and the asymmetry parameter η_Q. The composite quadrupolar parameter P_Q is given by the isotropic chemical shift and the chemical shift projection in the isotropic dimension of the MQMAS spectrum and the asymmetry parameter η_Q was assumed to be equal to 0.5. This first method is frequently used in the literature[16, 17] but may result, for large Qcc sites, in an imprecise estimation of the NMR parameters and an inaccurate estimation of the signal intensity (overestimation for weak quadrupolar signals and underestimation for strong quadrupolar signals). The second method is an alternative method[18] based on a computational estimation of the isotropic chemical shift and the quadrupolar coupling constant via the simultaneous decomposition of both [23]Na selective and quantitative MAS and [23]Na MQMAS spectra using the dmfit2007 program[14] taking the distribution of the interaction parameters into account. Due to the quadrupolar character of the sodium nucleus, Gaussian lineshapes could not be used to model sodium cations signals. As a consequence, Czjzek's lineshape was used[19, 20]. The isotropic chemical shift δiso, the quadrupolar coupling constant Qcc and asymmetry parameter η_Q were adjusted from the MQMAS spectrum. Then, the intensity of each signal was calculated from the MAS spectrum.

2.2.2 Porosity measurements. Textural properties as surface area and microporous volume are characterized by N_2 adsorption: at first, samples are outgassed under vacuum (10^{-5} torr) at 500°C for 12h and the adsorption isotherm of N_2 measured at 77K on a *Micromeretics ASAP 2000* device.

2.2.3 Procedure for gravimetric runs. Adsorption and thermal desorption experiments were carried out on a Setaram TAG 24 thermogravimetric system. Partially dehydrated samples at 300°C were saturated in situ with gaseous nC10 at 150°C and ex situ with liquid nC14. After then, they were heated from ambient temperature to 800°C with a heating rate of 10°C/min under helium flow. Temperature was maintained at 150°C for 2 hours to evaporate liquid nC_{10} or nC_{14}.

2.2.4 Breakthrough curves. Adsorption consisted in a continuous injection of n-paraffins on a 1 m length column filled by adsorbent previously partially dehydrated at 300°C and pre-saturated by a nC5. These n-paraffins were then desorbed by injecting continuously nC5 in the same column. We proceeded these steps at 175°C and 10 mL/min.

2.3 Results and discussion

2.3.1 Exchange level. From ^{27}Al NMR spectra decompositions, we get the proportion of aluminium surrounded by sodium and calcium cations and from those of ^{23}Na NMR spectra, the total quantity of sodium cations located into the zeolite network. Thanks to these values, we recalculated the exchange level of each zeolite and compared it to the one derived from bulk elemental analysis (Table 2). In the cases of Mg-5A and Sr-5A, we show a good concordance between all exchange levels values. On the contrary, exchange levels for Ca-5A mainly appear lower when calculated from NMR spectra. As these experiments concern only zeolite framework, such a lower values compared to the one obtained by elemental analysis suggest that not all the calcium cations are located on extra-framework positions of the zeolite. For example, cations could be involved in a second phase coexisting with zeolite crystallites. This hypothesis is supported some $CaCO_3$ detected on XRD diffraction diagrams (Figure 2). After zeolite calcination, all exchange levels become almost equal. We propose that during thermal treatment, calcium cations migrate into the zeolite network on cationic position. So this kind of treatment will be necessary to get the expected Ca-5A adsorbent.

Table 2 *Exchange levels of zeolites calculated from different results*

Zeolite	Thermal treatment	Exchange level (%)		
		ICP-AE	^{27}Al NMR	^{23}Na NMR
Mg-5A		43	42	42
		60	58	57
		68	70	68
Ca-5A		46		32
		62	65	59
		76	69	61
	500°C/1h	79	78	82
Sr-5A		88		90
	500°C/1h	88		88

2.3.2 Cationic distribution. Deeper study of ^{23}Na NMR spectra allows to determine the distribution of sodium cations between sites I and II (Table 3). For Mg-5A, cationic distribution on sites I and II is far from the theoretical one : there are more cations on sites II than expected and this will probable have consequence on molecules diffusion through the zeolite network as the supercage window is partially blocked. For Ca-5A, the same phenomenon is observed but less pronounced and for Sr-5A, we finally reach the distribution predicted by energy calculations. This variation of cationic distribution is probably related to the cation size and to their polarisability. As these measurements are carried out on hydrated solids, the biggest cation is Mg^{2+} (because of its weaker polarisability, it will be the one having the bigger hydration sphere). So, it is probably better stabilized on site II that is less confined. However, to explain the migration of cations Ca^{2+} and Sr^{2+} from site I to site II during thermal treatment, we have to consider

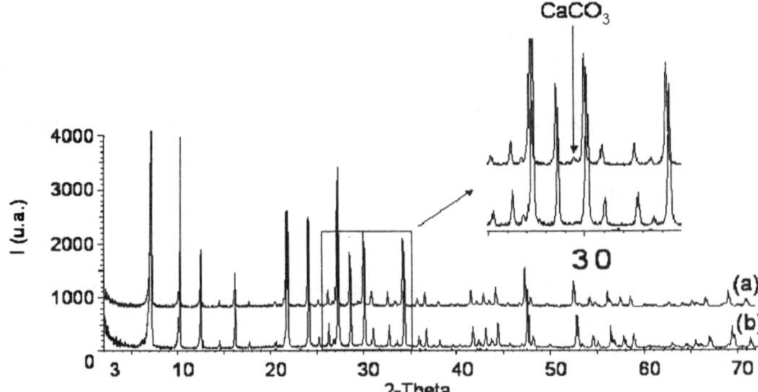

Figure 2 *XRD diagrams of Ca-5A before (a) and after (b) thermal treatment*

Table 3 *Cationic distribution of zeolites partially exchanged by divalent cations*

Zeolite	Exchange level (%) ICP-AE	Thermal treatment	%Site I (I+Ib)		%Site II	
			theo.	real	theo.	real
Mg-5A	43		85	64	15	36
	60		95	62	5	38
	68		100	60	0	40
Ca-5A	46		85	73	15	27
	62		95	82	5	18
	76		100	78	0	22
	79	300°C/1h	100	50	0	50
	79	500°C/1h	100	25	0	75
Sr-5A	88		100	100	0	0
	88	500°C/1h	100	75	0	25

another phenomenon. Activation dehydrate cations so that their coordination sphere is no more compete. To compensate the loose of coodinative water molecule, they migrate from the six-membered ring (site I) to the eight-membered ring (site II). This also justify that this depopulation of site I in favor of sites II is even enhanced when heating temperature is increased.

2.3.3 Formation of extra-framework aluminium species. Characterisation of 5A zeolites by ^{27}Al NMR evidence the formation of some extra-framework species for some of them, namely the Ca-5A and Sr-5A, that is favored by thermal treatment : a new thin lineshape appears near 78 ppm as it has already been seen in the literature.[3] Their origin and mechanism of formation are not really known but it is probably once more related to cation size and polarisability (cf. strength to extract an aluminium from the framework). If they are located in the supercage, they will probably hinder the diffusion of molecules and limit the porosity. However, they will probably play a little role on zeolite adsorption properties as they represent at least 5% of the total aluminium content. And if they are positioned in the sodalite cage as it has been suggested in the literature, they will have even less influence on zeolite adsorption properties.

2.3.4 Adsorption and diffusion properties. Adsorption capacities of nC10 and nC14 are determined from thermodesorption on zeolites partially hydrated saturated by nC10 or nC14 (Table 4).Very similar values are obtained from breakthrough curves. They seem to reach an optimum for Ca-5A as the microporous volume measured by N_2 adsorption at

Table 4 *Microporous volumes and quantity of nC10 and nC14 adsorbed on zeolites*

Zeolite	Exchange level (%) ICP-AE	V_{micro} (mL/g)	n_{mol}(nC10)/uc	n_{mol}(nC14)/uc
Mg-5A	74	0,217	1,4	1,1
Ca-5A	83	0,243	**2,0**	**1,3**
Sr-5A	97	0,125	1,6	0,8
Ba-5A	97		0,4	0,5

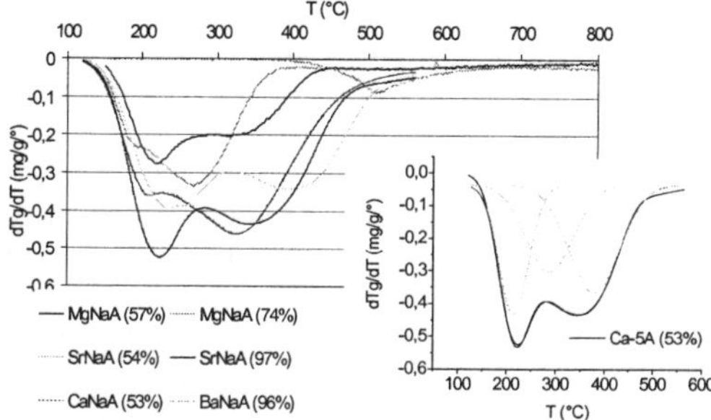

Figure 3 *Thermodesorption spectra of nC10 on different zeolites*

77K. We propose that Ca-5A correspond to the best compromise between cation size and site position. Smaller Mg^{2+} cations are probably localized on a greater scale on site II and larger Sr^{2+} and Ba^{2+} cations, even if proportionally more present on site I, have their site I more widely displaced towards the supercage.

This result also confirms that extra-framework aluminium species are not enough or positioned in sodalite cage and thus do not impact the adsorbent properties.

Comparing thermodesorption spectra (Figure 3), it appears that n-paraffins molecules are adsorbed on three types of sites. The strongest one is favored for the smaller cation and by the thermal treatment. We propose that this strongest site corresponds to the site II.

From adsorption on thermobalance and breakthrough curves (Figure4), we get information on diffusion properties of n-paraffins molecules into the zeolite. First of all, adsorption plateau seems to be more rapidly reached in the case of Ca-5A suggesting that diffusion depends on both cationic position and size of the cation. Indeed, in Ca-5A, there are more divalent cations positioned on site II than on Sr-5A but smaller ones. Comparing the desorption of Mg-5A and Ca-5A, we observe that this step occurs more easily for Ca-5A. This is probably a consequence of the fact that molecules are more strongly adsorbed on site II that are predominant in Mg-5A.

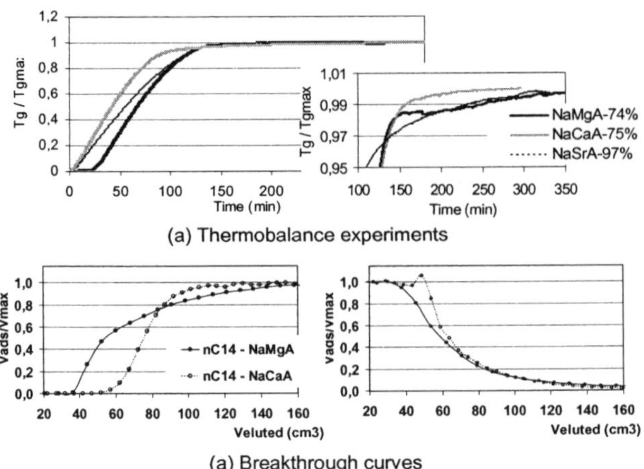

Figure 4 *Thermobalance experiments (a) and breakthrough (b) curves*

3 CONCLUSION

Aim of this study was to better understand parameters influencing the adsorption and diffusion properties of 5A zeolite depending of the nature and quantity of divalent cation. Comparing the evolution of cationic distribution, content of extra-framework aluminium species and results of adsorption/desorption experiments, it appears that diffusion limitations are mostly due to presence of cation on site II but also depends on cation size. Presence of extra-framework species do not seem to have any influence that support the hypothesis of their location in sodalite cage.

Therefore, Ca-5A appears as the optimized adsorbent. But, to get the expected exchange level, thermal treatment is necessary. However, it will induce a migration of some calcium cations from site I to site II that will be harmful to the adsorption performances. So to improve adsorbent, a solution has now to be found to induce a new migration of cations from site II to site I.

References

1 Breck D.W. in *Zeolite Molecular Sieves*, Wiley, New York, 1974.

2 J.J. Pluth and J.V. Smith, *J. Am. Chem. Soc.*, 1980, **102**, 4704.

3 D.R. Corbin, R.D. Farlee and G.D. Stucky, *Inorg. Chem.*, 1984, **23**, 2920.

4 D. Freude, J. Haase, H. Pfeifer, D. Prager and G. Scheler, *Chem. Phys. Letters*, 1985, **114**, 543.

5 Sherry H. S., Walton H. F. *J. Phys. Chem.* 1967, **71**, 1457.

6 K.Ogawa, M.Nitta, and K.Aomura. *J.Phys.Chem.*, 1979, **83**, 1235.

7 K.Ogawa, M.Nitta, and K.Aomura, *J.Phys.Chem.*, 1978, **82**, 1655.

8 T.Takaishi, Y.Yatsurugi, A.Yusa, and T.Kuratomi, *J. Chem. Soc. Far. Trans.*, 1975, **1**, 97.

9 M. Pruski, H. Ernst, H. Pfeifer and B. Staudte, *Chem. Phys. Letters,* 1985, **119**, 412.

10 E. Lippmaa, A. Samoson, M. Mägi, *J. Am. Chem. Soc.,*1986, **108,** 1730.

11 J.P. Amoureux, C. Fernandez and S. Steuernagel, *J. Magn. Reson. A*, 1996, **123**, 116.

12 D. Massiot, *J. Magn. Reson. A*, 1996, **122**, 240.

13 D.Massiot, F.Fayon, M.Capron, I.King, S.Le Calvé, B.Alonso,J-O.Durand, B.Bujoli, Z.Gan and G.Hoatson, *Magnetic Resonance in Chemistry* , 2002, **40**, 70.
14 L. Frydman and J.S. Harwood, *J. Am. Chem. Soc.,* 1995, **117**, 5367.
15 A. Medek, J.S. Harwood and L. Frydman, *J. Am. Chem. Soc.,* 1995, **117**, 12779.
16 M. Feuerstein, M. Hunger, G. Engelhardt and J.P. Amoureux, *Solid State NMR*, 1996, **7**, 95.
17 K.H. Kim and C. Grey, *J. Am.. Chem. Soc.*, 2000, **122**, 9768.
18 A.A. Quoineaud, V. Montouillout, S. Gautier, S. Lacombe and C. Fernandez, *Stud. Surf. Sci. Catal.*, 2002, **142**, 391.
19 G. Czjzek, *Phys. Rev. B*, 1982, **25**, 4908.
20 G. Czjzek, J. Fink, F. Götz, H. Schmidt, J.M. Coey, J.P. Rebouillat and A. Liénard, *Phys. Rev. B*, 1981, **23**, 2513.

ASSESSING GENERIC FORCE FIELDS TO DESCRIBE ADSORPTION ON METAL-ORGANIC FRAMEWORKS

David Fairen-Jimenez[†], Patricia Lozano-Casal[†], and Tina Düren[*]

Institute for Materials and Processes, School of Engineering and Electronics, University of Edinburgh, King's Buildings, Edinburgh EH9 3JL, United Kingdom
[†] These authors contributed equally.

1 INTRODUCTION

Metal-organic frameworks (MOFs) are fairly new porous materials with many potential industrial applications including adsorption separations, catalysis and gas storage.[1] MOFs are built of a metal corner unit (normally a metal ion or metal cluster such as Zn_4O) and organic linker units able to bridge the metal corners (e.g. dicarboxylate). The large variety of possible linker and corner units gives rise to a vast number of possible materials with cavities and pores of different shapes and tuneable sizes. [2-4]

Adsorption processes on MOFs are increasingly studied both experimentally and by molecular simulation.[5-7] Molecular simulation can not only be used to predict the macroscopic adsorption performance but also provide a detailed picture on the molecular scale, which is not readily accessible from experimental methods. This allows studying in detail how the structure influences the adsorption performance and therefore forms an essential part in the identification and design of promising MOF materials.

Yet, the prerequisite for accurately predicting the adsorption behaviour of a fluid in a MOF is the accurate description of the interactions between the fluid molecules and the MOF framework. So far no force fields have been developed specifically for MOFs. The only exception is the work by Tafipolsky and co-workers[8] who parameterised the MM3 force field for IRMOF-1. However, their work is specific to IRMOF-1 and cannot be transferred to other MOFs. Thus, in most simulation studies of adsorption and diffusion in MOFs generic force fields are used. As these force fields contain parameters for the whole periodic table or large subsets thereof, the interactions inside any MOF can be described simply by choosing the corresponding parameters for the framework atoms from a table. Yet, these force fields have been parameterised for certain systems (e.g. OPLS for proteins and small organic molecules), and it is unclear whether they accurately describe the interactions in materials of a completely different nature, such as porous MOFs.

In order to assess whether the widely used generic force fields are able to accurately describe the interactions, we studied the adsorption of methane in three different MOFs with three generic force fields: UFF (Universal Force Field),[9] Dreiding[10] and OPLS (Optimised Potentials for Liquid Simulations).[11] The UFF force field contains all atoms in the periodic table whereas the Dreiding force field is limited to a small number of metals. The OPLS force field describes a more limited number of atoms (i.e., no metals are included) but, in contrast with the other two, distinguishes between organic groups taking into account the different chemical nature of atoms within the organic linkers. All three force fields use the 12-6 Lennard-Jones (LJ) potential to describe interatomic interactions. UFF and Dreiding have been parameterised for selected organic, biological, main group inorganic, and transition metal complex structures whereas OPLS was parameterised for protein simulations and for small organic molecules.

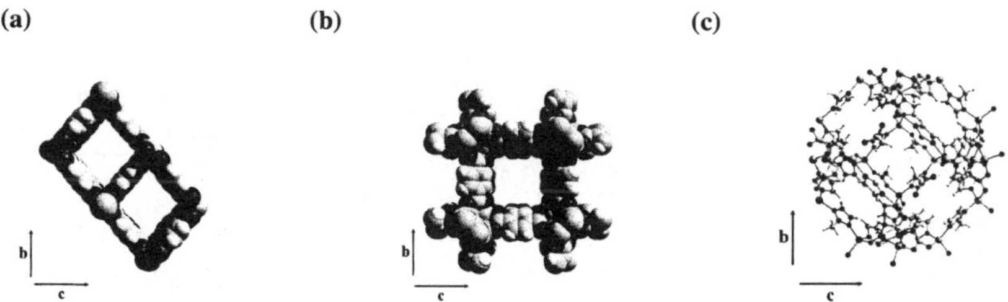

Figure 1: *View of (a) the 1D rhombic channels, which form the MIL-47 structure; (b) the interconnected porous structure for IRMOF-;1 and (c) the 3D crystal structure of ZIF-8, which comprises large cavities and small windows.*

In order to address the questions how accurately each of these force fields describes the interactions between a MOF and adsorbed methane molecules and whether one outperforms the others, we have chosen three MOF materials with different structures, pore shapes and sizes: MIL-47, IRMOF-1 and ZIF-8. MIL-47[12] contains 1D rhombic channels with a pore diameter of around 11 Å, each of which is delimited by four walls of benzenedicarboxylate (BDC) and four chains of corner-shared vanadium octahedra [Figure(1a)]. The structure of IRMOF-1[13] results from the interconnection of the coordination centres of Zn_4O with six BDC linkers which reticulates into a three-dimensional cubic structure [Figure 1(b)]. The framework gives rise to spherical pores of 15.1 Å and 11.0 Å[14]. Finally, ZIF-8[15] is a prototypical zeolitic imidazolate framework, exhibiting a topology formed by four- and six-ring ZnN_4 clusters interconnected by 2-methylimidazolate (IM) linkers. The ZnN_4 clusters give rise to pores with a diameter of 11.6 Å, with apertures between pores of 3.4 Å [Figure 1(c)]. For all of these MOFs experimental adsorption isotherms as well as data for the differential enthalpy of adsorption or the isosteric heat of adsorption is available from the literature which will allow us to asses whether generic force fields are suitable for simulating adsorption in MOFs.

2 COMPUTATIONAL DETAILS

Methane adsorption in the MOF materials was investigated using GCMC[16] simulations implemented in the multipurpose simulation code Music.[17] An atomistic model was used for the MOF frameworks with framework atoms kept fixed at the positions reported from crystallography. The standard 12-6 Lennard-Jones (LJ) potential was used to model all interatomic interactions. Three different force fields were used to simulate adsorption on the materials: UFF, Dreiding and OPLS. Due to the lack of potential parameters for certain atoms in the Dreiding and OPLS FFs, atoms missing from a particular force field (i.e. V from Dreiding Zn and V from OPLS) as well as the inorganic oxygen atoms present in MIL-47 and IRMOF-1 (i.e. in OPLS) were always described using UFF parameters, whereas all other atoms were described using the UFF, Dreiding and OPLS force field. All

                                                          Characterisation of Porous Solids VIII

Lennard-Jones parameters used in this study are summarised in Table 1. The model used for methane was derived by Goodbody et al.[18] from vapour liquid equilibrium data and uses a united-atom description of the methane molecules, i.e. one methane molecule is represented by a single sphere ($\sigma_{CH4} = 3.730$ Å, $\varepsilon_{CH4}/k_B = 148.000$ K). The Lorentz-Berthelot mixing rules were employed to calculate methane/framework parameters. Interactions beyond 18.650 Å were neglected.

Table 1: *Potential parameters of different force fields for MIL-47, IRMOF-1 and ZIF-8).*

	UFF		Dreiding		OPLS	
	σ_i / Å	(ε_i/k_B) / K	σ_i / Å	(ε_i/k_B) / K	σ_i / Å	(ε_i/k_B) / K
C	3.431	52.838	3.473	47.856		
C_{COO}					3.750	52.838
C_{CH}					3.550	35.225
O	3.118	30.193	3.033	48.158	N/A	N/A
O_{COO}					2.960	105.676
N	3.261	34.722	3.292	72.967	3.250*	85.548*
H	2.571	22.142	2.846	7.649	2.420	15.097
V	2.801	8.052	N/A	N/A	N/A	N/A
Zn	2.462	62.399	4.045	27.677	N/A	N/A

* These parameters are specific for imidazolate.

Depending on the temperature and MOF, $5 \cdot 10^6 - 3 \cdot 10^7$ number of Monte Carlo steps were performed carefully ensuring that equilibrium was reached. The first 40% were used for equilibration, and the remaining steps were used to calculate the ensemble averages. The Peng-Robinson equation of state was used to calculate the gas-phase fugacities for methane. In order to compare our simulation results with available experimental data, we carried out the simulations at 303 K for MIL-47, and 125 K, 200 K, 270 K and 300 K for IRMOF-1 and ZIF-8.[19] For the same reason, the differential enthalpy of adsorption was calculated for MIL-47 whereas the isosteric heat of adsorption was calculated for IRMOF-1 and ZIF-8. The differential enthalpy of adsorption, ΔH^D (Eq. 1), in MIL-47 was calculated from fluctuation theory during the GCMC simulation.[20]

$$\Delta H^D = H^R + RT - \left(\frac{\partial U}{\partial N}\right)_T = H^R + RT - \frac{\langle UN \rangle - \langle U \rangle \langle N \rangle}{\langle N^2 \rangle - \langle N \rangle^2} \tag{1}$$

where H^R is the molar residual enthalpy of the pure component in the real gas state, R is the universal gas constant, U is the average potential energy per molecule, and N is the average number of molecules adsorbed. The isosteric heats of adsorption, Q^{st}, in IRMOF-1 and ZIF-8 was calculated from the Clausius-Clapeyron equation (Eq. 2):

$$Q^{st} = -RT \left(\frac{\partial \ln P}{\partial T}\right) \tag{2}$$

Note that Qst is also directly related to the potential energy and can be calculated from ΔH^D by substracting RT.[21]

3 RESULTS AND DISCUSSION

Figures 2 shows the experimental as well as the simulated adsorption isotherms for the three different MOFs. All isotherms exhibit a type I shape, typical for microporous solids. Figure 2 (a) shows the isotherms for IRMOF-1 and ZIF-8 at two different temperatures. As expected, the amount adsorbed increases with decreasing temperature. By comparing the experimental and simulated isotherms, we found that in all cases the amount adsorbed was overpredicted with respect to the experimental results. There is very little difference between the different simulated curves indicating that the simulation results are not very sensitive to the choice of the force field. The deviation between the experimental and simulated adsorption uptakes is not unexpected. The input into the simulations is the perfect and completely activated crystal i.e. without defects and devoid of any solvent or precursor. However, the experimental uptake depends strongly on the synthesis and processing of the sample,[22] which can lead to defects in the crystal, as well as collapses in the porosity Additionally, experimental samples can retain some precursor and solvent molecules inside the pores. These two effects cause a reduction in the effective adsorption pore volume and therefore the maximum amount adsorbed. Thus, the amount adsorbed on its own is clearly insufficient to assess whether generic force fields are able to describe accurately the interactions between adsorbate molecules and the MOF frameworks.

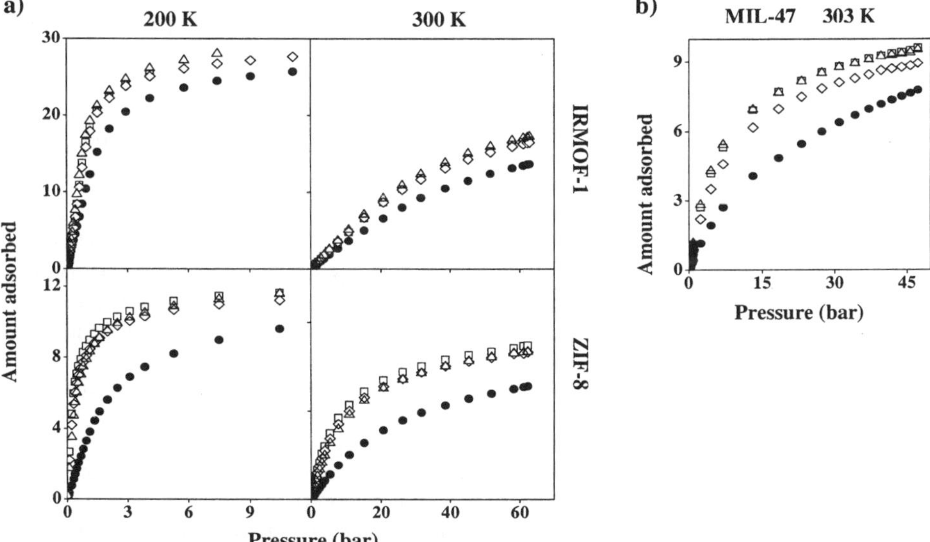

Figure 2: *Simulated and experimental adsorption isotherms of methane on* (**a**) *IRMOF1 and ZIF-8 (at 200 K, 300 K), and* (**b**) *MIL-47 (303 K) (□ – UFF; ◇ – Dreiding; △ – OPLS, ● – experimental).*

The differential enthalpy of ΔH^D as well as the isosteric heat of adsorption Q^{st} are directly related to the interaction energy (compare equation 1). Figure 3 shows how these two properties evolve as a function of the loading. For ZIF-8, the experimental Q^{st} values are overestimated by around 20 % for UFF and Dreiding and by 17 % for OPLS. The simulated values for IRMOF-1 are much closer to the experimental values (deviations of ~5% at intermediate covering). This result directly from the larger slope of the simulated isotherms compared to the experimental in the low pressure range. In the low pressure range, the amount adsorbed is proportional to the Henry's constant, another property that is directly related to the potential energy. Higher slopes imply stronger interaction in the adsorbent-adsorbate system. In contrast to this, the simulated values for ΔH^D in MIL-47 are slightly underestimated with respect to the experimental data (deviations of <13% at low loading and ~18% at higher loading), which could be within the experimental error. The results for all three force fields are rather similar with differences of 2 %. Additionally, we found that in all cases ΔH^D and Q^{st} increase with methane loading. This indicates that the adsorbate-adsorbate interactions in addition to adsorbate-adsorbent interactions become more important with higher loading.

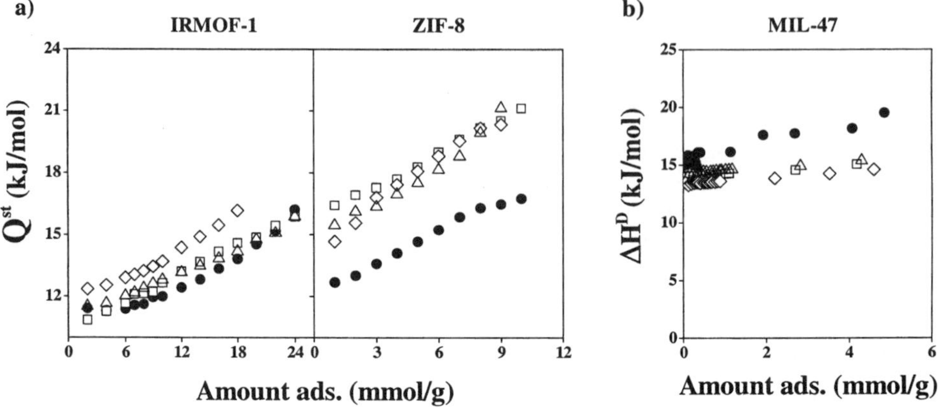

Figure 3: *Comparison between experimental (closed symbols) and simulated* (**a**) *isosteric heat of adsorption, Qst, for IRMOF-1 and ZIF-8, and* (**b**) *differential enthalpy ΔH^D forMIL-47 at 303 K (□ – UFF; ◇ – Dreiding; △ – OPLS, ● – experimental).*

To properly describe a given adsorption process, obtaining a good fit for the pore volume is not as important as achieving a good fit for the shape of the isotherm, which is related with the strength of the interactions. Adsorption uptake in simulated curves can be corrected by a scaling factor Φ. This scaling factor is the ratio of the experimental and the simulated maximum amount adsorbed thus taking into account that the real material has nonporous defects and inaccessible pores. Applying scaling factors, figure 4 illustrates that the shape of the experimental isotherms is captured to a different extent in the different MOFs investigated. In the case of MIL-47 (not shown) and ZIF-8, the isotherms exhibit a larger slope at low pressures for the simulated isotherms. In contrast, for IRMOF-1 the normalised simulated isotherms perfectly match the experimental curve. This indicates that the imperfections in the IRMOF-1 are most likely due to non-porous defects such as partial framework collapse leading to about 20 % of inaccessible pores. In addition to this, the imperfections in the MIL-47 and ZIF-8 samples alter the interaction between the methane and the MOF material itself and could result from defects in the crystal structure or solvent molecules left over in the pores due to imperfect activation.

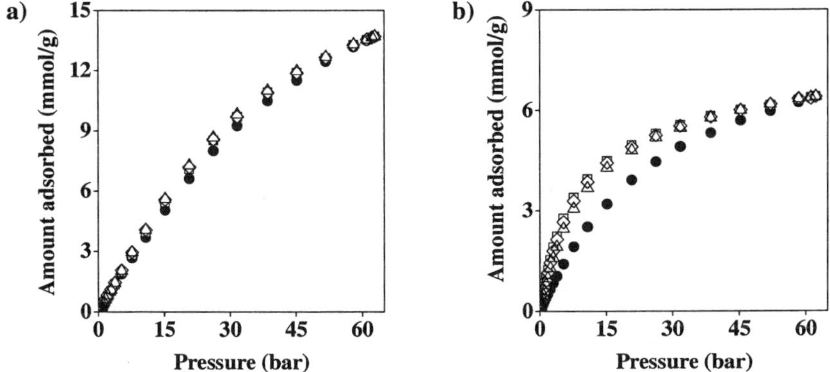

Figure 4: *Scaled simulated isotherms to fit the experimental data of* (**a**) *IRMOF-1 (Φ factors used for UFF: 0.82, Dreiding: 0.83, OPLS: 0.79) and* (**b**) *ZIF-8 (Φ factors used for UFF: 0.74, Dreiding: 0.77, OPLS: 0.77), ($\square$ – UFF; $\diamond$ – Dreiding; $\triangle$ – OPLS, $\bullet$ – experimental), T = 300K.*

We further quantified these observation by using the Dubinin-Radushkevich equation (DR),[23] which has been derived specifically to describe adsorption of vapours and can be applied directly to the methane isotherms obtained at 125 K. Kaneko *et al.*[24, 25] describe how the DR equation can be applied at supercritical conditions by estimating a quasi-saturation pressure, P_{0q}, for supercritical gases in micropores:

$$\ln\left(\frac{W_L}{W}\right)^{1/2} = \frac{RT}{\beta E_0}\left(\ln P_{0q} - \ln P\right) \tag{3}$$

where W is the gas uptake when the adsorption potential is A (with $A = RT \ln P/P_0$); W_L is the micropore capacity of the material and βE_0 is the characteristic adsorption potential of the material for the adsorbate in question. In order to compare the results obtained for the three different force fields, we calculated the values of P_{0q} from the experimental curve. The micropore capacity W_L was calculated by fitting either the Langmuir or Toth isotherm to the individual curves. With these values, we obtained the βE_0 values using equation (3).

The results obtained can be found in Table 2. From these data we can see that MIL-47 exhibits a slightly larger adsorption capacity than ZIF-8 at around 300 K. However, this value is increased by a factor of two in the case of IRMOF-1 at the same temperature. Again, these results are not very sensitive to the choice of the force field. The averaged ratios βE_0(experimental)/βE_0(simulated) for the three different force fields give an idea how well the simulation capture the shape of the experimental adsorption isotherms. For IRMOF-1 these ratios are 0.98, 0.98 and 0.95 at 300 K, 200 K and 125 K, respectively (lower temperatures values not shown); whereas the ratio for both MIL-47 and ZIF-8 is 0.89 at 300 K. This confirms the observations discussed previously that the simulations capture the shape of the isotherms in IRMOF-1 much better than in the two other materials. The deviations between the three force fields are negligible for IRMOF-1 (βE_0 deviation of ~2%). These deviations are more significant for MIL-47 and ZIF-8 around 12%). Overall, closest agreement is achieved with the Dreiding force field in MIL-47, whereas UFF gives the best agreement for IRMOF-1 and OPLS for ZIF-8.

Table 2: *Textural properties obtained from the application of Langmuir and Toth equations (W_L) and DR equation (βE_0 and P_0).*

	MIL-47		IRMOF-1		ZIF-8	
	303 K; P_{0q} = 435 MPa		300 K; P_{0q} = 1195 MPa		300 K; P_{0q} = 274 MPa	
	W_L (mmol/g)	βE_0 (kJ/mol)	W_L (mmol/g)	βE_0 (kJ/mol)	W_L (mmol/g)	βE_0 (kJ/mol)
Exp	11.65	11.55	19.63	11.66	8.68	10.61
UFF	11.02	13.10	20.70	11.82	10.24	12.05
Dreiding	10.59	12.76	21.41	11.86	9.85	11.93
OPLS	10.80	13.24	21.32	11.89	9.99	11.69

4 CONCLUSIONS

The purpose of this paper was to address the questions how accurately UFF, Dreiding and OPLS describe the interactions between a MOF and adsorbed methane molecules, and whether one outperforms the others. GCMC simulations were carried out for three MOFs with different structures, pore shapes and sizes: MIL-47, IRMOF-1 and ZIF-8. For each simulation, potential parameters were taken from the three different force fields to describe each atom involved in the fluid-fluid and fluid-framework interactions. Comparison with experimental enthalpy of adsorption or isosteric heat data clearly show that the generic force fields investigated here are able to correctly describe the interaction between adsorbate molecules and the MOF framework. There is little difference between the three different force fields, indicating that the simulation results (adsorption isotherm as well as isosteric heat and differential enthalpy data) are not very sensitive to the choice of the force field. Moreover, the Dreiding force field achieves the closest agreement with experimental data for MIL-47, whereas UFF gives the best fit for IRMOF-1 and OPLS for ZIF-8 thus showing that none of the force fields performs better than the others in all cases. We would like to point out, however, that the UFF force field contains all parameters for all elements in the periodic table which makes it easier to use in a consistent way i.e. without having to take parameters from other sources.

5 ACKNOWLEDGEMENTS

We thank Philip Llewellyn and colleagues (Université de Provence, France) for the experimental data on MIL-47 and Taner Yildirim and co-workers (University of Pennsylvania, USA) for the experimental data on IRMOF-1 and ZIF-8. We also thank the European Commission (DeSANNS (No FP6-SES6-020133)) and Universidad de Granada (Spain) for financial support.

References

1. U. Mueller, M. Schubert, F. Teich, H. Puetter, K. Schierle-Arndt, and J. Pastre, *Journal of Materials Chemistry*, 2006, **16**, 626.
2. O.I. Lebedev, F. Millange, C. Serre, G. VanTendeloo, and G. Ferey, *Chem. Mater.*, 2005, **17**, 6525.
3. H.K. Chae, D.Y. Siberio-Perez, J. Kim, Y. Go, M. Eddaoudi, A.J. Matzger, M. O'Keeffe, and O.M. Yaghi, *Nature*, 2004, **427**, 523.
4. G. Férey, C. Serre, C. Mellot-Draznieks, F. Millange, S. Surblé, J. Dutour, and I. Margiolaki, *Angewandte Chemie International Edition*, 2004, **43**, 6296.
5. S. Bourrelly, P.L. Llewellyn, C. Serre, F. Millange, T. Loiseau, and G. Ferey, *J. Am. Chem. Soc.*, 2005, **127**, 13519.
6. T. Düren, L. Sarkisov, O.M. Yaghi, and R.Q. Snurr, *Langmuir*, 2004, **20**, 2683.
7. W. Zhou, H. Wu, M.R. Hartman, and T. Yildirim, *J. Phys. Chem. C*, 2007, **111**, 16131.
8. M. Tafipolsky, S. Amirjalayer, and R. Schmid, *J Comput Chem*, 2007, **28**, 1169.
9. A.K. Rappe, C.J. Casewit, K.S. Colwell, W.A. Goddard, and W.M. Skiff, *Journal of the American Chemical Society*, 1992, **114**, 10024.
10. S.L. Mayo, B.D. Olafson, and W.A. Goddard, *Journal of Physical Chemistry*, 1990, **94**, 8897.
11. W.L. Jorgensen, D.S. Maxwell, and J. Tirado-Rives, *Journal of the American Chemical Society*, 1996, **118**, 11225.
12. K. Barthelet, J. Marrot, D. Riou, and G. Ferey, *Angewandte Chemie-International Edition*, 2001, **41**, 281.
13. H. Li, M. Eddaoudi, M. O'Keeffe, and O.M. Yaghi, *Nature*, 1999, **402**, 276.
14. L.D. Gelb and K.E. Gubbins, *Langmuir*, 1999, **15**, 305.
15. X.C. Huang, Y.Y. Lin, J.P. Zhang, and X.M. Chen, *Angewandte Chemie-International Edition*, 2006, **45**, 1557.
16. D. Frenkel and B. Smit, *Understanding Molecular Simulations: From Algorithms to Applications*, Academic Press, San Diego, 2002.
17. A. Gupta, S. Chempath, M.J. Sanborn, L.A. Clark, and R.Q. Snurr, *Molecular Simulation*, 2003, **29**, 29.
18. S.J. Goodbody, K. Watanabe, D. Macgowan, J. Walton, and N. Quirke, *Journal of the Chemical Society-Faraday Transactions*, 1991, **87**, 1951.
19. W. Zhou, H. Wu, M.R. Hartman, and T. Yildirim, *Journal of Physical Chemistry C*, 2007, **111**, 16131.
20. D. Nicholson and N.G. Parsonage, *Computer Simulation and the Statistical Mechanics of Adsorption*, Academic Press, London, 1982.
21. T. Vuong and P.A. Monson, *Langmuir*, 1996, **12**, 5425.
22. J. Liu, J.T. Culp, S. Natesakhawat, B.C. Bockrath, B. Zande, S.G. Sankar, G. Garberoglio, and J.K. Johnson, *J. Phys. Chem. C*, 2007, **111**, 9305.
23. M.M. Dubinin, E.D. Zaverina, and L.V. Radushkevich, *Zhurnal Fizicheskoi Khimii*, 1947, **21**, 1351.
24. K. Kaneko, S. Ozeki, and K. Inouye, *Colloid and Polymer Science*, 1987, **265**, 1018.
25. K. Kaneko and K. Murata, *Adsorption-Journal of the International Adsorption Society*, 1997, **3**, 197.

A METHOD FOR PREDICTING THE DIFFUSION COEFFICIENT BASED ON THE PHYSICAL CHARACTERISATION OF POROUS CARBONS BY ADSORPTION

Q. Cai, A. Buts, M.J. Biggs and N.A. Seaton

Institute for Materials and Processes, University of Edinburgh, King's Buildings, Mayfield Road, Edinburgh, EH9 3JL, Scotland, United Kingdom. M.Biggs@ed.ac.uk.

1 INTRODUCTION

Diffusion is important in many applications of nanoporous carbons, including in gas separation, catalysis, gas storage, controlled release and electrochemical energy storage. It is, therefore, important to be able to model diffusion of fluids in nanoporous carbons when seeking to design them and associated processes in these and other contexts. The long standing approach to modelling diffusion in porous solids involves averaging a pore-level transport model over the pore size distribution (PSD) and then using the so-called tortuosity[1] to account for pore system topology and other features omitted from the averaging (see, for example, *ref.* 2). Diffusion experiments are the main means used to determine tortuosities. This has a number of disadvantages – there is currently no commercially available apparatus, the experiments are challenging and potentially time consuming, and a sufficient quantity of the carbon must exist. By combining molecular simulation at the single-pore level with a pore network (PN) model whose physical characteristics are determined using equilibrium experiments, these issues can be largely circumvented. This paper briefly describes such a method along with its initial assessment.

2 OUTLINE OF METHOD

The method developed and assessed here for the determination of transport diffusion coefficients of fluids in nanoporous carbons on the basis of their physical characteristics derived from equilibrium experiments is illustrated in Figure 1. External field non-

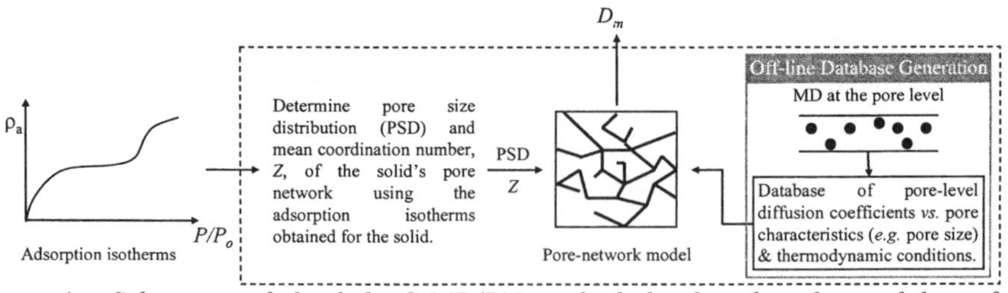

Figure 1 *Schematic of the hybrid MD/PN method developed and tested here for determining the diffusion coefficients of nanoporous carbons.*

equilibrium molecular dynamics[3,4] (EF-NEMD) is used to determine the transport diffusion coefficient of the fluid of interest within single pores as a function of the thermodynamic conditions and pore characteristics (*e.g.* pore width if the pore is to be modelled by the standard graphitic slit pore as done here[5]); these are stored in a database for subsequent look-up. The PSD and mean coordination number of the pore network, Z, which is defined as the average number of pores meeting at a junction, are separately determined from adsorption isotherms for the solid of interest using the approaches of Seaton and co-workers.[6-8] Both the physical characteristics of the solid and the MD-derived database of single-pore diffusion coefficients are then brought together to construct a series of pore network models whose averaged diffusion coefficient are then determined using the renormalized effective medium approximation (REMA) approach of Zhang and Seaton[9], which is an extension of the original REMA approach[9] to the case where the diffusion coefficients of the pores of the network are distributed. Further details beyond those given here may be found in Cai *et al.*[11]

3 OUTLINE OF APPROACH FOR ASSESSING METHOD

Clearly, one means of assessing the method illustrated in Figure 1 is to compare its predictions with experimental data. However, such a comparison cannot, by itself, serve to validate the approach. If the predictions agree with experiment, this might be the result of a fortuitous cancellation of errors that may not apply under other conditions. Similarly, if the predictions are poor, it will not be clear which aspects of the approach are at fault. These issues can be circumvented if we use a very well characterised nanoporous material whose pore space and processes therein can be probed unambiguously and in detail. Unfortunately, such materials and experimental methods do not exist in the case of nanoporous carbons (if at all). We have, therefore, used Virtual Porous Carbons (VPCs), which are complex computer-based model solids that capture elements of real carbons,[12] as illustrated in Figure 2 to initially assess the MD/PN method.

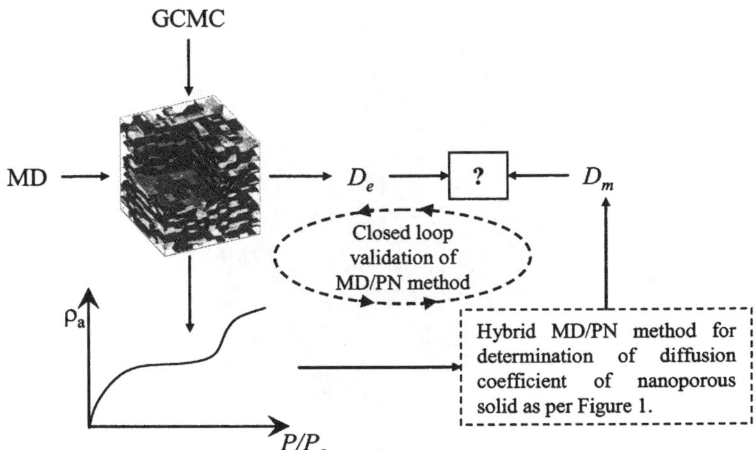

Figure 2 *Absolute assessment strategy for evaluating the MD/PN method for determining the transport diffusion coefficient of nanoporous carbons. The VPC shown on the left hand side of this figure is in the form of an isoenergy map of a microporous region of size 88.42×76.57×80.50 Å³ whose corner has been removed to show the nature of the pore space; the solid and pore space are coloured black and grey respectively.*

This process first involves use of grand canonical Monte-Carlo (GCMC) simulation and EF-NEMD to determine the sorption isotherms, N, and *actual* transport diffusion coefficients, D_a, of the VPC respectively – example isotherms and flow field through the VPC considered here are shown in Figure 3 and Figure 4 respectively. The GCMC-derived isotherms are then used to determine the PSD and Z of the VPC as outlined in §2, which are in turn used to determine the MD/PN-based transport diffusion coefficient, D_e, of the VPC as also outlined in §2. This is then compared with D_a. If this comparison reveals any differences, detailed probing of the VPC structure and diffusion process therein, and their comparison with the underlying elements of the MD/PN model (*e.g.* PSD, Z, pore-level diffusion), can be carried out to elucidate the reason for the mismatch and suggest routes to improving the MD/PN method.

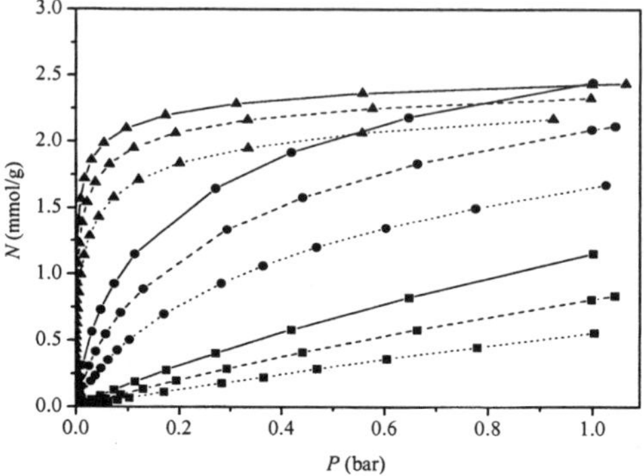

Figure 3   *GCMC-determined isotherms for CH_4 (square), CF_4 (circle) and SF_6 (triangle) on the VPC at 258 K (solid line), 275 K (dash line) and 296 K (dotted line). These were used to determine the PSD and Z of the VPC pore network (see §2). See Cai et al.[13] for further details.*

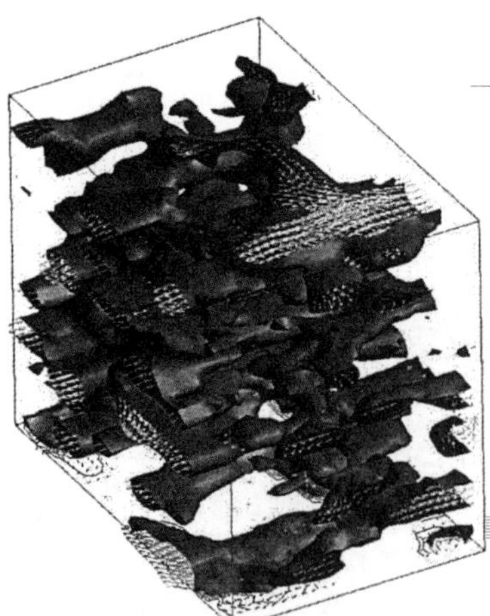

Figure 4   *EF-NEMD determined flow envelope for CH_4 in the VPC considered here at 298 K and 1 bar. The forcing is applied from right to the left and is sufficiently small for the system to be in the linear response regime. The size and shade of the arrows revealed by cutting away part of the flow envelope is related to the local speed. See Cai et al.[11] for further details.*

RESULTS AND DISCUSSION

Figure 5 shows the actual transport diffusion coefficient of the VPC considered here, D_a, and the diffusion coefficient predicted for the VPC by the MD/PN method, D_e, for a wide pressure range. This figure shows that the MD/PN-derived diffusion coefficients are generally in very good agreement with the actual coefficient both quantitatively and qualitatively. It should be emphasised that this comparison involves no adjustable parameters. The fact that the MD/PNM approach gives good results for the VPC suggests that it would give quantitatively accurate predictions for transport diffusion in real carbons.

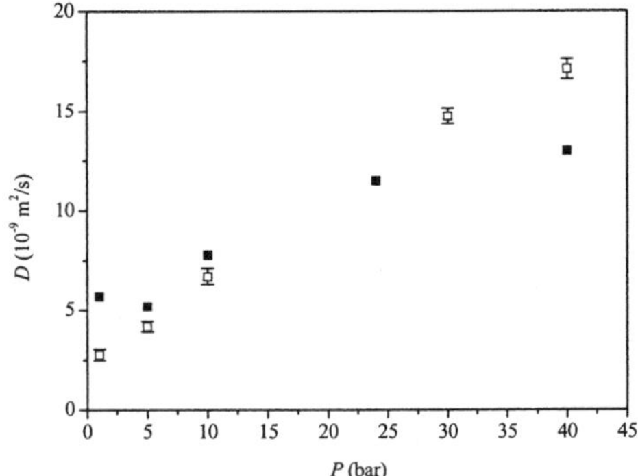

Figure 5   *Comparison of the actual transport diffusion coefficients of methane in the VPC, D_a, (solid squares) with those predicted by the MD/PN method, D_e, (open squares). The error bars associated with the latter are $\pm\,\sigma$, where σ is the standard deviation. Error bars are not shown when smaller than the symbols.*

Although the comparison in Figure 5 is very encouraging, the MD/PN method under- and over-predicts the transport diffusion coefficient at the lower and higher ends of the pressure range considered here respectively. In order to understand the origin of these differences and gain greater insight into the working of the MD/PN method, we investigate here the influence of the PSD and Z on the MD/PN-predicted diffusion coefficient.

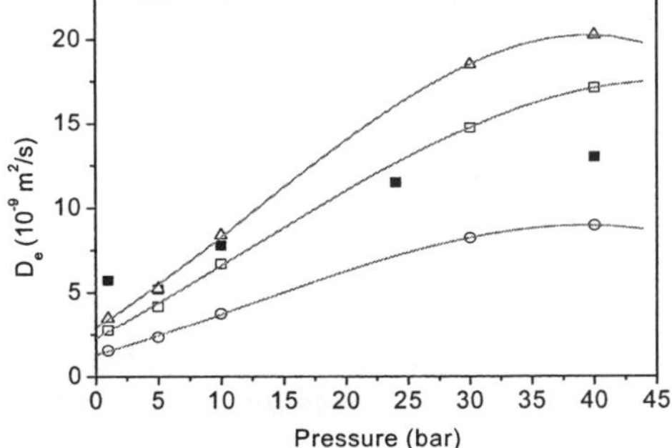

Figure 6   *Comparison of the actual diffusion coefficients of CH_4 in the VPC (closed squares) with the MD/PN-derived effective diffusion coefficients (open symbols) using the adsorption-derived PSD and the actual mean coordination number ($Z = 2.06$; squares) and slightly smaller ($Z = 1.9$; circles) and larger ($Z = 2.2$; triangles) mean coordination numbers. The lines are a guide to the eye only.*

Figure 6 shows a comparison between the actual diffusion coefficient of the VPC and the diffusion coefficients predicted by the MD/PN method using the PSD of the VPC with mean coordination numbers slightly higher and lower than the actual mean coordination number. This figure shows that even slight variations in the coordination number leads to substantially different predictions by the MD/PN method, especially at higher pressures. This suggests that the method of López-Ramón *et al.*[7] yields mean coordination numbers that are suitable for the prediction of diffusion within pore networks. It should be stressed, however, that Z is not necessarily "correct" in any fundamental sense – in practice it is impossible to calculate a unique value of Z for the pore network of a real carbon or a VPC as this can only be done on the basis of an arbitrary decomposition of the pore space in terms of a set of discrete pores.[13] Rather, we have demonstrated that the method of López-Ramón *et al.*[7] gives a measure of the connectivity that is a useful predictor of transport through the pore network.

The likely origin of the differences seen in Figure 5 can be appreciated by comparing the PSD determined from the adsorption isotherms with the actual (geometric) PSD of the VPC obtained by direct analysis, Figure 7. As discussed in our earlier work,[13] the adsorption-derived PSD analysis allocates pore volume to larger pores around 30 Å that do not exist in reality. At low pressure, there is little adsorption in the larger pores, and the smaller pores in the PN will carry the bulk of the flux. As Figure 7 shows, however, the number of these pores is underestimated because some of the porosity has been incorrectly assigned to (non-existent) larger pores that make little contribution at low pressure to the flux. This leads to under-prediction of the diffusion coefficient at low pressures. At high pressures, on the other hand, the larger pores are more highly loaded and carry a higher proportion of the flux, thus leading to an over-prediction of the diffusion coefficient. The errors in the MD/PN predictions appear, therefore, to originate with the incorrectly determined PSD. Similar analysis using a VPC whose PSD is much better predicted, which is published elsewhere,[11] further supports this conclusion.

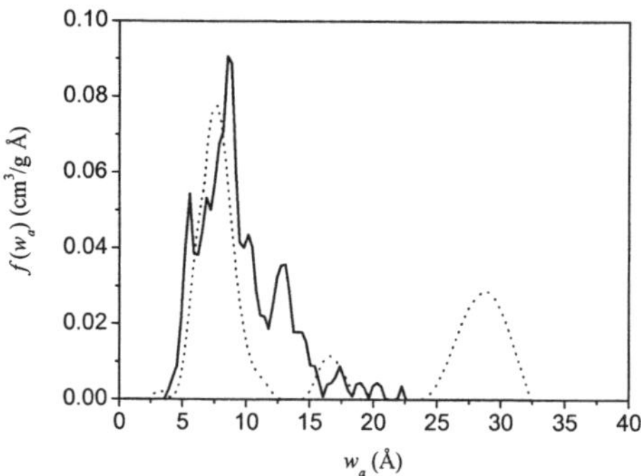

Figure 7 *Comparison of the geometric PSD of the VPC (solid line) with the adsorption-derived PSD (dashed line). The pore size, w_a, is defined as the distance between the surface of the pore surface carbons. See Cai et al.[13] for further details on the determination of these PSDs.*

5 CONCLUSIONS AND FUTURE WORK

Using an absolute assessment methodology based on Virtual Porous Carbons (VPCs), we have shown that by bringing together single pore diffusion coefficients derived from molecular dynamics (MD) with a pore network (PN) model constructed using adsorption-derived pore size distribution (PSD) and mean coordination number, Z, we can accurately predict transport diffusion coefficients of nanoporous carbons without input from non-equilibrium experiments. It was further shown that the mean coordination number determined for nanoporous carbons by the method of López-Ramón *et al.*[7] are suitable for the prediction of diffusion within pore networks. It was, however, also shown that the PSD obtained from adsorption can be a significant source of error in the MD/PN-predicted diffusion coefficients (this is despite it being suitable for prediction of adsorption isotherms[13]) – this suggests that further improvements are required in current methods for PSD prediction. Such improvements can be assessed using the absolute assessment process described here and elsewhere.[11,13,14] Once satisfactory, further comparison of the MD/PN method with data for real carbons is warranted.

Acknowledgements

QC gratefully acknowledges an Overseas Research Scholarship (ORS) from Universities UK. The UK Engineering and Physical Sciences Research Council (GR/R87178/01) is acknowledged for the financial support of this work.

References

1   N. Epstein, *Chem. Engng. Sci.*, 1989, **44**, 777.
2   C. Feng and W.E. Stewart, *Ind. Eng. Chem. Fundam.*, 1973, **12**, 143.
3   E.J. Maginn, A.T. Bell and D.N. Theodorou, *J. Phys. Chem.* 1993, **97**, 4173.
4   S. Chempath, R. Krishna and R.Q. Snurr, *J. Phys. Chem. B*, 2004, **108**, 13481.
5   Q. Cai, M.J. Biggs and N.A. Seaton, *PCCP*, 2008, **10**, 2519.
6   N.A. Seaton, *Chem. Engng. Sci.*, 1991, **46**, 1895.
7   M.V. López-Ramón, J. Jagiełło, T.J. Bandosz and N.A. Seaton, *Langmuir*, 1997, **13**, 4435.
8   H. Liu, L. Zhang and N.A. Seaton, *Chem. Engng. Sci.*, 1992, **47**, 4393.
9   L. Zhang and N.A. Seaton, *AIChE J.*, 1992, **38**, 1816.
10 M. Sahimi, B.D. Hughes, L.E. Scriven, and H.T. Davis, *Phys. Rev. B*, 1983, **28**, 307.
11 Q. Cai, A. Buts, N.A. Seaton and M.J. Biggs, *Chem. Engng. Sci.* in press.
12 M.J. Biggs and A. Buts, *Mol. Sim.*, 2006, **32**, 579.
13 Q. Cai, A. Buts, N.A. Seaton and M.J. Biggs, *Langmuir* 2007, **23**, 8430.
14 M.J. Biggs, A. Buts and D. Williamson, *Langmuir*, 2004, **20**, 7123.

POSITRON LIFETIME SPECTROSCOPY ON CONTROLLED PORE GLASS

S. Thränert[1], D. Enke[2], G. Dlubek[3] and R. Krause-Rehberg[1]

[1]Martin-Luther-Universität Halle-Wittenberg, Naturwissenschaftliche Fakultät II,
Institut für Physik, 06099 Halle, Germany
[2]Martin-Luther-Universität Halle-Wittenberg, Naturwissenschaftliche Fakultät II,
Institut für Chemie, 06099 Halle, Germany
[3]ITA - Institut für innovative Technologien GmbH, Köthen, Außenstelle Halle,
Wiesenring 4, 06120 Lieskau, Germany

1. INTRODUCTION

Mesoporous glasses (pore size 2 - 50 nm) are of interest for various applications (low-dielectric thin films, catalysis, molecular filters). For their utilization it is necessary to characterize their properties, especially the pore size, using capable methods. Numerous techniques for the determination of the pore size are established such as mercury intrusion, low temperature nitrogen adsorption and electron microscopic investigations. However, these techniques have one problem in common: their sensibility is limited to a minimum pore size D (diameter) of $\sim 2 - 4$ nm. Furthermore, mercury intrusion and nitrogen adsorption are applicable only for open pore systems.

Positronium annihilation lifetime spectroscopy (PALS) provides just in the range 0.5 - 10 nm a high sensitivity to the pore size and furthermore allows a non-destructive measurement, also on closed pore systems. In dielectric amorphous material a part of the implanted positrons may form a bound state with an electron which is called positronium (Ps) [1]. The singlet spin state of Ps (*para*-positronium, *p*-Ps) decays very rapidly due to self-annihilation via the emission of 2 γ rays with a mean intrinsic lifetime of $\tau_S = 0.125$ ns and is not suitable to deliver information about the pores. The long-living triplet spin state of Ps (*ortho*-positronium, *o*-Ps) is, however, a suitable probe for measuring the pore sizes [2-4]. In vacuum *o*-Ps decays with a mean lifetime of $\tau_T = 142$ ns via emission of 3 γ rays. But in matter the lifetime of *o*-Ps can be reduced markedly by pick-off annihilation, a quenching process of the *o*-Ps caused by interaction with electrons of suitable spin (opposite direction to the Ps's positron spin) leading to a 2 γ decay. Ps formed near the pore or at its internal surface will be localized at the pore. The driving force for the localization is the exchange interaction of the Ps's electron with the electrons of the molecules surrounding the pore. So the pick-off annihilation lifetime, which can be extracted from the lifetime spectra, contains information about the pore size.

A typical lifetime spectrum of CPG consist of four different exponential lifetime components which have its origin in the annihilation of the *p*-Ps ($\tau_1 \sim 0.125$ ns), the annihilation of free positrons which have not formed Ps ($\tau_2 \sim 0.5$ ns), the annihilation of *o*-Ps in micropores (subnanometersize holes appearing in the disordered structure of the glass, $\tau_3 \sim 1.5$ ns) and in micro- or mesopores of different sizes ($\tau_4 > 1.5$ ns to 142 ns).

As mentioned, *o*-Ps decays in matter via the pick-off process and self-annihilation with the rates λ_{po} and $\lambda_T = 1/\tau_T = 1/142$ ns^{-1}, respectively. If *o*-Ps is localized at a pore it will show a probability $P < 1$ to be found outside the pore. Here, in the wall of the pore, it undergoes pick-off annihilation with the rate λ_A which is assumed to correspond to the spin averaged vacuum annihilation rate $\lambda_A = (\lambda_S + 3\lambda_T)/4 = 2.00$ ns^{-1} [5]. With $\lambda_{po} = \lambda_A P$ the total *o*-Ps decay rate is given by

$$\frac{1}{\tau_{o-Ps}} = \lambda_{o-Ps} = \lambda_A P + \lambda_T (1 - P) \tag{1}$$

The traditional Tao-Eldrup model [5-7] provides a simple relation which allows to calculate the pore size from the lifetime of *o*-Ps localized at the pore. A spherical pore characterized by a rectangular potential well of the radius R and a finite potential barrier V is described by a well with an infinite potential barrier V and a size $R+\Delta R$ [5]. By this construction the wave function of the localized Ps overlaps with bulk (electron layer in the wall of the pore) in the range between R and $R+\Delta R$. This model was originally applied for small pores, $R < 1$ nm. Here the second term in eq. (1) can be neglected so that $\lambda_{o-Ps} = \lambda_A P$. Moreover, *o*-Ps will occupy only the ground state in the pore. The wave function for this state is the spherical Bessel function $j_0(r)$. The *o*-Ps lifetime is then related to the pore radius R via

$$\lambda_{TE}(R) = \lambda_A \left[1 - \frac{R}{R + \Delta R} + \frac{1}{2\pi} \sin\left(\frac{2\pi R}{R + \Delta R} \right) \right] \tag{2}$$

where ΔR was determined empirically to be 0.166 nm by measuring vacancy-rich molecular solids and zeolites with known hole sizes [6-8]. A corresponding equation may be derived for cuboid geometry [9, 10].

For larger pores, *o*-Ps will populate higher energy levels when the temperature is above 0 K. The population of higher levels increases with the temperature which will cause a distinct temperature dependency of the *o*-Ps lifetime. The consideration of excited states complicates the calculation of the annihilation rate. Gidley et al. extended the Tao-Eldrup model assuming rectangular pore geometrie [11, 12]. The Lublin group has extended the Tao-Eldrup model and published calculations of the *o*-Ps lifetime assuming spherical or cylindrical pore geometries [2, 3, 10, 13, 14]. In the calculations integrals of spherical or cylindrical Bessel functions appear [15]. It is assumed that at finite temperatures *o*-Ps populates the energy levels in the potential well following the Boltzmann statistics. The overlap of the *o*-Ps wave function and the bulk is given by δ, a constant parameter, analogous to ΔR in the TE model, which depends somewhat on the assumed pore geometry. By comparison with experimental pore sizes, this parameter was determined to be $\delta = 0.18$ nm [11, 16] for rectangular geometry and $\delta = 0.19$ nm for cylindric geometry [17].

The aim of this work is to find a calibration curve for the dependence of the longest lifetime on the pore size using one of the extended TE models. We pay attention on lifetime experiments on controlled pore glasses (CPG) of various pore sizes at room temperature. A major advantage using CPG samples is the free choosable pore size. Furthermore the temperature dependence of the o-Ps lifetime in selected CPG samples is measured and compared with theoretical calculations to prove the universality of the model. Furthermore we spent particular attention for getting information on the pore size distributions. For the analysis of the lifetime spectra we employed the routine LifeTime, version 9.0 (LT9.0) [18]. This routine is extremely useful for the analysis of hole size distributions in amorphous polymers [19, 20] and was also applied for porous glasses [13]. In the literature conventionally routines are used which assume that the lifetime spectrum consists of a sum of discrete exponential terms. These routines are generally useful but do not provide information on pore size distributions. Several authors employed the routines CONTIN [21] or MELT [22] which assume a priori a distribution of lifetimes. It was found, however, that the results of both programs are not unambiguous and may show artefacts [23-26]. Thus, a broad lifetime distribution may be mirrored as two, more narrow, neighbour distributions. In LT9.0 the number of components and the shape of the distribution are assumed from the beginning. These constraints lead to distinct decrease in the number of freedom in the analysis and reduce the danger of artefacts (see chapter 3). We use the results from LT9.0 and show a way to transfer lifetime distributions in pore size distributions.

2. CONTROLLED PORE GLASSES – CONVENTIONAL CHARACTERIZATION

The mesoporous glasses investigated in this study were prepared and characterized in the laboratory of one of us (D. Enke). For a review of porous glasses please refer to [27]. For our experiments we used CPG powder (porous micro-spheres with a particle size of 80 µm) with a mean pore size D from 2 nm to 70 nm. The pore size D was estimated as mean free path within the pore which is defined, independently of the shape of the pores, by $D = 4V/S$. Here V is the volume and S the internal surface of the pores, both values can be determined from nitrogen adsorption and/or mercury intrusion technique. Nitrogen sorption measurements were performed by using a Sorptomatic 1990 apparatus (ThermoFinnigan). Surface areas can be determined from the linear part of the Brunauer-Emmett-Teller (BET) equation in a relative pressure range (p/p_0) of the adsorption isotherms between 0.05 and 0.25 [28]. The total pore volume was estimated from the amount of gas adsorbed at the relative pressure $p=p_0$ = 0.99 assuming that pores were filled subsequently with condensed adsorptive in the normal liquid state. The mercury intrusion measurements were carried out on a Pascal 440 apparatus (ThermoFinnigan). The mean pore diameters D of samples with pore sizes below 5 nm were calculated from the total pore volume V (nitrogen adsorption) and the BET surface Area S according to $D = 4V/S$ or using the BJH-method [29] and the adsorption branch of the isotherm. The BJH-method also allows a calculation of the pore size distribution. Mercury intrusion was used for pore sizes > 5 nm. Here, the mean pore diameter was calculated by applying the Washburn equation [30].

For comparing the results of porosimetry with those from PALS one has to assume the shape of the pores. For spheres of diameter d one gets $D = 4V/S = 2d/3$ while for cylindric, infinite capillars of diameter d the relation $D = d$ is valid.

3. EXPERIMENTS AND DATA ANALYSIS

At first we investigated the correlation of the pore size and the o-Ps pick-off annihilation lifetime at T = 300 K. In a second step, the temperature dependence of lifetime spectra in the range between 50 K and 500 K was measured for several glasses with selected pore sizes. The PALS measurements were done using a fast-fast coincidence system [1, 31, 32] with a time resolution of 250 ps (FWHM, ^{22}Na source), an analyzer channel width of 121.5

ps, and a number of 8000 channels corresponding to a maximum delay time of 972 ns. Spectra with a total count number of 4×10^6 coincidence counts were recorded. The activity of the used positron source was 4×10^6 Bq (3.2 µCi). Such a weak source is required to obtain the longest lifetime τ_4 in a sufficient way [33] because the intensity of τ_4 in our CPG samples is rather weak ($I_4 \sim 5\%$). By using a stronger source the background due to random coincidences would disturb a proper analysis of the long lifetime. To avoid cryo-condensation effects the sample chamber was evacuated to 10^{-8} mbar. Furthermore the sample was heated up to room temperature after every low temperature (< 300 K) measurement before cooling down to the next lower temperature. The temperature control system consists of thermo couples and a PID regulation system unit with an accuracy of 0.1 K.

For analysis of the spectra the routine Life-Time [34] in its version 9.0 was used [18] assuming four exponential lifetime terms, $s(t) = \Sigma(I_i/\tau_i)\exp(-t/\tau_i)$ ($i = 1, 2, 3, 4$; τ - lifetime, $\lambda = 1/\tau$ annihilation rate), with the mean lifetimes τ_i and the relative intensities I_i ($\Sigma I_i = 1$). Nine fit parameters appear in the analysis: four τ_i, three I_i, the time zero t_0, and the background B. For reducing the statistical scatter of fit parameters we have fixed the p-Ps lifetime to its vacuum value $\tau_1 = 0.125$ ns. The resolution function used in the spectrum analysis was determined as a sum of two Gaussians. The source correction was estimated from a measurement of defect-free silicon ($\tau_{Si} = 219$ ps). τ_2 (e+ annihilation) varied between 0.4 ns and 0.5 ns, τ_3 (*o*-Ps annihilation from intrinsic holes in the glass structure) between 1.5 and 2.0 ns, and τ_4 (*o*-Ps annihilation in mesopores) from 21 ns to 134 ns depending on the pore size. With increasing pore size the relative intensity of this component, I_4, decreased from 14 % to 3 %. The routine Excited Energy Levels and Various Shapes (EELViS) [35] was used to calculate the theoretical correlation of the pore size and the longest lifetime τ_4. Furthermore the temperature dependency was calculated to compare it with our experimental results.

4a. RESULTS AND DISCUSSION – THE CALIBRATION CURVE

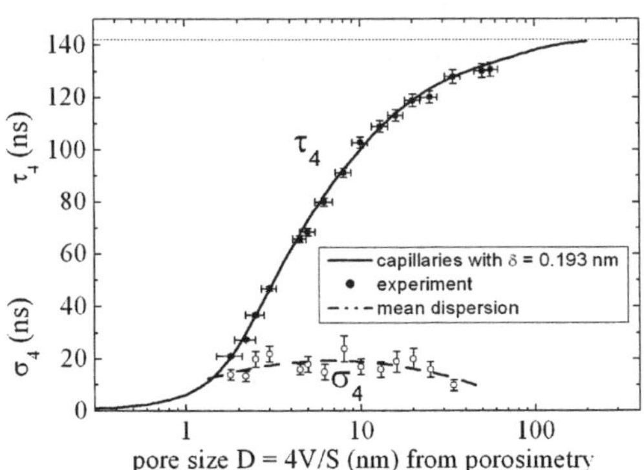

Fig. 1: The mean *o*-Ps lifetime τ_4 as a function of the pore size D (mean free path $D = 4V/S$) from porosimetry for CPG. The closed circles represent our experimental data at 300 K, the line correspond to the best fit of our experimental data on theoretical results of the ETE model for capillaries with $\delta = 0.193$ nm (here $D = d$ is the diameter of the cylindrical capillaries). σ_4 is the mean dispersion (square-root of variance) of the *o*-Ps lifetime distribution. For mean lifetimes τ_4 larger than 100 ns the value of σ_4 could not be correctly determined.

Figure (1) shows the mean *o*-Ps lifetime τ_4 as a function of the pore size $D = 4V/S$ (mean free path) from porosimetry for the CPG's at room temperature (300 K). The line corresponds to theoretical results of the ETE model for infinitely long cylinders of diameter $d = D$ with the overlap value δ as a parameter. The spherical model can not describe the experimental data for small pores, so it was not possible to achieve a calibration curve. For the cylindrical model δ was determined so that the sum of the squares of deviations between experiment and theory weighted with the error of experiments gives a minimum value. The estimated δ = (0.193±0.005) nm for cylindrical geometry is in agreement with the data of Ciesielski et al. [16], δ = 0.19 nm, obtained for porous Vycor glass while Zaleski et al. [3] determined recently δ = 0.18 nm for mesoporous silica sieves MCM-41. The error of 0.005 nm is due to the errors in porosimetry and in lifetime experiments.

This data can be used as a CPG calibration curve for the correlation of the *o*-Ps lifetime attributed to mesopores and the pore size *D*. At room temperature the extended TE model is a useful porosimetry tool for a characterisation of porous media.

Three selected CPG samples were used for temperature-dependent measurements to prove the universality of the extended TE model. The given pore sizes in Fig. (2) were calculated from the measured long lifetime using the cylindrical extension of TE model with δ = 0.193 nm. The theoretical temperature dependency was calculated using the same model. In our experiments we found three different scenarios: only the measurements done on the sample with a pore size of 4 nm agree reasonably to the theoretical calculations in the entire investigated temperature range from 60 K to 500 K. The measurements done on smaller pores (here represented by the 2.5 nm curve) show too long *o*-Ps lifetimes for low temperatures (250 K and beyond). However, for high temperatures the data agrees well to the model. Regarding the experimental data for the larger pores (here represented by the 20 nm curve) one observes a contrary temperature behaviour in comparison to the small pores. For temperatures below 300 K the long lifetime τ_4 is too short compared to the theoretical calculations whereas the measurements for higher temperatures show too long lifetimes. For large pores we only find at room temperature an agreement to the model.

So the model assumptions seem to be too simple to explain the temperature dependence of the *o*-Ps lifetime in the entire temperature range. The behaviour of the *o*-Ps lifetime in larger pores at low temperatures is probably caused by a localization of Ps at the internal surface of the pore due to the strong van der Waals interaction of the highly polarizable Ps with the glass molecules [36, 37]. The polarization of Ps is eight times larger than of the hydrogen atom since the reduced mass of Ps is only $m_e/2$ (m_e – electron mass). It was estimated that the van der Waals energy between the Ps and the SiO_2 molecule is about 1 eV at a distance of 0.2 nm [36]. The localization in the surface potential is stronger at lower temperature which increases the pick-off rate and decreases the lifetime of *o*-Ps. This effect is not considered in the ETE model which assumes a simple square-well potential.

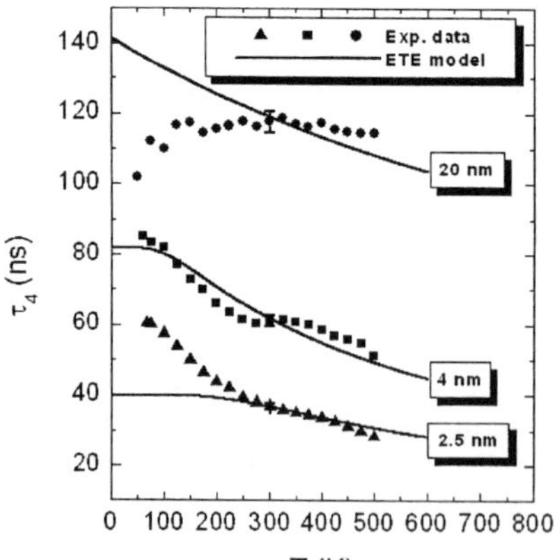

Fig. 2: Temperature dependence of the o-Ps lifetime for different pore sizes (symbols) in comparison with the ETE model calculations (lines).

The increase of the *o*-Ps lifetime with decreasing temperature for small pores is more difficult to understand. An analogous unexpected behaviour was observed for Vycor glasses [17], Silica gels [3, 38], and mesoporous silica fibers [2, 3, 39]. The nature of this behaviour seems to be still unknown. Recently a model was developed which attributes this effect to the thermal activation of surface atoms (-OH groups) at medium and higher temperatures and its closer attachment to the pore surface at low temperatures [40]. This effect would increase the effective pore size with decreasing temperature and would be particular strong for small pores. The model fitted well the temperature dependence of the *o*-Ps lifetime. Other groups [41] have shown that a chemical modification (trimethylsilylation, for example) of pore surfaces may change the *o*-Ps lifetime. Within the ETE model this effect may be treated as a change in the overlap parameter δ caused by the surface chemistry. Further research seems to be required to enlighten these important problems.

4b. RESULTS AND DISCUSSION – THE PORE SIZE DISTRIBUTION

In order to get information on the size distribution of mesopores we have used the routine LT9.0 in its distribution mode. The routine LT9.0 assumes that the function $\alpha_4(\lambda)$ follows a log normal distribution where $\alpha_4(\lambda)$ is the distribution of the o-Ps annihilation rate in the pores [34]. From $\alpha_4(\lambda)$ the lifetime distributions $\alpha_4(\tau) = \alpha_4(\lambda)\lambda^2$ and its mean (first momentum), $<\tau_4>$, and standard deviation (square-root of second momentum), σ_4, can be calculated. We have analyzed truncated spectra, as frequently done in the literature [13].

We assumed a single component with distributed lifetimes, fixed the time zero of the spectrum to the value from the complete analysis, and determined the parameter τ_4 and σ_4. These values are used for the further discussion. For mean lifetimes $\tau_4 > 100$ ns the value of σ_4 exhibited dramatically increasing statistical errors since the flat lifetime curve merges more and more with the background from random coincidence events. Therefore, we limited this analysis to the data from the range $\tau_4 < 100$ ns. The analyzed lifetime distributions show mass centers $<\tau_4>$ which agree with the lifetimes τ_4 from the discrete term analysis within the statistical error limits. The standard deviations of the distributions, σ_4, are shown in Fig.1. They show only a weak variation with the pore size D.

Four samples with pore sizes $D = 4V/S$ of 1.8 nm, 2.5 nm, 4.5 nm, and 6.2 nm (from porosimetry) are chosen for analyzing the pore size distribution. The corresponding τ_4/σ_4 values are 21.1/14.8, 46.9/17.6, 65.9/18.9, and 80.0/19.3 (all values in ns). These width parameters σ_4 which are used for calculation of distributions are values obtained from smoothing the experimental data by fitting a quadratic function (dash-dotted line in Fig.1). The lifetime distributions are of roughly Gaussian shape with a tail at the side of larger lifetimes.

The routine LT assumes that $\alpha_4(\lambda)$ follows a logarithmic Gaussian function which determines the shape of $\alpha_4(\tau) = \alpha_4(\lambda)\lambda^2$. One problem of this kind of analysis is that the fitting procedure does not take into account that τ_4 can not exceed 142 ns. Therefore, the tail of the distributions in case of larger lifetimes can be overestimated and also exceed the physical limit of the *o*-Ps lifetime in vacuum. From the distribution of *o*-Ps lifetimes, $\alpha_4(\tau)$, one can calculate the distribution of the hole diameters n(d) taking into account the non-linear character of this transformation [21,42]

$$n(d) = \alpha_4(\tau)\, d\tau_4/dd. \tag{3}$$

For this calculation we need an analytical function which describes the dependency $\tau_4 = \tau_4(d)$ and which can be differentiated. For this purpose we fitted a numerical calibration curve for $\delta = 0.193$ nm and 300 K (more in detail: soon to be published).

Figure (3) shows the distribution (pdf) $n(d)$ of pore diameters d for selected samples. The arrows show the d-values directly calculated from the mean o-Ps lifetime, $d = d(\tau_4)$. $n(d)$ should be considered as an apparent distribution which comes from the true variation in the diameter of cylindrical pores but contains also the effect of the irregular, non-linear character of pores. The long tail in the distribution for larger pores should be considered as overestimation which comes from an overestimation of $\alpha_4(\tau)$ and from the nonlinear character of the τ_4 vs. d relation for larger lifetimes.

For two samples we also show the (volume weigthed) pore size distribution obtained from the BJH-method. For comparison those curves are normalised to the same height and maximum. The BJH-curve for the 2.5 nm sample shows general agreement to the pore size distribution obtained from PALS. For the 4.5 nm sample the pore size distribution obtained from PALS is more narrow compared to the BJH-method. Also the longer tail in the PALS distribution is obvious. The BJH-method is limited to a minimum pore size of 2 nm, so it is not possible to use this method for micropores. This limitation leads also to the problem that the distribution can only be measured particularly, as shown in Fig. (3). However, PALS is not limited and a powerful tool to measure pore size distributions for micropores and smaller mesopores.

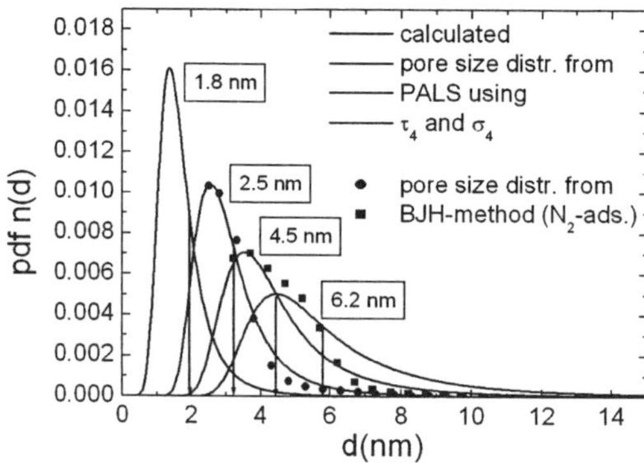

Fig. 3: Distribution $n(d)$ of pore diameters d for selected samples. The distribution is normalized to 1. The arrows show the d-values directly calculated from the mean o-Ps lifetime, $d = d(\tau_4)$, (1.77 nm, 3.09 nm, 4.38 nm, and 5.8 nm). The porosimetry delivered pore sizes D = 4V/S of 1.8 nm, 2.5 nm, 4.5 nm, and 6.2 nm. The closed symbols show the pore size distribution obtained from the BJH method and the adsorption branch of the isotherm for the 2.5 nm and 4.5 nm samples.

5. SUMMARY

The lifetime of o-Ps annihilating from localized states at mesopores of controlled pore glasses can be described by the extended Tao-Eldrup (ETE) model assuming a cylindrical pore geometry (diameter d, infinite length). The pore size determined as mean free path, D, by N_2 adsorption and Hg intrusion techniques varied between 1.8 and 64 nm. Correspondingly, the mean o-Ps lifetime increased from 21 and 134 ns. From a last-squares fit to the experimental data the overlap parameter δ of the ETE model was determined to be $\delta = 0.193 \pm 0.005$ nm at 300 K assuming $d = D$. An analytic (modified logistic) function fitted to the theoretical ETE model curve can be used for the calculation of the pore size directly from the o-Ps lifetime. This analytic calibration curve makes the application of PALS as routine porosimetry tool easier. In this context the particular sensitivity of PALS for small pores is to emphazise. Using the calibration curve the o-Ps lifetime distribution, analyzed in this work by employing the routine LifeTime9.0, can be converted into a pore size distribution.

It was found that the ETE model describes the temperature dependence of the o-Ps lifetime correctly only for a medium pore size such as $D = 4$ nm. For larger pores and low temperatures a smaller o-Ps lifetime than predicted by the model was measured which might have its origin in the strong van der Waals interaction of the highly polarizable Ps with the glass molecules at the pore surface. For small pores the o-Ps lifetime shows an unexpected increases at low temperatures. Possible reasons for this behavior are briefly discussed.

ACKNOWLEDGEMENT

The authors wish to thank Dr. R. Zaleski for supplying the routine EELViS and Dr. J. Kansy for making available the routine LT9.0.

REFERENCES

[1] O. E. Mogensen, *Positron annihilation in chemistry* (Springer, Berlin, 1995).
[2] R. Zaleski, J. Goworek, A. Kierys, phys. stat. sol. (c) 4, 3814 (2007).
[3] R. Zaleski, J. Wawryszczuk, J. Goworek, Rad. Phys. Chem. 76, 243 (2007).
[4] D. W. Gidley, R. S. Vallery, M. Liu, H.-G. Peng, phys. stat. sol. (c) 4, 3796 (2007).
[5] S. J. Tao, J. Chem. Phys. 56, 5499 (1972).
[6] P. Kirkegaard, M. Eldrup, O. E. Mogensen, J. Pedersen, Comp. Phys. Comm. 23, 307 (1981).
[7] M. Eldrup, D. Lightbody, J. N. Sherwood, Chem. Phys. 63, 51 (1981).
[8] Y. C. Jean, Microchem. J. 42, 72 (1990).
[9] B. Jasinska, A. E. Koziol, T. Goworek, J. Radioanal. and Nucl. Chem. 210, 617 (1996).
[10] T. Goworek, K. Ciesielski, B. Jasinska, J. Wawryszczuk, Chem. Phys. 230, 305 (1998).
[11] T. L. Dull, W. E. Frieze, D. W. Gidley, J. N. Sun, A. F. Yee, J. Phys. Chem. B 105, 4657 (2001).
[12] D. W. Gidley, W. E. Frieze, T. L. Dull, A. F. Yee, E. T. Ryan, Phys. Rev. B 60, R5157 (1999).
[13] T. Goworek, K. Ciesielski, B. Jasinska, J. Wawryszczuk, Chem. Phys. Lett. 272, 91 (1997).
[14] T. Goworek, Radiat. Phys. Chem. 68, 331 (2003).
[15] R. Zaleski, *Tables of Pick-off model calculations* (Lublin, Poland, 2002).
[16] K. Ciesielski, A. L. Dawidowicz, B. Jasinska, T. Goworek, J. Wawryszczuk, Chem. Phys. Lett. 289, 41 (1998).
[17] T. Goworek, K. Ciesielski, B. Jasinska, J. Wawryszczuk, Rad. Phys. Chem. 58, 719 (2000).
[18] J. Kansy, *LT for Windows, Version 9.0* (Silesian University, Katowice, Poland,).
[19] G. Dlubek, A. Sengupta, J. Pionteck, R. KrauseRehberg, Macromolecules 37, 6606 (2004).
[20] G. Dlubek, E. M. Hassan, J. Pionteck, R. Krause-Rehberg, Phys. Rev. E 73, 031803 (2006).
[21] R. B. Gregory, Nucl. Instr. and Meth. A 302, 496 (1991).
[22] A. Shukla, M. Peter, L. Hoffmann, Nucl. Instr. and Meth. A 335, 310 (1993).
[23] G. Dlubek, Ch. Hübner, S. Eichler, Nucl. Instr. and Meth. B 142, 191 (1998).
[24] M. P. Petkov, C. L. Wang, M. H. Weber, K. G. Lynn, J. Phys. Chem. B 107, 2725 (2003).
[25] M. H. Weber, K. G. Lynn, in *Principles and Applications of Positron and Positronium Chemistry* (edited by Y. C. Jean, World Scientific, Singapore, 2003).
[26] G. P. Babu, V. Manobar, K. Sundarshan, P. K. Pujari, S. B Manohar, A. Goswami, J. Chem. Phys. B 106, 6902 (2002).
[27] D. Enke, F. Janowski, W. Schwieger, Microporous and Mesoporous Materials 60, 19 (2003).
[28] K. S. W. Sing, J. Rouquerol, in *Handbook of Heterogenous Catalysis* (edited by G. Ertl, H. Knozinger, J. Weitkamp, Wiley-VCH, Weinheim, 1997).
[29] A. B. Shelekhin, A. G. Dixon, Y. H. Ma, J. Membr. Sci 83, 181 (1993).
[30] E. Washburn, Proc. Natl. Acad. Sci. USA 115, 7 (1921).

[31] Y. C. Jean, P. E. Mallon, D. M. Schrader, editors, *Principles and Application of Positron and Positronium Chemistry* (World Scientific, Singapore, 2003).
[32] R. Krause-Rehberg, H.S. Leipner, *Positron annihilation in semiconductors* (Springer, Berlin, 1999).
[33] S. Thraenert, E. M. Hassan, R. Krause-Rehberg, Nucl. Instr. and Meth. B 248, 336 (2006).
[34] J. Kansy, Nucl. Instr. and Meth. A 374, 235 (1996).
[35] R. Zaleski, *Excited Energy Levels and Various Shapes (EELViS)* (Poland, 2007).
[36] S. M. Kim, W. J. L. Buyers, J. Phys. C: Solid State Phys. 11, 101 (1978).
[37] A. L. R. Bug, T. W. Cronin, P. A. Sterne, Z. S. Wolfson, Rad. Phys. Chem. 76, 237 (2007).
[38] M. Sniegocka, B. Jasinska, J. Wawryszczuk, R. Zaleski, A. Derylo-Marczewska, I. Skrzypek, Acta. Phys. Polonica 107, 868 (2005).
[39] B. Jasinska, R. Zaleski, M. Sniegocka, R. Reisfeld, E. Zigansky, phys. stat. sol. (c) 4, 3985 (2007).
[40] B. Ganguly, D. Gangopadhyay, D. Dutta, S. Chatterjee, T. Mukherjee, B. D. Roy, J. Phys. Chem. 76, 263 (2007).
[41] C. He, T. Oka, Y. Kobayashi, N. Oshima, T. Ohdaira, A. Kinomura, R. Suzuki, Appl. Phys. Lett. 91, 024102 (2007).
[42] D. W. Gidley, W. E. Frieze, T. L. Dull, J. Sun, A. F. Yee, C. V. Nguyen, D. Y. Yoon, Appl. Phys. Lett. 76, 1282 (2000).

DYNAMIC MEAN FIELD THEORY FOR FLUIDS CONFINED IN POROUS MATERIALS: APPLICATION TO AN INKBOTTLE PORE GEOMETRY

P. A. Monson

Department of Chemical Engineering, University of Massachusetts, Amherst, MA 01003, U.S.A.

1 INTRODUCTION

Classical density functional theories are important tools in understanding the molecular level behavior of fluids confined in porous materials and are now widely regarded as a key component in characterization of porous materials by adsorption.[1,2] These theories allow us to calculate the free energy and density distribution for fluids inside porous materials for equilibrium states of the system. They also allow us to study metastable states associated with hysteresis loops in adsorption/desorption isotherms. It is also of interest to study the dynamics in the approach to equilibrium and metastable equilibrium for fluids in pores. This has recently been done using molecular dynamics and dynamic Monte Carlo simulations[3]. In this paper we discuss a recently developed theoretical approach to studying the dynamics that makes use of a lattice gas treatment of the system[4]. The theory yields the time evolution of the density distribution in the system and the equilibrium or metastable equilibrium states from mean field density functional theory emerge as the long time limit in the dynamics.

Our approach is based on a lattice gas description of fluids in pores, which has been used widely in recent density functional studies of fluids confined in porous materials.[5-7] The theory is constructed as an approximation to the average behavior over an ensemble of Kawasaki dynamics Monte Carlo simulations of the model in which mean field approximations are used for the transition probabilities.[8] The theory builds on the work of Martin[9] and of Penrose[10] on the dynamic mean field theory for the Ising and binary alloy models, respectively. That work was originally motivated by an interest in developing an extension of the Cahn-Hilliard approach[11-13] to phase separation dynamics that would include a microscopic theory of the local mobility. It is also equivalent to the theory of diffusion recently presented by Aranovich and coworkers.[14] In addition to being a mean field approximation ot a dynamic Monte Carlo simulation of the system the present approach can be interpreted as a theory of diffusion or as a dynamic density functional theory.[15]

Our first application of this approach was to a slit pore model.[4] Here we describe initial results from an application to the inkbottle geometry. This geometry consists of a larger cavity joined with two smaller cavities, which are in contact with the bulk. A visualization of the lattice model of the inkbottle is shown in figure 1 where the large cavity has a wall separation of 12 sites and the small cavities have wall separations of 6 sites. Since the density is uniform in the *y*-direction the problem becomes two-dimensional. The inkbottle geometry occupies a special place in the literature on adsorption hysteresis literature because it is a very simple model of a pore network and illustrates the concept of pore blocking. Since the pore condensation and evaporation in the small cavities occurs at lower pressure than that in the large cavities, on desorption the evaporation from the large cavities may be delayed until the small cavities can empty. Recently, however, using molecular simulations it was discovered that the large pores could empty even though the small cavities remain filled.[16] Subsequent calculations have indicated that the behavior encountered depends on the size of the small cavities[17] and also on the temperature.[18] Here we use dynamic mean field theory to study the dynamic mechanisms of adsorption and desorption in a processes that start with the system in an

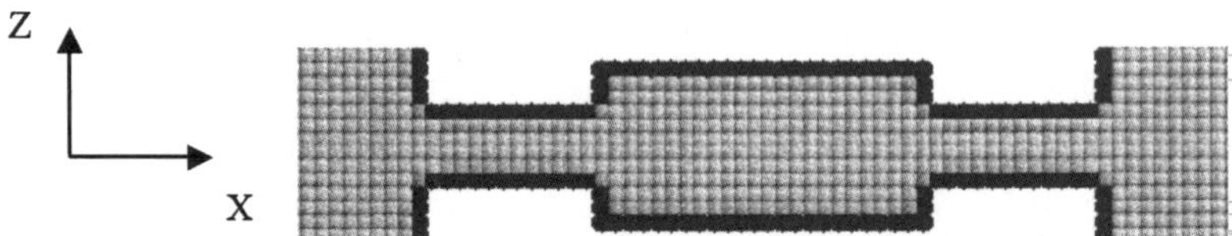

Figure 1 *Visualization of the inkbottle geometry used in this work for a low pressure state where the fluid forms a monolayer on the pore walls. The fluid density is represented by shading of the lattice sites and darker shading indicates higher density. The density is uniform in a direction normal to the plane of the figure. Periodic boundary conditions are applied in the x-direction.*

equilibrium state at a given chemical potential and proceed by quenching the chemical potential to a higher or lower value.

In the next section of the paper we briefly describe the model for the inkbottle pore geometry used here as well as the static and dynamics mean field theories. In section 3 we present results for the static and dynamic behavior of the system. Section 4 gives a summary of our results and conclusions.

2 MODEL AND METHODS

2.1 Lattice Gas Model and Thermodynamics in Mean Field Theory

We use a single occupancy, nearest neighbor lattice gas model of a fluid confined in a porous material. The fluid molecules occupy the site of a simple cubic lattice and the Hamiltonian can be written[5,7]

$$H = -\frac{\varepsilon}{2}\sum_{i}\sum_{a} n_i n_{i+a} + \sum_{i} n_i \phi_i \qquad (1)$$

where ε is the nearest neighbor interaction strength, n_i is the occupancy of site **i**, with **i** denoting a set of lattice coordinates. ϕ_i is the external field at site **i** emanating from the pore walls, which here is taken to be a nearest neighbor attraction with strength -3ε, k is Boltzmann's constant and T is the absolute temperature.

In the mean field approximation the Helmholtz energy is given by

$$F = -kT\sum_i \left[\rho_i \ln \rho_i + (1-\rho_i)\ln(1-\rho_i)\right] - \frac{\varepsilon}{2}\sum_i \sum_a \rho_i \rho_{i+a} + \sum_i \rho_i \phi_i \tag{2}$$

where ρ_i is the average occupancy (density) at site i. For a system at fixed N, V and T (canonical ensemble) we can determine the density distribution by minimizing F subject to the constraint of a fixed number of molecules. This leads to the equations

$$kT \ln\left[\frac{\rho_i}{(1-\rho_i)}\right] - \varepsilon \sum_a \rho_{i+a} + \phi_i - \mu = 0 \qquad \forall\ i\ ; \qquad \sum_i \rho_i = N \tag{3}$$

where μ is a Lagrange multiplier which coincides with the chemical potential. We can also implement the mean field theory in the grand ensemble (fixed μ, V and T) by solving equations (3) without the fixed N constraint.

2.3 Dynamic Mean Field Theory

We have followed the presentation of Gouyet et al.[8] in the development of the dynamic mean field theory and full detail is given in our previous paper. We focus on the average occupancy for site i at time t defined by

$$\rho_i(t) = \langle n_i \rangle_t = \sum_{\{n\}} n_i P(\{n\},t) \tag{4}$$

where $P(\{n\},t)$ is the probability of observing a given set of occupation numbers, $\{n\}$, at time t. The evolution of $\rho_i(t)$ is determined from a material balance in terms of net flux site into site i from its nearest neighbor sites expressed in terms of Kawasaki dynamics with Metropolis transition probabilities. After replacing site occupancies by their ensemble averages we express this as

$$\frac{\partial \rho_i}{\partial t} = -\sum_a \left[w_{i,i+a}\rho_i(1-\rho_{i+a}) - w_{i+a,i}\rho_{i+a}(1-\rho_i) \right] \tag{5}$$

where $w_{i,j}$ is the transition probability for a molecule to move from site i to site j. The other factors in this expression take into account the occupancy restrictions for the lattice model. The transition probabilities are given by[14]

$$w_{i,j} = \begin{cases} w_0 & E_j < E_i \\ w_0 \exp[-(E_j - E_i)/kT] & E_j > E_i \end{cases} \tag{6}$$

where w_0 is the transition rate in the absence of interactions, which sets the timescale for the system, and

$$E_i = -\varepsilon \sum_a \rho_{i+a} + \phi_i \tag{7}$$

To implement the method for the geometry in figure we begin by adding an additional layer of sites in the bulk at one end of the system. The density in this layer is held constant at the value consistent with the bulk chemical potential (calculated in mean field theory) for the state into which we want to quench the system. The initial density elsewhere in the system is set to values associated with the equilibrium density distribution at the initial chemical potential obtained by solving equation (3) in the grand ensemble. Using equations (3) and (7) we can readily relate the transition probabilities to the chemical potential at each site. In this way we can show that the chemical potential is uniform in the system when $\partial \rho_i / \partial t = 0 \ \forall \ \mathbf{i}$.[4] The system evolves to this state for long times and the density distribution is then given by the static mean field theory in the grand ensemble.

Equation (5) can be integrated with Euler's method, so that the density evolves via

$$\rho_i(t) = \rho_i(t + \Delta t) - \Delta t \sum_{\mathbf{a}} \{w_{i,i+a}(t)\rho_i(t)[1 - \rho_{i+a}(t)] - w_{i+a,i}(t)\rho_{i+a}(t)[1 - \rho_i(t)]\} \qquad (8)$$

We find that dimensionless time steps less than about $w_0 \Delta t \sim 0.2\text{-}0.3$ are sufficient to achieve stable solutions. Higher order integration schemes may lead to improved stability

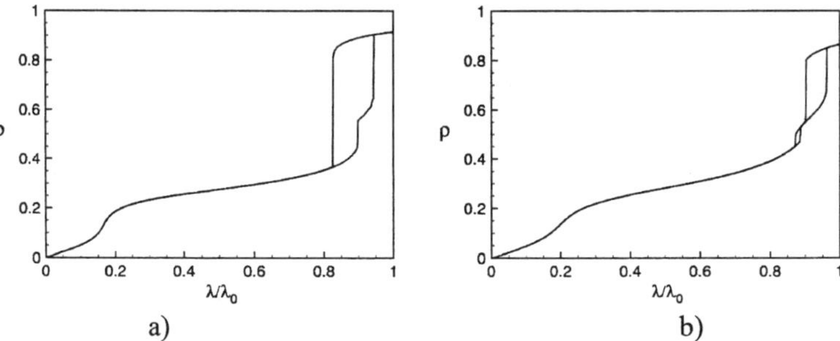

a)   b)

Figure 2 *Adsorption/desorption isotherms for the inkbottle pore system shown in figure 1: a) $kT/\varepsilon = 1.125$ ($T/T_c = 0.75$); b) $kT/\varepsilon = 1.275$ ($T/T_c = 0.85$).*

and the potential to use longer time steps, but this should be weighed against the increased computer time for each step of the integration.

3 RESULTS

3.1 Static Behavior

To begin with we discuss the static behavior of the system in mean field theory, which was also recently studied by Libby and Monson.[18] Figure 2 shows adsorption/desorption isotherms at two temperatures. We plot the density versus the relative activity, λ/λ_0, where $\lambda = \exp(\mu/kT)$ and λ_0, is the value for the bulk saturated vapor. The

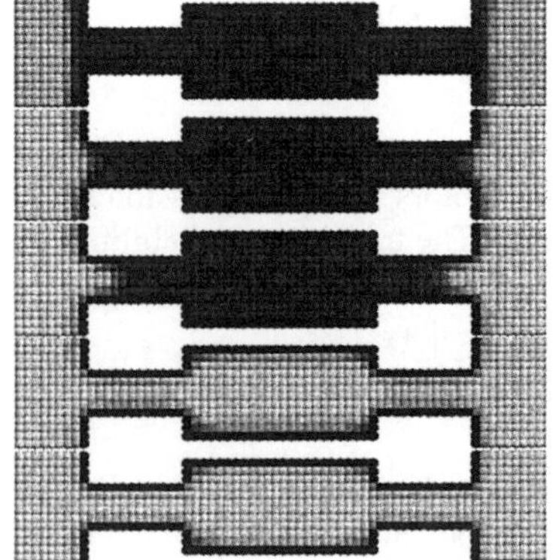

Figure 3 *Visualizations of states on the desorption isotherms shown in figure 5a.*

relative activity is close in value to the relative bulk pressure except for effects of gas imperfection. In both cases the adsorption isotherms involve two steps associated with the filling of the small and large cavities. In addition there is an inflection point at low activity associated with monolayer-like adsorption on the pore walls. The desorption branches of the isotherms for the two temperatures differ in that the evaporation from the pore at the lower temperature is a single step process but a two step process at the higher temperature.

Visualizations of the density distributions for states on the desorption isotherms are shown in figure 3 and 4. In figure 3 we see that desorption proceeds starting from the formation of a meniscus at the entrance to the small cavity. As the chemical potential is reduced the meniscus recedes into the cavity until a point is reached where the dense fluid state becomes unstable. An additional decrease in the chemical potential leads to evaporation of liquid from both cavities. This behavior is reminiscent of the traditional pore-blocking picture, where evaporation from the large cavity cannot occur until the small cavity can empty. Looking at the isotherms in figure 2 one sees that this behavior could be also interpreted in terms of the increased range of metastability of the liquid in the large and small cavities at the lower temperature. Figure 4 shows that desorption involves evaporation of liquid from the large cavity while the small cavity remains filled, followed by evaporation from the small cavity. This remarkable behavior was first seen in Monte Carlo and molecular dynamics simulations of an off-lattice model for this pore geometry.[16]

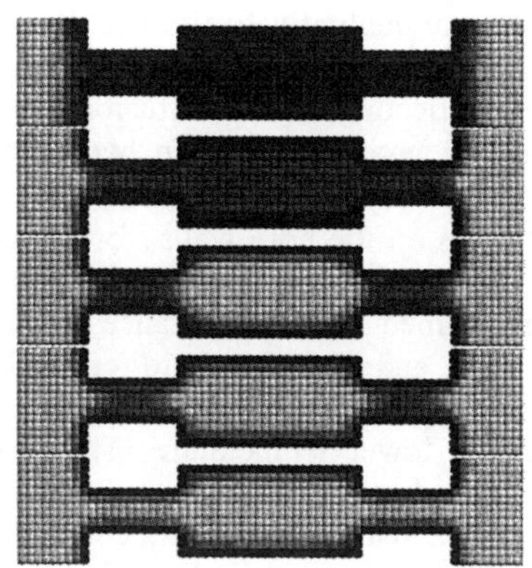

Figure 4 *Visualizations of states on the desorption isotherms shown in figure 5b.*

3.2 Dynamics

We have thus far considered three types of chemical potential quench. In the first case we make a large quench from a chemical potential corresponding to very low bulk pressure, where little fluid is adsorbed in the pores, to a value to close to that of the bulk saturated vapor, where the pores are essentially filled with liquid. In the second case we make a quench of the same magnitude but starting at the high chemical potential state and

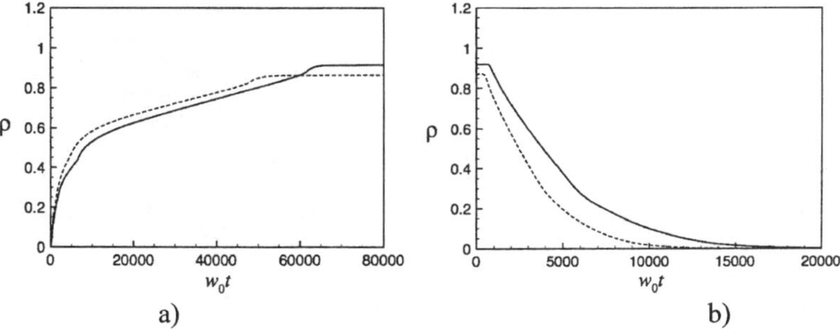

a)   b)

Figure 5 *Density vs. time plots from dynamic mean field theory for two temperatures (full line: $kT/\varepsilon = 1.125$ ($T/T_c = 0.75$); dashed line $kT/\varepsilon = 1.275$ ($T/T_c = 0.85$)): a) adsorption; b) desorption.*

quenching to the low chemical potential state. In addition we have considered a smaller desorption quench for $kT/\varepsilon = 1.275$ between states either side of the desorption step in figure 3b during which the large cavity is emptied.

Figure 5 shows the average density in the system plotted as a function of time (dynamic uptake plots) for the large adsorption and desorption quenches at each temperature. The curves for the two temperatures are qualitatively similar. We see that equilibration is faster for the higher temperature reflecting higher diffusion rates. The adsorption process occurs in two broad stages. First we have the formation of adsorbed layers on the pore walls, for $w_0 t$ up to about 5000. Around that time liquid bridges are formed near the entrances of the small cavities and this leads to cusp-like behavior in the uptake curve, particularly evident at the lower temperature. The next stage is the filling of the small and large cavities. This is a slower process requiring mass transfer of the fluid through the liquid bridges in the small cavities. The filling concludes with the disappearance of a vapor bubble in the center of the large cavity indicated in step-like behavior in the uptake curve at long times. Visualizations of the states appearing at key stages of the dynamics for the lower temperature are shown in figure 6.

Figure 6 *Visualizations from adsorption dynamics at $kT/\varepsilon = 1.125$.*

Figure 5b shows the density versus time in the corresponding desorption quenches. The first things to note is the difference in time scales between desorption and adsorption. The process of emptying the system is roughly four times faster than the process of filling. The reason for this is evident in the visualizations shown in figure 7 for the lower temperature. We see that the desorption dynamics proceeds by evaporation from the liquid surfaces with menisci retreating into the small cavities from the external surface and finally merging in the large cavity. The mass transfer to the external surface takes place in the vapor-like regions of the system. Notice that in contrast to what is seen in a quasistatic desorption process, i.e. following states on the desorption isotherm in figure 3b, the irreversible desorption mechanism is the same at the two temperatures with no emptying of the large cavity while the small cavities remaining filled.

Finally in figure 8 we show visualizations from a desorption quench at the higher temperature between states just either side of the step associated with desorption from the large cavity. This allows us to see the dynamic mechanism of the emptying of the large cavity with the small cavity remaining filled. We see first a loss of

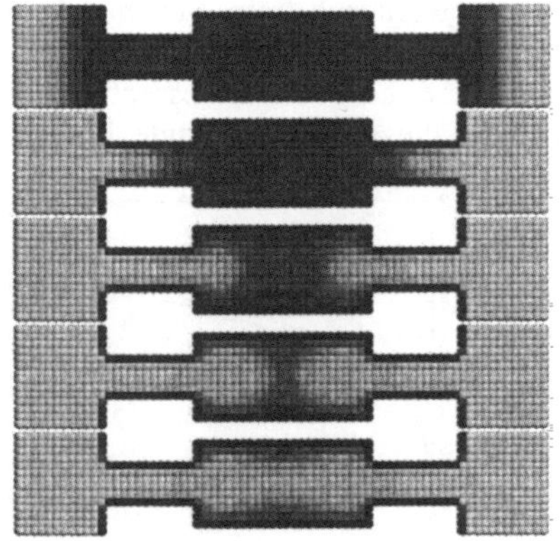

Figure 7 *Visualizations from the desorption dynamics at $kT/\varepsilon = 1.125$.*

density in the liquid throughout the system and then particularly in the large cavity. This is followed by the formation of a vapor bubble in the large cavity that gradually increases in size. Notice how the menisci near the entrances of he small cavity start to recede into the pore but then move out again after the bubble appears. We are seeing the stretching and breakup of the liquid in the system, reminiscent to the stretching and breakage of an elastic material.

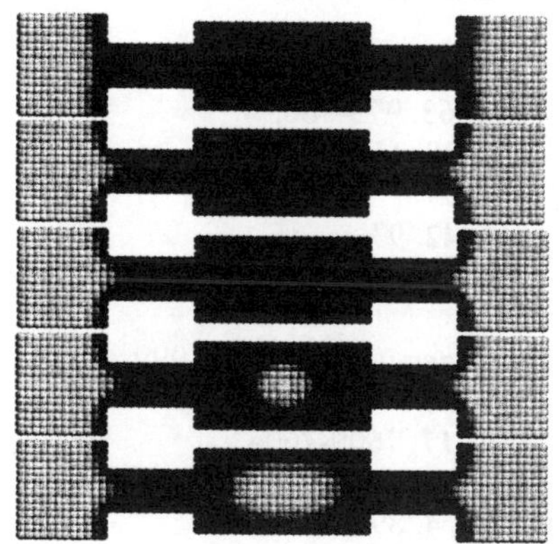

Figure 8 *Visualizations from desorption dynamics at kT/ε = 1.275 for a quench between states just either side of the larger desorption step in the isotherm in figure 2b.*

4 SUMMARY AND CONCLUSIONS

We have described a dynamic mean field theory for fluids confined in porous materials. The theory is based on a lattice gas model of a confined fluid and reduces to the mean field (density functional) theory for the system in the equilibrium limit. The approach can be used to describe several aspects of the dynamics of confined fluids, including the dynamics of adsorption and desorption. The theory can be viewed as a dynamic density functional theory, a theory of diffusion or an approximation to the average behavior over an ensemble of Kawasaki dynamics simulations.

As an illustrative example we have considered the case of an inkbottle pore geometry, a simple model for a pore network. The results show the importance of network effects in influencing equilibration rates for adsorption and desorption. In addition visualizations of the density during the dynamics provide insights into the nature of the density redistributions occurring during the dynamics. We believe the dynamic mean field theory should be a valuable tool in understanding the relaxation dynamics for fluids confined in porous materials.

5 ACKNOWLEDGMENTS

This work was supported by a grant from the national Science Foundation (Grant No. CBET-0649552).

6 REFERENCES

1.   R. Evans, *Journal of Physics-Condensed Matter*, 1990, **2**, 8989-9007.
2.   L. D. Gelb, K. E. Gubbins, R. Radhakrishnan and M. Sliwinska-Bartkowiak, *Reports on Progress in Physics*, 1999, **62**, 1573-1659.
3.   R. Valiullin, S. Naumov, P. Galvosas, J. Karger, H. J. Woo, F. Porcheron and P. A. Monson, *Nature*, 2006, **443**, 965-968.
4.   P. A. Monson, *Journal of Chemical Physics*, 2008, **128**, 084701.

5.   E. Kierlik, P. A. Monson, M. L. Rosinberg, L. Sarkisov and G. Tarjus, *Physical Review Letters*, 2001, **8705**, 055701.
6.   A. P. Malanoski and F. van Swol, *Physical Review E*, 2002, **66**, 041602.
7.   H. J. Woo and P. A. Monson, *Physical Review E*, 2003, **67**, 041207.
8.   J. F. Gouyet, M. Plapp, W. Dieterich and P. Maass, *Advances in Physics*, 2003, **52**, 523-638.
9.   G. Martin, *Physical Review B*, 1990, **41**, 2279-2283.
10.   O. Penrose, *Journal of Statistical Physics*, 1991, **63**, 975-986.
11.   J. W. Cahn, *Journal of Chemical Physics*, 1959, **30**, 1121-1124.
12.   J. W. Cahn, *Acta Metallurgica*, 1961, **9**, 795-801.
13.   J. W. Cahn, *Journal of Chemical Physics*, 1965, **42**, 93-&.
14.   D. Matuszak, G. L. Aranovich and M. D. Donohue, *Journal of Chemical Physics*, 2004, **121**, 426-435.
15.   U. M. B. Marconi and P. Tarazona, *Journal of Chemical Physics*, 1999, **110**, 8032-8044.
16.   L. Sarkisov and P. A. Monson, *Langmuir*, 2001, **17**, 7600-7604.
17.   A. Vishnyakov and A. V. Neimark, *Langmuir*, 2003, **19**, 3240-3247.
18.   B. Libby and P. A. Monson, *Langmuir*, 2004, **20**, 4289-4294.

CHARACTERIZATION OF SINGLE WALL CARBON NANOHORNS RESTRUCTURED WITH MECHANICAL TREATMENT

Koki Urita[1], Shinya Seki[1], Hideyuki Tsuchiya[1], Hiroaki Honda[1], Shigenori Utsumi[1], Tomonori Ohba[1], Hirofumi Kanoh[1], Masako Yudasaka[2], Sumio Iijima[2,3] and Katsumi Kaneko[1]

[1]Department of Chemistry, Graduate School of Science, Chiba University, Inage, Chiba 263-8522
[2]Japan Science and Technology Corporation, NEC Corporation, 34 Miyukigaoka, Tsukuba, Ibaraki 305-8501, Japan
[3]Department of Physics, Meijo University, 1-501 Shiogamaguchi, Tenpaku, Nagoya 468-8502, Japan.

1   INTRODUCTION

One of key sciences for environmentally friendly technologies is chemistry on nanoporous materials that have been widely applied to adsorption, separation, catalysis, storage, removal, sensing, etc. Zeolites and activated carbons are the representative nanoporous materials that have been widely applied to various technologies. The environmental aspects have led to new attempts to get traditional nanoporous materials of better performance by addition of mesopores and morphology control [1,2]. On the other hand, we need to find out new functions in newly developed nanoporous materials such as carbon nanotubes, ordered mesoporous silicas and metal organic frameworks (MOFs). The intrinsic merits of the new nanoporous materials have been expected to be applied to construction of environmentally friendly technologies. Carbon nanotube families such as single wall carbon nanotube (SWCNT), double wall carbon nanotube (DWCNT), and multiwall carbon nanotube (MWCNT) of high electronic conductivity, which are different from MOFs and ordered mesoporous silicas, have great application potentials. In particular, all carbon atoms of SWCNT are exposed to both interfaces, and thereby interfacial properties of single wall carbons with the relevance to clean fuel gas storage and electrical charge storage have gathered great attention. However, preparation of highly pure SWCNT of 100 mg order is still difficult. Ordinary SWCNT contains several wt. % metallic catalysts. The reliable study on interfacial chemistry of SWCNT needs pure samples of 100 mg order. Also recently, importance of defects in SWCNT has been indicated with high-resolution transmission electron microscope (HR-TEM) and theoretical studies [3,4]. Hence, the relationship between interfacial properties and defective structure should be elucidated experimentally by using pure SWCNT, being quite difficult at this stage. Single wall carbon nanohorn (SWCNH), which is produced by CO_2 laser ablation of pure graphite without any catalysts under an atmosphere of Ar at room temperature, consists of graphene tube and horn structures [5]. The internal porosity can be controlled by donation of nanoscale windows on the wall by oxidation; the optimum oxidation conditions for formation of nanoscale windows were determined [6]. The preceding studies with Raman spectroscopy and electrical conductivity responses on gas adsorption led that SWCNH is essentially defective, exhibiting n-type behavior being different from SWCNT [7]. Consequently, SWCNH is appropriate for study on the defect-associated structural changes of single wall nanocarbons.

In this study, we control the defect structure of SWCNH with the repeated mechanical treatment using grinding and compression. We characterize thus-treated SWCNHs with different techniques.

2 EXPERIMENTAL

SWCNH supplied by NEC was partially oxidized at 823 K in O_2 flow of 10 mL/min for 10 min(the details were reported in [6]); it is denoted by ox-SWCNH. SWCNH and ox-SWCNH assemblies were dispersed ultrasonically in ethanol, and then they were dried at 343 K to give highly packed assemblies. The SWCNH samples of 120 mg were ground with an agate mortar in air for three min and they were compressed under 500 kg/cm^2 in vacuum for 10 min. The grinding-compression treatment was repeated until nine times. The SWCNH and ox-SWCNH assemblies treated n-times will be denoted n-SWCNH and n-ox-SWCNH, respectively in this article. The morphological changes of SWCNH samples with the mechanical treatment were observed with FE-SEM and HR-TEM.

N_2 adsorption isotherms of SWCNH and ox-SWCNH samples at 77 K were measured to evaluate their nanoporosity with a volumetric apparatus (Autosorb 1: Quantachrome) after pretreatment at 423 K and 1 mPa for 2 h. Also the superwide pressure range adsorption (SWPA) isotherms from 10^{-10} of P/P_0 were gravimetrically measured with a laboratory designed UHV system, which includes high vacuum system, pressure-gauges, and laser detector-assisted quartz balance system. The buoyancy curve of the SWCNH samples as a function of He pressure up to 10 MPa was measured to evaluate the precise particle density after preheating. The samples were also examined by thermogravimetry (TG) and differential thermal analysis (DTA). TG-DTA measurements were carried out in the oxygen flow of 100 mL/min over the temperature range of ambient temperature to 1073 K with the heating rate of 3 K/min. The samples also examined by X-ray photoelectron spectroscopy (XPS: JPS-9010MX) and Raman spectroscopy (NRS-1000) with the excitation laser wavelength of 532 nm. We also measured their dc-electrical conductivity changes of the compressed SWCNH pellet by using van der Pauw method.

3 RESULTS AND DISCUSSION

3.1   Structural and morphological changes with mechanical treatment

Figure 1 shows FE-SEM images of as-received SWCNH and three to nine times-treated SWCNH.   SWCNH before the compression treatment is an assembly of separated spherical colloids. After three times-treatment, the distance between spherical colloids becomes shorter, suggesting formation of a mutually-associated structure of the SWCNH colloids. The nine times-treated SWCNH has an almost continuous structure. The ox-SWCNH gave a more continuous assembly structure than SWCNH with the mechanical treatment. The HR-TEM images of the non-treated SWCNH and nine times-treated one are shown in Figure 2. The images clearly show a remarkable structure change; the successive mechanical treatment makes each particle round and gives a dense packing.

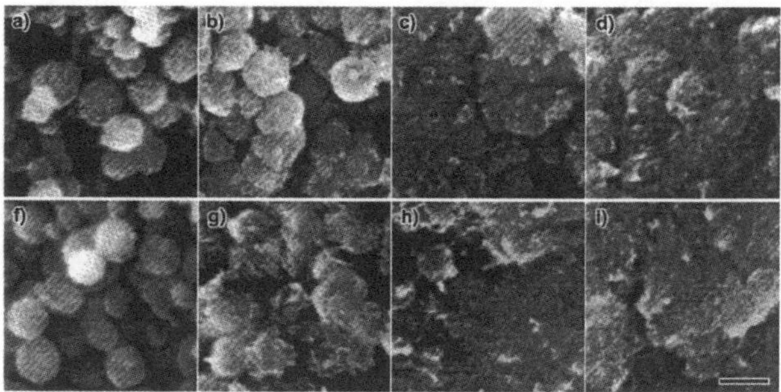

Figure 1   *SEM images (a–d) SWCNH and (f–i) ox-SWCNH. Mechanochemical treating number is (a,f) zero, (b,g) three, (c,h) five and (d,i) nine. Bar length is 100 nm.*

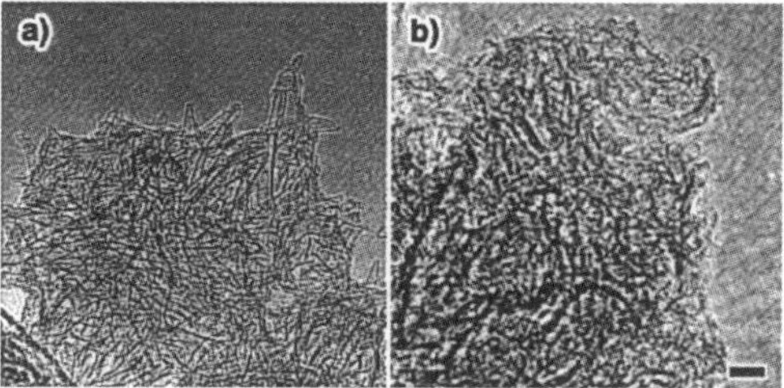

Figure 2   *TEM images of (a) SWCNH and (b) 9-SWCNH. Bar length is 10 nm.*

Raman spectra of both SWCNH and ox-SWCNH had characteristic of less ordered graphitic carbons such as nanoporous carbons. That is, the intensity of the D-band at 1330 cm^{-1} was comparable to that of G-band at 1580 cm^{-1}, being completely different from SWCNT. Therefore, SWCNH samples have more defective structures than SWCNTs. The G-band intensity of both of SWCNH and ox-SWCNH decreases remarkably with the treatment up to the five times-treatment, as shown in Figure 3. The intensity of G- and D-bands of ox-SWCNH was smaller than that of SWCNH. This is because ox-SWCNH even without the mechanical treatment is more defective due to addition of nanoscale windows on the single wall.

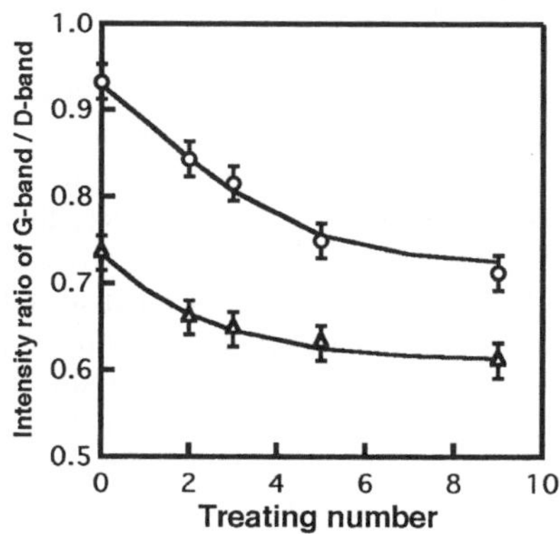

Figure 3   *D/G ratio of SWCNH (circle), ox-SWCNH (triangle) against treating number (The bar indicates the reproducibility).*

The preceding study on the analysis of C1s peak of XPS showed that SWCNH has sp^3 carbon atoms that are associated with the defective structures [8]. Therefore, the changes in the peak intensity of sp^3 and sp^2 carbon atoms with the mechanical treatment were examined for SWCNH and ox-SWCNH. Figure 4 shows clear increases in the percent of the sp^3 carbon atom with the mechanical treatment. In particular, the sp^3 carbon atoms become predominant in the surface region of ox-SWCNH. Accordingly a serious structure transformation on the mechanical treatment is associated with the change in the carbon-carbon bonding.

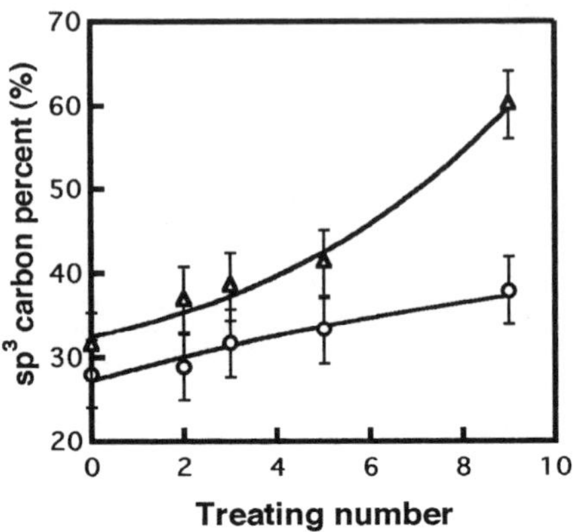

Figure 4   *The percent of sp^3 carbon in SWCNH (circle) and ox-SWCNH (triangle) against treating number (The bar indicates the accuracy).*

3.2   Nanoporosity change with mechanical treatment

The weight change with He pressure for SWCNH assemblies as a function of the compression-treatment frequency decreases linearly due to the buoyancy effect; the particle density is reversely proportional to the slope of the weight-pressure linear relationship that becomes small with the mechanical treatment. Therefore, the particle density becomes larger with the mechanical treatment from 1.25 g / mL to 1.98 g / mL. The largest density is slightly smaller than the density of bulk graphite (ρ_{gra} = 2.27 g/mL). The ratio of ($\rho - \rho_{cl}$) to ($\rho_{gra} - \rho_{cl}$) gives a measure of opening of the internal tube spaces through the mechanical treatment. Here, ρ and ρ_{cl} are the particle densities of the treated and non-treated SWCNH samples, respectively. Table 1 collects changes in the particle density and the opening ratio of wall in SWNH particles; the opening ratio of SWCNH rapidly increases with the mechanical treatment. The particle density of the 9-SWCNH is close to that of ox-SWCNH. Consequently, the mechanical treatment opens the SWCNH walls. However, the opening mechanism should be completely different from oxidation. The high temperature oxidation donates nanoscale windows on the wall of SWCNH particles, which were evidenced by molecular sieve effect [9] and HR-TEM observation [10]. On the contrary, the mechanical treatment must break the tube and/or horn structure.

Figure 5(a and b) shows N_2 adsorption isotherms of the treated SWCNH and ox-SWCNH. The N_2 adsorption isotherms of non-treated SWCNH is of IUPAC II type. The mechanical treatment increases adsorption uptake below 0.1 of P/P_0, indicating the predominant contribution of micropores. In particular, the successive treatment transforms the adsorption isotherm type from Type Ⅱ to Type I; the micropores become major pores in SWCNH assemblies by the intensive treatment. Figure 5(c) shows the N_2 adsorption isotherms over the logarithm of P/P_0 range from 10^{-10} to 10^{-6} as a function of the treating time. The initial uptake pressure in the extremely low P/P_0 range suggests the narrowest pore width. Here, the initial uptake pressure shifts from 10^{-8} to 10^{-10} of P/P_0 with the increase of the mechanical treatment. If we assume that newly developed ultramicropores with the mechanical treatment are slit-shaped or wedge-shaped, we can refer the relationship between the rising pressure and average pore width in the previous study [11]. The nine times treatment gives rise to slight amount (0.3 % of the whole microporosity) of small pores of less than 0.5 nm in width. On the other hand, the mechanical treatment of ox-SWCNH shifts the rising pressure to a lower pressure side than SWCNH and the initial uptakes were about three times-larger than SWCNH; the 9- and 4-ox-SWCNH gave ultramicropores of less than 0.4 nm and 0.45 nm, respectively; the initial uptakes corresponded to only 1 % of the whole nanoporosity. Production of ultramicropores of less than 0.5 nm in width suggests the presence of strong particle-particle contacts on the successive mechanical treatments.

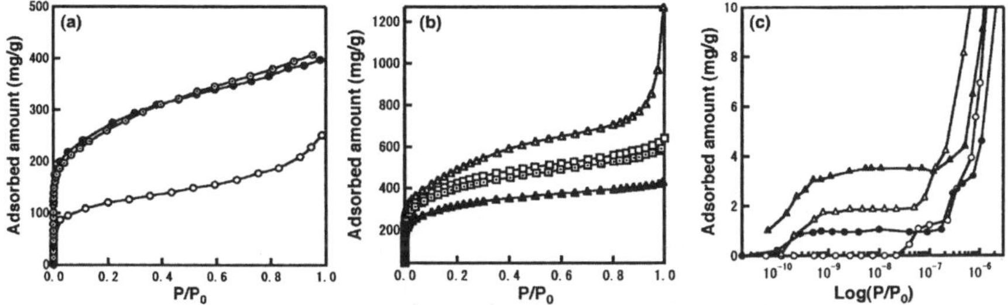

Figure 5 *N_2 Adsorption isotherms of (a) SWCNH, (b) ox-SWCNH and (c) logarithmic pressure plot of SWCNH and ox-SWCNH. Mechanochemical treatment number is zero (open circle and triangle), three (square), four (circle with dot), five (square with dot) and nine (closed circle and triangle).*

Table 1 *Particle density of n-SWCNH and ox-SWCNH*

Sample	Particle density (g/mL)	Opening ratio
SWCNH	1.25	0
2-SWCNH	1.85	0.59
4-SWCNH	1.86	0.60
9-SWCNH	1.98	0.66
ox-SWCNH	2.05	0.78
9-ox-SWCNH	2.09	0.82

3.3 sp³ carbon-associated structural changes with mechanical treatment

The above-mentioned Raman spectroscopic data indicate that the single carbon wall of SWCNH is highly defective and thereby the mechanical strength is much weaker than that of SWCNT. Hence the mechanical treatment produces the dramatic change in the chemical bonding evidenced by the increase of sp³ carbon atoms in ox-SWCNH, in particular. The chemical bonding transformation of carbon atoms from sp^2 to sp^3 needs a large enthalpy change (314 kJ/mol), which can be supplied by the bond dissociation during the mechanical treatment. Then, the structural change on the treatment should be associated with the sp^2–sp^3 bonding transformation; the sp^3 amount should be a good measure of the defective structure. The porosity changes are plotted against the sp^3 percent for SWCNH and ox-SWCNH, as shown in Fig. 6. The micropore volume increases with the sp^3 contribution for SWCNH, whereas the micropore volume steeply decreases with the sp^3 percent for ox-SWCNH. Both tendencies at the common sp^3 percent region are mutually different each other. Accordingly, the reasons for both porosity changes should be different from each other. The mechanical treatment opens the internal tube spaces of closed SWCNH, leading to the observed increase of the micropore volume. On the contrary, the mechanical treatment should close the open internal tube spaces of ox-SWCNH using the unstable carbon atoms produced on the mechanical bond dissociation.

The sp^2 carbon atoms form conjugated π–electron system, giving high electronic conductivity. On the contrary, the presence of sp^3 carbon atoms cut the conducting path of

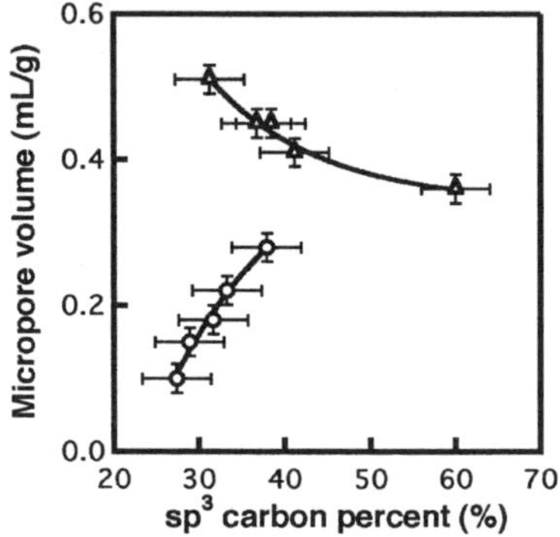

Figure 6   *Micropore volume changes against the percent of sp^3 carbon in SWCNH (circle) and ox-SWCNH (triangle).*

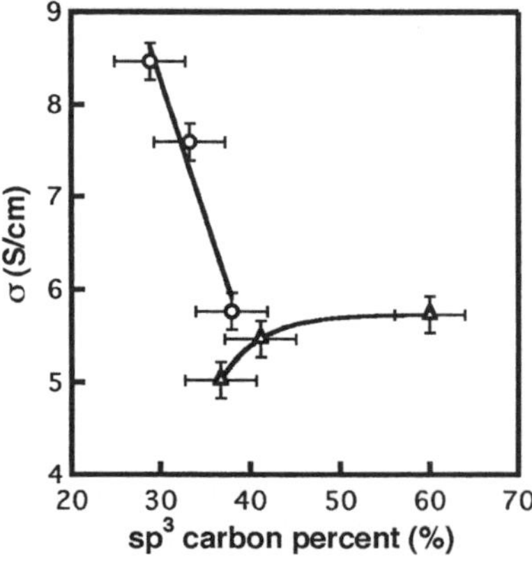

Figure 7   *Electrical conductivity changes against the percent of sp^3 carbon in SWCNH (circle) and ox-SWCNH (triangle).*

the conjugated π-electrons, giving rise to the electrical conductivity decrease observed, as shown in Fig. 7. The electrical conductivity of ox-SWCNH is almost constant irrespective of the mechanical treatment, because ox-SWCNH should have ill-conjugated π-electron structures due to addition of the nanowindows and the sp^3 carbons. On the other hand, the mechanical treatment reduces remarkably the electrical conductivity of SWCNH. The intensive mechanical treatment of SWCNH should induce the reconstruction between defective SWCNH particles through the sp^3 bonding. The theoretical study on the reconstruction of SWCNHs by Kawai et al suggested the formation of the junction structure through the sp^3 carbon atoms[4]. Formation of the junction structure in SWCNH particles should block the opening sites, reducing the micropore volume, as shown for ox-SWCNH. At the same time, such junction structures should produce very narrow spaces between the reconstructed SWCNH particles. The increase of ultramicropores inducing adsorption uptake below $P/P_0 = 10^{-6}$ in the SPA data should stem from the junction formation.

Conclusion

The mechanical treatment induces not only porosity changes, but also the sp^2 to sp^3 bonding transformation in SWCNH and ox-SWCNH. The formation of the junction structure between SWCNH particles through the sp^3 carbon atoms is strongly suggested.

Acknowledgements

This work was funded by Grant-in-Aid for Scientific Research S (No.15101003) from the Japanese Government.

References

1 Y. Tao, H. Kanoh, L. Abrams and K. Kaneko, *Chem. Rev.*, 2006, **106**, 896.
2 S. Lei, J. Miyamoto, H. Kanoh, Y. Nakahigashi and K. Kaneko, *Carbon*, 2006, **44**, 1884.
3 M. Terrones, F. Banhart, N. Grovert, J.-C. Charlier, H. Terrones, and P. M. Ajayan, *Phys. Rev. Lett.*, 2002, **89**, 075505-1.
4 T. Kawai, S. Okada, Y. Miyamoto, and A. Oshiyama, *Phys. Rev. B*, 2005, **72**, 035428.
5 S. Iijima, M. Yudasaka, R. Yamada, S. Bandow, K. Suenaga, F. Kokai, and K. Takahashi, *Chem. Phys. Lett.*, 1999, 309, 165.
6 S. Utsumi, J. Miyawaki, H. Tanaka, Y. Hattori, T. Itoi, N. Ichikuni, H. Kanoh, M. Yudasaka, S. Iijima, and K. Kaneko, *J. Phys. Chem. B*, 2005, **109**, 14319.
7 K. Urita S. Seki, S. Utsumi, D. Noguchi, H. Kanoh, H. Tanaka, Y. Hattori, Y. Ochiai, N. Aoki, M. Yudasaka, S. Iijima and K. Kaneko, *Nano Lett.*, 2006, **6**, 1325.
8 S. Utsumi, H. Honda, Y. Hattori, H. Kanoh, K. Takahashi, H. Sakai, M. Abe, M. Yudasaka, S. Iijima and K. Kaneko, *J. Phys. Chem. C*, 2007, **111**, 5572.
9 K. Murata, K. Hirahara, M. Yudasaka, S. Iijima, D. Kasuya, and K. Kaneko, *J. Phys. Chem. B*, 2002, **106**, 12668.
10 K. Ajima, M. Yudasaka, K. Suenaga, D. Kasuya, T. Azami, and S. Iijima, *Adv. Mater.*, 2004, **16**, 397.
11 M. Sunaga, T. Ohba, T. Suzuki, H. Kanoh, S. Hagiwara and K. Kaneko, *J. Phys. Chem. B*, 2004, **108**, 1065.
12 K. Urita, S. Seki, H. Tsuchiya, H. Honda, S. Utsumi, C. Hayakawa, H. Kanoh, T. Ohba, H. Tanaka, M. Yudasaka, S. Iijima and K. Kaneko, *J. Phys. Chem. C* **112**, 8759-8762 (2008).

ADSORPTION CHARACTERISATION OF ENERGY GASES ON MICROPOROUS COORDINATION POLYMERS

A. Matsumoto and T. Ito

Department of Materials Science, Toyohashi University of Technology, Toyohashi 441-8580, JAPAN

1 INTRODUCTION

Microporous coordination polymers (MPCP), or metal organic frameworks (MOF), constructed by transition metal ions or metal oxide clusters and bridging organic ligands are promising materials for energy gas storage.[1-5] Since the pore wall of MOF consists of the framework of the ligands, MOF has higher porosity and surface area than conventional porous materials such as activated carbon, aluminosilicate zeolites and mesoporous silica. The frameworks of some MOF are also known to exhibit high flexibility and its void swells by adsorption of guest adsorbates.[6-8] Because of such unique pore characteristics, MOF is expecting as a high-capacity adsorbent for energy gases such as hydrogen and methane.[1-5]

The present study focused on two kinds of MOF materials, MOF-5 and two-dimensional layered MOF (denoting LPC or latent porous complex). MOF-5 (IRMOF-1) of composition $Zn_4O(BDC)_3$ (BDC=1,4-benzenedicarboxylate) is consisted by Zn_4O-cluster linked to six carboxylates of a 1,4-benzenedicarboxylate, and is one of well-known MOF with a cubic three dimensional pore structure. MOF-5 is expected as storage medium of methane and hydrogen.[1-5] LPC is composed by a stack of layers composed by organo copper complex of $[Cu(FBF_3)_2(bpy)_2]$ (bpy=4,4′-bipyridine). The layer has a two-dimensional grid structure and the size of the grid is ca. 0.77 nm × 0.77 nm so that it is intrinsically microporous. But each layer was stacked with a shift of a half period for a and b directions, the microporous voids on each layer were blocked by the $Cu(FBF_3)_2$ parts of neighbouring layers. Therefore, LPC does not usually show a porous character. However, the layer distance expands by pressurizing gas adsorptive, and then LPC can adsorb the gas.[8] Adsorption of energy gases, hydrogen and methane, has been studied for these materials however no adsorption microcalorimetric analysis has been reported yet. In this study, adsorption microcalorimetry is conducted to characterize the adsorption behaviour of the energy gases, methane and hydrogen, on MOF-5 and LPC.

2   EXPERIMENTAL

2.1 Sample Preparation

MOF-5 was prepared by the method reported in ref. 1.[1] LPC was prepared by a modified procedure based on the method in ref.6. In this preparation, methanol and dichloromethane were used instead of acetonitrile and diethyl ether in ref.6, respectively.

2.2 Physicochemical Characterization

X-ray diffraction (XRD) patterns of the samples were obtained using a Rigaku RINT2000 instrument with CuKα radiation. Nitrogen adsorption isotherms were measured volumetrically using an automatic nitrogen adsorption apparatus (Autosorb 1M, Quantachrome) at 77 K. Each sample was treated in vacuo (0.1 mPa) at 373 K for 11 h before the nitrogen adsorption measurement.

2.3 Methane and Hydrogen Adsorption

The high-pressure adsorption isotherms and the differential energies of adsorption of methane and hydrogen were measured at isothermal conditions, 203, 243 and 298 K by a twin conduction type microcalorimeter (Setarum, BT-2.15) attached to a volumetric adsorption apparatus. Each adsorption isotherm was consecutively measured with adsorption microcalorimetry. Hydrogen of 99.99999 % purity (G1 grade, Taiyo-Nippon Sanso) and methane of 99.999 % purity (G1 grade, Taiyo-Nippon Sanso) were used in the research without further purification. Sample was evacuated at 1 mPa and 423 K for 12 h (MOF-5) or 2 h (LPC) before each measurement.

3   RESULTS AND DISCUSSION

3.1 X-ray Diffraction and Nitrogen Adsorption

X-ray diffraction patterns of the prepared samples, MOF-5 and LPC, are identical with their reported patterns, indicating successful formation of these materials. Figure 1 shows the nitrogen adsorption isotherms of MOF-5 and LPC. The isotherm of MOF-5 showed the type Ia character in the IUPAC classification which is typical type for microporous materials because MOF-5 has a jungle-gym-type framework with uniform microporous voids. Since "pore wall" of MOF-5 is consisted by $[Zn_4O]^{6+}$ and coordinating ligands, the specific surface area and pore volume estimated by the nitrogen adsorption were extremely high, 3550 m^2/g and 0.98 mL/g, respectively. Pore size estimated by the DFT method was quite uniform 1.2 nm. On the other hand, the isotherm of LPC showed a unique character which could not classify into any IUPAC types: the adsorption uptake was almost zero at relative pressure P/Po < ca.0.35 but it suddenly increased at P/Po=0.38. This phenomenon is called the gate effect where an interlayer distance of two dimensional sheets of $Cu(bpy)_2(BF_4)_2$ expand by diffusion of gas molecules at a given pressure or gate pressure which is P/Po = 0.38 in the present case.[6] Pore characteristics such as specific surface area, pore volume and pore size were difficult to analyze based upon conventional adsorption theories, because the pore structure is flexible and the equilibrium pressure does not directly relate to pore diameter.

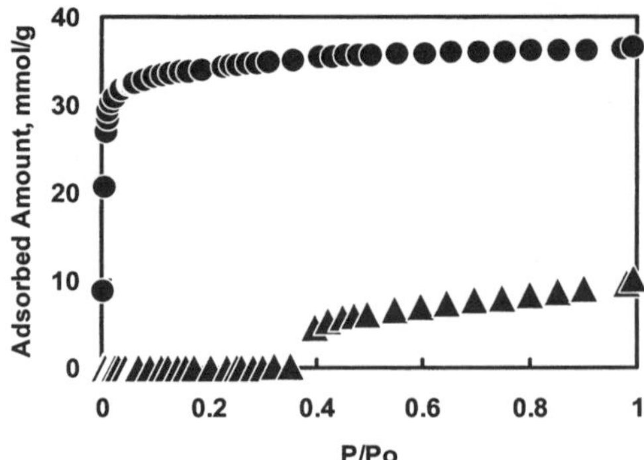

Figure 1 *Nitrogen adsorption isotherms. Keys: circle, MOF-5; triangle, LPC*

3.2 Methane Adsorption

Adsorption isotherms of methane on MOF-5 were of Henry type at 243 and 298 K, but that at 203 K showed rather type Ib character as shown in Figure 2. Adsorption uptakes (surface excess) at lower adsorption temperature were higher: the uptake at ca. 3.2 MPa increased from 9.4 mmol/g (13 mass%) to 29.2 mmol/g (32 mass%) with decreasing the adsorption temperature from 298 K to 203 K. The density of adsorbed methane in the pore of MOF-5 was estimated to be 0.40 g/mL at 3.2 MPa and 203 K by the adsorption uptakes and the evaluated pore volume, which is higher than the density of methane at the critical point (190.5 K) of 0.16 g/mL, and was comparable to that in a liquid phase at its boiling point (111 K), 0.42 g/mL. This result indicates that methane is densely adsorbed in the micropore of MOF-5 on this condition.

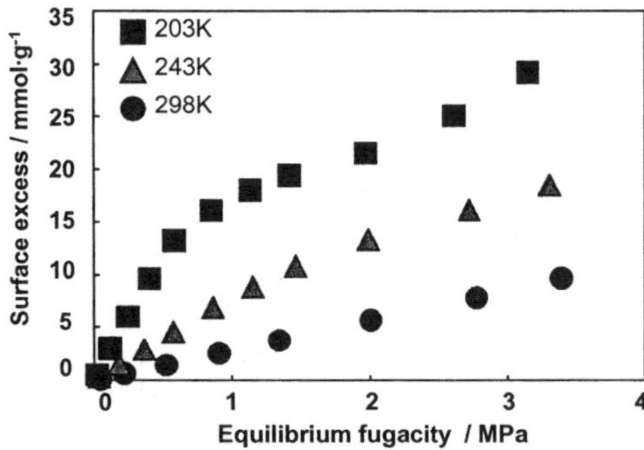

Figure 2 *Adsorption isotherms of methane on MOF-5 at different temperature*

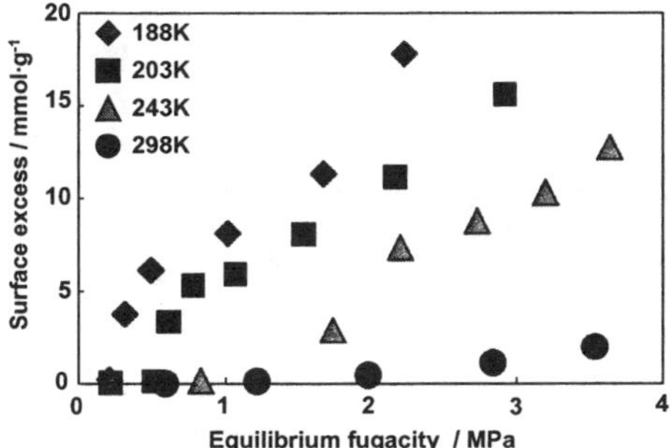

Figure 3 *Adsorption isotherms of methane on LPC at different temperature*

LPC did not adsorb methane at all until the pressure (fugacity) range of 0.6 MPa but the adsorption uptake gradually increased above 0.6 MPa at 298 K, as shown in Figure 3. The narrow void distance between *ab* planes suddenly expands at the gate pressure 0.6 MPa, thereby the adsorption capacity increases as well as nitrogen and carbon dioxide adsorption (the gate effect).[6,8] The gate pressure decreased with decreasing the adsorption temperature and was 0.3 MPa at 188 K. The adsorption amount increased, and "knee" of the isotherm became clear with decreasing the adsorption temperature. The adsorption amount at 2.5 MPa was 9.4 mmol/g (13 mass %) at 188 K. Because the framework of LPC would swell with adsorption, the density of adsorbed methane could not be estimated in the case of MOF-5.

The differential adsorption energy $\Delta_{ads}u$ can be calorimetrically evaluated as the differential adsorption heat q_{diff} (>0). They are related by the expression: $\Delta_{ads}u = -q_{diff}$.[9] The differential energy of adsorption, $\Delta_{ads}u$, of methane on MOF-5 was shown in Figure 4. An ordinate in Figure 4 shows an absolute value of $\Delta_{ads}u$ which corresponds to q_{diff}. The differential adsorption heat, q_{diff} was 12 - 16 kJ/mol at the initial stage of adsorption and became almost constant 10 - 12 kJ/mol at the uptakes of ca.5 - ca.15 mmol/g. There are five different kinds of adsorption sites in MOF-5 for Ar and N_2 adsorption.[10] And these sites would also be active ones for the methane adsorption. The strongest one is the $\alpha(CO_2)_3$ site, where adsorbate interacts with two oxygen atoms of a carboxylate ion. The number of the $\alpha(CO_2)_3$ site is 5.5 sites/unit cell, which coincides with the adsorption uptake giving high heat evolution at 203 K in Figure 4. Therefore the initial high heat would be mainly attributable to the interaction with such strong interactive sites.

The differential adsorption energy at lower adsorption temperature showed higher value. The adsorption energy $\Delta_{ads}u$ is written by the differential enthalpy of adsorption $\Delta_{ads}h$ as the following equation: $\Delta_{ads}u = \Delta_{ads}h + RT$ (R: gas constant, T: temperature).[10] According to the relation, the differential heat at 298 K would be 0.8 kJ/mol higher than those at 203 K. This value coincides well with the difference in q_{diff} at 298 and 203 K, as shown in Figure 4. The differential heat of adsorption q_{diff} at the uptake of ca.5 - ca.15 mmol/g was higher than the vaporization enthalpy of methane 8.2 kJ/mol, suggesting that the adsorption would take place by the specific interaction between the framework of MOF-5 and methane. Because of the microporous voids of MOF-5, the enhancement of adsorption by overlapping of the potential fields also contributes the strong interaction in the adsorption process. The isosteric heat of adsorption q_{st} ($q_{st} = -\Delta_{ads}h$) was estimated by applying the Clausius-Clapeyron equation, $q_{st} = -R[\partial \ln P/\partial(T^{-1})]_\Gamma$ (P, equilibrium pressure; Γ, adsorbed amount per unit surface area) on the adsorption isotherms. Here we note that the present methane adsorption is measured above the critical temperature where the phase transition never takes place although the Clausius-Clapeyron equation is based on a phase transition

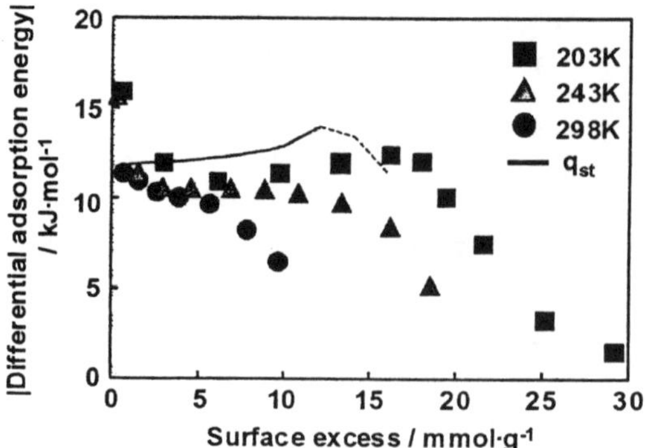

Figure 4 *Differential energy of adsorption of methane on MOF-5*

between liquid and solid phases. Nevertheless the evaluated q_{st} for MOF-5 was 12 - 14 kJ/mol which was similar to the experimental values. These results also suggest that methane adsorbed densely in the microporous voids of MOF-5.

The differential adsorption heat of LPC was very unique as shown in Figure 5. The heat q_{diff} was not detected below the gate pressure, but it immediately increased with adsorption at the gate pressure. However, different from the adsorption on MOF-5, q_{st} abruptly decreased with further adsorption in spite of continuous increase of the adsorption uptake. The initial heat by adsorption just at the gate pressure would be attributable to direct interaction between the framework and methane. The steep decrease in q_{diff} would be consumption of energy by layer expansion of LPC above the gate pressure. Since the total heat flow from the adsorption system can be measured by adsorption microcalorimetry, the method cannot separately detect the adsorption energy and the heat for expansion.

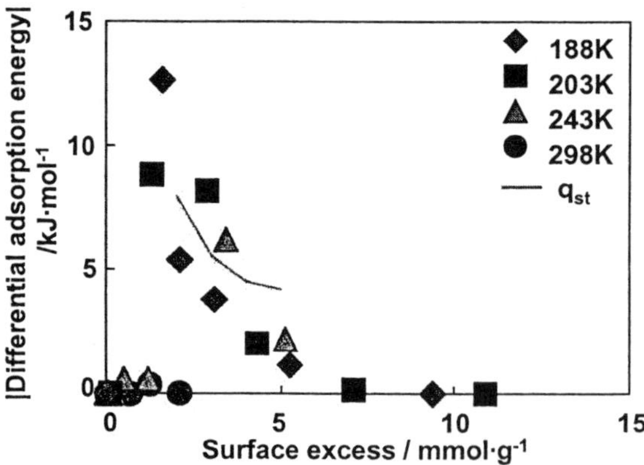

Figure 5 *Differential energy of adsorption of methane on LPC*

3.3 Hydrogen adsorption

Figure 6 shows adsorption isotherms of hydrogen on MOF-5 at different temperature. The isotherms measured at 203-298 K were of Henry type at the ranges of equilibrium pressure less than ca. 5 MPa; the uptakes increased linearly with equilibrium pressure, and they gradually increased over 5 MPa. The adsorption amount increased with decreasing the adsorption. The uptakes at these temperature were not significant; e.g. 6-8 mmol/g (1.2-1.6 mass%) at 6 MPa. The differential adsorption heat at these temperatures was 5-7 kJ/mol at very initial stage of adsorption, however, it immediately decreased with increasing the adsorption amount (Figure 7). The initial higher heat would be direct interaction between the framework of MOF-5 and hydrogen, but the abrupt decrease of the heat suggests the interaction between hydrogen and the framework is not enough to give rise to strong adsorption at these temperatures.

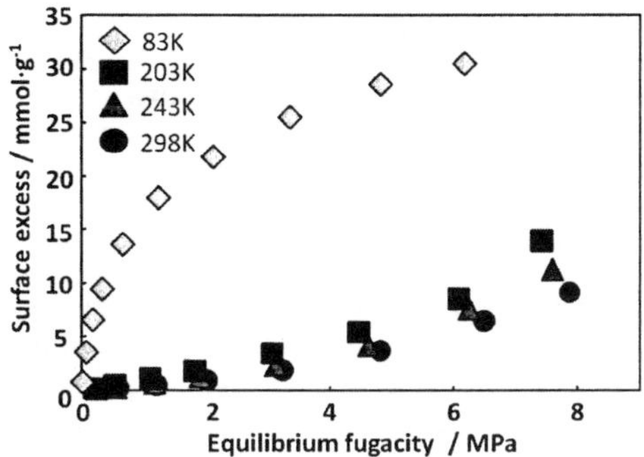

Figure 6 *Adsorption isotherms of hydrogen on MOF-5*

On the other hand, the adsorption isotherm at 83 K was very different from other ones. The isotherm is rather of type Ib, indicating that strong adsorption takes place at 83 K. The adsorption amount at 6MPa was 30.4 mmol/g (6 mass %), which is comparable to a target value of hydrogen storage for automobile fueling by the Department of Energy of the U.S. government.[2] This result is interesting that MOF-5 has such high adsorption activity even over the boiling point of nitrogen. q_{diff} was 6.1 kJ/mol at initial stage of adsorption and it decreased gradually with increasing the uptake as shown in Figure 7. But different from the adsorption at higher temperature, the high heats of 3.7-4.9 kJ/mol were continuously observed until the adsorption amount of 21mmol/g. These results suggest that MOF-5 would be a promising material for hydrogen storage.

Hydrogen adsorption on LPC was also measured. But different from MOF-5, significant adsorption activity was not observed (Figure 8) so that the differential heats were not detected.

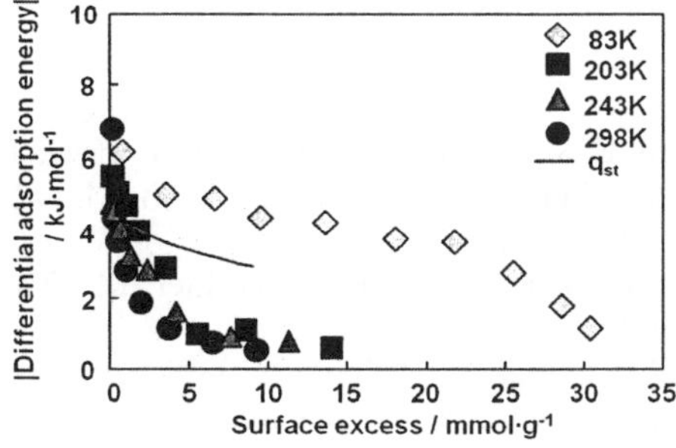

Figure 7 *Differential adsorption energy of hydrogen on MOF-5*

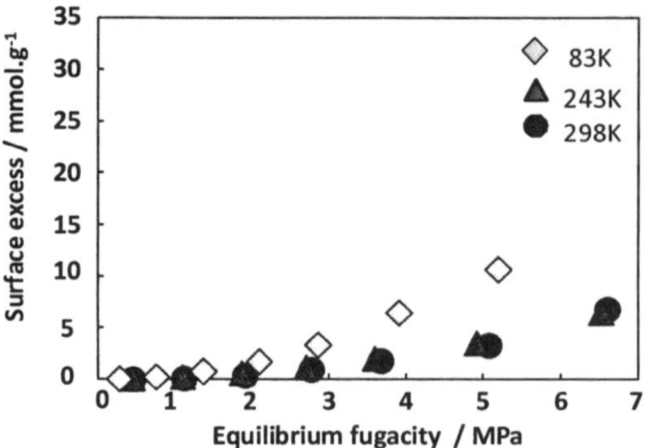

Figure 8 *Adsorption isotherms of hydrogen on LPC*

4 CONCLUSION

The adsorption of methane and hydrogen on metal organic frameworks MOF-5 and LPC, were characterized by adsorption microcalorimetry. The adsorption amount of methane and hydrogen increased with decreasing adsorption temperature. MOF-5 showed high adsorption capacity, which was 29.2 mmol/g (32 mass%) at 3.2 MPa and 203K. Adsorption microcalorimetry measurements revealed that methane adsorbed strongly and densely in micropore of MOF-5 at 203 K; the density of adsorbed phase (0.48 g/mL) was even higher than that of liquid phase at boiling point (0.42 g/mL). The gate adsorption, where the adsorption suddenly takes place at certain pressure, was observed in methane adsorption on LPC. The pressure at which adsorption begins (the gate pressure) shifted to lower pressure by decreasing adsorption temperature, and it was 0.3 MPa in the adsorption at 188 K. The change in the differential adsorption heat of methane on LPC indicates that the framework of LPC less interact with methane than that of MOF-5. The hydrogen adsorption activity of MOF-5 was rather high at 83 K; the adsorbed amount was 30.4 mmol/g (6 mass %) at 6 MPa, which is comparable to a target value of hydrogen storage for automobile fueling by the Department of Energy of the U.S. government. On the other hand, LPC did not adsorb enough amount of hydrogen and the adsorption heat was hardly observed.

Acknowledgement
A part of this work was supported by the Daiko Foundation.

References

1   H. Li, M. Eddaoudi, M. O'Keefe and O. Yaghi, Nature, 1999, **402**, 276.
2   N.L. Rosi, J. Eckert, M. Eddaudi, D.T. Vodak, J. Kim, M. O'keeffe and O.M. Yaghi, Science, 2003, **300**, 1127.
3   M. Eddaudi, J. Kim, N. Rosi, D. Vodak, J. Wachter, M. O'keeffe and O.M. Yaghi, Science, 2002, **295**, 469.
4   W. Zhou, H. Wu, M.R. Hartman and T. Yildirm, J. Phys. Chem. C, 2007, **111**, 16131.
5   U. Mueller, M. Schubert, F. Teich, H. Puetter, K. Schierle-Arndt and J. Pastré, J. Mater. Chem., 2006, **16**, 626.

6   D. Li and K. Kaneko, Chem. Phys. Lett., 2001, **335**, 50.
7   S. Onishi, T. Ohmori, T. Ohkubo, H. Noguchi, L. Di, Y. Hanzawa, H. Kanoh and K. Kaneko, Appl. Surf. Sci., 2002, **196**, 81.
8   A. Kondo, H. Noguchi, S Ohnishi, H. Kajiro, A. Tohdoh, Y. Hattori, W-C Xu, H. Tanaka, H. Kanoh and K. Kaneko, 2006, **6**, 2581.
9   For example, S.J. Gregg and K.S. W. Sing, Adsorption Surface Area and Porosity, 2nd Ed, Academic Press, London, 1982, F. Rouquerol, J. Rouquerol and K. Sing, Adsorption by Powders and Porous Solids, Academic Press, London, 1999.
10   J.L.C. Rowsell, E.C. Spencer, J. Eckert, J.A.K. Howard, O.M. Yaghi, Science, 2005, **309**, 1350.

POST-SYNTHESIS MODIFICATION OF A METAL-ORGANIC FRAMEWORK WITH SILVER IONS

G. A. Emberger,[1] Y.-S. Bae,[1] S. T. Nguyen,[2] J. T. Hupp,[2] L. J. Broadbelt,[1] and R. Q. Snurr[1]

[1] Department of Chemical and Biological Engineering, Northwestern University, Evanston, IL 60208, USA
[2] Department of Chemistry, Northwestern University, Evanston, IL 60208, USA

1 INTRODUCTION

Recently, metal-organic frameworks (MOFs) have garnered much interest as a new class of porous, crystalline solids with potential applications in gas storage, catalysis, and sensing.[1-3] In addition, the tailorability of MOFs may lead to new breakthroughs in adsorption-based separation processes.[4-6] Studies to date have indicated that MOFs have potential in air separations,[7] CO_2 separations,[8, 9] and hydrocarbon separations.[10]

Olefin/paraffin separations constitute an important process in the chemical and petrochemical industries and are currently accomplished through highly energy-intensive cryogenic distillation. One promising alternative to distillation is π-complexation of olefins to Ag^+, which has been used to achieve facilitated transport of olefins in polymer membranes[11, 12] and selective adsorption of olefins in nanoporous materials.[13, 14] A recent study showed that Ag^+ could complex to carbonyl groups in a poly(vinyl methyl ketone) membrane and undergo reversible complexation with olefins.[15] We reasoned that Ag^+ could be introduced post-synthetically into a MOF and complexed to the pore walls through interaction with the π bonds in the MOF linker molecules. The presence of the Ag^+ ions could then enhance the selectivity of the MOF for olefins over paraffins. Given that post-synthesis modification of MOFs has been demonstrated only in a limited number of studies to date,[16-18] another objective of this work was to illustrate a new method for post-synthesis modification of MOFs. As a starting point, we chose to post-synthetically modify the MOF Zn(NDC)(4,4'-Bpe)$_{0.5}$ (**1**) (NDC = 2,6-naphthalenedicarboxylate, 4,4'-Bpe = 4,4'-*trans*-bis(4-pyridyl)ethylene (see Figure 1)).[19]

2 METHOD

2.1 Sample Preparation

The synthesis of **1** has been reported previously.[19] Here, we present an alternative preparation: H$_2$NDC (0.216 g, 1.0 mmol), 4,4'-Bpe (0.182 g, 1.0 mmol), and Zn(NO$_3$)$_2$·6H$_2$O (0.298 g, 1.0 mmol) were dissolved in *N,N*-dimethylformamide (DMF)

Figure 1 *4,4'-trans-bis(4-pyridyl)ethylene (4,4'-Bpe, left) and 2,6-naphthalenedicarboxylate (NDC, right)*

(HPLC grade, 100 ml). The solution was divided into 5-ml portions in 20 40-ml volatile organic compound (VOC) vials, which were heated at 80 °C for 24 h in an oven and then cooled to room temperature over a period of 45 min. The reaction mixtures from the 20 vials were combined together and filtered to afford a white powder. This product was washed with DMF (3 × 10 ml) and dried under aspiration for 10 min. The isolated powder (150 mg) was ground with a mortar and pestle to make the sample uniform. All chemicals used in this synthesis and in the following procedure were obtained from Sigma-Aldrich unless otherwise noted.

2.2 Post-synthesis Modification with Silver Triflate

The MOF was first immersed in ACS-reagent grade methanol (MeOH, EMD Chemicals, 4 ml) for at least 4 days to exchange out the DMF, which has been known to reduce silver ions.[20] The resulting MeOH-exchanged MOF (**1·MeOH**) was then filtered, washed with MeOH (3 x 10 ml), and dried under aspiration for 10 min. **1·MeOH** was subsequently evacuated at 100 °C for 24 h to generate the evacuated MOF 1_{evac}.

The silver modification procedure follows: Anhydrous MeOH was first obtained by refluxing ACS-reagent grade MeOH over CaH_2 (Fluka) overnight and distilling the MeOH under N_2. In a dark dry box, silver triflate (AgOTf) was dissolved in this anhydrous MeOH to form a 3.89 mM solution. This clear solution was added to 1_{evac} in a ratio of Ag^+ to Zn^{2+} of 1.7 and left in the dark for 24 h. (While stirring of the sample was found to give comparable modification results to the no–stirring case, mechanical degradation of 1_{evac} was found to occur under stirring conditions.) After 24h, the resulting silver-modified MOF (**1·Ag$^+$**) was isolated from the reaction mixture via filtration. This slightly orange powder was washed with MeOH (3 x 10 ml) and dried under aspiration for 10 min.

2.3 Characterization

Powder X-ray diffraction (PXRD) was carried out with a Rigaku XDS 2000 diffractometer using nickel-filtered Cu Kα radiation ($\lambda = 1.5418$ Å). The scanning angle ranged from $2\theta = 5°$ to $2\theta = 50°$ with a step size of 0.1°. Each step used a 1-s counting time. Inductively coupled plasma (ICP) spectroscopy was performed using a Varian ICP spectrometer. Samples (1-3 mg) were dissolved in neat HNO_3 and heated at 100 °C in an oil bath until no particles were left and water vapor was no longer produced. The acid solution was diluted to 2% in deionized H_2O and analyzed for Zn (202.548 nm) and Ag (328.068 nm). Thermogravimetric analysis (TGA) was performed by heating samples (3-5 mg) from 25 °C to 700 °C at a rate of 10 °C/min under N_2 on a Mettler-Toledo TGA/SDTA851e. Infrared spectra of KBr pellets containing **1·MeOH** and **1·Ag$^+$** were collected with a resolution of 8 cm^{-1} in the range of 4000-400 cm^{-1} with a Thermo-Nicolet Nexus 870 spectrometer. N_2 adsorption/desorption isotherms were obtained with a Quantachrome Autosorb 1-MP. Adsorption of C_2H_4, C_2H_6, CH_4, and CO_2 was measured volumetrically as detailed elsewhere.[9] Ultra-high-purity grade N_2 was used while C_2H_4,

C_2H_6, and CH_4 were chemically pure grade and CO_2 was 99.9%+. Activation of the samples was accomplished by outgassing *in vacuo* overnight at 150 °C.

3 RESULTS AND DISCUSSION

The PXRD patterns of as-synthesized **1** and **1·MeOH** are shown in Figure 2A. The PXRD pattern of the as-synthesized MOF matched the simulated PXRD pattern of **1**.[19] After solvent exchange, the low-angle peaks in the PXRD pattern shifted to higher 2θ values. It has been suggested that this results from framework shrinking due to ligand flexibility in response to new guest molecules.[19] Upon modification of **1**$_{evac}$ with AgOTf, new peaks appeared in the PXRD patterns at $2\theta = 10.5°$ and $2\theta = 21.5°$ (Figure 2B). The presence of these new peaks strongly suggested that Ag^+ has indeed been introduced into the pores of **1**$_{evac}$ with some regularity. Additional peaks were sometimes present in the PXRD patterns of modified samples as shown in Figure 2B for **1·Ag$^+$** with Ag:Zn = 0.87:1 where a prominent new peak appeared at $2\theta = 7.1°$. The variation in the location of new peaks for different samples of **1·Ag$^+$** may be due to different modes of interaction between AgOTf and various functional groups in the MOF. Another notable result was the decrease in intensity in the PXRD patterns upon introduction of AgOTf into the pores. This suggests that the framework was degraded during the post-synthesis modification.

ICP results showed that the silver loading varied significantly although the same post-synthesis modification procedure was used in three different experiments. Sample loadings obtained had ratios of Ag:Zn of 0.38, 0.59, and 0.87. This variation could be due to different particle sizes in the samples or pores near the outside of the crystal becoming blocked with anions at different times during the modification.

TGA experiments also demonstrated that AgOTf had been successfully introduced into the pores (Figure 3). For **1·MeOH**, a large step in the TGA trace above 425 °C indicated the degradation of the MOF. TGA traces of samples of **1·Ag$^+$** showed a step prior to degradation of the framework near 400 °C. The percent mass loss in the first step of these TGA traces matches the molar mass equivalent of present triflate, suggesting that this step corresponds to the removal of triflate species. Additional TGA and PXRD

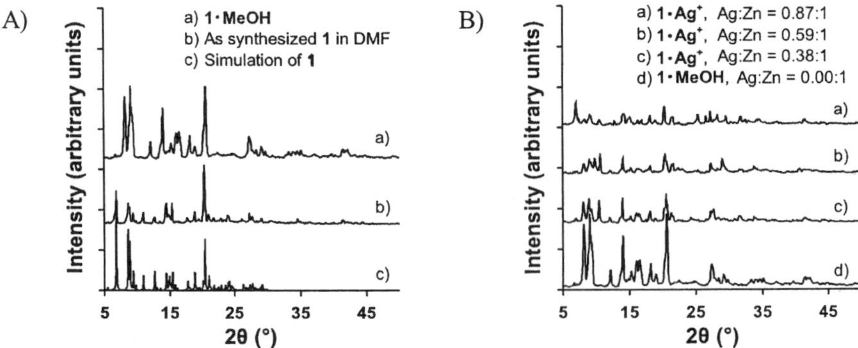

Figure 2 *PXRD patterns of A) as-synthesized 1 and 1·MeOH compared to the simulated PXRD pattern of 1 and B) 1·MeOH and 1·Ag$^+$ at various Ag$^+$ loadings*

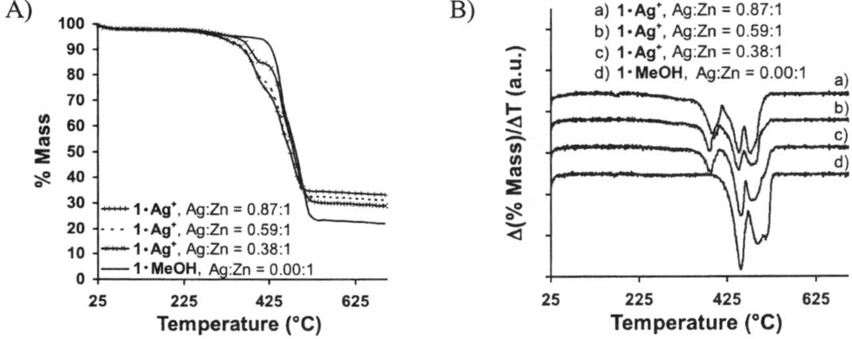

Figure 3 *A) TGA traces and B) derivatives of the TGA traces for 1·MeOH and 1·Ag⁺*

analysis revealed that the Ag^+-modified MOF is stable in MeOH for up to two weeks.

To determine if Ag^+ formed a π-complex with the MOF framework, IR spectra of **1·MeOH** and **1·Ag⁺** were collected (Figure 4). If complex formation of Ag^+ with olefins occurs in **1·Ag⁺**, the C=C stretching frequency (1600-1700 cm⁻¹) in **1·Ag⁺** should be red-shifted by 50-60 cm⁻¹ from that in **1·MeOH**.[21] Interestingly, only the **1·Ag⁺** sample with the highest loading ratio, Ag:Zn = 0.87:1, exhibited such a shifted peak (1537 cm⁻¹). The presence of this characteristic stretch may be attributed to Ag^+ interacting with the C=C bonds in the framework and is consistent with the additional diffraction peak detected at $2\theta = 7.1°$ in the PXRD pattern of **1·Ag⁺** with Ag:Zn = 0.87:1. However, it raises questions regarding the nature of the interaction between Ag^+ and the framework in samples with lower Ag:Zn ratios. Does addition of Ag^+ lead to framework-disruptive interactions with the carboxylate and pyridyl groups? To what extent does the charge-balancing inclusion of the triflate anions affect the pore environment inside the MOF?

Characteristic features in the IR spectra of **1·MeOH** and **1·Ag⁺** at various loadings of Ag^+ (Figure 4) supported two conclusions regarding the structure of **1·Ag⁺**: 1) the triflate anion is present inside the pore of the MOF and 2) below the 0.87:1 Ag:Zn loading, Ag^+ interacts with the MOF framework primarily through non-C=C-π–bonded modifications. Stretching frequencies associated with the triflate anion appeared as a broad peak at 1260

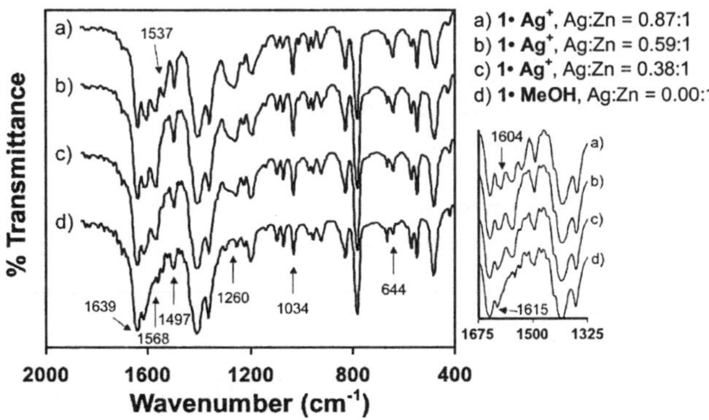

Figure 4 *IR spectra for 1·MeOH and 1·Ag⁺ at various Ag⁺ loadings*

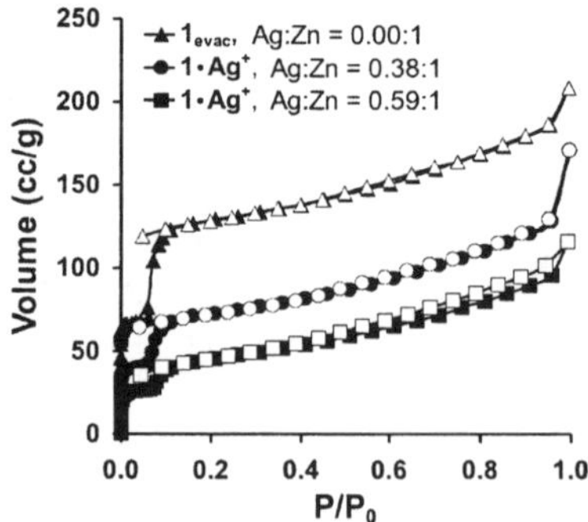

Figure 5 *N_2 sorption isotherms at 77 K for 1_{evac} and $1 \cdot Ag^+$ at various Ag^+ loadings. Filled symbols represent adsorption data, and open symbols represent desorption points.*

cm^{-1} and as growth in peaks at 1192, 1142, 1034, and 644 cm^{-1},[22, 23] all of which became larger with increasing silver loading. In addition, the peaks at 1639 and 1615 cm^{-1} decreased slightly while new peaks at 1604 and 1568 cm^{-1} appeared. The peak at 1639 cm^{-1} can be assigned to the Zn-carboxylate groups by comparison to the IR spectra of IRMOF-8, which is composed of the NDC ligand and Zn_4O corners,[24] and by inference from DFT calculations on MOF-5.[25] As the anti-symmetric stretch of carboxylic acid salts are known to appear in the 1610-1560 cm^{-1} range,[26] we attributed the peaks at 1604 and 1568 cm^{-1} to newly formed interactions between Ag^+ and the carboxylate groups in the framework, at the expense of the original Zn_2-carboxylate bonds. Based on a recent study on $Zn(BPE)X_2$, where X is a halide,[27] and on a control experiment where only the 4,4′-Bpe ligand was modified, the decrease in the peak at 1615 cm^{-1} in $1 \cdot Ag^+$ with increasing silver loading and the corresponding appearance of the peak at 1604 cm^{-1} could indicate that Ag^+ also coordinated to the pyridyl groups. The growth in the peak at 1497 cm^{-1}, which was observed in the IR spectrum of 4,4′-Bpe in this study and another study,[27] additionally supports the disruption of Zn^{2+} coordination to the pyridyl groups. Loss of coordination of the metal corners to the organic linkers explains the decreased intensity of the PXRD patterns observed after modification of 1_{evac} with AgOTf.

The adsorption capacities of 1_{evac} and $1 \cdot Ag^+$ were measured using N_2 at 77 K. As shown in Figure 5, the adsorption capacity was reduced significantly when AgOTf was in the pores. The unmodified MOF had a Langmuir surface area of 299 m^2/g (Table 1), in agreement with previous measurements.[19] The Langmuir surface areas of $1 \cdot Ag^+$ were 181 and 117 m^2/g for Ag:Zn ratios of 0.38 and 0.59, respectively (Table 1). The Dubinin-Radushkevich (DR) pore volumes of these two samples were only 60% and 40% of the

Table 1 *Surface areas, pore sizes, and volume data obtained from N_2 sorption isotherms at 77 K for 1_{evac} and $1 \cdot Ag^+$. See text for definitions of V_2/V_1 and V_{large}/V_{small}.*

Ag:Zn	BET Pressure Range (P/P₀)	BET Surface Area (m²/g)	Langmuir Surface Area (m²/g)	DR Pore Volume (cc/g)	HK Pore Size (Å)	V_2/V_1	V_{large}/V_{small}
0.00	0.007 − 0.06	287	299	0.11	4.8, 12.9, 14.6	0.83	0.89
0.38	0.007 − 0.05	171	181	0.064	4.9, 14.5	0.78	0.80
0.59	0.007 − 0.05	110	117	0.042	4.9, 15.2	0.70	0.69

pore volume of 1_{evac}. These results were consistent with a reduction of pore volume due to the presence of the bulky triflate anion. It should be noted that while the TGA results indicated that triflate may be removed from the pores at 400 °C, activating $1 \cdot Ag^+$ at this temperature overnight was not done to avoid potential MOF degradation. The decreased capacity in $1 \cdot Ag^+$ could also be caused by disruption of the framework upon modification.

Interestingly, the N_2 sorption isotherms exhibited two steps in the adsorption curve and hysteresis in the desorption curve, suggesting the presence of super micropores. The pore sizes were determined from the desorption curves using the Horvath-Kawazoe (HK) method (Table 1). In all cases, there were pores of ~4.9 Å, as expected from the crystal structure, and pores of 14-15 Å. For 1_{evac} and $1 \cdot Ag^+$ at different silver loadings, the ratios of the volume adsorbed in the second step to that in the first step of the N_2 isotherm (V_2/V_1) are in good agreement with the ratios of the volume of the larger pore to that of the smaller pore obtained from integration of the pore size distribution (V_{large}/V_{small}) (Table 1). These results indicate that the samples probably contained super micropores. In addition, the isotherms (Figure 5) showed an linear increase in adsorption capacity after the two steps, indicating of the presence of mesopores, possibly resulting from Ostwald coarsening of some of the micropores.[28]

To determine the effect of post-synthesis modification of 1_{evac} with Ag^+ on selectivity for olefins over paraffins, single-component C_2H_4 and C_2H_6 adsorption isotherms were measured at 298 K for 1_{evac} and $1 \cdot Ag^+$ (Figure 6A). A decrease in adsorption capacity was observed for $1 \cdot Ag^+$, in agreement with the N_2 adsorption results. Smaller adsorbate molecules, such as CO_2 and CH_4, were also studied (Figure 6B) to investigate if they could penetrate into triflate-containing pores more readily than C_2H_4 and C_2H_6. Whereas 1_{evac} adsorbed three times more C_2H_4 and C_2H_6 than $1 \cdot Ag^+$ with Ag:Zn = 0.59:1, it only adsorbed twice as much CO_2 and CH_4 as the Ag^+-modified MOF. Thus, the triflate anions were not as effective at blocking the entrance of small adsorbate molecules into the pores.

The single-component isotherms were used as inputs to ideal adsorbed solution theory (IAST) calculations[29] to predict the selectivities for equimolar mixtures (Figure 7). No difference in selectivity for C_2H_4 over C_2H_6 was observed between 1_{evac} and $1 \cdot Ag^+$, each with selectivities of 1.3 at 1 bar. The selectivity for CO_2 over CH_4 was 3.0 for 1_{evac} and 2.7 for $1 \cdot Ag^+$ at 1 bar. These results show that the introduction of Ag^+ into the MOF did not enhance selectivity for C_2H_4 or CO_2, possibly due to triflate anions blocking sorbate molecules from interacting with Ag^+. Over the pressure range studied, the

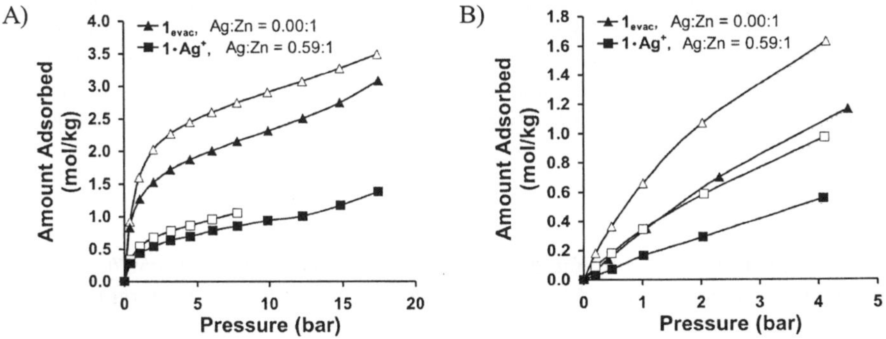

Figure 6 *Isotherms at 298 K of 1_{evac} and $1 \cdot Ag^+$ with Ag:Zn = 0.59:1 for A) C_2H_4 (open symbols) and C_2H_6 (filled symbols) and B) CO_2 (open symbols) and CH_4 (filled symbols)*

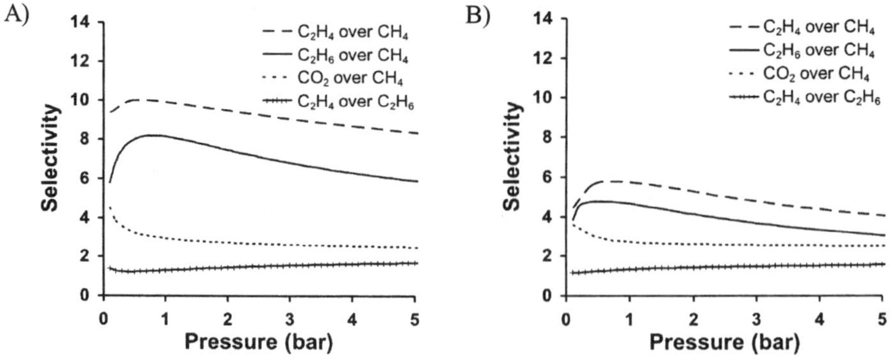

Figure 7 *IAST selectivity plots for C_2H_4/CH_4, C_2H_6/CH_4, CO_2/CH_4, and C_2H_4/C_2H_6 equimolar mixtures at 298 K for A) 1_{evac} and B) $1 \cdot Ag^+$ with Ag:Zn = 0.59:1*

selectivities for C_2H_4 and C_2H_6 over CH_4 were determined to be less in $1 \cdot Ag^+$ than in 1_{evac}. The greater selectivity found for 1_{evac} in these cases agrees with the observation that 1_{evac} absorbed as much as three times the amount of C_2H_4 and C_2H_6 as $1 \cdot Ag^+$, but only twice as much CO_2 and CH_4 as $1 \cdot Ag^+$.

4 CONCLUSION

The post-synthesis modification of 1_{evac} with AgOTf was shown to result primarily in the interaction of Ag^+ with the functional groups originally coordinated to the Zn^{2+} corners of the MOF. These interactions caused some disruption of the framework as determined by PXRD and IR spectroscopy. Investigation of the adsorption capacity of 1_{evac} and $1 \cdot Ag^+$ illustrated that increased silver loading led to decreased capacity, likely due to the inclusion of triflate anions inside the pores, in addition to framework disruption. Single-component isotherms and parallel IAST calculations demonstrated that the presence of Ag^+ in the pores did not enhance selectivity for olefins over paraffins. The relatively bulky triflate anions may have prevented π–donor sorbates from interacting with Ag^+ in the confined space of the pores. The use of silver salts with smaller anions may lead to a Ag^+-modified MOF that is more selective for olefins. Preliminary PXRD and IR studies have indicated that modification of 1_{evac} with $AgNO_3$ yielded a Ag^+-modified MOF that is similar to $1 \cdot Ag^+$ but with smaller nitrate anions inside the pores. Another vein for future work is the ion-exchange of Ag^+ into MOFs with negatively charged linker molecules.[30]

Acknowledgements

This research was supported by the National Science Foundation (CTS-0507013). YSB thanks the Korean Government (MOEHRD) for a Korea Research Foundation Fellowship (KRF-2006-352-D00040). GAE is a National Defense Science and Engineering Graduate Fellow. JTH additionally acknowledges support from the U.S. Department of Energy (grant No. DE-FG02-1ER15244). The authors thank Dr. Omar Farha, Ms. Tendai Gadzikwa, and Ms. Karen Mulfort for helpful discussions.

References

1. O. M. Yaghi, H. Li, C. Davis, D. Richardson and T. L. Groy, *Acc. Chem. Res.*, 1998, **31**, 474.
2. S. Kitagawa, R. Kitaura and S.-I. Noro, *Angew. Chem. Int. Ed.*, 2004, **43**, 2334.
3. G. Ferey, *Chem. Soc. Rev.*, 2008, **37**, 191.
4. Q. M. Wang, D. Shen, M. Bülow, M. L. Lau, S. Deng, F. R. Fitch, N. O. Lemcoff and J. Semanscin, *Microporous Mesoporous Mater.*, 2002, **55**, 217.
5. R. Q. Snurr, J. T. Hupp and S. T. Nguyen, *AIChE J.*, 2004, **50**, 1090.
6. S. Ma, D. Sun, X.-S. Wang and H.-C. Zhou, *Angew. Chem. Int. Ed.*, 2007, **46**, 2458.
7. Y. Li and R. T. Yang, *Langmuir*, 2007, **23**, 12937.
8. D. N. Dybtsev, H. Chun, S. H. Yoon, D. Kim and K. Kim, *J. Am. Chem. Soc.*, 2004, **126**, 32.
9. Y.-S. Bae, K. L. Mulfort, H. Frost, P. Ryan, S. Punnathanam, L. J. Broadbelt, J. T. Hupp and R. Q. Snurr, *Langmuir*, 2008, **in press**, DOI: 10.1021/la800555x.
10. B. Chen, C. Liang, J. Yang, S. S. Contreras, Y. L. Clancy, E. B. Lobkovsky, O. M. Yaghi and S. Dai, *Angew. Chem. Int. Ed.*, 2006, **45**, 1390.
11. R. B. Eldrige, *Ind. Eng. Chem. Res.*, 1993, **32**, 2208.
12. S. Hess, C. Staudt-Bickel and R. N. Lichtenthaler, *J. Membr. Sci.*, 2006, **275**, 52.
13. J. Padin and R. T. Yang, *Chem. Eng. Sci.*, 2000, **55**, 2607.
14. C. A. Grande, J. D. P. Araujo, S. Cavenati, N. Firpo, E. Basaldella and A. E. Rodrigues, *Langmuir*, 2004, **20**, 5291.
15. H. S. Kim, J. H. Ryu, H. Kim, B. S. Ahn and Y. S. Kang, *Chem. Commun.*, 2000, 1261.
16. K. L. Mulfort and J. T. Hupp, *J. Am. Chem. Soc.*, 2007, **129**, 9604.
17. Z. Wang and S. M. Cohen, *J. Am. Chem. Soc.*, 2007, **129**, 12368.
18. M. J. Ingleson, J. P. Barrio, J. Bacsa, C. Dickinson, H. Park and M. J. Rosseinsky, *Chem. Commun.*, 2008, 1287.
19. B. Chen, S. Ma, F. Zapata, E. B. Lobkovsky and J. Yang, *Inorg. Chem.*, 2006, **45**, 5718.
20. I. Pastoriza-Santos and L. M. Liz-Marzan, *Langmuir*, 1999, **15**, 948.
21. M. A. Bennett, *Chem. Rev.*, 1962, **62**, 611.
22. P. A. Yeats, J. R. Sams and F. Aubke, *Inorg. Chem.*, 1971, **10**, 1877.
23. G. A. van Albada, W. J. J. Smeets, A. L. Spek and J. Reedijk, *J. Chem. Crystallography*, 1998, **28**, 427.
24. M. Eddaoudi, J. Kim, N. Rosi, D. Vodak, J. Wachter, M. O'Keeffe and O. M. Yaghi, *Science*, 2002, **295**, 469.
25. B. Civalleri, F. Napoli, Y. Noël, C. Roettia and R. Dovesia, *Crystal Eng. Commun.*, 2006, **2006**, 364.
26. J. L. Lambert, H. F. Shurvell, D. A. Lightner and R. G. Cooks, *Organic Structural Spectroscopy*, Prentice-Hall, Upper Saddle River, New Jersey, USA, 1998, p. 194.
27. Z. Ozhamam, M. Yurdakul and S. Yurdakul, *J. Mol. Struct. (THEOCHEM)*, 2006, **761**, 113.
28. C. S. Tsao, M. S. Yu, T. Y. Chung, H. C. Wu, C. Y. Wang, K. S. Chang and H. L. Chen, *J. Am. Chem. Soc.*, 2007, **129**, 15997.
29. A. L. Myers and J. M. Prausnitz, *AIChE J.*, 1965, **11**, 121.
30. Y. Liu, V. C. Kravtsov, R. Larsen and M. Eddaoudi, *Chem. Commun.*, 2006, 1488.

OPTIMIZING H_2 VOLUMETRIC STORAGE CAPACITY OF CARBON MATERIALS FOR MOBILE APPLICATIONS

L. Zubizarreta[1], J. Jagiello[2], A. Arenillas[1], CO Ania[1], J.B. Parra[1], J.J Pis[1]

[1] Instituto Nacional del Carbón, CSIC, Apartado 73, 33080 Oviedo, Spain
[2] Micromeritics Instrument Corporation, 1 Micromeritics Dr. Norcross GA 30093, USA.

1 INTRODUCTION

Hydrogen storage is recognized as the critical point to enable the widespread use of H_2 as an energy source in mobile applications. The United States Department of Energy (DOE) and the European Union (EU) have established their targets for on-board hydrogen storage systems, including the minimum gravimetric and volumetric capacity (6 wt% and 45 g H_2 L^{-1} for 2010). Among the numerous works published dealing with the H_2 storage in solid materials in the last decade, very few mention the volumetric capacity of the materials studied.[1-5] Considering the limited volume available in automobile applications the volume is the critical factor for use of these solids as storage systems. Therefore, it is important to consider storage data on both volume and weight capacity basis. Optimizing the volumetric storage capacity presents a challenge because increasing carbon pore volume for higher gravimetric adsorption capacity reduces its density. It follows that only "good" pores of certain sizes are beneficial to the improved volumetric capacity. The benefits of the simultaneous analysis of adsorption data of multiple gases to achieve a more complete characterization of microporous carbon materials were recently demonstrated.[6] The objective of this work is to gain a better understanding of the effects of carbon pore structure on the H_2 gravimetric and volumetric capacity under different temperature and pressure conditions. It is attempted to find the relationships between carbon pore size distribution and its volumetric capacities measured at different pressures and various temperatures.

2 METHODS

2.1. Synthesis of the carbon materials

A series of porous carbon materials have been prepared and investigated as candidates for hydrogen storage: carbon xerogel, chemically activated carbon xerogels and carbon nanospheres. Detailed experimental synthetic routes have been reported and discussed elsewhere.[7-9] A commercially available activated carbon Maxsorb (from Kansai Coke and Chemicals) was also used for comparative reasons.

Briefly, an organic xerogel was obtained from the sol-gel polymerization of resorcinol and formaldehyde in water (ratio resorcinol:catalyst = 300) and further drying under reduced pressure (0.012 bar).[8] The organic gel was chemically activated with KOH (ratio activated agent to carbon 3:1) at 1023 K to generate porosity (OX-Act sample). Sample CX was prepared by carbonization (1073 K, 2 hours) of the organic gel; further chemical activation of the carbon xerogel yielded CX-Act sample.[7] Carbon nanospheres (CS) were synthesised by a two-step polymerisation of furfuryl alcohol, drying and carbonisation at 923 K for 5 hours.[9]

2.2 Characterization and H$_2$ storage

Nitrogen and hydrogen adsorption measurements at 77 K were performed with a Micromeritics ASAP 2010M. Before the experiments, the samples were outgassed under vacuum (10^{-6} bar) at 523 K overnight (17 hours). The N$_2$ isotherms were used to calculate the specific surface area, S_{BET}, total and micro pore volumes, V_{TOTAL}, and pore size distributions (PSD). In this work, the slit pore model was assumed for carbon pores and the calculations of N$_2$ model adsorption isotherms (kernels) were performed following the implementation of Tarazona's nonlocal density theory (NLDFT).[10] Details of model parameters and calculations were given and discussed elsewhere.[6,10,11]

The high pressure hydrogen adsorption isotherms were performed at room temperature (298 K) up to 90 bars using a high pressure gravimetric analyzer from VTI Scientific Instruments; the manifold volume and buoyancy effect were corrected by carrying out helium isotherms. Leaks and background noise were also evaluated. Before performing the isotherms, the samples were outgassed in situ at 523 K under vacuum. The hydrogen storage capability of the materials was evaluated from the isotherms at cryogenic and room temperature, and at different pressures (1 and 90 bar). Hydrogen was supplied by Air Products with an ultrahigh purity (i.e., 99.9992%). The packing density of the materials was determined by pressing a given amount of activated carbon (0.250 g approximately) in a mould at a pressure of 550 kg cm^{-2}. The measurements were repeated several times. The densities obtained have an error smaller than 3%.

3 RESULTS AND DISCUSSION

The studied carbon materials possess different morphologies, powders and monoliths, due to the synthetics routes employed. According to the large number of studies reported in the literature in recent years,[12-16] in order to obtain high gravimetric storage capacities, highly microporous materials with a large specific surface area must be used, and morphological or structural characteristics have little or no influence on the gravimetric storage capacity of the samples. Therefore, we can assume that the textural properties of these carbons will be the only factors affecting their hydrogen storage capability. Hence, the selection of the materials employed in this work was based on their different textural features. This allows us to investigate the relationship between carbon pore size distributions and the volumetric H$_2$ storage capacities measured at various pressures and temperatures.

There is a general understanding that microporous materials are the most appropriate for hydrogen adsorption. It has been also shown that the hydrogen storage capacity strongly depends on the micropore width.[17,18] For this reason, it is very important to obtain as much as information as possible about the full microporosity range, including micropore sizes. Table 1 compiles the main textural parameters of the studied carbon materials, obtained from the nitrogen adsorption data at 77 K.

Table 1 Main textural parameters of the studied carbons materials obtained from N_2 adsorption at 77 K

	S_{BET} $[m^2g^{-1}]$	V_{TOTAL} $[cm^3g^{-1}]$	Cum. Vol. [< 20 Å] $[cm^3g^{-1}]$	Cum. Vol. [< 50 Å] $[cm^3g^{-1}]$	Vol. [20<w< 50 Å] $[cm^3g^{-1}]$
CX	636	0.436	0.168	0.242	0.073
CX-Act	1540	0.863	0.559	0.660	0.102
OX-Act	2037	1.271	0.753	1.256	0.503
CS	543	0.269	0.205	0.213	0.008
Maxsorb	3494	1.651	0.874	1.591	0.717

The corresponding N_2 adsorption isotherms and the cumulative pore size distributions (PSD) are shown in Figures 1(a,b). The PSDs of our carbons were calculated from N_2 adsorption isotherms measured at 77 K using numerical algorithm SAIEUS[19] with the appropriate N_2 kernel[6]. The studied carbons represent a variety of surface areas, and pore volumes; also, the cumulative PSD curves show significant differences in the pore structures, particularly in the narrow micropore range.

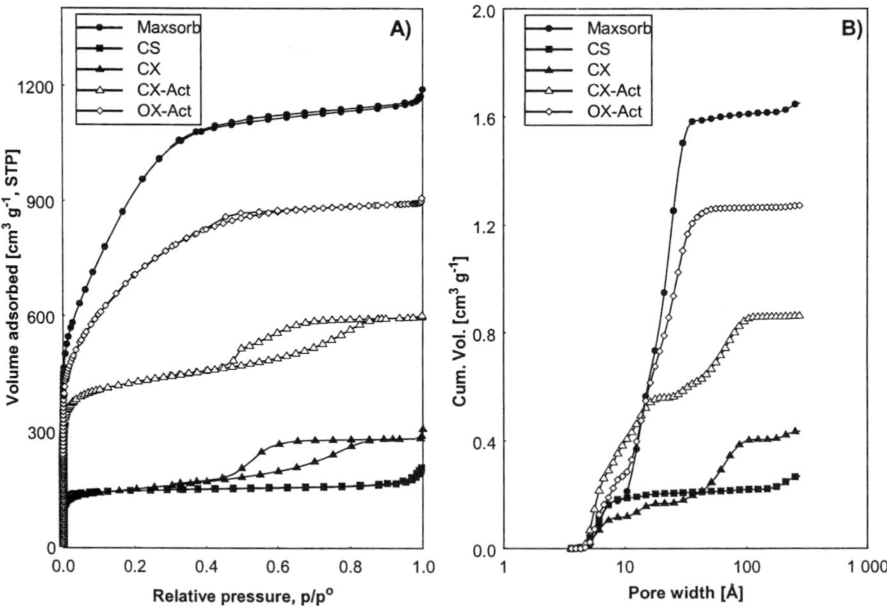

Figure 1. *A) Experimental N_2 adsorption isotherms at 77 K and B) calculated cumulative PSD curves for the studied carbons.*

Based on Table 2 one may conclude that the overall trend in the H_2 uptake at 77 and 298 K (Table 2) follows the sequence of increasing specific surface areas and pore volumes for our samples: CS<CX<CX-Act<OX-Act<Maxsorb. However, this general observation is not sufficient for describing how pores with different sizes affect the H_2 capacity of carbon samples.

It is interesting to note that the activated xerogels CX-Act and OX-Act present similar hydrogen gravimetric uptakes at low temperature (Table 2) to those obtained for Maxsorb carbon, which possesses much higher surface area and pore volumes: between 1.3 and 2 times larger than the mentioned xerogels (Table 1). It appears that a large fraction of the porosity in Maxsorb carbon is underused for hydrogen adsorption.

To explain this observation based on structural parameters of CX-Act and OX-Act samples, we note that the common feature of these samples is their large volume of micropores smaller than 10-15 Å (Figure 1b). Indeed, the volume of pores below 10 Å is almost 2 times higher in CX-Act than in Maxsorb. This does not mean that only pores below 15 Å contribute to the overall H_2 capacity. Larger pores also contribute in smaller extent, but they cannot be neglected. To show the contribution of pores larger than 10 Å to H_2 capacity we compare CS and Maxsorb carbons. Pore volumes of these carbons are very similar below 10 Å (Figure 1b), but the content of larger micropores (10-20 Å) is much higher for Maxsorb and as a result, this carbon has significantly higher gravimetric H_2 capacity than CS sample.

Table 2. *Gravimetric and volumetric hydrogen storage capacities of the carbons studied in this work, obtained at different pressures and temperatures*

SAMPLE	Packing density (g cm⁻³)	Gravimetric H₂ uptake (% wt)		Volumetric H₂ uptake (kg m⁻³)	
		77 K, 1bar	298 K, 90bar	77 K, 1bar	298 K, 90 bar
CS	**1.88**	1.2	0.3	**23.2**	**5.8**
CX	1.02	1.3	0.2	12.9	2.4
CX-Act	0.24	2.4	0.5	5.7	1.1
OX-Act	0.63	2.7	0.6	16.8	4.0
Maxsorb	**0.47**	**2.8**	**0.8**	13.6	3.7

The gravimetric capacity at room temperature and high pressure follows the same general trend observed at cryogenic temperatures (Table 2 and Figure 2). Again, the porosity of Maxsorb seems to be underused, given the relatively high storage capacity of the activated xerogels. At room temperature, an almost linear correlation was found between the gravimetric storage and the volume of pores below 10 Å.

In the earlier study[20] it was shown that the cryogenic low pressure H_2 adsorption data can be used to predict the high pressure adsorption of this gas. Here, the PSDs calculated from N_2 adsorption were used in conjunction with the appropriate H_2 NLDFT kernel to predict the excess adsorption isotherms at 298 K. Selected values of this kernel are shown in Figure 3. A semi-quantitative agreement between the calculated and experimental high pressure isotherms was achieved (Figure 2). A good prediction is obtained for Maxsorb and CS, whereas the largest deviations were found for the activated CX-Act and OX-Act samples, which are characterized by a large contribution of narrow micropores, stressing the importance of an accurate and reliable characterization of the materials. This result also confirms that in the case of physical adsorption the knowledge of the carbon PSD allows prediction/estimation of high-pressure H_2 capacity, as shown in previous investigations.[20]

It is important to realize that different pores "collect" different amounts of hydrogen. Theoretical illustration of the capacity of pores with different sizes for H_2 storage is given in Figure 3 in terms of H_2 excess density vs. pore width. For the comparison, bulk hydrogen gas densities evaluated for different pressures at 298 K using the Younglove equation of state[21] are also included in this figure. The calculated densities in pores smaller than 10 Å are much higher than the corresponding densities in gaseous phase for each pressure. It is interesting to note that the calculated absolute H_2 density (excess + bulk gas) in 3 Å pores approaches the liquid density of H_2 (15.5 mmol cm^{-3}) at the critical point (33 K). The most important observation that can be made here is that the H_2 excess density drops very quickly with increasing pore size. This trend is similar for low and high pressures (from 1 up to 100 bar). The excess density of adsorbed hydrogen reaches high values for pores below 10 Å and drops for pores wider than 20 Å. If we calculate the decrease of the excess density of adsorbed hydrogen vs the maximum value at 3 Å, there is a 5, 10 and 20-fold fall for pores below 10, 20 and 50 Å, respectively. This indicates that the pore volume for larger pores is underused. In addition, the existence of larger pores decreases the density of the carbon material. It follows that the optimal pore size range for maximizing the hydrogen storage is below 10-15 Å. Similar observations were recently reported by Yushin et al.[22]

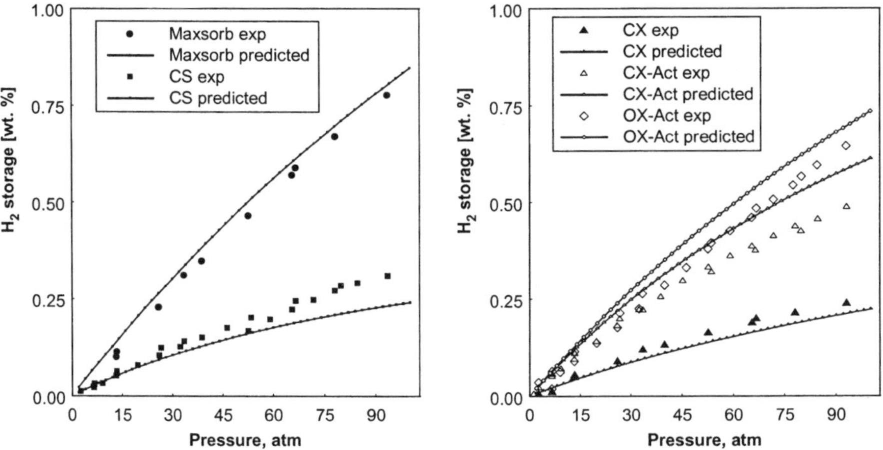

Figure 2. *Experimental (symbols) and predicted (symbols with lines) high-pressure excess H_2 adsorption isotherms at 298 K*

A suitable hydrogen storage system must not only reach high gravimetric but also high volumetric storage capacities. The volumetric storage capacity is especially important in mobile applications where one of the crucial factors to be taken into account is the volume of the system. The volumetric storage capacity is directly related to the packing density i.e. the amount of material that it is possible to pack per unit volume. Whereas it is the porosity alone which determines the gravimetric storage capacity.

For instance, CS sample displays a homogeneous pore size distribution, having most of the pores below 10 Å (Figure 1b). The presence of such large contribution of narrow micropores is beneficial for both high packing density and high density of H_2 adsorbed in the pores. Therefore, in spite of small gravimetric storage capacity this sample having the highest packing density has also the highest volumetric capacity (see bold values in Table 2). One can say that this sample has only the "good" pores for the H_2 storage.

It is interesting to compare CX and CS samples whose gravimetric capacities are similar but the packing density of CX is ca. 8 times lower than that of CS sample. As a result, the volumetric storage capacity of CX is much lower than that of CS. In this case, the mesoporosity of the non activated carbon xerogel (CX) is responsible for the decrease in the packing density value, pointing out the negative effect of large pores in the storage ability.

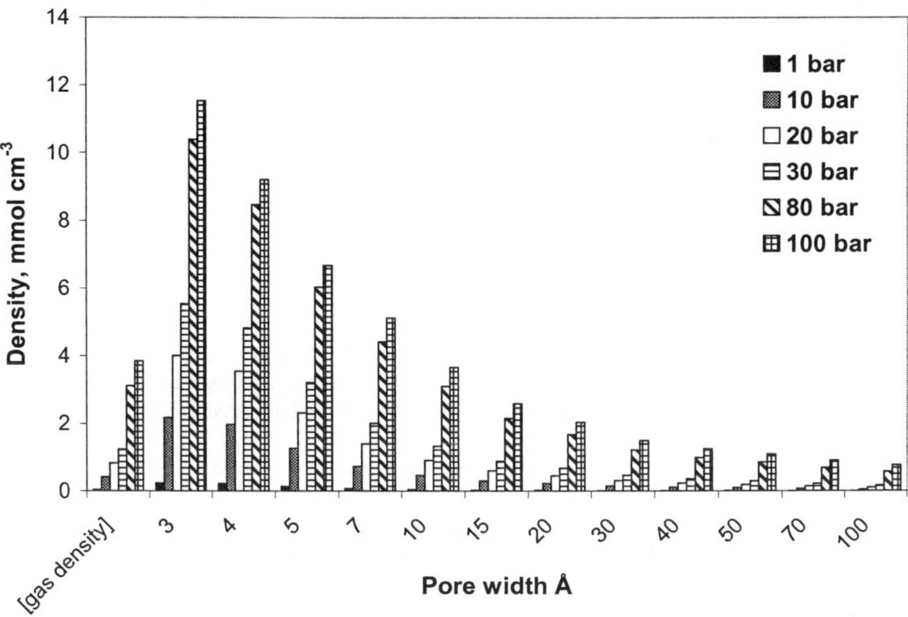

Figure 3. *Calculated excess densities of adsorbed H₂ at 298 K as a function of pore width and pressure and bulk H₂ gas densities at 298 K calculated using Younglove equation [21] for different pressures.*

After activation, the carbons undergo a remarkable decrease in the packing density, as a consequence of the creation of new porosity (i.e., opening porosity plus creation of new porosity during activation). This diminution is higher for CX-Act than for the organic xerogel OX-Act. This behaviour must be explained in terms of the different structure of these materials. The initial carbon xerogel prior to activation is composed of interconnected almost-spherical nodules constituting a three dimensional network. After chemical activation, this structure remains. The activation process does not alter the characteristic interconnected structure of carbon gels: both samples still display interconnected spherical nodules of the same size (~ 150 nm). The mesoporosity formed during the gel synthesis remains in the activated carbon xerogel and the morphology of the chemically activated samples is the same as in the non-activated ones. This indicates that, during chemical activation of carbon xerogels, the increase in microporosity occurs without modifying the general interconnected structure (interconnections between the carbon nodules are strong enough to avoid the network's breaking up, probably because the C-C bonds created during pyrolysis simultaneously to H and O removal from the polymer network are numerous enough to overcome the chemical aggression).

Unlike the chemically activated carbon xerogels, the morphology of OX-Act does not consist in interconnected almost-spherical nodules. A severe erosion of the surface can be observed. The structure still seems interconnected but the spherical nodules disappeared to form a melt-like structure, leading to the destruction of most of the mesoporosity. Thus, the resulting sample is mainly a microporous carbon. Therefore, OX-Act presents higher volumetric capacity than CX-Act.

Carbon materials described in this work do not reach the targets established by DOE and EU and to our knowledge, no successful results of meeting these targets have been reported for adsorption systems at ambient temperature. However, our objective was not to reach those targets but to gain a better understanding of the effects of carbon pore structure on its H_2 capacity and, specifically, relationships between the carbon PSD and its gravimetric and volumetric capacities. We selected our samples to demonstrate and discuss these relationships in conjunction with theoretical (DFT) description of H_2 adsorption in the pores. Our results may be useful in designing future storage systems with variable temperature and pressure.

4 CONCLUSIONS

In this work, we have shown that the hydrogen gravimetric storage capacity of activated carbons depends strongly on their pore size distribution. Pores of different sizes "collect" different amounts of hydrogen and hence do not contribute equally to the carbon storage capacity. It follows that the total pore volume and surface area are not sufficient to describe that capacity.

Calculations of the amount adsorbed in the pores of different sizes show that H_2 density drops very quickly with increasing pore size. "Good" pores that adsorb H_2 at high density are those that are smaller than 10-15 Å. Pores larger than 20 Å do not contribute much to the storage capacity and may be considered underused.

The volumetric storage capacity depends on the gravimetric storage capacity and on the material packing density. Large volume of wider pores is detrimental to the volumetric capacity in two ways: (i) these pores do not adsorb H_2 effectively; (ii) they contribute to the volume of carbon material and thus reduce its density. A carbon having the lowest gravimetric storage capacity (e.g. CS sample) may have the highest volumetric capacity if it contains only "good" pores that adsorb H_2 at high density. Carbon materials with high pore volume of pores smaller than 10-15 Å are the best candidates to reach the high targets of storage capacity.

Acknowledgements

L.Z. acknowledges the support from the CSIC I3P Programme cofinanced by the European Social Fund. J.J. acknowledges the support of the Center for Applied Energy Research, University of Kentucky. C.O.A. thanks the Spanish MEC for a Ramon y Cajal Research Contract.

5 REFERENCES

1. P. Bénard, R. Chachine, *Scripta Materialia*, 2007, **56**, 803.
2. B. Buczek, L. Czepirski J. Zietkiewicz, *Adsorption*, 2005, **11**, 877.
3. J.B. Parra, C.O. Ania, A, Arenillas, F. Rubiera, J.M. Palacios, J.J. Pis, *J Alloys Compounds*, 2004, **379**, 280.
4. H. Tagaki, H. Hatori, Y. Soneda, N. Yoshizawa, Y. Yamada, *Mater. Sci. Eng. B-Solid State Mater. Adv. Technol.*, 2004, **108** (1-2), 143.
5. L. Zhou, Y. Zhou, Y. Sun, *Int. J. Hydrogen Energy*, 2004, **29**, 319.
6. J. Jagiello, C.O. Ania, J.B. Parra, L. Jagiello, J.J. Pis, *Carbon*, 2007, **45**, 1066.
7. L. Zubizarreta, A. Arenillas, J.J. Pis, *Appl. Surf. Sci.*, 2008, **254**, 3993.
8. L. Zubizarreta, A. Arenillas, A. Dominguez, J.A. Menendez, J.J Pis, *J. Non-Crystal. Solids*, 2008, **354**, 817.
9. L. Zubizarreta, A. Arenillas, J.P. Pirard, J.J. Pis, N. Job, *Micropor. Mesopor. Mater.*, 2008, DOI 10.1016/j.micromeso.2008.02.023
10. P. Tarazona, U. Marini Bettolo Marconi, R. Evans, *Mol. Phys.*, **60**, 1987, 573.
11. C.O. Ania, J.B. Parra, F. Rubiera, A. Arenillas J.J. Pis, *Stud. Surf. Sci. Cat.*, 2006, **160**, 319.
12. L. Zubizarreta, E.I. Gomez, A. Arenillas, C.O. Ania, J.B. Parra, J.J Pis, *Adsorpt.*, 2008, DOI 10.1007/s10450-008-9116-y.
13. B. Panella, M. Hirscher, S. Roth, *Carbon*, 2005, **43**, 2209.
14. A. Zütel, *Materials Today*, 2003, **6**, 24.
15. R. Strobel, L. Jorissen, T. Schliermann, V. Trapp, W. Schutz, K. Bohmhammel, G. Wolf, J. Garche, *J. Power Sources*, 1999, **84** (2), 221.
16. L.L. Vasiliev, L.E. Kanonchik, A.G. Kulakov, D.A. Mishkinis, A.M. Safonova, N.K. Luneva, *Int. J. Hydrogen Energy*, 2007, **32**, 5015.
17. R. Gadiou, N. Texier-Mandoki, T. Piquero, S. Saadallah, J. Parmentier, J. Patarin, *Adsorpt.*, 2005, **11**, 823.
18. M. Rezpka, P. Lamp, M.A. de la Casa-Lillo, *J. Phys. Chem. B,* 1998, **102**, 10894.
19. J. Jagiello, *Langmuir*, 1994, **10**, 2778.
20. J. Jagiello J, A. Ansón A, M.T. Martínez, *J. Phys. Chem. B*, 2006, **110**, 4531.
21. B.A. Younglove, *J. Phys. Chem. Ref. Data*, 1982, **11**, 1.
22. G. Yushin, R. Dash, J. Jagiello, J. E. Fischer, Y. Gogotsi, *Adv. Funct. Mater.* 2006, **16**, 2288.

TAILORING CARBON GELS FOR THEIR USE IN THE ENERGY SECTOR

L. Zubizarreta[1], A. Arenillas[1], J.A. Menéndez[1], J.J. Pis[1], P.J.M. Carrott[2], F.L. Conceição[2] and M.M.L. Ribeiro Carrott[2]

[1] Instituto Nacional del Carbón, CSIC, Apartado 73, 33080 Oviedo, Spain
[2] Centro de Química de Évora and Departamento de Química, Universidade de Évora, Colégio Luís António Verney, 7000-671 Évora, Portugal

1 INTRODUCTION

Porous solids are industrially very important materials in the energy sector for applications such as gas separation, gas storage, catalysis and electrochemistry. Among the family of porous solids, zeolites, silicas and activated carbon are the most used materials. However, in order to increase their efficiency in specific applications there is a need to design new materials with controlled and specific properties (i.e., textural and morphological). Recently, various novel techniques have been developed to control the pore structure in carbon materials: template-based techniques, sol-gel methods, etc.[1] However, the processes required for the preparation of these new porous carbon materials involve, in some cases, long times and high costs. In order to make these carbon materials more competitive than traditional carbon materials, there is a need to optimise the manufacturing steps.

Carbon gels are novel porous materials that have received considerable attention in the literature over the last decade. They are usually made by sol-gel polymerisation of resorcinol and formaldehyde followed by a drying step, and carbonisation of the resulting organic gel. In the preparation of this type of carbon materials a large number of variables are involved and their variation allows a wide range of different carbon gels with different textural properties, especially in the range of the mesopores, to be obtained.[2] One of the most important variables in the preparation is the drying process required before the carbonisation of the material. Generally, it is the step which takes longer and requires more sophisticated techniques; so, in order to make carbon gels competitive with activated carbons it is necessary to find a fast and cheap alternative, maintaining the possibility of tailoring the properties of the material. Thus, the optimisation of the drying method is one of the most important challenges in this field. Another attractive characteristic of carbon gels is that they can be prepared in different forms, i.e., as powder, monoliths, microbeads, etc., with high packing densities. This facilitates their use in a wider range of applications. Finally, in order to broaden the range of properties of carbon gels, it could be very interesting to increase the microporosity of this type of carbon materials by different activation techniques and to perform a detailed characterisation of the microporosity created. Therefore, they can be used in more applications, such as gas storage, gas separation, fuel cells and supercapacitors, where high micropore volumes are required.

In this work, a wide variety of different carbon gels have been obtained, by using different drying methods (vacuum, microwave and supercritical drying). Some of them were chemically activated and the textural development was evaluated using different techniques. A wide variety of carbon xerogels and aerogels with textural development at the mesopore and micropore level that could cover a wide range of applications in the energy sector have been obtained.

2 MATERIALS AND METHODS

2.1 Synthesis of Carbon and Activated Carbon Gels

Organic gels were synthesised by polycondensation of resorcinol (R) and formaldehyde (F) in basic, sodium carbonate (C), solution. The R/F molar ratio was equal to the stoichiometric value (0.5). The conditions used in the preparation of gels for vacuum and microwave drying were the following: the dilution ratio, D, (*i.e.,* the total solvent/reactants molar ratio) was 5.7 and R/C molar ratios of 300, 500, 750 and 1000. For supercritical drying, the conditions used were D of 20.2 and R/C ratio of 100.

Resorcinol (VWR International, 99%) and sodium carbonate (UCB, 99.5%) were first solubilised in deionised water in a sealable flask under magnetic stirring. After dissolution, formaldehyde (Aldrich, 37 wt% in water, stabilised by 10-15 wt% methanol) was added and the mixture was stirred until a homogeneous solution was obtained. The solution was then placed in an oven at 85 °C during 72 h for gelation and ageing.

The organic gels obtained were dried by vacuum, microwave or supercritical drying. In the first case, the samples were kept at 60 °C and the pressure progressively reduced from 10^5 Pa to 1200 Pa. The drying procedure was performed over 20 h. The samples were then heated to 150 °C (1200 Pa) and kept overnight. Microwave drying was performed in a unimode cavity oven using a power of 1000 W during 30 min, as preliminary experiments had shown that 30 min was sufficient to reach constant weight.[3] For supercritical drying the organic gels were filtered and placed in pure acetone, which was changed twice a day for 3 days. The CO_2 supercritical drying was carried out in a Polaron E3100 apparatus.

After drying, the xerogels were pyrolysed at 800 °C under nitrogen flow in a tubular oven. The following heating program was used: (i) ramp at 1.7 °C min^{-1} to 150 °C and hold for 15 min; (ii) ramp at 5 °C min^{-1} to 400 °C and hold for 60 min; (iii) ramp at 5 °C min^{-1} to 800 °C and hold for 120 min; and (iv) cool slowly to room temperature. The aerogels were heated directly to 800 °C at a rate of 5 °C min^{-1} and held for 60 min before cooling.

Finally, some of the carbon gels obtained were chemically activated. An organic and carbon xerogel synthesised with R/C 750 and vacuum dried were selected for chemical activation with KOH. The mixture of precursor gel and hydroxide was continuously stirred at 85 °C until dryness and then pyrolysed at 750 °C under nitrogen flow. The heating rate was 5 °C min^{-1} and the flow rate was 800 cm^3 min^{-1}. The samples were maintained at 750 °C for 2 h and then cooled down slowly by natural convection. The samples were then washed repeatedly with a 5 M solution of HCl, followed by distilled water until a final pH of 6. Finally the samples were dried at 110 °C overnight.[4] Additionally, an organic aerogel was impregnated by placing the dried gel in a H_3PO_4 solution, for two hours at 25 °C with agitation, and then heated at 110 °C until complete dryness. The impregnated gel was pyrolysed at 450 °C in a horizontal furnace in nitrogen with a flow rate of 85 cm^3 min^{-1}. The heating rate was 5 °C min^{-1} and the dwell time 60 min. The pyrolysed sample was washed with distilled water until constant pH and dried at 110 °C.

The samples are designated as follows: O is the first letter to indicate the organic gel, the following letter or letters give the type of drying method (MM for microwave drying, R for vacuum drying, and S for supercritical drying); the following number gives the value of the R/C ratio. For example, the sample MM500 corresponds to a carbon gel prepared with R/C = 500, dried by the microwave method and carbonised. For activated samples the letter A and the impregnant/precursor ratio were added to the nomenclature of the corresponding precursor as the first and final letter respectively.

2.2 Characterisation of the Samples

Textural characterisation of the samples was carried out by carbon dioxide adsorption-desorption isotherms at 0 °C in a TriStar 3000 from Micromeritics (xerogels) or a manual manometric apparatus (aerogels), and nitrogen adsorption-desorption isotherms at -196 °C in a ASAP 2010 from Micromeritics (xerogels) or CE Sorptomatic 1990 (aerogels). The use of carbon dioxide adsorption at 0 °C for a proper textural characterisation of the narrow micropores has been established as an effective procedure.[5,6] Indeed, carbon dioxide adsorption occurs in pores smaller than 0.7 nm, while nitrogen does not easily enter such small pores. Thus, the combination of nitrogen and carbon dioxide adsorption isotherm data provides complementary and relevant information about the full micropore range. The Dubinin-Radushkevich (DR) method[7] was applied to the carbon dioxide and nitrogen adsorption isotherms, in order to obtain the characteristic energy E_0, the mean micropore size, L_0, the narrow micropore volume, W_0-CO_2, wider micropore volume, W_0-N_2, and micropore surface area, S_{mi}. The BET surface area was also evaluated from the nitrogen adsorption isotherms.[8] The maximum pore diameter, d_p, (i.e. the pore diameter limit below which smaller pores represent 95% of the total pore volume), was deduced from pore size distributions provided by the Broekhoff-de-Boer method applied to the adsorption branch of the nitrogen isotherm.[9] The pore volume at saturation, V_p, was also calculated. In the case of microporous or mesoporous materials, V_p corresponds to the total pore volume of the material. The so-called external surface area, S_e, was obtained from the comparison of the adsorption isotherm with a reference isotherm determined at the same temperature on a non-porous carbon black (Sing's classical α_s-plot for N_2 at -196 °C, also extended to other vapours). In addition, real surface area S_{TOT} was evaluated as $S_{TOT}=S_{mi}+S_e$.[10]

3 RESULTS AND DISCUSSION

3.1 Textural Properties of Carbon Gels

The N_2 adsorption isotherms determined on the carbon gels are shown in Figure 1 and the results of analysis of the N_2 and CO_2 isotherms are given in Table 1. It can be seen from the form of the isotherms and the V_p and d_p values in Table 1, that increasing the R/C ratio of the xerogels increases the mesopore size. In general, the mesopore size obtained using the microwave drying is smaller, but the variation is only around 1-3 nm. It is also evident that the mesoporosity of the aerogel is qualitatively similar to that of the vacuum dried gels but with, as expected, higher total pore volume. On the other hand, the microporosity created during the carbonisation of the dried organic gels is remarkably similar. Table 1 shows the micropore volumes obtained by applying the Dubinin-Radushkevich equation to the N_2 and CO_2 adsorption isotherms. It can be seen that all of the carbon xerogels present almost the same micropore volume, and if W_0-N_2 and W_0-CO_2 are compared very similar values are found, indicating that independently of the drying method used the carbon xerogels obtained present a similar narrow micropore size distribution. The carbon aerogel apparently has higher microporosity. However, the value obtained by direct application of the DR Equation is overestimated due to the very high mesopore surface area.[11] The value

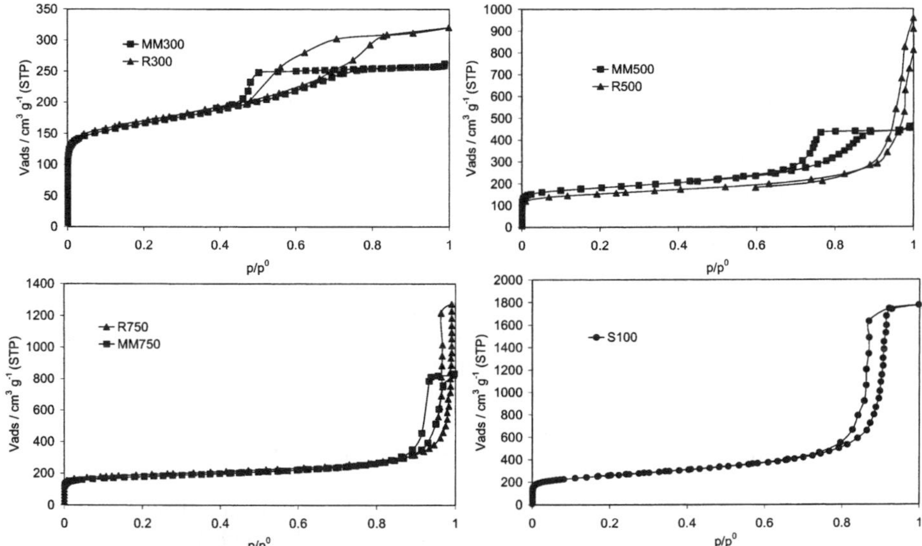

Figure 1 *N_2 adsorption-desorption isotherms of carbon gels dried by vacuum, microwave and supercritical drying and synthesised with different R/C.*

can be corrected using the nonane pre-adsorption method and the value obtained is 0.25 cm^3 g^{-1}, which is effectively the same as that of the xerogels.

Taking into account that a wide range of synthesis conditions and drying methods were used in this work, we can conclude that the micropore volume of RF carbon gels is not significantly influenced by the synthesis and drying procedures used.

The vacuum drying required one week to obtain the dried gels, supercritical drying took 3 days, and the microwave drying method only needed 30 minutes. In the case of using supercritical or vacuum drying with all R/C ratios used, the monolithicity of the samples was maintained. However in the case of microwave drying some samples broke into pieces. This can be explained taking into account the fact that in supercritical drying there

Table 1 *Textural properties of carbon gels synthesised with different R/C and dried by different methods*

Samples	S_{BET} (m^2 g^{-1}) ± 5	W_0-N_2[a] (cm^3 g^{-1}) ± 0.01	W_0-CO_2[a] (cm^3 g^{-1}) ± 0.01	V_p[b] (cm^3 g^{-1}) ± 0.05	d_p[c] (nm) ± 1
MM300	620	0.24	0.24	0.40	8
MM500	670	0.27	0.25	0.70	23
MM750	656	0.26	0.27	1.58	44
R300	636	0.26	0.26	0.58	10
R500	571	0.25	0.23	1.18	26
R750	723	0.28	0.27	1.49	45-50
S100	920	0.36	0.37	2.74	22

[a] obtained by applying Dubinin-Radushkevich method to N_2 and CO_2 adsorption isotherms
[b] calculated from N_2 adsorption at saturation
[c] obtained by applying Broekhoff-de Boer theory to N_2 adsorption isotherms

are no surface tension forces, while in vacuum drying these forces can be minimised by controlling the temperature and the pressure during the drying process. However, in the case of microwave drying, it is more difficult to control the operating conditions. In this work, using a power of 1000 W some carbon gels preserved the monolithicity, while others broke into pieces. Improved control of the power during the microwave drying may allow the monolithicity to be preserved in all cases. Hence, microwave drying may prove to be a good alternative in order to minimise the time required in the preparation of the material.

After carbonisation of the organic gels, the microporosity of the carbon gels is not very well developed. Other work has shown that physical activation in CO_2 does not result in a very significant improvement.[11] The mesoporosity of the gels could be applied in the separation of big molecules, such as organic compounds, peptides, etc. However, for their use as molecular sieves in the energy sector, i.e, for the separation of H_2/CO_2, CH_4/CO_2, it would be necessary to increase the inherent microporosity created during the synthesis process to at least 0.5 cm^3 g^{-1} and to control the pore size.

3.2 Textural Properties of Chemically Activated Carbon Gels

In order to increase the microporosity of the carbon gels, different chemical activation techniques were evaluated. In the case of chemical activation with KOH, two different precursors were activated: an organic xerogel and a carbon xerogel. Figure 2 shows the N_2 adsorption isotherms of non-activated and activated samples. As can be observed, the chemical activation with KOH (Figure 2a) produces a notable increase of the microporosity in both cases (activation of organic and carbon xerogel), leading to an increase of the N_2 adsorption at low relative pressures. But there is a big difference between the activated carbon xerogels obtained from different precursors, mainly in the mesopore range. It can be seen that the activated carbon xerogel obtained from the carbon xerogel preserves the mesoporosity created during the synthesis having a type IV isotherm in the BDDT classification. However, in the case of the activated carbon xerogel obtained from organic xerogel it is mainly a microporous material with a type I isotherm in the BDDT classification. Similar behaviour is found when the organic aerogel is activated with phosphoric acid. The N_2 isotherms are shown in Figure 2b; it can be seen that the isotherm shape also changes from Type IV to Type I, and that there is a large increase in the uptake of N_2 at lower pressures. The most significant difference between KOH and phosphoric acid activation is that in the first case the isotherm reaches a plateau at about 0.2 p^o while the latter only reaches the plateau at about 0.6 p^o, indicating the existence of much wider pores.

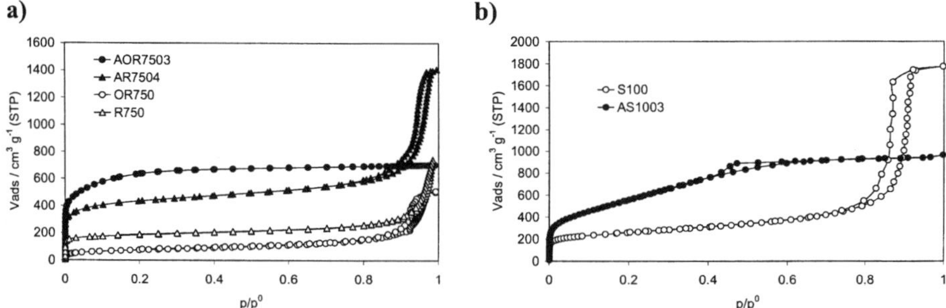

Figure 2 *N_2 adsorption isotherms of different carbon gels and their chemically activated counterparts.*

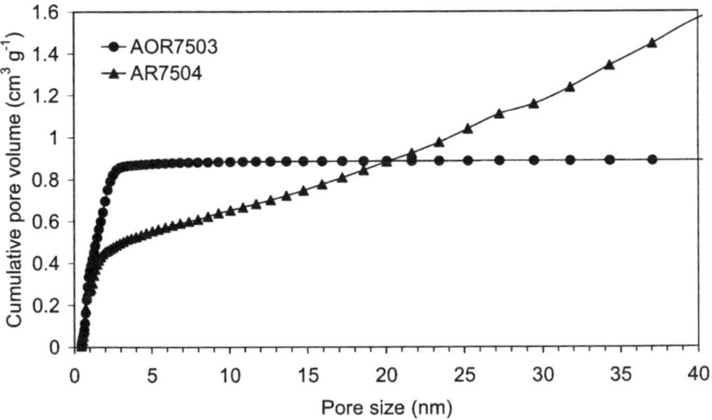

Figure 3 *Pore size distribution applying DFT method to N_2 adsorption isotherm.*

These results show that activated carbon gels with very different textural properties in the micro and mesopore range can be obtained by chemical activation, and this method therefore offers the possibility of producing adsorbents which can be used in different applications in the energy sector.

The pore size distribution of some samples are shown in Figure 3. It can be seen that AR7504 has a bimodal pore size distribution, with a high micropore volume, but at the same time large mesopore volume, which could facilitate gas diffusion inside the carbon bed reducing the pressure drop. However, for gas storage applications, like CH_4 and H_2, it could be better to use activated carbon xerogel obtained from organic xerogel, due to its high micropore development with a unimodal pore size distribution and with a higher packing density, in comparison with AR7504, due to the absence of large pores. High micropore volume and higher packing density allow both high gravimetric and volumetric storage capacities, required for a suitable gas storage system, to be obtained.

Additional information about the microporosity of the samples is given in Table 2, which shows the textural properties of a non-activated carbon xerogel (R750) and two corresponding activated carbon xerogels, and of a non-activated carbon aerogel (S100) and a corresponding activated carbon aerogel. As was mentioned above, it is possible to obtain carbon gels with very different textural properties in the mesopore range by varying the synthesis conditions. However, as can be seen from Table 2, activated carbon gels with different microporosity can be obtained depending on the precursor used for activation. The non-activated carbon xerogel and aerogel present comparatively low micropore volumes centred on an average pore size of about 0.8-0.9 nm. Additionally, they present high external surface areas. When the carbon xerogel is chemically activated it can be seen that the average pore size increases from 0.82 to 1.15 nm and the micropore volume increases up to 0.61 cm^3 g^{-1}. This sample also presents a high external surface area. However, in the case of activation of the organic xerogel and aerogel, the increase of the microporosity is higher (0.81 and 1.41 cm^3 g^{-1}). For the organic xerogel and corresponding activated carbon xerogel the average micropore size is very similar: 1.17 nm. On the other hand the activated carbon aerogel has a wider pore size than the original organic aerogel. In both cases the external surface area of the activated gel is very low, and both are mainly microporous materials.

The differences in the external surface area can be explained taking into account that pores of width, L_0, beyond 2 nm correspond to classical meso- and macropores, in which

Table 2 *Textural properties of different carbon gels, obtained from N_2 adsorption isotherm*

Sample	R750	AR7504	AOR7503	S100	AOS1003
Micropore volume, W_o ($cm^3 g^{-1}$)	0.28	0.61	0.81	0.25[a]	1.41
Characteristic Energy, E_o ($kJ mol^{-1}$)	24.6	20.8	20.6	18.4	15.6
Average micropore size, L_o (nm)	0.82	1.15	1.17	0.87[a]	2.55
Microporous surface, S_{mi} ($m^2 g^{-1}$)	658	1061	1385	575	1106
External (non-microporous) surface, S_e ($m^2 g^{-1}$)	207	387	9	719	39
Total surface area ($m^2 g^{-1}$), $S_{TOT} = S_{mi}+S_e$	865	1448	1394	1294	1145
S_{BET} ($m^2 g^{-1}$)	723	1472	2199	920	2070

[a] estimated on the basis of nonane pre-adsorption results

capillary condensation takes place and where the surface to volume ratio decreases rapidly. Due to the reduced adsorption potential, the surface of these pores may be regarded as equivalent to the surface of the corresponding non-porous material (e.g. carbon black).

This so-called external surface area, can be obtained from the comparison of the adsorption isotherm with a reference isotherm determined at the same temperature on a non-porous carbon black (Sing's classical α_s-plot for N_2 at -196 °C, also extended to other vapours).[8]

The combination of the different techniques leads to an assessment of the micropore volume W_0, the micropore surface area S_{mi}, the external surface area S_e and the total available surface $S_{TOT}=S_{mi}+S_e$. These concepts have well defined meaning and S_{TOT} may be compared with the approach based on the classical BET theory usually applied to nitrogen adsorption at -196 °C. If S_{TOT} and S_{BET} of the samples of Table 2 are compared it can be seen the lowest difference corresponds to AR7504. However, in the case of the samples AOR7503 and AOS1003 there is a big difference between these two values. According to other authors[8] in carbon materials with a pore width between 0.8 and 1.1 nm there is a good agreement between these two surface areas. However, in samples with higher pore width like AOR7503 and AOS1003, the value of S_{BET} increases faster than S_{TOT}.

3 CONCLUSIONS

A wide range of synthesis conditions and drying methods were used to prepare mesoporous resorcinol-formaldehyde organic gels, and different carbonisation or activation procedures were used in order to produce new carbon gels with controlled microporosity. As is well known, the mesoporosity of the gels is dependent on the synthesis conditions and type of drying. On the other hand, the results presented here indicate that the microporosity generated during carbonisation is not significantly influenced by the synthesis and drying procedures used, but can be significantly improved by resorting to chemical activation.

On carbonisation, low micropore volumes and mean pore widths of about 0.25 cm^3 g^{-1} and <1 nm, respectively, are obtained. KOH activation of a previously carbonised gel increased the micropore volume twofold and increased the micropore width from 0.82 to 1.15 nm, while retaining the mesoporosity of the sample. Even larger increases in micropore volume were obtained by KOH or H_3PO_4 activation of the non-carbonised organic gels, although with the disappearance of mesoporosity. The pore size was also increased, to 1.17 nm and 2.55 nm in the case of KOH and H_3PO_4, respectively.

The work reported here has stated that it is possible to obtain a variety of different carbon xerogels and aerogels with different microporosity and mesoporosity characteristics, by KOH and H_3PO_4 chemical activation under different conditions.

Acknowledgements

L.Z. acknowledges the support from the CSIC I3P Programme cofinanced by the European Social Fund. The work was partially supported by the Fundação para a Ciência e a Tecnologia (PhD grant SFRH/BD/19681/2004, Plurianual Finance Project Centro de Química de Évora (619) and project Nº PTDC/EQU-EQU/64842/2006) with national and European community (FEDER) funds. T.A. Centeno is also kindly acknowledged for her contribution in the textural characterisation of the samples.

References

1 M. Inagaki and J.M.D. Tascón in *Activated Carbon Surfaces in Environmental Remediation*, ed. T. Bandosz, Academic Press, vol. 7, 2006, p 49.
2 S.A. Al-Muhtaseb and A. Ritter, *Adv. Mater.*, 2003, **15**, 101.
3 L. Zubizarreta, A. Arenillas, A. Domínguez, J.A. Menéndez and J.J. Pis, *J. Non-Cryst. Solids*, 2008, **354**, 817.
4 L. Zubizarreta, A. Arenillas, J.P. Pirard, J.J. Pis and N. Job, *Microporous Mesoporous Mater.* 2008, doi:10.1016/j.micromeso.2008.02.023.
5 J. Jagiello and M. Thommes, *Carbon*, 2004,**42**, 1227.
6 D. Lozano-Castello, D. Cazorla-Amoros and A. Linares-Solano, *Carbon*, 2004, **42** 1233.
7 M.M. Dubinin in *Progress in Surface and Membrane Science, vol. 9,* ed. J.F. Danielli, M.D. Rosenberg and D. Cadenhead, Academic Press, New York, 1975, p 1.
8 J.B. Parra, J.C. de Sousa, R.C. Bansal, J.J. Pis and J.A. Pajares, *Adsorpt. Sci. Technol.*, 1995, **12**, 51.
9 A.J. Lecloux in *Catalysis Science and Technology, vol. 2*, ed. J.R. Anderson and M. Boudart, Springer, Berlin, 1981, p 171.
10 F. Stoeckli and T.A.Centeno, *Carbon*, 2005, **43**, 1184.
11 P.J.M. Carrott, F.L. Conceição and M.M.L. Ribeiro Carrott, *Carbon*, 2007, **45**, 1310.

INVESTIGATION ON THE STRUCTURE OF SILICA/POLYURETHANE NANOCOMPOSITES

K. Rübner[1], H Goering[1], P. Klobes[1] and H. Tschritter[2]

[1] BAM Federal Institute for Materials Research and Testing, 12200 Berlin, Germany
[2] Chemiewerk Bad Köstritz GmbH, Heinrichshall, 07586 Bad Köstritz, Germany

1 INTRODUCTION

The production of inorganic-organic composites from silica and polyurethane (PUR) or other polymers has been a topic of applied polymer research for many years. Commonly, the silica is supplied with water glass or solid highly disperse silicas, aerosils, aerogels, zeoliths and silicates or organic silicon compounds. The results of syntheses are often disappointing because the incorporation of silica into the polymer comes along with a drastically deterioration of properties due to leaning effects and poor workability of the reaction mixture. Furthermore, only small amounts of silica can be incorporated into the polymer.[1-6]

A new approach is the inorganic-organic nanocomposite, which is synthesised according to EP 1414880.[1] With a polyethylene glycol based organosol, which contains colloidal silica (silicic acid), and diphenyl methane diisocyanate silica/polyurethane nanocomposites are produced as soft or hard foamed and compact materials. The nanocomposites can incorporate up to 35 % silica. Their properties differ considerably from those of similar produced polyurethanes without silica. To explain these macroscopic changes, pore structure measurements and dynamic mechanical analyses are used.

2 EXPERIMENTAL

2.1 Synthesis of Nanocomposites

The starting material was an aqueous silica colloid produced at Chemiewerk Bad Köstritz. The aqueous silica sols contain long term stable polysilicic acids (silicas), which are amorphous and highly reactive. The spherical silica particles have a size of about 60 nm. Their specific surface area varies between 60 and 330 m²/g depending on the production process.

The nanocomposites were synthesised according to EP 1414880.[1] First the aqueous alkaline silica sol was slightly acidified. Then the water was substituted by polyethylene glycol with an average molar mass of 600 g/mol. An organosol containing colloidal silica was formed (Figure 1). Due to reaction between the organosol and diphenyl methane diisocyanate silica/PUR nanocomposites were made as soft and hard foams as well as soft and hard compact materials. They contained different contents of silica up to 25 % by mass (hard polymer) and 35 % by mass (soft polymer). An almost same viscosity of the reaction mixture was provided by variation of the concentration of foaming and cross-linking agents to produce sample series with rising silica contents but with nearly same bulk densities (see Tables 1 and 3). Thus a better comparability of nanocomposite samples was ensured.

The measurements of macroscopic properties and pore structure were carried out with hard foams. All types of nanocomposites were studied by dynamic mechanical analysis.

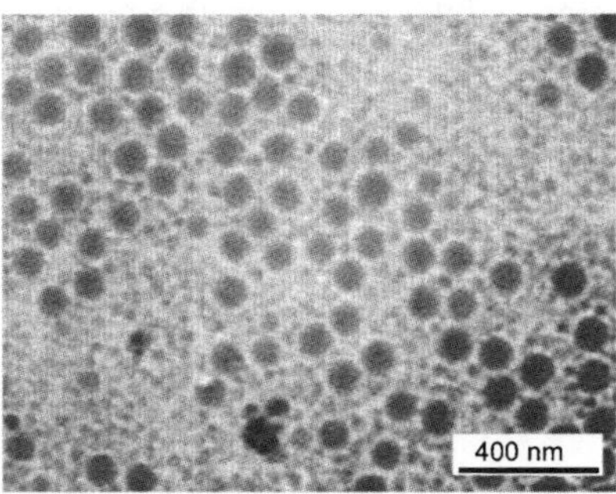

Figure 1 *TEM transmission micrograph of the system silica/polyethylene glycol containing solid particles of about 30 % by mass*

2.2 Methods

The compressive strength and the flexural strength of the nanocomposites each were measured on six measuring samples with a size of 20 mm x 20 mm x 40 mm and 10 mm x 10 mm x 60 mm using the universal testing device ToniNORM (ToniTechnik Berlin). The load measuring range was 2 to 10 kN for the compressive strength and 0.2 kN for the flexural strength.

The total porosity of the nanocomposites was calculated from the ratio of bulk density and density according to DIN 66137-1.[7] The bulk density was determined from weighing and measuring the geometrical size of the samples. The density analysis was performed with a gaspycnometer ACCUPYC 1330 (Micromeritics) using helium according to DIN 66137-2.[8] The samples were cut in pieces of 10 mm x 10 mm x 5 mm size for that purpose. The mass of the measuring samples was 10 g.

The flammability of the nanocomposites was studied by a laboratory test. Five measuring samples with the size of 15 mm x 50 mm x 50 mm were centrically exposed to the blue flame of Bunsen burner (600-700 °C). The times of exposition were consecutively 5 and 30 seconds. After cooling residues of the sample were weighed. The flammability was characterised by mass loss, after burning time and visual inspection.

The pore structure of the nanocomposites was measured by mercury intrusion porosimetry (MIP) with a Pascal 140/240 (POROTEC - Thermo Electron) according to ISO 15901-1:2005.[9] Beside the standard sample holders a ultramacropore kit was used in order to enlarge the measuring range. The mass of the measuring samples was 0.5 g. A measuring range from 0.002 to 200 MPa was used. However, the pressure-volume curves as well as the comparison between densities from MIP and helium pycnometry had shown that at pressures > 5 MPa a compression of the material takes place instead of a pore filling. Therefore, the evaluation of MIP data was restricted to pressures ≤ 5 MPa. By assuming a contact angle of 140° and a mercury surface tension of 0.48 N/m, a range of pores and pore entrances, respectively, from about 370 µm to 150 nm pore radius was measurable.

Water vapour isotherms of the nanocomposites were measured with a dynamic vapour sorption analyser DVS (POROTEC - SMS London) at 298 K. The mass of the measuring samples was 80 mg. To evaluate the isotherms, the ESW theory by Adolphs was used.[10-12] It provides a modeless way of calculating surface energies and specific surfaces areas directly from sorption isotherms. Thermodynamically, the excess surface work (ESW) is the sum of the surface free energy and the isobaric isothermal work of sorption. Physically, it means that each adsorbed molecule decreases the surface energy and at the same time increases the isothermal isobaric work of sorption. The ESW function Φ is defined as product of the adsorbed amount n_{ads} and the change of chemical potential $\Delta\mu$:

$$\Phi = n_{ads} \cdot \Delta\mu \tag{1}$$

It is assumed that the change in the chemical potential $\Delta\mu$ during isothermal adsorption is expressed by the ratio of pressure p to saturation vapour pressure p_s. It follows:

$$\Delta\mu = RT \cdot \ln\left(\frac{p}{p_s}\right) \tag{2}$$

with the gas constant R = 8.314 J/mol · K and the absolute temperature T in K. The measured isotherms are transformed into ESW isotherms by calculation of Φ according to (1) and plotting Φ versus the adsorbed amount n_{ads}. The ESW function (1) shows at least one minimum at the first layer, if the surface free energy and the work of sorption are balanced. The corresponding sorption energy Φ^*_{mono} can be correlated with the loss of degrees of freedom of the sorptive molecules, which is caused by adsorption. Thus the binding strength of the adsorption layer can be estimated. Additionally, a specific surface area A_{ESW} can be calculated from the adsorbed mass n_{mono} at the first minimum of the ESW function:

$$A = n_{mono} \cdot N_A \cdot A_{mol} \tag{3}$$

with the Avogadro constant $N_A = 6{,}023 \cdot 10^{23}$ mol^{-1} and the cross section area of an adsorbed water molecule $A_{mol} = 0.106$ nm².

Furthermore, images of organosols were taken with a transmission electron microscope (TEM) and of nanocomposites with a scanning electron microscope (SEM). Elements in the investigated sample area, such as Si, could be determined by the EDX detector (Energy Dispersive X-ray) installed inside the SEM.

The structure of the polymer matrix of the nanocomposites was characterized by dynamic mechanical analysis using the torsion-pendulum method according to DIN EN ISO 6721-2:1996.[13] An ATM 3 equipment (Myrenne) for free damped torsional oscillations was used. The measuring samples had a length of 80 mm, a width of 10 mm and a thickness 2-3 mm. The complex shear modulus was determined at the fixed frequency of 1 Hz depending on the temperature. The samples were rapidly cooled down to about -196 °C and then heated up to 180 °C using a heating rate of 2 K/min. The results are the shear modulus G', which characterises the stiffness of a sample, and the mechanical loss factor tan δ, which describes the damping behaviour.

3 RESULTS AND DISCUSSION

3.1 Macroscopic Properties

The macroscopic properties, such as bulk density, density, total porosity and strength, of two series of hard nanocomposite foams are summarised in Table 1. As intended, the samples of each series show nearly the same bulk density and total porosity. The density of the PUR is increased by the addition of silica. The nanocomposites show a considerable increase in compressive strength and flexural strength in comparison to the silica-free PUR. A strength maximum is reached for silica additions between 6 and 13 % by mass.

The flammability of the silica/PUR nanocomposites is significantly improved in comparison to the silica-free samples. The results are shown in Table 2. All nanocomposites are self-extinguishing. They show the tendency to be non-combustible. The mass loss after fire exposition decreases with growing silica content. It is two-thirds smaller than those of the silica-free PUR.

Even up to high-silica contents of 25 % the samples of the two series does not show a loss in macroscopic properties. Quite the contrary, the strength and the flammability of PUR are improved by the silica additions. This indicates that the silica particles actively take part in the development of the polymer network.

Table 1: *Properties of hard nanocomposite foams depending on the silica content (two series with different bulk density*

Silica content (% by mass)	Bulk density (g/cm³)	Density (g/cm³)	Total porosity (%)	Compressive strength (MPa)	Flexural strength (MPa)
0	0.16	1.37	88.4	1.30	1.80
6.5	0.17	1.56	89.1	1.75	2.25
12.5	0.14	1.52	90.8	1.65	1.43
19.7	0.16	1.50	89.3	1.63	1.60
24.5	0.17	1.54	89.0	1.44	1.68
0	0.20	1.28	84.4	1.35	1.81
12.2	0.21	1.45	85.5	3.01	3.64
24.5	0.22	1.45	84.8	1.55	1.82

Table 2: *Flammability parameters of hard nanocomposite foams (bulk density 1.6 g/cm³) depending on the silica content*

Silica content (% by mass)	Afterburning time (s)	Mass loss (% by mass)	General observations
0	20	13.9	burning, sooting
6.5	3	8.2	self-extinguishing
12.5	4	9.4	self-extinguishing
19.7	0	5.2	without burning
24.5	0	4.6	without burning

3.2 Pore Structure

The results of pore structure measurements give further evidence for the assumption that the silica influences the PUR network. SEM micrographs of hard nanocomposite foams show that the PUR system is able to incorporate up to 35 % silica nanoparticles. Samples containing less than 15 % SiO_2 are characterised by a homogeneous distribution of silica in the polymer. Composites with higher SiO_2 contents have a more inhomogeneous silica distribution with silica agglomerates in lamellas, strips and nodes of the polymer foam as shown in Figure 2.

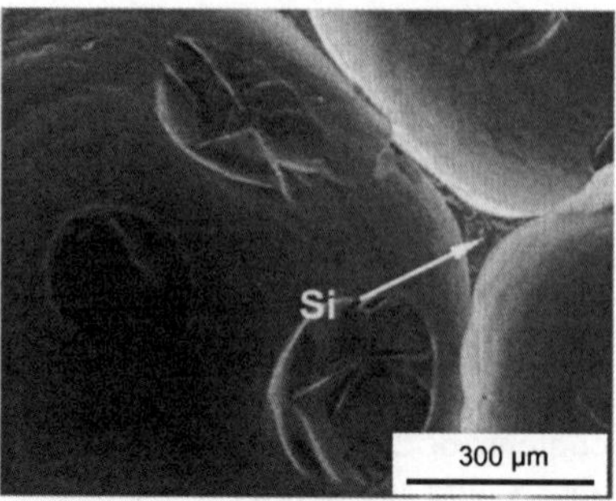

Figure 2 *SEM micrograph of hard nanocomposite foam containing 17 % silica; detection of SiO_2 particles by EDX*

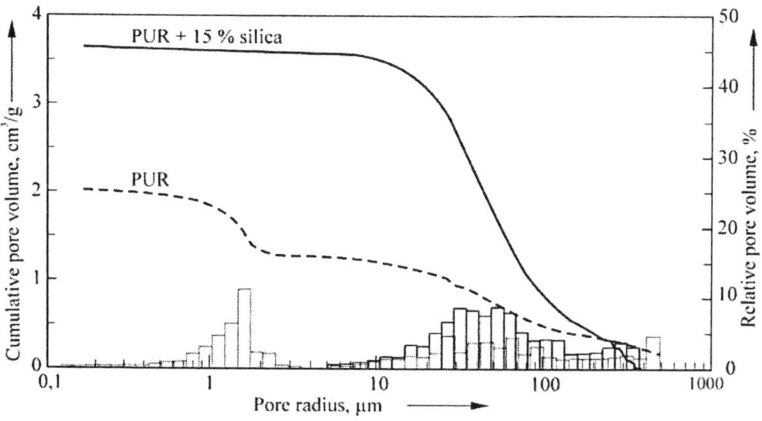

Figure 3 *Pore size distribution of foamed PUR with and without silica addition from mercury intrusion measurements*

The main pore size distributions curves measured by mercury intrusion porosimetry on a third series of hard polyurethane foams with and without silica addition are shown in Figure 3. The single parameters obtained from MIP are summarised in Table 3. The silica-free material shows a bimodal pore size distribution with a wide maximum at pore radii in the range of 35 to 70 µm and a second maximum at pore radii of about 1.5 µm. The peak at 1.5 µm decreases with growing silica content in the polymer and vanishes at 15 % silica. Furthermore, the addition of silica leads to an increase in total pore volume and in porosity, especially for pore radii > 20 µm. These findings are not fully understood yet. Probably, the silica addition influences both, the PUR network and the foaming process. For example, a silica layer formed on the polymer lamellas could reduce the formation of small pores. Additionally, the viscosity of the polymer mixtures grows due to silica additions. That interferes with the foaming process and could inhibit the development of a fine pore system.

According to ISO 15901-1:2005 the specific surface area could not be calculated on the basis of MIP data because the material is considered to be compressible, small pores < 150 nm pore radius are not detected and the pore system probably consist of irregular non-cylindrical pores.[9]

The comparison of the values from density measurements with densities determined by MIP shows the familiar differences. The bulk densities from MIP are higher than those from geometrical size measurements because the pores until 7.4 µm pore radius are already filled with mercury and do not contribute to the sample volume. The densities from MIP and helium pycnometry are in quite good agreement if the material heterogeneity with regard to the small sample mass and the compressibility are taken into account. This confirms the decision to evaluate the MIP data for pressures ≤ 5 MPa only.

Table 3: *Results of density measurements and mercury intrusion porosimetry (MIP)*

SiO_2	ρ_{bulk}	ρ	P_{tot}	V_{tot}	r_1	r_2	$\rho_{bulk, MIP}$	ρ_{MIP}	P_{MIP}
0	0.20	0.73	72.6	1.85	1.5	55.2	0.49	0.69	29.0
8	0.20	0.80	75.0	2.36	1.4	47.4	0.46	0.92	50.0
12	0.18	1.05	82.9	3.40	1.2	40.3	0.43	0.99	81.8
15	0.16	1.27	87.4	3.63	--	48.8	0.41	1.32	87.9

SiO_2 silica content (% by mass)

ρ_{bulk} bulk density (g/cm³), ρ density (g/cm³), P_{tot} total porosity (%) from density measurements

V_{tot} total pore volume at 5 MPa (cm³/g), r_1 and r_2 most frequent pore radii (µm), $\rho_{bulk,MIP}$ bulk density at 0,1 MPa (g/cm³), ρ_{MIP} density at 5 MPa (g/cm³), P_{MIP} porosity (%) from MIP

Figure 4 shows the water vapour isotherms on PUR and silica/PUR nanocomposite. These are type III isotherms that are characteristic for hydrophobic polymeric materials.[14] The silica addition only marginally influences the water adsorption on the PUR foam. The silica particles are obviously embedded in the PUR that the silica surface is not accessible to the water vapour. Furthermore, according to Gregg and Sing an outstanding feature of the adsorption of water vapour on silica gel is its sensitivity to the degree of hydroxylation of the silica surface.[14] A silica gel with a high or medium concentration of hydroxyl groups shows type II isotherms and a mostly dehydroxylated silica gel shows type III isotherms. So the type III isotherms of the nanocomposite could also be explained in terms of participation of the silica particles in the polyurethane synthesis via their superficial hydroxyl groups.

According to the classical models of BET and BJH further evaluation of the type III isotherms is not possible. To derive more information from the isotherms, the ESW theory is used. The transformation into ESW isotherms and the calculated parameters, minimum of sorption energy Φ^*_{mono} and specific surface area A_{ESW}, are shown in Figure 6. The minimum of the ESW isotherms corresponds to a loss of degrees of freedom of 1.6 RT/2 and 1.8 RT/2, respectively. This describes a medium strong physisorption of the water molecules on the surface. However, the deeper and sharper ESW minimum of the nanocomposite shows that the nanocomposite better adsorbs the water. Possibly, the silica addition slightly reduces the hydrophobic properties of the PUR foam. Further evaluation of ESW isotherm yields a specific surface area of about 6 m²/g for both samples.

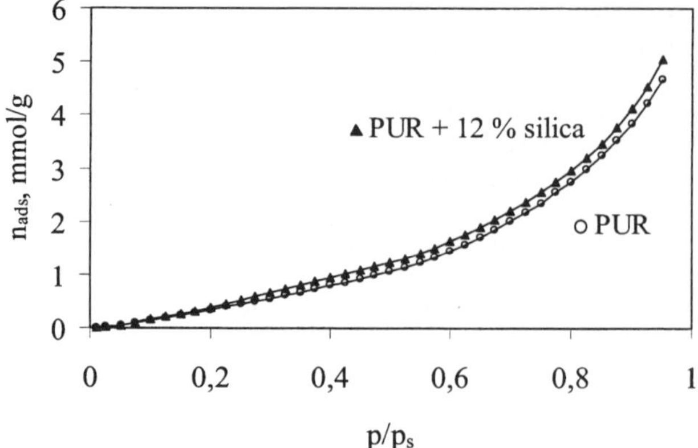

Figure 4 *Water vapour isotherms at 298 K on foamed PUR with and without silica addition from DVS measurements*

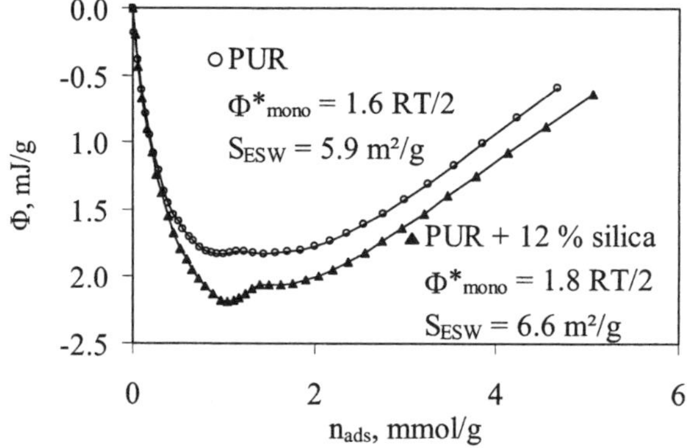

Figure 5 *ESW diagrams of foamed PUR with and without silica addition calculated from water vapour isotherms in Figure 4*

3.3 Structure of Polymer Matrix

The results of the dynamic mechanical analysis of the nanocomposites also confirm the assumption that the silica nanoparticles are involved in the formation of the polymer network. The shear modulus and the mechanical loss factor of compact and foamed samples containing different silica contents are shown in Figures 6 and 7. (For better clarity. the curves of the hard foams, which show the same effects like the soft ones, are left in Figure 7.) The silica nanoparticles do not show negative filler effects on all studied PUR systems. The molecular dynamics are only marginally influenced up to high silica additions. This is proved by nearly unchanged glass transition temperatures. However, the glass transition range is slightly broadened for higher silica contents above the glass temperature. The foamed nanocomposites even show a second relaxation phase there. This second phase appears in conjunction with locally high particle concentrations as identified by SEM. Furthermore, the polymer system is little reinforced by the silica particles as shown by the small change of the plateau modulus. The temperature dependence of the plateau modulus indicates that the SiO_2 nanoparticles stabilise the polymer network showing better entropy elasticity.

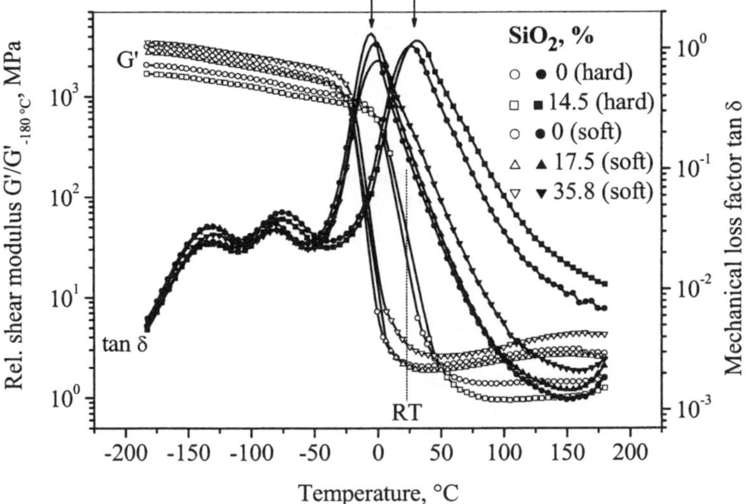

Figure 6 *Dynamic mechanical analysis of hard and soft compact nanocomposites containing different silica contents (arrows indicate glass transition temperatures)*

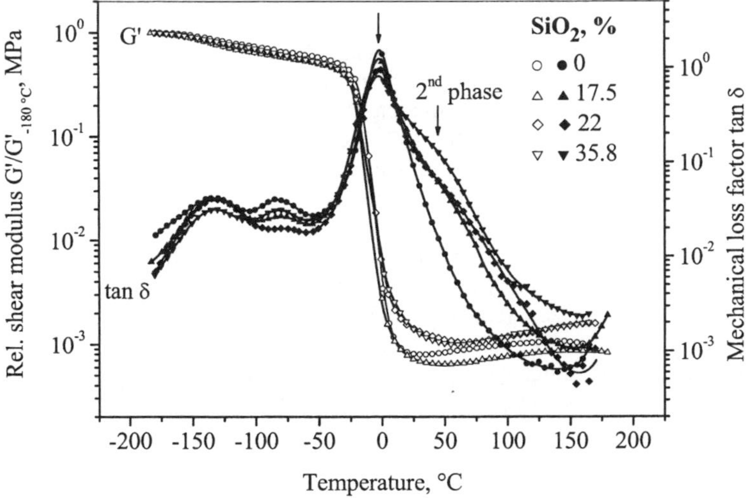

Figure 7 *Dynamic mechanical analysis of soft nanocomposite foams containing different silica contents (arrows indicate glass transition temperatures)*

4 CONCLUSION

By synthesis according to EP 1414880, the PUR system is able to incorporate silica nanoparticles up to 25 % (hard nanocomposite) and 35 % (soft nanocomposite) without drastically deterioration of properties. The properties are even superior to those of silica-free PUR.

The results of the different characterisation methods lead to the conclusion that the silica particles take part in the synthesis of polyurethane via their superficial hydroxyl groups and adsorbed water molecules on the surface of the colloidal silica. An inorganic-organic silica/PUR nanocomposite is formed.

By influencing the polymer network and the foaming process of PUR, rising silica additions cause only slight changes of the structure of the polymer matrix. the pore structure and the hydrophobicity of the material. However, the strength and the flammability of the nanocomposites are improved with growing silica contents.

Acknowledgments

The provision of nanocomposite samples and the helpful discussions by J. Rübner, formerly Department of Chemistry of Technische Universität Berlin, are gratefully acknowledged with thanks. Many thanks are also due to K. Oppat and J. Götze, Division Building Materials of BAM, as well as C. Prinz, Division Structural Analysis of BAM, who carried out the material tests and the pore structure measurements.

References

1   H. Tschritter, G. Knölle, H. Oehlert, M. Ottow and J. Rübner, European Patent EP1414880, 06.05.2004.
2   D. Dieterich, Angew. Makromol. Chem., 1979, **76/77**, 79.
3   P.H. Markusch, I. Iliopulos and D. Dietrich, Adv. Urthane Sci. Technol., 1981, **8**, 263.
4   R.K. Iler, United States Patent US2739076, 20.03.1956.
5   R. Becker, Polyurethane, Fachbuchverlag Leipzig, Leipzig, 1973, 67.
6   T. Adebahr, Farbe und Lack, 2000, **107**, 192.
7   DIN 66137-1, Beuth-Verlag, Berlin, November 2003.
8   DIN 66137-2, Beuth-Verlag, Berlin, Dezember 2004.
9   ISO 15901-1:2005, ISO International Organization for Standardization, 2005.
10 J. Adolphs and M.J. Setzer, J. Coll. Interface Sci., 1996, **180**, 70.
11 J. Adolphs and M.J. Setzer, J. Coll. Interface Sci., 1996, **184**, 443.
12 J. Adolphs, Appl. Surface Sci., 2007, **253**, 5645.
13 DIN EN ISO 6721-2:1996, Beuth-Verlag, Berlin, Dezember 1996.
14 S.J. Gregg, K.S.W. Sing, Adsorption, Surface Area and Porosity, Academic Press, London, 1982, 250 and 269.

NARROW MICROPOROSITY CHARACTERIZATION IN POLYARAMID-DERIVED CARBONS BY ADSORPTION OF N₂, CO₂ AND ORGANIC VAPOURS

M.C. Almazán-Almazán[1], M. Pérez-Mendoza[2], I. Fernández-Morales[2], M. Domingo-García[2], F.J. López-Garzón[2], A. Martínez-Alonso[3], F. Suárez-García[3], J.M.D. Tascón[3]

[1]. Department of Chemical Engineering. Univ. of Liège. Liège, Belgium
[2]. Departamento de Química Inorgánica, Universidad de Granada, 18071 Granada, Spain
[3]. Instituto Nacional del Carbón, CSIC, Apartado 73, 33080 Oviedo, Spain

1 INTRODUCTION

Activated carbon fibers (ACFs) exhibit certain advantages over conventional (powdered or granular) activated carbons, such as rapid adsorption/desorption rates and narrow pore size distributions. Moreover, they are easy to handle, can be molded, have no packing or channelling problems and can be used in form of cloths or felts. The use of high-cristallinity fibrous polymers, such as Kevlar [poly (p-phenylene terephthalamide)] and Nomex [poly (m-phenylene isophthalamide), henceforth Nomex], as precursors for ACFs was initially proposed by Freeman[1]. Later on, ACFs have been prepared from these polymers by physical and chemical activation[2,3]. The main advantage of polyaramids as precursors is the ability to yield ACFs with a very uniform pore size distribution (PSD) in the micropore range, which is little sensitive to the activation method, activating agent, variety of precursor and/or experimental conditions[4]. In this regard, polyaramid-based ACFs compare favourably with ACFs from more conventional (non-crystalline) precursors, whose PSD, albeit narrow, is easily widened by activation to include mesopores. A further (indirect) proof for the outstandingly narrow character of the PSDs of Nomex-based ACFs is the ability of these materials to yield carbon molecular sieves with unpreceded selectivity for the highly demanding O_2/N_2 separation[5].

The use of adsorbates with different molecular sizes has been suggested[6] as the best possible way to analyze the pore size distribution in solids with very narrow microporosity, including carbon molecular sieves. Moreover, the so-called molecular probe method provides useful information on the capacity of a given material to adsorb volatile organic compounds.

In this work, ACFs were prepared from Nomex, either alone or pre-impregnated with different amounts of phosphoric acid and subsequently activated in CO_2 to different burn-off (BO) degrees. The aim is to compare the results of characterizing the microporosity of these solids by adsorption of nitrogen (at 77 K), carbon dioxide (at 273 K) and of several organic vapours with different molecular dimensions and geometries (benzene, n-hexane, cyclohexane and 2,2-dimethyl butane, at 303 K).

2. METHOD AND RESULTS

2.1 Experimental

The ACFs were prepared from crystalline Nomex tow 2.2 dtex, either pure, or pre-impregnated with different amounts of H_3PO_4. The precursor was pyrolyzed at 850 °C in argon and then activated at 800 °C in CO_2 to different BOs. The activated samples are identified as N-BO for the ACFs prepared from pure Nomex, and as NPX-BO for those prepared from H_3PO_4-impregnated Nomex (X is the impregnation ratio, either 3 or 7 wt.%). Further details about samples preparation can be found elsewhere[2].

Gas adsorption isotherms were measured in volumetric adsorption analysers Micromeritics ASAP 2010 (for N_2) and Quantachrome NOVA 1200 (for CO_2). Prior to the adsorption experiments, the samples were outgassed at 523 K overnight under vacuum. The adsorption of benzene, hexane, cyclohexane and 2,2-dimethyl butane (2,2-DMB) was carried out by a static technique, using a conventional gravimetric system. The adsorption isotherms of all probes were measured a 303 K, and 24 hours of equilibrium time was allowed for each point. For the present investigations, it was assumed that the adsorbed compounds were in liquid state. The DR equation has been applied to the adsorption data to determine the micropore volume and the mean pore width (L_0), the last one from the adsorption energy (E_0) using either the Dubinin equation when $E_0 < 20$ kJ/mol:

$$L_0(nm) = \frac{24}{E_0} \qquad (1)$$

or the equation proposed by Stoeckli when $20 < E_0 < 42$ kJ/mol:

$$L_0(nm) = \frac{10.8}{E_0 - 11.4} \qquad (2)$$

2.2 Results and discussion

2.2.1 N_2 and CO_2 adsorption

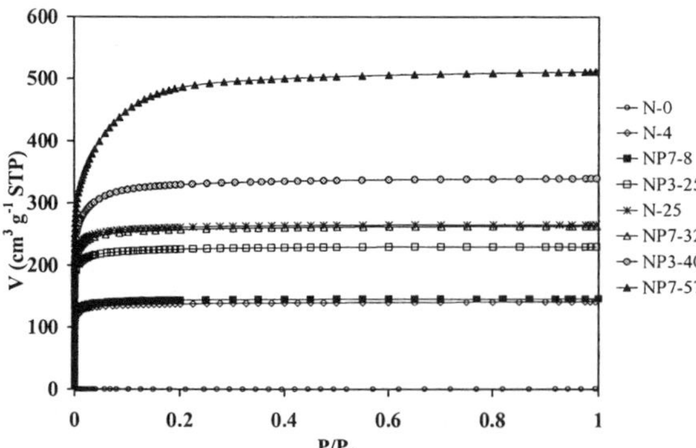

Figure 1 *High-resolution N_2 adsorption isotherms.*

The isotherm (Figure 1) for pyrolysed and non-activated Nomex (sample N-0) is type II according to the IUPAC classification[7]. The rest of the isotherms belong to type I, although differences can be established as the BO increases[8]. At early stages of activation the isotherms are type Ia with a rather sharp knee, reflecting the existence of narrow microporosity (pore widths smaller than 0.7 nm) and a homogeneous pore size distribution. As the BO increases the isotherms change to type Ib, which is characterized by a rounded knee due to the existence of supermicroporosity (pore sizes between 0.7 and 2 nm). The isotherm for the material activated to the largest BO (57 wt.%) is type Ib with a sloping (non-horizontal) linear branch at high relative pressures ($P/P_0 > 0.2$) and a more rounded knee, related to larger and heterogeneous microporosity, plus some small mesoporosity.

The α_S-plots derived from the N_2 adsorption data (Figure 2) for all ACF samples activated to BOs ≤ 32 % show filling swing (FS) at $\alpha_S < 0.5$, with no evidence of cooperative swing (CS), further confirming that ultramicropores are the prevalent pores in these materials. The FS is less marked in the samples activated at highest BOs (40, 57%), suggesting a decrease in the relative amount of ultramicropores. Only the sample activated to the highest BO (NP7-57) shows a slight deviation for $\alpha_S > 0.5$, indicative of some supermicropores.

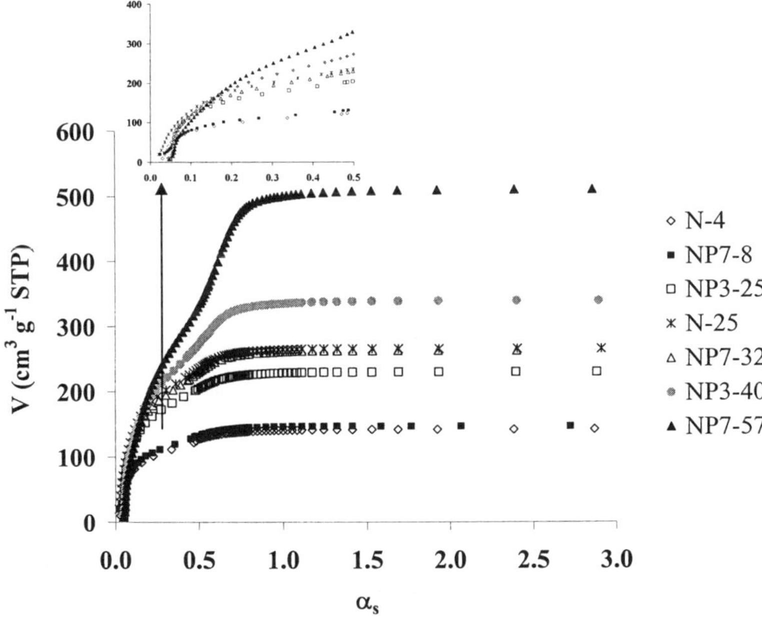

Figure 2 *α_S-plots derived from the N_2 adsorption data*

The textural parameters calculated from the N_2 isotherms are collected in Table 1. All of them increase systematically with increasing BO except the ultramicropore volume ($V_{u\mu p}$). This applies to samples prepared from both, fresh and pre-impregnated Nomex. A comparison of the total pore volume V_p (N_2) with the micropore volumes $V_{\mu p}$ (α_S) and V_0 (DR, N_2) shows that their values practically coincide for all samples. At equivalent BOs, V_p (N_2) is always higher for the ACFs prepared from pure Nomex, probably due to a more thorough activation (especially at low BOs) when using the latter precursor. The relative contribution of ultramicroporosity to the total microporosity can be estimated from the ratio between the values of $V_{u\mu p}$ (α_S) and $V_{\mu p}$ (α_S) reported in Table 1. This ratio decreases from 71% for N-4 to 26% for NP7-57. This trend is the same of the progressive weakening of the FS in the curves of Fig. 2 as the BO increases. Accordingly, the system becomes more heterogeneous and the mean pore width, L_0 (DR, N_2), rises (Table 1) as the BO increases.

Table 1. *Textural parameters from N_2 and CO_2 adsorption*

| | N_2 | | | | | | | CO_2 | |
| | | | DR method | | α_S method | | | | |
Sample	S_{BET} $m^2\,g^{-1}$	V_p $cm^3\,g^{-1}$	V_0 $cm^3\,g^{-1}$	L_0 nm	$V_{u\mu p}$ $cm^3\,g^{-1}$	$V_{\mu p}$ $cm^3\,g^{-1}$	$V_{u\mu p}/\,V_{\mu p}$ (%)	V_0 $cm^3\,g^{-1}$	L_0 nm
N-0	1	0.00	-	-	-	-	-	0.19	0.49
N-4	555	0.22	0.22	0.75	0.15	0.21	71	0.24	0.51
NP7-8	584	0.23	0.23	0.67	0.15	0.22	68	0.24	0.49
NP3-25	898	0.36	0.36	0.74	0.21	0.35	60	0.34	0.60
N-25	1040	0.41	0.41	0.72	0.24	0.39	61	0.37	0.61
NP7-32	1030	0.41	0.41	0.86	0.22	0.39	56	0.40	0.68
NP3-40	1286	0.53	0.50	1.05	0.18	0.50	36	0.45	0.75
NP7-57	1809	0.79	0.71	1.23	0.19	0.74	26	0.47	0.84

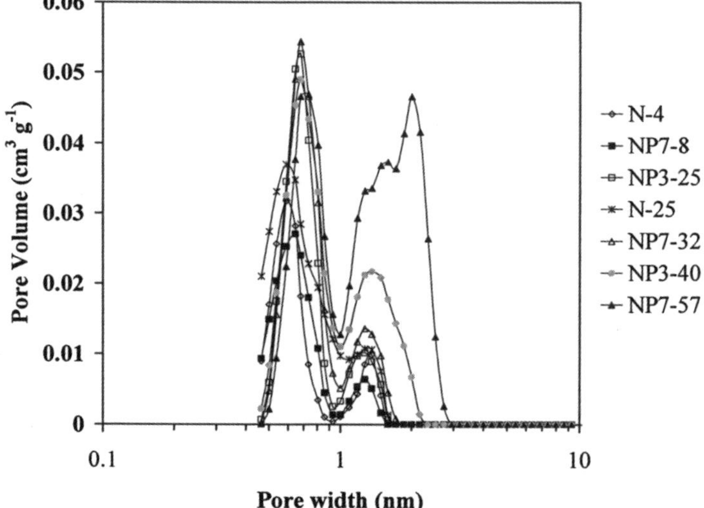

Figure 3 *The PSDs distributions, by application of the NLDFT*

The PSDs distributions, obtained by application of the non local density functional theory (NLDFT), for all samples (Figure 3) have two peaks in the micropore region: the first one, with a maximum at 0.6-0.7 nm, and the second changing from around 1.3-1.4 nm (for BOs<40%) to 2.0 nm for the sample with the highest BO degree. The PSDs clearly show a widening in the micropore size distribution with increasing BO. This takes place in a restricted pore size range below 2.0 nm, i.e. in the microporosity range, for all the samples except the one activated at 57 % BO, for which some contribution of pores between 2.0 and 2.9 nm is observed. In summary, all the ACFs are essentially microporous and contain ultramicropores. In addition, the micropore volume and the mean size increase upon activation.

Carbon dioxide adsorption is a useful tool for characterizing porous solids with very narrow microporosity where N_2 cannot access[9]. This adsorbate covers only the microporosity range from around 0.3 to 0.9 nm. The Dubinin-Radushkevich parameters, V_0 (CO_2) and L_0 (CO_2), are collected in Table 1. It is seen an increase of V_0 with BO, although a slight pore widening also occurs with activation. A comparison of the micropore volumes calculated from CO_2 and N_2 adsorption shows that for samples with low BO (up to 32%) the value of V_0 (CO_2) is very close to V_0 (N_2), which suggests a very narrow and homogeneous microporosity. For larger BO (40 and 57%), V_0 (N_2) is significantly larger than the V_0 (CO_2), what is typical of highly activated carbons with wide micropore distributions. Thus, the activation develops new micropores together with the widening of the existing ones.

2.2.2 Organic vapours adsorption

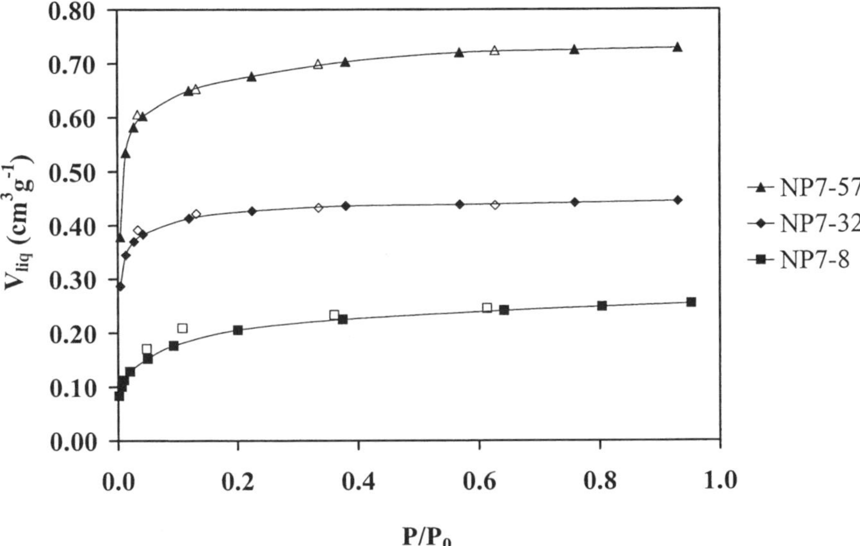

Figure 4 *Adsorption-desorption isotherms of benzene on selected ACF samples.*

The uptake of benzene on pyrolyzed Nomex is very small, but the isotherm (not shown) is much closer to type I compared to N_2. Therefore, benzene adsorption at 303 K approaches equilibrium better than N_2 at 77 K on the pyrolysed material. For all the activated materials, the benzene isotherms are type I. A small low pressure hysteresis (LPH) loop appears for the less activated samples, and completely disappears for BO larger than 32%. The fact that the LPH loop disappears with increasing activation means that this phenomenon is associated with adsorption either into pores whose widths are very close to the critical dimension of benzene, or into wider pores that are only accessible through constrictions similar in size to this molecule.

                                    Characterisation of Porous Solids VIII

Table 2. *Textural parameters from organic vapours adsorption*

Sample	C_6H_6 V_0 $cm^3 g^{-1}$	V_p	L_0 nm	C_6H_{14} V_0 $cm^3 g^{-1}$	V_p	L_0 nm	C_6H_{12} V_0 $cm^3 g^{-1}$	V_p	L_0 nm	2,2-DMB V_0 $cm^3 g^{-1}$	V_p	L_0 nm
N-0	0.02	0.02	-	0.01	0.01	-	-	-	-	-	-	-
N-4	0.14	0.15	1.41	0.14	0.15	1.75	0.08	0.08	1.80	0.02	0.02	2.03
NP7-8	0.19	0.25	1.50	0.19	0.21	1.48	0.07	0.07	1.46	0.05	0.05	1.93
NP3-25	0.20	0.25	0.77	0.26	0.28	0.88	0.12	0.12	1.22	0.11	0.12	1.60
N-25	0.31	0.39	1.22	0.37	0.38	1.17	0.26	0.27	1.14	0.24	0.25	1.17
NP7-32	0.44	0.45	1.02	0.67	0.70	0.90	0.34	0.34	0.98	0.32	0.33	1.00
NP3-40	0.44	0.44	1.15	0.53	0.54	1.20	0.48	0.48	1.18	0.46	0.47	1.22
NP7-57	0.70	0.73	1.19	0.72	0.73	1.17	0.68	0.70	1.19	0.68	0.69	1.22

The characteristics curves of some selected samples are shown in Figures 5. Those in Figures (5a), (b) and (c) are representative of samples with low (N-4), medium (N-25) and high (NP7-57) degree of activation, respectively. It is seen that the adsorption of the four organic probes is smaller than that of nitrogen in samples with low degree of activation (Figure (5a)). Moreover, the characteristics curves do not coincide, in such a way that the adsorbed amount decreases as the molecular dimension rises, suggesting molecular sieve behaviour. This is probably related to constrictions that hinder the access of the molecules depending on the molecular size. This statement is based on the values of the pore width obtained from nitrogen that are larger than the molecular dimensions (L_0, Table 1). The slope of the linear part of the characteristic curves in Figure (5a) increases with the molecular dimension of the adsorbents. This means that the molecular probes are adsorbed in ranges of pores of different mean sizes, i.e. the larger the molecular dimension the wider the pore size where those molecules are adsorbed. This is supported by the trend of L_0 values of the less activated samples in Table 2. It has been reported that the smaller benzene adsorption compared to that of N_2 is related to the density values used for calculations. Thus, the V_0 (DR, N_2) values could be overestimated if the adsorption occurred in very narrow micropores in which the density may be higher than the bulk density due to strong molecular interactions[10]. Nevertheless the characteristic curves in Figure (5a) and the values of L_0 (Table 2) suggest that the smaller benzene adsorption is related to a gate effect for the more geometrically unequal molecule of benzene[11]. This gate effect could also be the reason of the small LPH loop in the isotherms (Figure 4). The deviation of the linearity of the characteristic curves of n-hexane and cyclohexane at low relative pressures is probably related to diffusion problems produced by constrictions at the entrance of the micropores. The less adsorbed molecule is 2,2-DMB in accordance to its largest molecular dimension. In addition, its adsorption occurred in large micropores (near 2 nm in width) with no diffusion restrictions.

Figure (5b) is representative of the characteristics curves of the ACF with medium activation degree. It is seen that in this case the adsorption of the organic probes have increased and it is closer to that of nitrogen. In spite of this, the characteristics plots are different for every adsorbate and the adsorption increases as the molecular dimension decreases except for the n-hexane. This means that the medium degree of activation has developed the microporosity that now is more accessible to all the adsorbates, although the molecular sieve character still remains. The fact that n-hexane is adsorbed in larger extension than benzene could be due to two concurrent effects related with geometrical and kinetic considerations. The first one deals with the several configurations that n-hexane can adopt, which permit a more efficient adsorption. The second one is related to the higher vapour pressure of n-hexane and hence with its higher kinetic energy. As a consequence of these two factors, n-hexane adsorption is more efficient and faster. The characteristics curves in Figure (5b) are parallel having a slope a bit larger than this of nitrogen. This suggests that the organic vapours are adsorbed in micropores of the same dimension, near 1.2 nm, in all cases (Table 2). This means that the pore dimension is larger than the molecule size and, therefore, the molecular sieve effect has to be related with the geometry of the molecules[11].

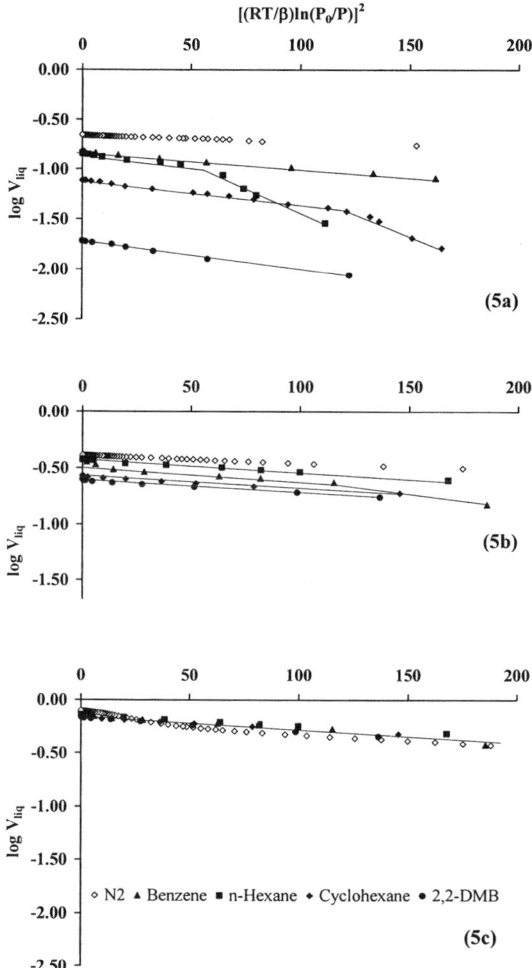

Figure 5 *The characteristics curves of some selected samples. 5a sample N-4, 5b sample N-25 and 5c sample NP7-57*

The characteristics curves in Figure (5c) are representative of samples with high degree of activation. There is an increase in the adsorption capacity for all the adsorbates that is similar to that of nitrogen. In addition, all the organic probes have the same characteristic curve not coincident to that of nitrogen, although the linear part of this is parallel to that of the organic probes. Similar trends, reported for active carbons with high degree of activation[6] have been explained as a consequence of broad microporosity, produced by the very high degree of activation. The micropore volume accessible to every organic probe and the mean pore sizes (Table 2) are in agreement with the above statement. Moreover, it is worth mentioning the large increase in the adsorption capacity of all these samples, without substantial modification of L_0, with the highest activation degree.

3. CONCLUSIONS

From the N_2 and CO_2 adsorption results it can be concluded that: i) The volumes of micropores increase with the BO, ii) The mean pore size is also larger as the BO rises, and iii) The volumes of ultramicropores accessible to carbon dioxide increase in small extension with the activation. From the vapours adsorption it can be concluded that the samples with small activation degree have molecular sieve effect. This is due to constrictions that disappear as the activation increases in such a way that the sample with the highest activation does not show this effect.

Acknowledgments

The support of Junta de Andalucía (Proyecto P06-FQM-01585 and Grupo de Investigación RNM342) is acknowledged. M.C.A.A is grateful to Ministerio de Educación y Ciencia (MEC) and Fundación Española para la Ciencia y la Tecnología (FCYT) for the financial support as a postdoctoral contract. M.P.M. thanks MEC for a Ramón y Cajal contract.

References

1. J.J. Freeman, J.B. Tomlinson, K.S.W. Sing, C.R. Theocharis, *Carbon* 1993, **31**, 865.
2. F. Suárez-García, J.I. Paredes, A. Martínez-Alonso, J.M.D. Tascón, *J. Mater. Chem.* 2002, **12**, 3213.
3. F. Suárez-García, A. Martínez-Alonso, J.M.D. Tascón, *Microp. Mesop. Mat.* 2004, **75**, 73.
4. S. Villar-Rodil, F. Suárez-García, J.I. Paredes, A. Martínez-Alonso, J.M.D. Tascón. *Chem. Mater.* 2005, **17**, 5893.
5. S. Villar-Rodil, R. Denoyel, J. Rouquerol, A. Martínez-Alonso, J.M.D. Tascón. *Chem. Mater.* 2002, **14**, 4328.
6. J. Garrido, A. Linares-Solano, J.M. Martín-Martínez, M. Molina-Sabio, F. Rodríguez-Reinoso, R. Torregrosa. *J. Chem. Soc., Faraday Trans. 1* 1987, **83**, 1081.
7. K.S.W. Sing, D.H. Everett, R. Hauel, L. Moscou, J. Rouquerol, T. Sieminieska, *Pure and Applied Chemistry* 1985, **57**, 603.
8. F. Rodríguez-Reinoso, A. Linares-Solano, in *Chemistry and Physics of Carbon*, ed., P.A. Thrower, Marcel Dekker, New York, 1989, vol 21, Chapter 1, p 1.
9. R.V.R.A. Rios, J. Silvestre-Albero, A. Sepúlveda-Escribano,M. Molina-Sabio, F. Rodríguez-Reinoso, *J. Phys. Chem. C* 2007, **111**, 3803.
10. F. Derbyshire, M. Jagtoyen, R. Andrews, A. Rao, I. Martín-Gullón, E.A. Grulke, in *Chemistry and Physics of Carbon*, ed., L.R. Radovic, Marcel Dekker, New York, 2001, vol 27, Chapter 1, p 1.
11. I. Fernández-Morales, M.C. Almazán-Almazán, M. Pérez-Mendoza, M. Domingo-García, F.J. López-Garzón, *Microp. Mesop. Mat.* 2005, **80**, 107.

MONOLITHS: SYNTHESIS AND CHARACTERIZATION USING LIGNOCELLULOSIC PRECURSOR

Diana Paola Vargas[1], L. Giraldo[1], J.C. Moreno-Piraján[2], Yolanda Ladino[3] and Karim Sapag[4]

[1] Department of Chemistry, Universidad Nacional de Colombia, Ciudad Universitaria, Bogotá, Colombia.
[2] Department of Chemistry, Universidad de los Andes, Carrera 1A No. 18A-10, Bogotá, Colombia.
[3] Department of Chemistry, Universidad Pedagógica Nacional, Calle 72 N° 11-86, Bogotá, Colombia.
[4] Department of Physics, Universidad Nacional de San Luis, San Luis, Argentina.

1 INTRODUCTION

In recent years, rapid industrial growth has generated serious environmental problems that are a threat to the sustainability of the planet. As an alternative solution, porous materials are now widely used in the field of adsorption in both gas and liquid phase; one of these materials is activated carbon, which has excellent properties for the adsorption of different pollutant substances.[1] The monolith type refers to compact structures; the most common are the disc and honeycomb-type monoliths. The latter are structural units crossed by parallel lengthwise[2] canals. Honeycombs allow the passage of gases with a very small loss of load; they have a great geometric surface per weight or volume unit; they are easy to handle and have a series of properties to be improved, which make them effective adsorbents and catalytic supports to be used in environment decontamination. One of the main objectives in adsorption studies is to maximize the adsorbent density, minimizing the spaces that are not useful in the storage of certain gases and pollutants (meso-and macroporosity and inter particle spaces), while maintaining a high volume of micropores. In order to reduce these spaces, proper compaction is essential. This can be carried out by using mixtures of grains of various sizes or uniaxially compressing powders (with or without additives), or by preparing compact monoliths using a binder. As has been highlighted in several studies [3-4-5-6-7-8-9], the last option, which involves preparing activated carbon monoliths (MCA), is a way to reduce the space between particles, maximizing the material density.

The aim of this work was to prepare and study MCA monoliths through the chemical activation of coconut shell with zinc chloride, without the use of any binder. The structures were characterized by nitrogen adsorption at 77K, scanning electron microscopy and benzene immersion calorimetry. Finally, the monoliths were tested by adsorption of phenol in an aqueous solution.

2 METHODS AND RESULTS
2.1   Synthesis of discs and honeycombs

The coconut shell was ground and sieved for the synthesis using a particle size of 38 μM. It was then impregnated with a desiccant agent, in this case zinc chloride (1 gram of precursor by 2 ml solution) for 7h at 85° C followed by drying at 110° C for about 2 hours. After this, it was put in an axial press at varying pressure levels; pressing at 150° C formed the molding, obtaining discs and honeycombs by using two different types of moulds. These structures were carbonized in an oven at a temperature of 500° C, at N_2 flow of 85 ml/min and warming at 1°C/min for 2 hours. Finally, the samples were washed with hydrochloric acid 0.1 M and distilled water to neutral pH to eliminate the remnants of the chemical agent used in the impregnation.[6]

Different sets of monolith-type disc (MFD1) and honeycombs (MFH2) were prepared using $ZnCl_2$ concentrations (20% p/v, 32% p/v, 48% p/v) called: MFD120, MFD132, MFD148, MFH220, MFH232, MFH248 while maintaining the other conditions. The degree of impregnation (XgZn/g precursor) obtained for this series was 0.19, 0.3 and 0.46 respectively.

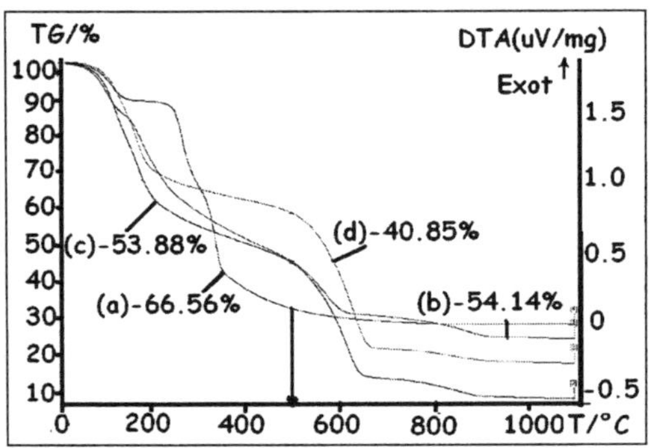

Figure 1 *Thermogravimetric Analysis for coconut shell raw and impregnation with $ZnCl_2$ (a) Raw (b) 20% $ZnCl_2$ (c) 32% $ZnCl_2$ and (d) 48% $ZnCl_2$.*

To assess how the temperature of the raw material and the impregnated sample affected behavior, samples were submitted to a thermogravimetric analysis (TGA). Figure 1 represents the thermogravimetric analysis of the lignocellulosic precursor. After the TGA analysis, the coconut shell was impregnated at the concentrations used in the study and the analysis focused on the temperature of 500°C. It was noted that, in general, the larger the concentration of the impregnating agent, the lower the loss of mass and, as a result, the more room the structure had for the process of carbonization. This situation is evidenced by the rate of mass loss at this point (crude - 66.56% [20%] - 54.14% [32%] - 53.88 % [48%] - 40.85%). Also, the process of dehydration increases sharply with the impregnation and there is a minor release of volatile materials under increasing concentrations of $ZnCl_2$. At temperature of 500°C, the rate of mass loss is about 50%, which makes it possible to obtain a porous solid with good characteristics.

2.2   Characterization and application of the monoliths

All the activated carbon monoliths were characterized by physical adsorption of N_2 at 77 K using a QUANTACHROME, Autosorb 3-B. The microporous volume was calculated by applying the Dubinin-Radushkevich equation and surface area was obtained by the BET method. The samples were also characterized by immersion calorimetry in benzene (0.37nm) using a Calvet[10] in conjunction with scanning electron microscopy (SEM). The implementation of the monoliths was tested for the adsorption of phenol in aqueous solutions of 500ppm-1500ppm. The residual concentration of phenol in the solutions after the adsorption was determined by a UV-VIS Milton Roy Co. Spectronic Genesys SN spectrophotometer.

2.3 Characteristics of the synthesized series

Figure 2 shows the isotherms for honeycombs and discs; type I isotherms are characteristic of microporous solids, according to IUPAC classification. The isotherms show that the specific surfaces are between 725 m^2/g^{-1} and 1523 m^2/g and the microporous volume between 0.39 cm^3/g and 0.79 cm^3/g, satisfactory data comparable to those reported in the bibliography.

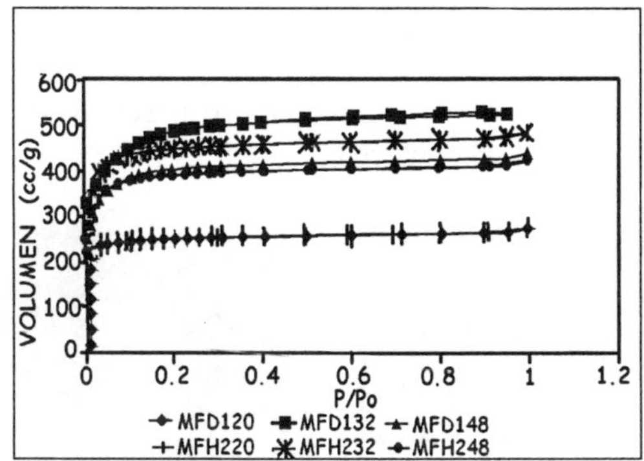

Figure 2 *Adsorption Isotherms of N₂ at 77 K for honeycombs and discs with different ZnCl₂ concentrations*

Table 1 shows the characteristics of the MCA. The yield of the carbonization process increases with a higher concentration of the impregnating agent. This is ratified by the TGA of the precursor at the concentrations used in the study because as the concentration increases in the impregnation agent, the residual mass available for carbonization also rises, due to the larger amount of retained volatile matter. The warming that results from the pressing reduces the loss of volatile matter, increasing resistance. During the heating treatment, the impregnating agent also acts as a drying agent, favoring the release of hydrogen and oxygen to the interior in the form of H_2O and COx or hydrocarbons. The inhibition of the loss of volatile matter and tar fixes the carbon and improves the solid yield. This agent penetrates the interior of the particles, producing a partial fragmentation of the cellulose and other vegetal biopolymers structures such as lignin and hemicellulose. Dehydration and condensation then occur, leading to the production of more aromatic compounds like tars, which remain on the impregnated particle surface and act as a binder.[11] This means that no material need be used to agglomerate the precursor particles, so the porosity of the final material is not affected by the presence of binders after the carbonization. Due to the impregnation of the precursor with $ZnCl_2$, the mass can be deformed with the pressure, reducing space between particles and maximizing the density of the material.

Table 1
Characteristics of the synthesized MCA

Monolith Series	Impregnation Relationship	Microporous Volume DR cm^3/g	BET Area m^2/g	YIELD
MFH220	0.19	0.39	726	45.94%
MFH232	0.30	0.72	1 315	50.83%
MFH248	0.46	0.62	1 170	53.71%
MFD120	0.19	0.45	821	35.09%
MFD132	0.30	0.79	1 523	45.05%
MFD148	0.46	0.65	1 206	51.93%

On the other hand, the microphotographs of the selected samples are shown in figures 3 and 4, where the development of the porosity in both internal and external structures can be seen. It is evident that the discs (Figure 3a) have a more compact structure than the honeycombs (Figure 4b), which show no space between particles. The images are comparable to those obtained in other reports.[12-13]

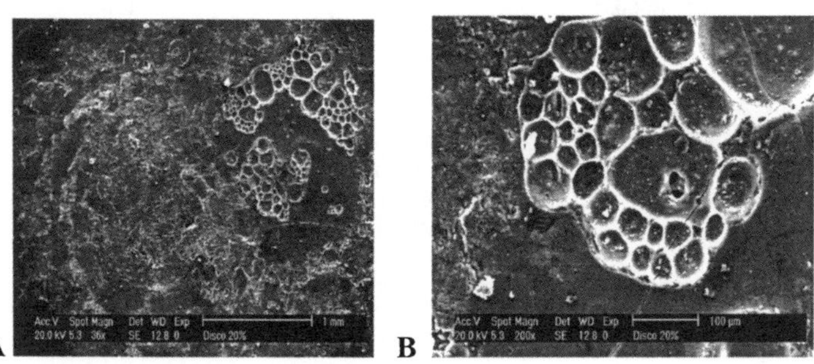

Figure 3 *Microphotographs SEM of discs.*

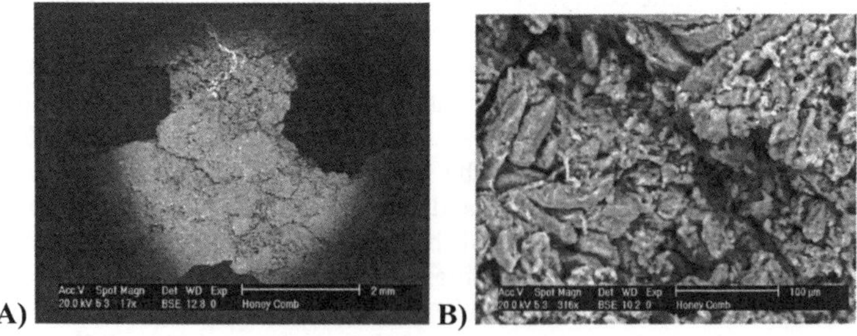

Figure 4 *Microphotographs SEM of honeycombs*

Table 2 shows the data obtained for the MCA immersion enthalpy into benzene, which are among 83.52 J/g-164.21 J/g. The immersion enthalpy is indicative of the solid- immersion liquid interaction. This value corresponds to the total amount of heat generated in the solute adsorption process and therefore we can conclude that there is interaction between the solid and the solvent. The results show an increase in the enthalpy of samples with the BET area (Figure 5), which is consistent with the bigger surface that can interact with the adsorbate. The correlation coefficients in this figure show that the data are consistent with the linearity.

Table 2.
Characteristics of synthesized MCA (immersion calorimetry and adsorption of phenol).

Monoliths Series	- ΔHinm (J/g)	Phenol Uptake (mg/g)	1500 ppm	500 ppm
MFD120	90.25		24.04	7.38
MFD132	164.21		78.05	32.41
MFD148	132.04		55.52	22.32
MFH220	83.52		39.69	20.09
MFH232	144.13		108.61	35.91
MFH248	115.53		93.61	31.33

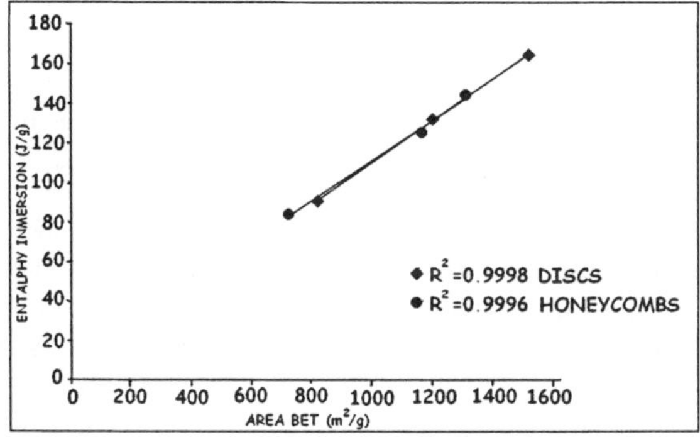

Figure 5 *Immersion Enthalpy into Benzene of the Monoliths at different BET Areas.*

2.4 Adsorption of phenol in aqueous solution

The MCA were tested as potential phenol adsorbents in the aqueous phase. Data (Table 2) show that the honeycomb-type monolith series has a greater phenol retention capacity than disc-type monoliths. This can be supported by the fact that the honeycomb channels facilitate the adsorption of this molecule as they reduce the transport problems in solutions. Figure 6 show that a larger surface area increases the retention of the phenol; in addition, the data show a clear linear performance. A relationship can be established between the surface chemistry of the MCA and the adsorption capacity of phenol. As the adsorption of phenolic compound in the monoliths occurs because the monoliths favor certain interactions that generate a change in the system enthalpy, the adsorbed quantity increases and immersion enthalpy occurs. Both of these processes are affected by the total content of acidic and basic groups in the activated carbon surface. When the total content of acidic groups decreases, the adsorbed amount of the phenolic compound increases, since this compound interferes less with the π electrons of the graphenic layers on the surface of the activated carbon and instead interacts with the π electrons of the aromatic ring of the solutes[14].

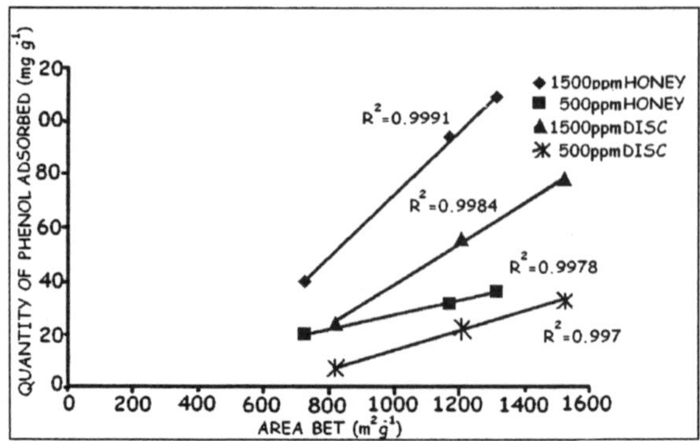

Figure 6 *Phenol adsorbed vs. Monoliths BET Area*

3 CONCLUSIONS

$ZnCl_2$ can be used as an impregnating agent for the synthesis of activated carbon monoliths using coconut shells as the precursor material. The obtained structures have good mechanical and textural properties achieved by the heating that takes place in the different production steps. The structures could be used for the adsorption of different compounds.

The molding temperature in the pressing process is a key aspect in the synthesis of the structures. 150°C is the appropriate temperature because of the scarce loss of volatile matter that favors its resistance. Six series of activated carbon monoliths (discs and honeycombs) were prepared, obtaining BET areas between 702 m^2/g and 1523 m^2/g, together with a micropore volume between 0.38 cm^3/g and 079 cm^3/g. Better surface and adsorption characteristics for phenol in aqueous solution were achieved for the MFD132 (discs) and MFH232 (honeycombs) series, which were obtained under the same conditions, differing only in the shape of their structures.

The influence of impregnation was evaluated in the BET area and in terms of the microporous volume; a maximum value was reached in the development of the microporosity and surface area, reaching 0.3 grams of Zn per coconut shell gram. An increase was also observed in the carbonization process yield, which led to a rise in impregnation attributed to a greater residual mass available for the carbonization that comes from a larger amount of retained volatile matter as evidenced by thermogravimetric analysis.

The immersion calorimetry into benzene for the samples show enthalpies among 83.52J/g-164.21J/g, which are proportional to the BET area, where higher heats of wetting were obtained for the series MFD132 and MFH232.

Finally, the synthesized activated carbon monoliths were found to affect the phenol adsorption in aqueous solution, with retained amounts between 7.38mg/g and 35.91mg/g for phenol solutions with 500ppm and between 24.02 mg/g and 108.61 mg/g for phenol solutions with 1500ppm. Despite the fact that the honeycombs retain greater quantities of phenol than discs with a similar area, both reduce the concentration of this molecule in the solution.

ACKNOWLEDGEMENTS

The authors wish to thank the Master Agreement established between the "Universidad de Los Andes" and the "Universidad Nacional de Colombia" and the Memorandum of Understanding entered into by the Departments of Chemistry of both universities.

References

1 Dinesh. Mohan, Kunwar. P. Singh and Vinod. K. Journal of Hazardous Materials., 2008, **152**, 1045-1053.

2 M. Yates, J. Blanco, P. Avila and M.P Martin. Microporous and Mesoporous Materials., 2000, **37**, 201-208.

3 José Manuel Gatica, José María Rodríguez Izquierdo, Daniel Sánchez; Tarik Chafik, Sanae Harti, Hicham Zaitan and Hilario Vidal. Comptes Rendus Chimie., 2006, **9**, 1215-1220.

4 J.Blanco; P. Ávila and Martin. MP. Microporous. and Mesopprous Materials., 2000,**37**, 201-208.

5 Krisztina László, György Onyestyák, Cyrille Rochas and Erik Geissler. Carbon., 2005, **43**, 2402-2405.

6 Y. Nakagawa, M. Molina-Sabio and F. Rodríguez-Reinoso. Microporous and Mesoporous Materials., 2007, **103**, 29-34.

7 Laishuan. Liu, Zhenyu. Liu, Jianli. Yang, Zhanggen. Huang and Zenghou Liu. Carbon., 2007, **45**, 2836-2842.

8 Agustín. F, J. Pérez-Cadenas, Freek. Kapteijn, Martijn M.P, Zieverink and Jacob A. Moulijn. Catalysis Today., 2007,**128**, 13-17.

9 Laishuan. Liu, Zhenyu. Liu, Zhanggen. Huang, Zenghou. Liu and Pingguang and Liu. Carbon., 2006, **44**, 1598-1601.

10 J. Silvestre-Albero, C. Gómez de Salazar, A. Sepúlveda-Escribano, F. Rodríguez-Reinoso, Colloid. and Surf. A: Phys. and Eng., 2001, **187-188**, 151-165

11 M. Molina Sabio, C. Almansa, F. Rodríguez-Reinoso. Carbon., 2003, **41**, 2113-2119

12 D. Lozano-Castelló, D. Cazorla-Amorós, A. Linares-Solano, and D.F. Quinn. Carbón., 2007, **40**, 2817-2825.

13 Ruth. Ubago-Pérez, Francisco. Carrasco-Martín, David. Fairén-Jiménez, Carlos and C Moreno-Castilla. Microporous and Mesoporous Materials., 2006, **92**, 64-70.

14 C. Moreno-Castilla. Carbon., 2004, **42**, 83-94.

MICROPOROSITY OF ACTIVATED CARBONS OBTAINED FROM PET

M.C. Almazán-Almazán[a], M. Domingo-García[b], M. Pérez-Mendoza[b], I. Fernández-Morales[b], F.J. López-Garzón[b]

[a]Dep. Chemical Engineering, University of Liège, Sart-Tilman, B-4000, Liège, Belgium
[b]Dep. Inorganic Chemistry, University of Granada, Fuentenueva s/n, 18071 Granada, Spain

1 INTRODUCTION

A high percentage of solid wastes in the world is constituted by plastics. PET (polyethylene terephthalate) represents more than 20% of the whole of these plastic residues. Hence, its recycling is prioritised. One of the traditional ways to re-use PET is the chemical recycling by solvolysis, i.e. by hydrolysis[1], although it has also been well described its use as precursor of carbonaceous materials[2,3].

In previous works, it was proposed a new method for the preparation of active carbons by the combination of these two ways of recycling, i.e. the basic hydrolysis with KOH and the subsequent pyrolysis of the resulting products[4,5]. In this way, the organic salts obtained by the de-polymerisation process are the carbon precursors, while the alkali-metal hydroxides behave as chemical activation agents during the pyrolysis[4,5]. This method allows avoiding some of the problems that appear during the direct pyrolysis of PET and subsequent physical activation with CO_2, such as the formation of organic volatile compounds, i.e. benzoic and terephthalic acids, which are released to the atmosphere. In these works, the reactions taking place during the thermal treatment, which explain the formation of the carbonaceous materials and their activation, were proposed. It was also shown that the activated carbons so prepared present high adsorption capacity and a relatively homogenous and narrow microporosity. Nevertheless, it has not been carried out a complete analysis of the influence of all experimental parameters involved in the carbonisation and activation processes.

In the present work, it has been performed an exhaustive evaluation of these experimental parameters, i.e. the KOH/residue ratio, the nitrogen flow rate, the soak time and the final temperature of the thermal treatment. The main objective is to deeply analyse the influence of these parameters on the evolution of the porous texture of the activated carbons obtained and, finally, to chose those conditions which optimise their textural characteristics.

2 EXPERIMENTAL

2.1 Experimental set-up

The de-polymerisation process of PET was carried out under the conditions previously described[5]. Four different series of samples have been obtained by varying the carbonisation conditions. In the first one, the carbonisation temperature was fixed at 700 °C, the residence time was 1 hour and the carrier gas (N_2) flow rate was 100 $cm^3 \cdot min^{-1}$, while the KOH/residue ratio (solid residue after depolymerisation) was varied: 1/1, 2/1 and 3/1. In the second series, the KOH/residue ratio was fixed to 1/1 and the N_2 flow rate was varied (100, 200 and 300 $cm^3 \cdot min^{-1}$), keeping the rest of the conditions unchanged. The third series was prepared by setting the temperature to 700 °C, the KOH/residue ratio to 1/1, the N_2 flow rate to 300 $cm^3 \cdot min^{-1}$ and the residence time to 1, 4 and 8 hours. Finally, a fourth series was obtained in the same way as the third one but using a carbonisation temperature of 800 °C instead.

The so prepared samples were labelled as follows: HKT-t-F. Where HK indicates the nature of the precursor, i.e. obtained by basic hydrolysis with KOH, T is the final temperature of the carbonisation process in Celsius degrees, t represents the soak time of this process and F refers to the nitrogen flow rate. When the KOH/residue of the hydrolysis ratio was changed, the nomenclature was given by HKT-t-R, where R corresponds to this ratio.

The textural characteristics of the AC were derived through the application of Dubinin-Radushkevich (DR) equation to the adsorption isotherms of N_2 at 77 K, and CO_2 at 273 K. These isotherms were obtained in a Micromeritics ASAP 2020 apparatus and in a classical volumetric system, respectively.

2.2 GCMC simulation of gas adsorption to generate a representative PSD

The characterisation of the AC under study was completed obtaining a representative pore size distribution (PSD). With this objective, the experimental adsorption data were combined with those obtained by Grand Canonical Monte Carlo (GCMC) simulation method[6-8].

Monte Carlo simulation in the grand canonical ensemble can be used to generate fluid adsorption information representative of real adsorption experiments. This method keeps the chemical potential, the temperature and the volume of the system constant while the number of molecules in a certain simulation cell, considered as a pore model of defined geometry, is allowed to fluctuate until equilibrium is reached, as occurs in experimental measurements.

Nitrogen adsorption isotherms on slit shaped pores of various widths (from 0.3 to 4.2 nm) were obtained by this method. This set of simulated isotherms was combined with the experimental isotherms, following the implementation proposed by Davies and Seaton[6]. This permits to solve the adsorption integral equation by numerical inversion and, hence, to obtain a discrete representation of the PSD that fits the real isotherm.

3 RESULTS AND DISCUSSION

3.1 Adsorption of N_2 and CO_2

The first variable analysed in the carbonisation process was the KOH/residue ratio. It is worth to note that the choice of an optimum KOH/residue ratio is a determinant factor because the study on the influence of the rest experimental parameters will be carried out on samples prepared with this ratio. Table 1 collects the textural parameters obtained by applying the DR equation to the adsorption isotherms of N_2 and CO_2 for this series of samples. The values of micropore volume, V_0, obtained by nitrogen adsorption, correspond to the application of the equation to the 10^{-5}-10^{-2} pressure range. The saturation volumes, V_S, for N_2 adsorption have been calculated at 0.9 relative pressure.

Table 1 *DR parameters for the samples obtained at various KOH/residue ratios*

Sample	Adsorbate	V_0 (cm^3 g^{-1})	V_S (cm^3 g^{-1})	E_0 (kJ mol^{-1})	L_0 (nm)
HK700-1-1/1	Nitrogen	0.18	0.23	24.13	0.85
	CO_2	0.16		32.50	0.51
HK700-1-2/1	Nitrogen	0.43	0.68	17.98	1.34
	CO_2	0.20		26.99	0.69
HK700-1-3/1	Nitrogen	0.48	0.72	18.61	1.29
	CO_2	0.44		26.39	0.72

Table 1 shows how the micropore adsorption capacity of both adsorbates is much higher when the KOH/residue is raised from 1/1 to 3/1. However, this increase produces a wider and heterogeneous micropore system as it can be concluded from the increase in L_0 values ($L_0(N_2)$ = 1.3 nm and $L_0(CO_2)$ = 0.7 nm) and in the important difference between the volume determined by DR for the N_2 adsorption, V_0, and the saturation volume, V_S. It is well known that in a chemical activation process coexist two mechanisms: the micropore formation and the widening of these created micropores[9]. In the table, it can be seen that a KOH/residue ratio of 3/1 produces these two effects in a large extension, although the pore widening is more significant. Nevertheless, there is an important decrease in carbon yield from 15.2% for the 1/1 sample to 6% in the 3/1 sample. Therefore, the 1/1 ratio was chosen as precursor to carry out the posterior studies.

Once the KOH/residue ratio was fixed, the effect of the nitrogen flow rate during the pyrolysis was studied. It was varied from 100 cm^3·min^{-1} to 300 cm^3·min^{-1}. The textural characteristics of the resulting samples are presented in Table 2:

Table 2 *DR parameters for the samples at various nitrogen flow rate*

Sample	Adsorbate	V_0 (cm^3 g^{-1})	V_S (cm^3 g^{-1})	E_0 (kJ mol^{-1})	L_0 (nm)
HK700-1-100	Nitrogen	0.18	0.23	24.13	0.85
	CO_2	0.16		32.50	0.51
HK700-1-200	Nitrogen	0.41	0.51	23.93	0.86
	CO_2	0.26		29.52	0.59
HK700-1-300	Nitrogen	0.38	0.42	27.85	0.66
	CO_2	0.34		30.64	0.56

It can be observed that an increase in the nitrogen flow rate produces an improvement of the adsorption capacity of N_2 and CO_2, without any widening of the existent microporosity, presenting small differences between V_S and V_0, specially in the case of the sample prepared at 300 $cm^3 \cdot min^{-1}$, which implies a quite homogeneous microporous system. This could be explained taking into account two factors: i) the faster removal of gases evolved during the carbonisation process favours the creation of narrow micropores[9], ii) when a higher nitrogen flow rate is used, the potassium formed in the reactions which take place between KOH and the carbon matrix can be removed. All these reactions are explained in reference 5 and references therein. This removal involves a displacement of the equilibrium, provoking an increase in the extension of the reaction between the carbon and KOH. On the other hand, the carbon yields for the three samples decreases slightly from 15.2% for the sample obtained at 100 $cm^3 \cdot min^{-1}$ to 10.5% for the sample obtained at 300 $cm^3 \cdot min^{-1}$. Finally, the flow rate of 300 $cm^3\ min^{-1}$ was considered as the optimum one.

Once the KOH/residue ratio and the N_2 gas flow rate were fixed, and the temperature of pyrolysis kept constant, a series of samples varying the soak time was obtained. The textural characteristics of these carbons are shown in Table 3.

Table 3 *DR parameters for the samples at various soak times*

Sample	Adsorbate	V_0 ($cm^3\ g^{-1}$)	V_S ($cm^3\ g^{-1}$)	E_0 (kJ mol^{-1})	L_0 (nm)
HK700-1-300	Nitrogen	0.38	0.42	27.85	0.66
	CO_2	0.34		30.64	0.56
HK700-4-300	Nitrogen	0.64	0.77	22.32	0.99
	CO_2	0.48		25.24	0.78
HK700-8-300	Nitrogen	0.60	0.68	22.40	0.98
	CO_2	0.44		25.80	0.75

Table 3 shows that when the temperature is kept constant at 700 °C, an increase of the soak time from 1 to 4 hours produces a development of not only the wider microporosity measured with N_2, but also of the narrow micropores accessible to CO_2. This fact suggests that the increase of the soak time creates narrow micropores and widens the existent ones, obtaining a more heterogeneous system, as it can be deduced from the increase in the difference between V_S and V_0 values. Nevertheless, longer times do not improve the textural characteristics of the samples, since the data in Table 3 of sample HK700-8-300 are very similar to those of sample HK700-4-300. It is possible that, under the reaction conditions, the activating agent KOH can be completely consumed after four hours, and remaining only the K_2CO_3 as a by-product of the reaction between the KOH and the carbon matrix[5,10]. The thermal decomposition of the carbonate into K_2O (which also reacts with the carbon matrix) can start at temperatures as low as 700 °C in the presence of carbon, but under these conditions is still very slow. Hence, it seems that at this temperature, the amount of K_2O produced is not high enough[5] to react with the carbon during the longest soak time.

Finally, when the temperature of pyrolysis was 800 °C (Table 4), a treatment during 1 hour is enough to obtain a carbon with better textural characteristics than those prepared at 700 °C and 4 hours. This can be explained considering that the decomposition of K_2CO_3 into K_2O commences to be important at 800 °C so the reaction between the carbon and KOH and K_2O is activated, enhancing the gasification of the carbon matter and the formation of potassium. The potassium formed, whose boiling point is below 800 °C, can intercalate into the carbonaceous matrix increasing the structure distortion and improving the volume of micropores[5,10]. As it can be observed in Table 4, the increase in residence time from 4 to 8 hours has a nearly negligible effect on the micropore volumes determined by applying the DR method to N_2 adsorption and on those determined by CO_2 adsorption.

Nevertheless, the total pore volumes, V_S, determined at relative pressure 0.9, increase markedly, indicating very heterogeneous pore networks.

Table 4 DR parameters for the samples at different temperature and soak times

sample	adsorbate	V_0 (cm^3 g^{-1})	V_S (cm^3 g^{-1})	E_0 (kJ mol^{-1})	L_0 (nm)
HK800-1-300	nitrogen	0.70	0.87	22.60	0.96
	CO_2	0.48		24.95	0.80
HK800-4-300	nitrogen	0.72	0.99	20.99	1.13
	CO_2	0.47		24.39	0.83
HK800-8-300	nitrogen	0.78	1.35	18.44	1.30
	CO_2	0.47		23.41	0.90

3.2 PSDs obtained by GCMC simulation of gas adsorption

More information to complete the study on the evolution of the porous texture of the ACs prepared by KOH hydrolysis of PET can be drawn out from their PSD curves. These have been obtained by fitting the simulated nitrogen isotherms on single-sized pore models to the experimental ones following the method proposed by Davies et al.[6] The PSDs corresponding to the samples under study are represented in Figures 1a to 1d. The discussion will focus on the microporous range as this is the most interesting part of the PSDs for the textural characterisation of these samples and also the most reliable part of the PSD according to the method employed[6].

Figure 1a collects the PSD curves for the series of samples obtained varying the KOH/residue ratio. The increase in the heterogeneity of the microporous system in samples prepared with 2/1 and 3/1 ratios is clearly observed in this figure, in which three peaks are observed in the PSDs for these two samples. Those at higher sizes (1.3 and 2.0 nm) are more important. This also explains the difference between V_0 and V_S collected in Table 1. Meanwhile, the sample with 1/1 ratio does not show the peak at 2.0 nm and the one at 1.3 nm is very small.

The samples obtained by changing the nitrogen flow rate (Figure 1b) present two peaks, i.e., a bimodal pore system, although the peak corresponding to supermicropores (1.3 nm) is significantly less important than the first one for the three samples. When N_2 flow rate is increased, the peak corresponding to the narrower micropores increases and is maximum for the sample prepared at 300 cm^3·min^{-1}, while the second one (1.3 nm) diminishes. All these results agree with the differences between the DR micropore volumes and the V_S which appear in Table 2.

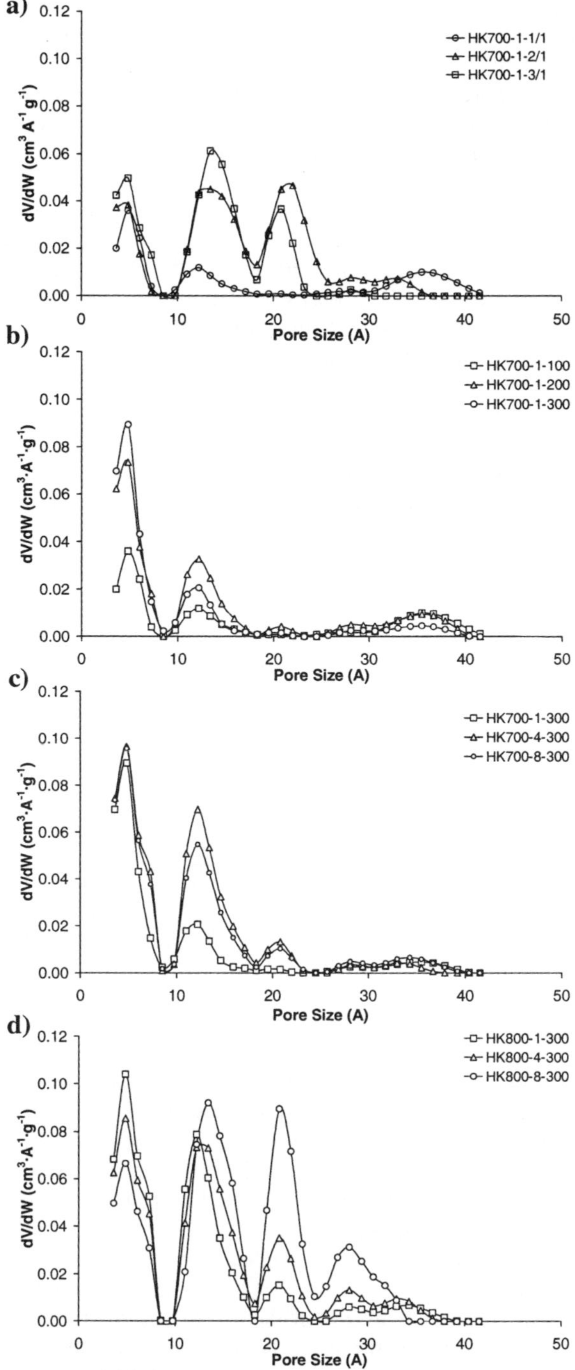

Figure 1 *PSDs obtained from N₂ adsorption at 77 K*

In the third series (varying the soak time), Figure 1c shows how the increase in residence time does not affect the volume of micropores smaller than 1 nm. Nevertheless, extending the time from 1 to 4 hours, the peak corresponding to wider micropores (1.3 nm) increases markedly, appearing also a third peak at sizes larger than 2.0 nm, in the small mesopores range. Increasing the time from 4 to 8 hours does not affect the PSDs curves, except for a slight decrease in the supermicropore volume. All this is also in agreement with data showed in Table 3.

Finally, when the soak time is studied at 800 °C, it has been obtained a series of samples with a clearly polymodal behaviour which becomes more remarked with the carbonisation time (Figure 1d). The curves point out the development of the micropore volume and the high heterogeneity of the porosity of this series of samples. The PSDs present three mean peaks at 0.5, 1.3 and 2.0 nm. It can be seen how the microporosity widens for longer times. Similarly as for the former series, these results are in good agreement with those collected in Table 4.

In Figure 6 it is represented, as an example, the fitting of the simulated isotherms based on the proposed PSDs to the experimental data corresponding to the samples obtained at 700 °C and different soak times. In this figure pressures are on logarithmic scale. It can be seen the considerable goodness of the fitting in the whole range of pressures.

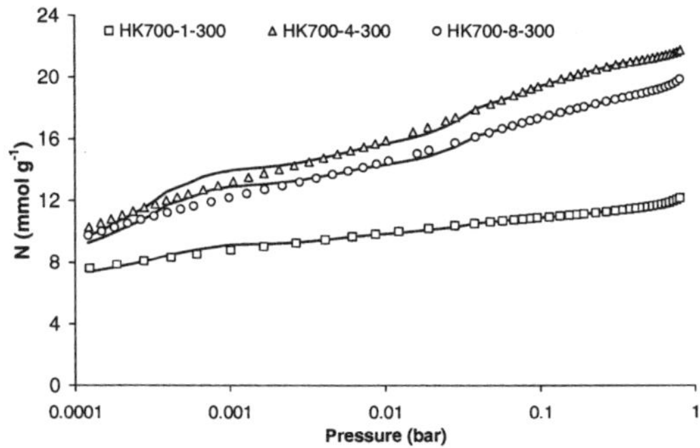

Figure 2 *Fitting of simulated adsorption isotherms (lines) to the experimental data*

4   CONCLUSIONS

The method previously proposed to obtain activated carbons from the products resulting of the de-polymerisation by the basic hydrolysis of PET has been optimised.

The KOH/residue ratio, the sweep gas flow rate, the temperature and the soak time have an important influence on the textural characteristics of the ACs prepared. The analysis of the experimental data and the results obtained by PSD simulated curves are in agreement in all cases, and they confirm that:

a)   The higher the KOH/residue ratio, the higher the micropore volume and the heterogeneity of the micropore size distribution, since an increase of this ratio produces a significant development and widening of the microporosity. It can be attributed to the double effect of the chemical activation process, favoured by the increasing amount of chemical reagents, i.e. the micropore formation and the opening of these created micropores.

b)   The study of the influence of the nitrogen flow rate reflects that it develops an homogeneous microporosity, probably due to that the faster removal of gases permits the creation of narrow micropores or to a displacement of the equilibrium of the reactions taking place in the carbon matrix.

c)   As the soak time passes from 1 to 4 h, keeping the temperature constant at 700 °C, there exists an enhancement of the narrow and wider microporosity. It suggests the creation of narrow micropores and the opening of the existent ones. Nevertheless, it seems that the gasification does not continue above 4 h of thermal treatment since the amount of chemical reagents is insufficient for longer times.

d)   At 800 °C, the extension of the reactions between the carbon and the activation agents is more important. An increase from 1 to 4 h provokes a large improvement of the microporosity, but when the soak time is prolonged, it is produced predominantly a widening of the pre-existent micropores.

After the complete study of the experimental parameters involved in the preparation of ACs by this method, it can be concluded that the carbon obtained with a KOH/residue ratio of 1/1 at 800 °C, with a nitrogen flow rate of 300 cm^3min^{-1} and during 1 hour presents interesting textural characteristics, improving those of the samples already described obtained by the same method[5] or by physical activation at 950 °C and 8 hours[2].

Acknowledgments

The authors acknowledge the economical support of Junta de Andalucía (Proyecto de Excelencia P06-FQM-01585 and Grupo de Investigación RNM342). M.C.A.A is grateful to Ministerio de Educación y Ciencia (MEC) and Fundación Española para la Ciencia y la Tecnología (FCYT) for the financial support as a postdoctoral contract. M.P.M. thanks MEC for a Ramón y Cajal contract.

References

1.   D. Paszun and T. Spychaj, Industrial Engineering Chemical Research, 1997, 36, 1373.

2.   I. Fernández-Morales, M.C. Almazán-Almazán, M. Pérez-Mendoza, M. Domingo-García and F.J. López-Garzón, Microporous and Mesoporous Materials, 2005, 80, 107.

3.   J.B. Parra, C.O. Ania, A. Arenillas, F. Rubiera and J.J. Pis, Applied Surface Science, 2004, 238, 304.

4.   F.J. López-Garzón, M. Domingo-García, I. Fernández-Morales, M. Pérez-Mendoza, M.C. Almazán-Almazán. International Patent n° WO2007/06778.

5.   M.C. Almazán-Almazán, M. Pérez-Mendoza, F.J. López-Domingo, I. Fernández-Morales, M. Domingo-García and F.J. López-Garzón. Microporous and Mesoporous Materials, 2007, 106, 219.

6.   G.M. Davies, N.A. Seaton and V.S. Vassiliadis, Langmuir, 1999, 15, 8235.

7.   M. Pérez-Mendoza, J. González, P.A. Wright and N.A. Seaton, Langmuir, 2004, 20, 7653.

8.   M. Pérez-Mendoza, C. Schumacher, F. Suárez-García, M.C. Almazán-Almazán, M. Domingo-García, F.J. López-Garzón and N.A. Seaton, Carbon, 2006, 44, 638.

9.   D. Lozano-Castelló, M.A. Lillo-Ródenas, D. Cazorla-Amorós and A. Linares-Solano, Carbon, 2001, 39, 741.

10. H. Teng and L.Y. Hsu, Industrial and Engineering Chemistry Research, 1999, 38, 2947.

LOW PRESSURE HYSTERESIS IN THE NITROGEN ADSORPTION ISOTHERM OF ZSM-5: EFFECT OF ACID TREATMENT

Elpiniki Panayi, and Charis R. Theocharis*

Porous Solids Group, Department of Chemistry, University of Cyprus, P.O. Box 20537, 1678, Nicosia, Cyprus

1. INTRODUCTION

Zeolites have long been of interest[1-3] in academic but also applied and industrial research, because of their significant adsorptive, catalytic, and ion-exchange properties. In particular, ZSM-5 has been of interest to our group[4,5] because of the unusual low-pressure hysteresis present in the nitrogen adsorption isotherm for this solid, at certain Si:Al ratios but not others. These solids have the general formula $Na_n[Si_{96n}Al_2O_{192}].16H_2O$ $(n<8)$[6,7] and can be synthesized with a wide range of possible Si:Al ratios. This ratio has a profound effect on the surface solid state properties of the solid, affecting among others, the number and strength of surface acid sites, and the charge of the framework and thus the number of exchangeable cations present in the channels. With increase in the Si:Al ratio, the thermal stability of the solid is reduced, and the nature of the surface changes from hydrophilic to hydrophobic. The catalytic behaviour of the solid is therefore influenced[8], and can at some extent be fine-tuned by adjusting the Si:Al ratio.

An important factor in making zeolites such an important class of adsorbents and catalysts is the fact that their microporosity is intracrystalline since it is formed as part of their crystal structure, and not as a result of aggregation or agglomeration of primary particles. However, in order to improve the mass flow characteristics of these solids in catalytic applications, it is often necessary to use small crystals, or more usefully create extra porosity in the solid[9-16]. The additional porosity is usually introduced by various post synthetic treatments, such as the suspension of the solid in acid or base, a treatment which leads to dealumination[13] in the former case, or Si removal in the latter[14-16].

In the present paper, we present an investigation of the effect of acid treatment on the adsorptive properties of samples of Na-ZSM-5, and especially on the shape and other properties of the low-pressure hysteresis loop. The impetus of the study was that the process of dealumination leads to a partial collapse of the zeolitic structure, and therefore, presumably, the introduction of structural faults in the structure. It would therefore be possible to examine the effect of such faults in the low-pressure hysteresis loop. The acids chosen for this study were HCl, H_2SO_4 and CH_3COOH. Both calcined and non calcined zeolite samples were used, as well as two different Si:Al ratios, namely 80 and 120. It has already been established that both the Si:Al ratio and the process of calcination influences the behaviour of zeolites vis-à-vis the presence of H^+ [17-20].

2. EXPERIMENTAL

The samples used in this study were commercially available, and were kindly provided by Degussa. In each experiment, 1 g of solid was suspended in 20 mL of the appropriate acid. After vigorous shaking for 5 min, the suspension was allowed to stand for various amounts of time, after which it was centrifuged, washed and dried at 373K for 24h. The parameters that were changed were acid concentration and contact time. Nitrogen adsorption isotherms were carried out using a Micromeretics ASAP 2000 automated apparatus at 77K, from a starting pressure of 0,13Pa. Prior to adsorption isotherm determination, samples were outgassed at 0,13Pa overnight at 373K. FTIR spectra were obtained using an 8500 Shimadzu Spectrometer.

3. RESULTS AND DISCUSSION

The Na-ZSM-5 samples with Si:Al 80 as provided, had still significant amounts of matrix within their pores, and therefore treatment with HCl or H_2SO_4 had no effect on their surface properties or pore volume. This is presumably because the matrix molecules would block access to the inner surface of the H^+ ions, and thus no dealumination would take place.

By contrast, acid treatment of a calcined sample of Na-ZSM-5 with Si:Al 80, a procedure known to lead to the removal of the matrix, led to noticeable differences in the N_2 adsorption isotherm of the samples, as seen in Figure 1. In the case of acetic acid, a change in the slope of the isotherm was observed at p/p° below 0.4, as well as a very significant drop in the BET surface area compared both with the HCl-treated and the non-acid treated samples. This could be attributed to the production of insoluble aluminum acetate $(CH_3COO)_3Al$ which precipitated in and on the sample and thus blocked the pores. The Al3+ ions released from the sample by the H+ ions formed the insoluble acetate.

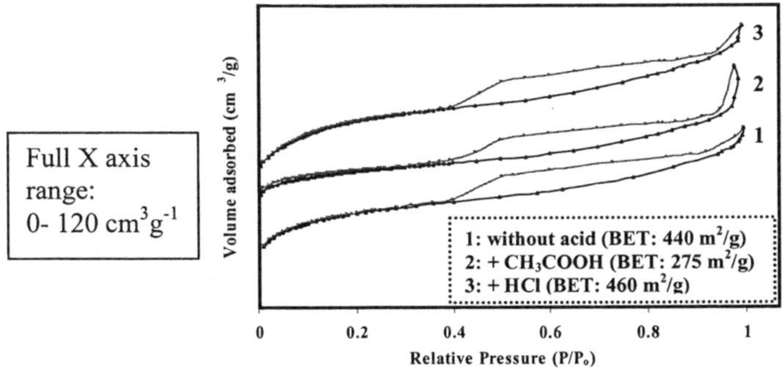

Figure 1: *Comparison of Nitrogen Adsorption Isotherms and BET surface area for (a) calcined Na-ZSM-5 (Si:Al 80), and after treatment (b) with HCl and (c) CH₃COOH.*

From previous work, it was know that Na-ZSM-5 with Si:Al 80 did not exhibit low-pressure hysteresis[5]. However, it is noticeable from examination of Figure 1 that acid treatment led to a change in the shape of the isotherm at very low partial pressures. Such changes cannot be due, in this case, to changes in pore size, but would be due to changes to the adsorbate-adsorbent interactions. Removal of Al^{3+} leaves behind imperfections in the walls of the zeolitic channel that may be able to act as sites of adsorption. On the other hand, Al ion removal would lead to a change in the overall charge carried by the framework. Both these factors would lead to a change in the shape of the isotherm. The difference in action between the two acids would be linked with delumination taking place to different extents due to pH differences.

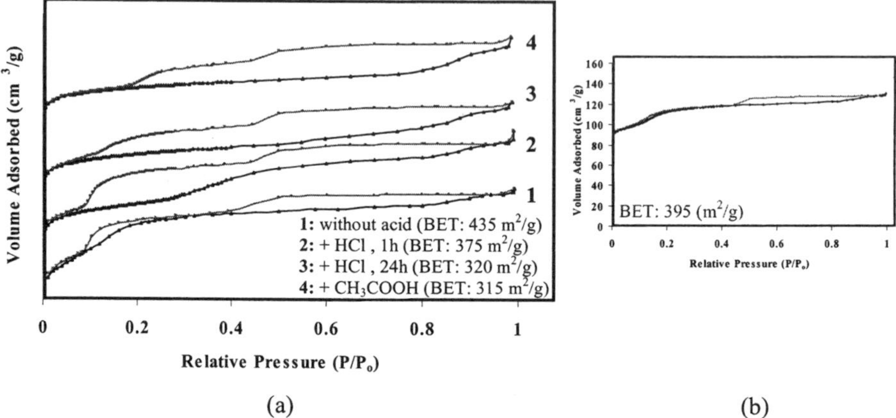

(a)   (b)

Figure 2: *Nitrogen adsorption isotherms of (a) uncalcined samples of Na-ZSM-5 (Si:Al 120) and of the same sample upon treatment with HCl and CH₃COOH, (b) uncalcined sample after treatment for 1h with HCl for 1h followed by outgassing at 673K. For (a), X axis range was 0-160 m^2g^{-1}*

For both the calcined and non-calcined Na-ZSM-5 with Si:Al 120 acid treatment had a very significant effect on their surface properties. This sample was known not to contain any significant amounts of matrix in its pores, as supplied by the manufacturer. From Figure 2, it can be seen that after acid treatment there was a significant distortion of the adsorption isotherm, as well as a significant reduction in the BET surface area. The extent of the distortion depended upon the length of the acid treatment. Upon re-evacuation of the sample at 673K, the hysteresis loop disappeared and the shape of the isotherm at low partial pressures was altered. After this treatment the BET surface area was restored to a value near that before treatment. Comparison of isotherms 1, 2 and 3 in Figure 2 indicates that the low pressure hysteresis moves to higher partial pressures upon acid treatment and is elongated with the result that it joins up with the loop at higher pressures which can be attributed to secondary mesopores created by aggregation of the zeolitic particles. The low-pressure hysteresis loop has been interpreted by Sing *et al*, Rouquerol *et al* and Unger *et al*[21-24] as being due to a reorganisation of the adsorbed molecules in the micropores. We previously argued[4,5,25] that the interaction of the adsorbed molecules with the channel walls would play a role in the occurrence of the hysteresis loop, and more specifically would be influenced by small changes in the local pore geometry and the adsorbent-adsorbate interaction forces. Such perturbations would be introduced by crystallographic faults in the walls, including missing cations, so-called dangling bonds, and surface groups such as –OH. It is suggested that the observations here are consistent with that interpretation, since it is well-established that acid treatment causes dealumination and loss of crystallinity in the zeolite. Water molecules would be strongly adsorbed at these faults, thus altering the local geometry of the pore. Outgassing at the higher temperature would remove the strongly held water molecules, causing another change in the shape of the isotherm, as shown in Figure 2b. This isotherm did not exhibit low-pressure hysteresis, as this sample was now H^+-ZSM-5 a system known not exhibit low-pressure hysteresis[5]. From isotherm 4 in Figure 2a, it can be gleaned treatment with acetic acid resulted in an even bigger change in the shape of the hysteresis loop, as well as a further shift to higher partial pressures. This may be attributed to the deposition of aluminium acetate in the pores and on the external surface of the sample. Given this enhanced effect further investigation of this system was carried out, and is presented later in this paper. Figure 3 shows the adsorption isotherms for acid treated and pristine calcined Na-ZSM-5 (Si:Al 120). The changes observed were similar to those seen in Figure 2.

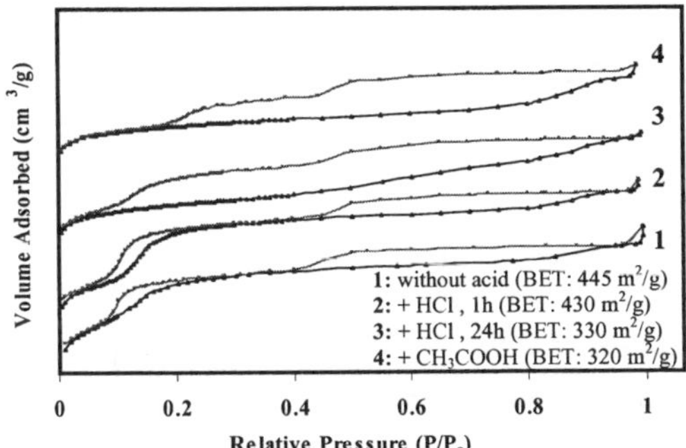

Figure 3 *Nitrogen adsorption isotherms of calcined samples of Na-ZSM-5 (Si:Al 120) and of the same sample upon treatment with HCl and CH₃COOH. X axis range 0-180 cm³g⁻¹*

Increasing the length of acid treatment of the zeolitic sample made the effects discussed previously, more enhanced. This can be seen, by comparing isotherms 2 and 3 in Figure 2. The effect was more marked in the case that obtains for acetic acid, where with more contact time, more aluminium acetate was deposited. This can be confirmed by examining the FTIR spectra, presented in Figure 4. The band at 3670 cm⁻¹ which has been attributed to structural OH groups reduces in intensity, presumably because of their binding with the insoluble acetate[26]. The production of aluminium acetate, is further confirmed by the peaks close to 1500 cm⁻¹ (C-H) and 1740 cm⁻¹ (C=O)[27]. Comparison of the spectra in Figure 4, confirms that more acetate is formed by increasing contact time with the acid.

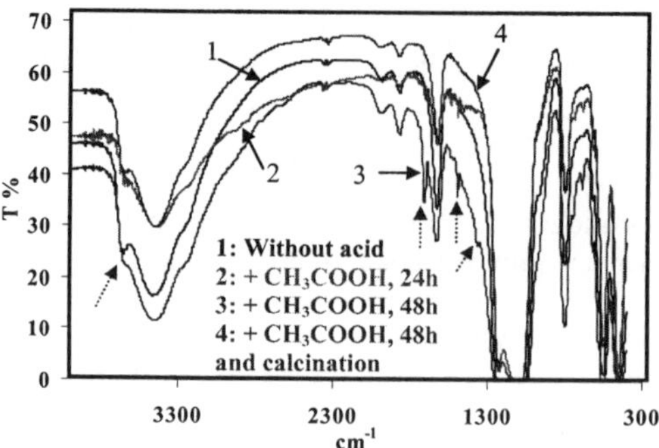

Figure 4: *FTIR spectra for Na-ZSM-5 (Si:Al 80) before and after treatment with CH₃COOH for 24h and 48h The sample was calcined prior to spectroscopy.*

Further examination was carried out of the effect of the concentration of the acid used to react with zeolite. These experiments were carried out on zeolite Na-ZSM-5 (Si:Al 120). The nitrogen adsorption isotherms are presented in Figure 5. For dilute acids up to 0.1M there is no change in the shape of the isotherm at low partial pressures, but there is a marked change in the shape and area of the low-pressure hysteresis loop, the general trend being towards a diminution of the size of the loop. For higher acid concentrations, specifically 1M and 17.4M (glacial) there is a marked change in the shape of the isotherm at low pressures towards a marked square shape, indicating stronger adsorbate-adsorbent interactions. The hysteresis loop has moved to higher partial pressures and expanded in size. In general, there is a lowering of the BET surface area, which intensified with increasing acid strength. These observations confirmed the inferences already made previously.

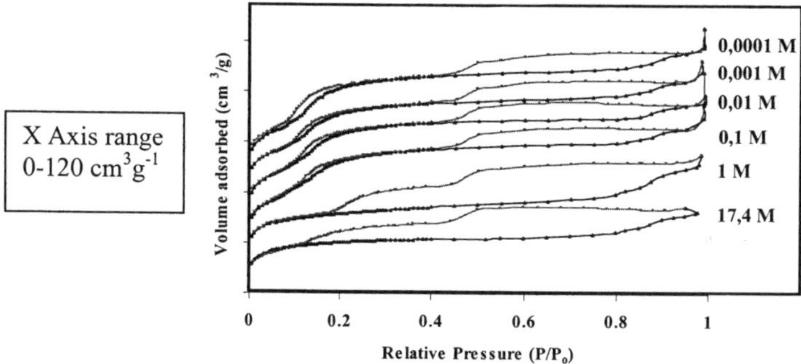

Figure 5: *N₂ adsorption isotherm for sample Na-ZSM-5 (Si:Al 120) after treatment with acetic acid solutions of varied concentrations from 0,0001 M to 17,4 M for 24h.*

In **Figure 6**, we present a comparison of XRD traces for Na$^+$-ZSM-5 with Si:Al=120 before and after acid treatment. Acids used were, hydrochloric, acetic and sulphuric acids.

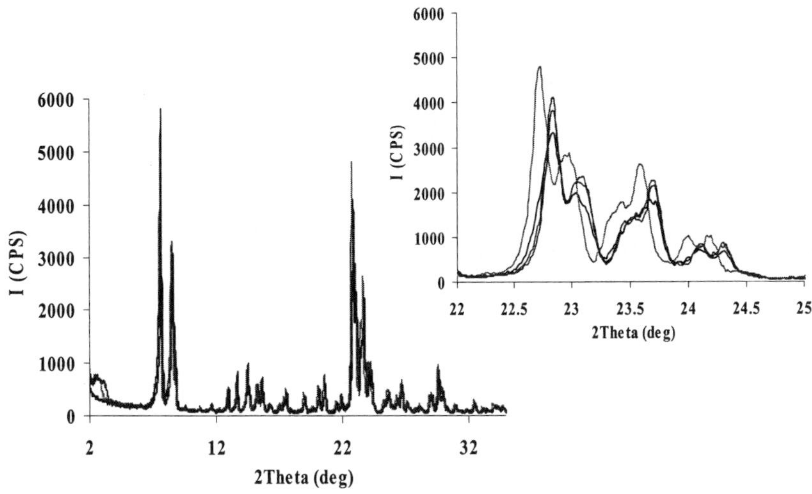

Figure 6 *XRD traces for Na$^+$-ZSM-5 with Si:Al =120 after acid treatment. The inset presents a magnification of the region 22 to 25° in 2Θ. Pristine acid, + HCl + H2SO4 + CH3COOH*

Comparison of the XRD traces shows that acid treatment under the conditions used here, does not lead to a significant loss of crystal integrity of the sample, although as it can be seen from the enlarged regions of the diffractogrammes in Figure 6, small displacements of the peaks do occur, whereas there is a more significant change in relative peak intensity, as well as the peak width. Both these changes are compatible with a change in sample composition brought about by dealumination. This is remarkable, given the relatively dilute aluminium content in these samples.

All the samples examined in the context of this studied also exhibited significant high-pressure hysteresis in their nitrogen adsorption isotherms. This can be attributed to pseudo-mesopores generated by the aggregation of the primary crystallites. The merging of the two loops observed in some isotherms displayed in Figure 3, Figure 4 and Figure 5 is probably simply due to the change in shape of the two loops, and has no further significance.

4. CONCLUSION

From the discussion above it can be concluded that the behaviour of ZSM-5 towards acid treatment and how this affects the shape and size of the low-pressure hysteresis loop confirmed previous inferences as to the reasons for the occurrence of this loop. It has been concluded that the changes observed could be attributed to the generation of additional faults in the pore walls. These would influence both the local geometry of the pore, as well as the strength of the adsorbate-adsorbent interaction. Furthermore, the presence of strongly held water molecules in the channels would further influence the local pore geometry. Given the commercially and scientifically very significant and widespread application of zeolites and specifically of the ZSM-5 structure type in catalysis and separation processes, the understanding of the factors influencing the characteristics of the solids as adsorbents and of the factors influencing theier behaviour is very significant. Acid treatment is often used as part of catalyst preparation and conditioning, and therefore a better understanding of the process, afforded by studies such as the present one are of great importance.

We acknowledge the financial support of the University of Cyprus. We are grateful to Degussa for the donation of the zeolite samples used in this study.

References

1.   W. Qu, Q. Zhou, Y. Wang, J. Zhang, W. Lan, Y. Wu, J. Yang, D. Wang, Polymer Degradation and Stability **2006**, 91, 2389.
2.   G. Manos, AA. Garforth, J. Dwyer, Ind. Eng. Chem. Res. **2000**, 39, 1198.
3.   P.L. Llewellyn, J.P. Coulomb, Y. Grillet, J. Patarin, H. Lauter, H. Reichert, H. Rouquerol, J. Rouquerol, Langmuir **1993**, 9, 1846.
4.   M.E. Eleftheriou, C.R. Theocharis, Characterisation of Porous Solids IV, Royal Society of Chemistry, London **1997**, 475.
5.   E. Panayi and C.R. Theocharis, "Studies in Surface Science and Catalysis", Elsevier Science Publishers, **2006**, 160, 273
6.   D.H. Olson, W.O. Haag, R.M. Lago, J. Catalysis **1980**, 61, 390.
7.   H. van Koningsveld, J.C. Janssen, H. van Bekkum, Zeolites **1990**, 10, 235.
8.   M.A. Ali, B. Brisdon, W.J. Thomas, Applied Catalysis A: General **2003**, 252, 149.
9.   J.C.Groen, L.A.A Peffer, J. Perez-Ramirez, Microporous and Mesoporous Materials **2003**, 60, 1.
10.   M. Ogura, S. Shinomiya, J. Tateno, Y. Nara, M. Nomura, E. Kikuchi, M. Matsukata, Appl. Catal. A. **2001**, 219, 33.
11.   A. Corma, A. Martinez, V. Matrinez-Soria, J. Catal. **2001**, 200, 259.
12.   X. Zhao, G.Q. Lu, G.J. Millar, Ind. Eng. Chem. Res. **1996**, 35, 2075.
13.   M. Muller, G. Harvey, R. Prins, Microporous Mesoporous Mater. **2000**, 34, 135.
14.   R.M. Dessau, E.W. Valyocsik, N.H. Goeke, Zeolites **1992**, 12, 776.
15.   R. Le Van Mao, S.T. Le, D. Ohayon, F. Caillibot, L. Gelebart, G. Denes, Zeolites **1997**, 19, 270.
16.   T. Suzuki, T. Okuhara, Microporous Mesoporous Mater. **2001**, 43, 83.
17.   J. C. Groen, L.A.A. Peffer, J.A. Moulijn, J. Perez-Ramirez, Chem. Eur. J. **2005**, 11, 4983.
18.   A. Cižmek, B. Subotić, I. Šmit, , A. Tonejc, R. Aiello, F. Crea, A. Nastro, Micropor. Mater. **1997**, 8, 159.
19.   T. Sano, Y. Nakajima, Z.B. Wang, Y. Kawakami, K. Soga, A. Iwasaki, Micropor. Mater. **1997**, 12, 71.

20.   J. C. Groen, J.A. Moulijn, J. Perez-Ramirez, J. Mater. Chem. **2006**, 16, 2121.
21.   U. Muller, H. Reichert, E. Robens, K.K. Unger, Y. Grillet, F. Rouquerol, J. Rouquerol, D. Pan and A. Mersmann, Fresenius Z. Anal. Chem., **1989**, 333, 433
22.   P.J.M. Carrott and K.S.W. Sing, Chem. & Ind., **1986**, 786
23.   H. Reichert, U. Muller, K.K. Unger, Y. Grillet, F. Rouquerol, J. Rouquerol, J.P. Coulomb, Studies in Surface Science and Catalysis, (Eds F. Rodriguez-Reinoso, J. Rouquerol, K.S.W. Sing and K.K. Unger), Elsevier, Amsterdam, **1991**, 62, 535
24.   U. Muller and K.K. Unger, Studies in Surface Science and Catalysis, (Eds K.K. Unger, J, Rouquerol, K.S.W. Sing and H. Kral), Elsevier, Amsterdam, **1988**, 39, 101
25.   G. Kyriacou and C.R. Theocharis, Studies in Surface Science and Catalysis, Elsevier Science Publishers **2002**, 144, 709.
26.   P.K. Tandon, S.B. Singh, M. Srivastava, Appl. Organometal. Chem. **2007**, 21, 264.
27.   Y. Zhang, I.J. Drake, A.T. Bell, Chemistry of Materials **2006**, 18, 2347.

IN SPECIES AND STRUCTURAL PROPERTIES OF IN-EXCHANGED ZSM5 AND MORDENITE

J. M. Zamaro[1], E. E. Miró[1], M. Yates[2], A. Martínez[3] and G. Fuentes[3]

[1] (Instituto de Investigaciones en Catálisis y Petroquímica) FIQ-UNL-CONICET. Santa Fe, Argentina
[2] Instituto de Catálisis y Petroleoquímica (CSIC). Madrid, Spain
[3] Universidad Autónoma Metropolitana Iztapalapa (UAMI). DF, México

1 INTRODUCTION

Nitrogen oxides (NOx) produced in the combustion processes of internal combustion engines in either stationary (e.g. energy plants) or mobile sources (e.g. automobiles) are major atmospheric pollutants. For over a decade now, zeolites have been studied as catalysts for the Selective Catalytic Reduction (SCR) of NOx to produce innocuous N_2, indium-zeolites being possible candidates.[1] Even though the catalytic behaviour of these solids has been extensively studied, further research is needed to define the nature of the indium species and their relationship with the catalytic performance. Different methods have been developed to add indium to zeolites, where the method selected determines the type of indium species generated. It is recognized that the SCR NOx active specie is InO^+ in the exchangeable zeolite site. Indium could also exist as In_2O_3 crystals and as highly dispersed non-stoichiometric oxides (In_xO_y).[3, 4] The active species could be generated by the solid state reaction at high temperature between zeolite protons and impregnated In_2O_3 as follows: $In_2O_3 + 2\ Z\text{-}H^+ \rightarrow 2\ Z\text{-}InO^+ + H_2O$. This treatment involves the impregnation of the solid, oxide generation at 500°C, and the subsequent reaction at 700°C. However, the zeolitic structures could be affected and more detailed knowledge on the consequences and limitations of such procedures is needed.

In this work we studied In-mordenite and In-ZSM5 exchanged samples produced through high-temperature treatments of $In(NO_3)_3$ impregnated solids. The transformations suffered by the zeolitic matrix and the exchanged indium species were also studied and their catalytic behaviour evaluated. For that purpose the indium-zeolites were characterised through Fourier Transform Infrared Spectroscopy (FTIR), Si and Al Magic Angle Nuclear Magnetic Resonance Spectroscopy (Si^{29}MASNMR and Al^{27}MASNMR), NO Temperature Programmed Desorption (NO-TPD), Thermogravimetric Analysis and Differential Thermal Analysis (TGA-DTA) techniques, and were evaluated in the SCR of NOx reaction using methane (CH_4-SCR) as reductant.

2 METHOD AND RESULTS

2.1 Experimental Setup and Results

2.1.1. Materials and Procedures

NH_4-mor Zeolyst (Si/Al = 10) and NH_4-ZSM5 Zeolyst (Si/Al=15) zeolites were used as supports. The supports were impregnated by the incipient wetness technique using a slight excess of solution (0.5 g/lt solution of In $(NO_3)_3$ Aldrich) and the solid allowed to dry at ambient temperature. Subsequently the solids were dried at 120°C overnight and then treated in air first for 12 h at 500°C and then for 2h at 700°C. Some samples were treated only at 500°C. The indium contents of the solids were 4 and 8 wt%.

Pelletised solids were characterised by FTIR using an ISRI chamber with CaF_2 windows, taking spectra of the samples at 350°C under N_2 flow. A Bruker IFS 66 equipment was used and the OH stretching range (3000-4000 cm^{-1}) was measured. Pellets treated with NO (5000 ppm) were also analysed by FTIR. They were first treated under vacuum at room temperature. Subsequently, temperature was steeply increased and the gas concentration (NO, NO_2 and N_2O) was measured with a Mattson Genesis II equipment using a cell for gas phase analysis.

Thermogravimetric experiments were carried out using a TGA-DTA 851e Mettler Toledo instrument, between 30 and 1000°C at a heating rate of 10°C/min in air.

The Si^{29} HDPEC-MASNMR and Al^{27}MASNMR spectra were acquired at 5 and 10 KHz, respectively, with a Bruker Advance II instrument (300 MHz).

TPD experiments were performed after flowing NO (5000 ppm) in He. A temperature ramp of 10°C/min was applied and the NO, NO_2, N_2O desorbed species were measured by FTIR using a cell for gas phase analysis.

The catalytic evaluation for NO_x SCR with CH_4 was performed using 1000 ppm NO, 1000 ppm CH_4 and 10% O_2, He balance, with GHSV = 15000 h^{-1} and between 300 and 600°C.

2.1.2. FTIR Characterisation

As mentioned above, the InO^+ exchanged in the zeolite framework is the active specie for the SCR of NO_x. However, apart from InO^+, other oxides such as In_2O_3 could also be formed.[3] In the high temperature exchange treatment of Indium impregnated NH_4-zeolites (as $In(NO_3)_3$), the generation of H^+ and, subsequently, its reaction with In_2O_3 at 700°C are produced.

In order to analyse the changes in OH^- groups, FTIR spectra were taken (Fig. 1 a and b). In the ammonium samples treated at 700°C a significant modification is observed, particularly in the case of mordenite. However, when In is exchanged, the decrease in the OH signal due to its reaction with In_2O_3 (producing exchanged InO^+ species) is higher for ZSM5 than for mordenite. These results indicate that while OH is more easily removed in mordenite, the extension of the In exchange (as InO^+) is lower than for ZSM5. This can be explained by taking into account that OH bridge groups are closer and that the Al acidity is lower in the mordenite sample. Probably the closer the OH's the higher extension of thermal dehydroxilation and also as aluminium atoms are closer they are less acid (Lewis) compared to ZSM5. This lower acidity could cause lower acidity (Brönsted) of the bridge OH (interchange sites). Moreover, a signal at 3650 cm^{-1} can also be observed (see indication in Fig. 1), which in the case of the mordenite sample is present after treatment at 700°C and can be assigned to $Al(OH)_2^+$ and $Al(OH)^{+2}$ species originated in the migration and hydration of Al of the zeolite framework. For ZSM5, this signal is also present in the fresh sample.

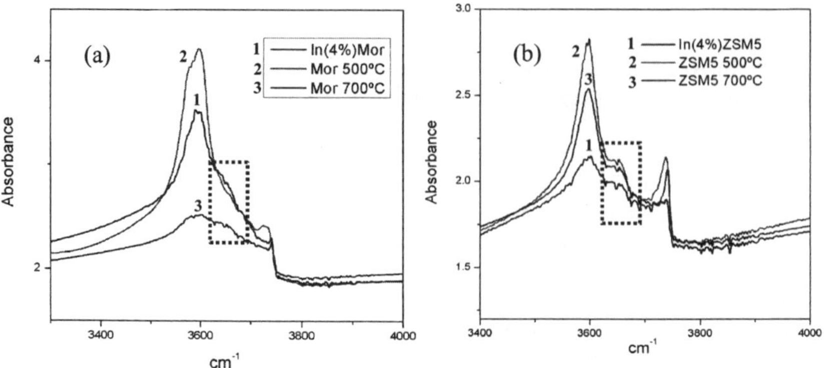

Figure 1 *FTIR spectra of NH₄ and In-zeolites pellets taken at 350°C with N₂ flow. a) mordenite b) ZSM5. In(4%)zeolite is denominated as the fresh sample.*

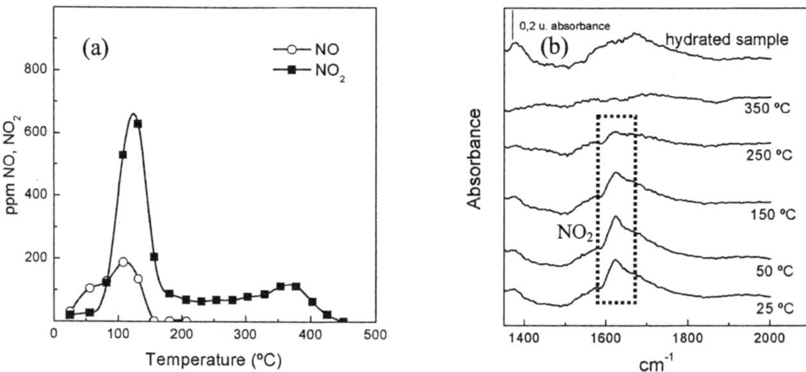

Figure 2 *a) NO desorption thermogram of In-mordenite. b) FTIR spectrum of dehydrated In-Mor pellet treated with NO and evacuated at different temperatures.*

2.1.3. NO-TPD Characterisation

Temperature-programmed NO desorption results for In(4%)-Mor are shown in Fig. 2a. NO_2 is the main gas evolved with two temperature zones. A small signal of NO at low temperature is also observed. NO_2 could be formed from the NO reaction with two differently exchanged InO^+ species, reducing the site to In^+, which are in turn reoxidised thus giving place to a redox cycle. The presence of adsorbed NO_2 was confirmed by FTIR of pellets treated in NO atmosphere (Fig. 2b), in which the only signal was at 1628 cm⁻¹, in agreement with the stretching of NO_2 bondings. The signal corresponding to the bonding of water is also in this zone, but with lower definition and shifted to higher wave numbers (see hydrated sample).

2.1.4. MAS NMR Characterisation

The thermal process involved in the In exchange caused structural changes in the zeolite, as observed by FTIR. To gain further insight into this aspect, Al^{27} and Si^{29} MASNMR spectra were analysed. Fig. 3.a shows the Si^{29} MASNMR of fresh and treated mordenites at 500°C and 700°C. A chemical shift through more negative values can be observed when the temperature increases, indicating a progressive dealumination process. The deconvolution of the signals allowed us to calculate the proportions of Siq_2, Siq_3 and Siq_4, as shown in Table 1. In the case of mordenite, the amount of Siq_2 and Siq_3 (surrounded by a higher number of Al atoms) decreases when the temperature increases. This is also verified with Al^{27} MASNMR, in which a significant octahedral Al signal (extra-framework Al) can be observed at 500°C. In these spectra, it can also be observed that the treatment at 700°C provokes a higher deformation of the Al environment, both for octahedral and tetrahedral coordination, as suggested by the tail of the Al^{27} MASNMR signal.

The NH_4 and In-ZSM5 samples show a slight modification of the Siq_2, Siq_3 and Siq_4 proportions, suggesting a high stability even at 700°C. This is shown in Table 1. The Al^{27} MASNMR spectra indicates a slight initial dealumination (fresh sample), which is slightly increased at 700°C.

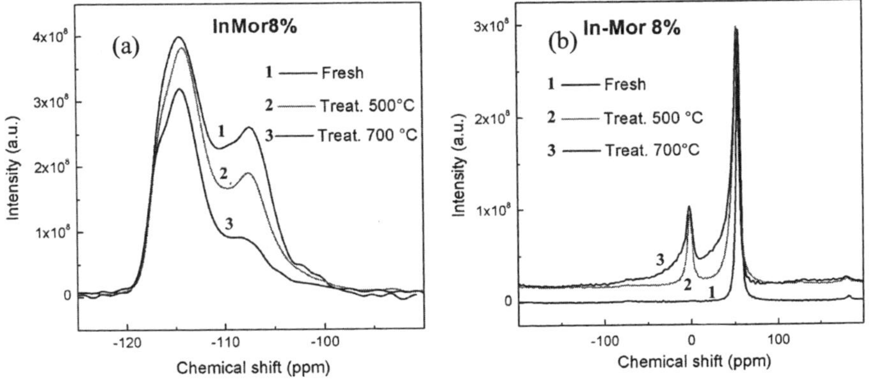

Figure 3 *a) Si^{29}MASNMR of fresh In-mordenite and treated at 500 and 700°C. b) Al^{29}MASNMR of fresh In-mordenite and treated at 500 and 700°C.*

Table 1 *MASNMR signals deconvolution*

Signal/Sample	InMor	InMor fresh	InZSM5	InZSM5 fresh
Siq_4	68.0	55.8	84	88.2
Siq_3	30.8	42	16	11.8
Siq_2	1.2	2.2	-	-
Al^{tet}/Al^{oct}	4.3	-	11.8	17.6

2.1.5. TGA-DTA Characterisation

This technique was useful to study the transformation of the functional groups and structures of the different zeolitic supports. NH$_4$-ZSM5 presents a weight loss at 420°C associated with NH$_3$ loss and another at 850°C due to dehydroxilation. This is confirmed by DTA which shows two endothermal processes at the same temperatures (Fig. 4a). In NH$_4$-Mor (not shown), the corresponding temperatures are 510°C (NH$_3$ loss) and 760°C and 900°C (dehydroxilation), suggesting that two types of OH are present.

2.1.6. XPS Characterisation

Surface In species were investigated by XPS. In Fig. 4b, the In3d XPS signal of In(4%)-Mor is shown. This sample was evaluated for the SCR of NO$_x$ yielding the results shown in Fig. 5. The 3d$_{5/2}$ signal is centred at 445 eV taking C as reference. This is in agreement with In (III). However, it is not possible to distinguish different species. From the In 3d$_{5/2}$ and Al 2p signal, a In/Al ratio of 0.23 is obtained which indicates a significant amount of surface In. The Si/Al ratio is 12.5, slightly different from the value of 10 reported by the manufacturer. The dealumination process could be the origin of the difference.

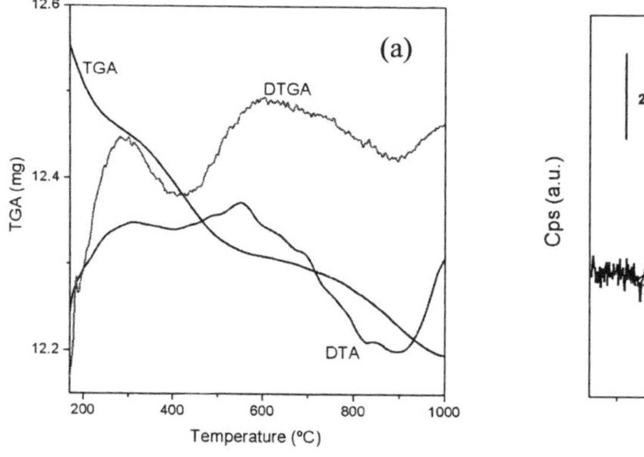

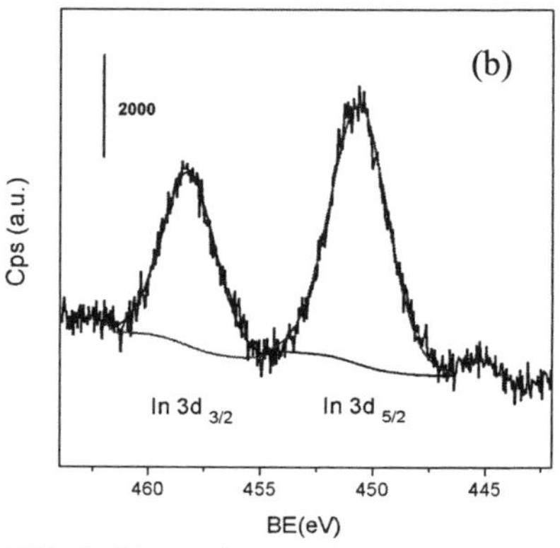

Figure 4 *a) TGA-DTGA and DTA of NH$_4$-ZSM5, b) In3d XPS signals present in the evaluated In(4%) Mor sample.*

2.1.7. Catalytic Characterisation

Fig. 5a shows NO and CH$_4$ conversion against temperature for the In-Mor catalyst. The maximum NO to N$_2$ conversion reaches 61% at 390°C. Fig. 5b shows that the selectivity (defined as NO$_x$ conversion/CH$_4$ conversion) presents a maximum value at 350°C. These values are in agreement with those previously reported by other authors.[2] However, an interesting aspect not previously reported is the significant production of CO obtained during reaction. In-Mor and In-ZSM5 catalysts presented similar behaviours. For In-Mor, the maximum CO production was about 15% of the initial CH$_4$ in the feed. This effect is more pronounced for In-ZSM5 (not shown).

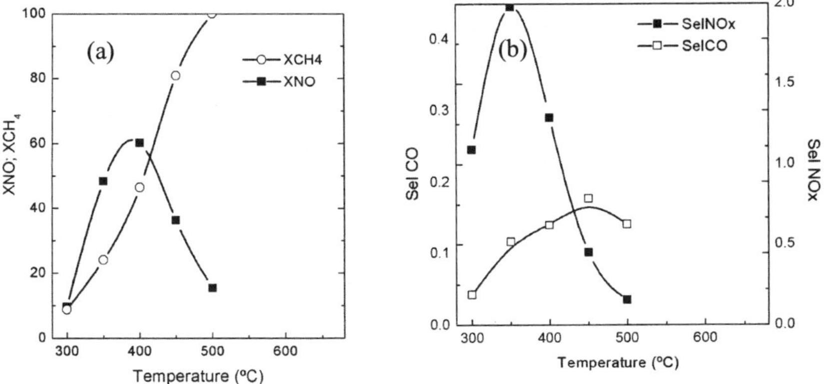

Figure 5 *In(4%)-Mordenite sample: a)NO and CH$_4$ conversion. b) CO production and NO to N$_2$ selectivity.*

3 CONCLUSION

The thermal activation process generates InO$^+$ species with two different environments. These species interact with NO in a redox cycle, producing NO$_2$ adsorbed at the catalyst surface. The In exchange performed at 700°C depends on the acidity of the exchange sites, being higher in the case of ZSM5 than for mordenite. A thermal dehydroxilation is produced at 700°C which is higher for mordenite (with lower Si/Al than ZSM5). Both processes (exchange and dehydroxilation) are accompanied by a loss of zeolitic structure, giving place to extra-framework Al formation. The distortion degree originated is bigger for mordenite.

In-Mor presents good activity for SCR of NO$_x$ with CH$_4$ but a significant amount of CO is produced. Future research is planned in order to obtain relationships between the chemical and structural properties of the catalysts with the catalytic behaviour.

Acknowledgments

The authors are grateful for the financial support received from: Project CIAM, Universidad Nacional del Litoral-CONICET (Argentina), Universidad Autónoma Metropolitana-CONACYT (México), and ANPCyT (Argentina).

References

1. X. Zhou, T. Zhang, Z. Xu, L. Lin, *Cat. Lett.*, 1996, **40**, 35.
2. M. Ogura, T. Ohsaki, E. Kikuchi, *Mic. Mes. Mat.*, 1998, **21**, 533.
3. E. E. Miró, L. Gutierrez, J. M. Ramallo López, F. G. Requejo, *J. Catal.*, 1999, **188**, 375.
4. M. Ogura, N. Aratani, E. Kikuchi, *Progress in Zeolite and Microporous Materials, Studies in Surface Science and Catalysis*, 1997, **105**, 1593.

THE PROPERTIES OF VIRTUAL SOLIDS USING MONTE CARLO INTEGRATION AND THEIR VALUE TO THE NANO-SCIENTIST

G.R. Birkett[1], L.F. Herrera[1] and D.D. Do[1]

[1]School of Engineering, The University of Queensland, St Lucia, QLD, 4072, Australia.

1 INTRODUCTION

The advent of molecular simulation and molecular models of adsorbents has led to an increased understanding of the adsorption and transport processes that occur in pores of molecular dimensions. With molecular models of solids it is possible, in principle, to measure all the properties of the solid. At the molecular scale however, properties such as surface area and pore size become ambiguous and we must impose some assumptions to be able to measure these quantities. In this paper we will present Monte Carlo integration techniques for measuring the surface area and pore size distribution of a porous solid which has a known structure or has had an atomistic model constructed from a technique such as Reverse Monte Carlo. These techniques are based upon the interaction potential of the adsorbent and the adsorbate and are unique to this binary system. To demonstrate the utility of these techniques a series of results will be presented for regular geometries for which we have some intuitive understanding of the pore size and area and also for disordered systems for which we cannot *a-priori* know these properties.

2 METHODS

2.1 Accessible Volume

The accessible volume (V_{acc}) is defined as the region of the pore space where the total potential between a probe particle and the solid is negative (Figure 1). V_{acc} was determined using the method proposed by Do and Do [1] which uses Monte Carlo integration. This inserts a probe particle into the simulation box at random and the potential between the particle and the solid is calculated. If the potential is negative, the insertion is counted as a success. This is repeated many times with the number of trial insertions, N_{ins}, and the number of successful insertions, N_{succ}, recorded. Every time a probe particle is inserted, V_{acc} is calculated as the fraction of successful insertions times the volume of the simulation box V_T (i.e. $V_{acc} = V_T (N_{succ} / N_{ins})$). The process is terminated when the difference between V_{acc} values for two consecutive insertions is less than the desired tolerance.

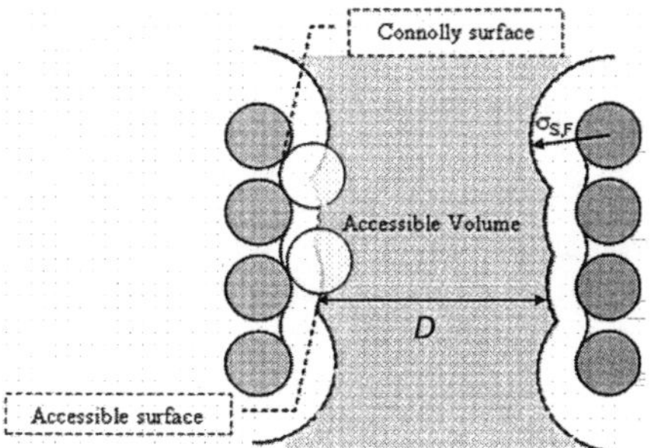

Figure 1 *The different definitions of the surface of a pore.*

2.2 Surface Area

In the literature there are many definitions of the surface for a porous material. Two such definitions are the Connolly surface and the accessible surface [2]. These are geometrical definitions of the surface and can be explained as follows: A single site LJ molecule is placed at a distance of $\sigma_{S,F}$ from the closest solid atom. It is moved about the surface such that the distance between the closest solid atom and the adsorbate molecule is always $\sigma_{S,F}$. The Connolly surface is that traced by the surface of an adsorbate molecule as it moves over the solid atoms. The accessible surface is defined as that traced by the centre of the adsorbate molecule. Both of these definitions are illustrated in Figure 1. The geometrical surface area is estimated with a method similar to that used for the accessible volume. Consider the surface of a solid atom as a sphere of radius $\sigma_{S,F}$ centred at the centre of the solid atom. Using this definition, the total area of a solid atom i is given by $A_i = 4\pi(\sigma_{S,F})^2$ and the total area of all solid atoms is $A = NA_i$, where N is the number of solid atoms. However, not all this surface is accessible to a probe particle. Therefore, the fraction of this area that is accessible for a fluid atom is determined using Monte Carlo integration (MCI).

In MCI, a solid atom is selected from the given solid configuration, and then a probe particle is placed on its surface at random. If the total potential between the probe particle and the solid is negative, the insertion is considered a success, otherwise it is a failure, then the probe particle is removed from the simulation box. This procedure is repeated a large number of times for each solid atom. Finally, the geometrical surface area is estimated as $A = \sum_{i=1}^{N} A_i S_i / (S_i + F_i)$, where S_i is the number of successful insertions on the surface of the solid atom i and A_i is the number of failed insertions on the surface of the same solid atom. This method also determines the solid atoms composing the surface of the solid with least one point where the total potential between a probe particle and the solid is negative.

2.3 Pore Size Distribution

The PSD method is based on the potential energy of probe molecules rather than purely by geometrical considerations. To achieve this, a method similar to that used by Gubbins and co-workers [2-5] is used and is based around the use of a spherical balloon particle. The balloon particle will be inserted, manipulated in accordance with the method outlined below, and deleted from the system many times over to calculate the properties of the porous solid. The method is outlined below:

<u>Step 1</u>: A balloon particle is placed at random in the simulation box, and the total potential is calculated. If the potential is positive, the insertion is counted as a failure and Step 1 is repeated. Otherwise, the potential is negative and the steps presented below are followed.

<u>Step 2</u>: The balloon particle is increased in size by Δr to r'. For each of the solid atoms comprising the surface (which were identified during the surface area determination) a probe particle is placed on the spherical surface of the balloon particle (with radius r') on the line joining the centres of the balloon particle and the solid atom k (line BC in Figure 2a). The potential energy ($\varphi_{k,S}$) between each probe particle and all solid atoms is computed. This results in a set of N_{surf} potential energies, $\{\varphi_{k,S}; k = 1, 2, ..., N_{surf},\}$ where N_{surf} is the number of solid atoms on the surface.

<u>Step 3</u>: If one or more of the potential energies ($\varphi_{k,S}$) are positive, a displacement is performed, otherwise the enlargement is accepted ($r' \rightarrow r$) and Step 2 is repeated. The displacement move is made over a random distance with the average displacement length chosen as $\Delta L = \Delta r/20$. A random, as opposed to fixed, displacement length is important because a fixed displacement length can result in particles becoming stuck between two opposing solid atoms. The balloon particle is moved along the vector sum, given by equation 1, between the solid atoms k and the centre of the balloon particle B (Figure 2a).

$$\vec{\zeta} = 2\eta\Delta L \frac{\sum\limits_{k=1}^{m} \vec{r}_{k,B}}{\left|\sum\limits_{k=1}^{m} \vec{r}_{k,B}\right|} \qquad \text{equation 1}$$

<u>Step 4</u>: The steps 2 and 3 are repeated until the balloon particle has reached the size of the pore. This is when it cannot move any further and the particle starts to move about the same spot or the maximum displacement, L_{max}, has been reached. The balloon particle cannot move farther than L_{max} from the initial insertion point to retain the ability to distinguish between two different, but connected, pores.

<u>Step 5</u>: Steps 1 to 4 are repeated with another initial random position for N times and a maximum radius of the balloon particle is obtained for each insertion point i.

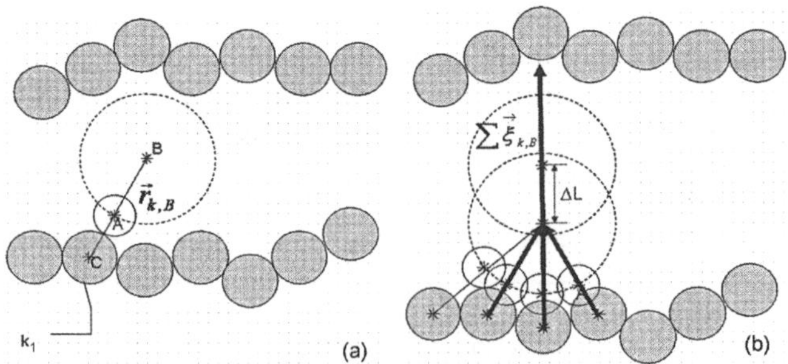

Figure 2 *a) The placement of a probe particle on the surface of the ghost particle and b) movement of the ghost particle.*

<u>Step 6</u>: The fraction of the pore volume with a diameter equal to D is determined as number of the balloon particles with diameter equal to D over the total number of successful insertions. As the unsuccessful insertions are counted in Step 1, this procedure calculates the accessible volume of the solid as a whole and for each pore size within the interval Δr. This data is used to create the Pore Size Distribution (PSD) and the Cumulative Probability Density (CPD), which are normalized using the total accessible volume.

2.4 Configurations

The configurations used to test the methods of determining the accessible volume, surface area and PSD are given below. These configurations have been selected to test different aspects of these techniques. All of the configurations have been, arbitrarily, constructed from a grid of carbon atoms with the exception of the graphite sheets (Figure 3a) and the nanotube (Figure 3e) which have the correct configurations for those forms of carbon. The Graphite sheet and Box, in Figure 3a and b, respectively, are shapes for which we expect the surface area to be easily calculated due to their simple geometry. The hollow cube, hollow sphere and nanotube provide simple geometries for which we should get a single pore size which we know from having constructed the solid. The constricted pore if the first step towards a real solid but again has a fairly simple geometry for which we would expect to measure two pore sizes. The disordered solids in Figure 4 have been constructed by removing atoms a randomly until the desired percentage of solid atoms are removed.

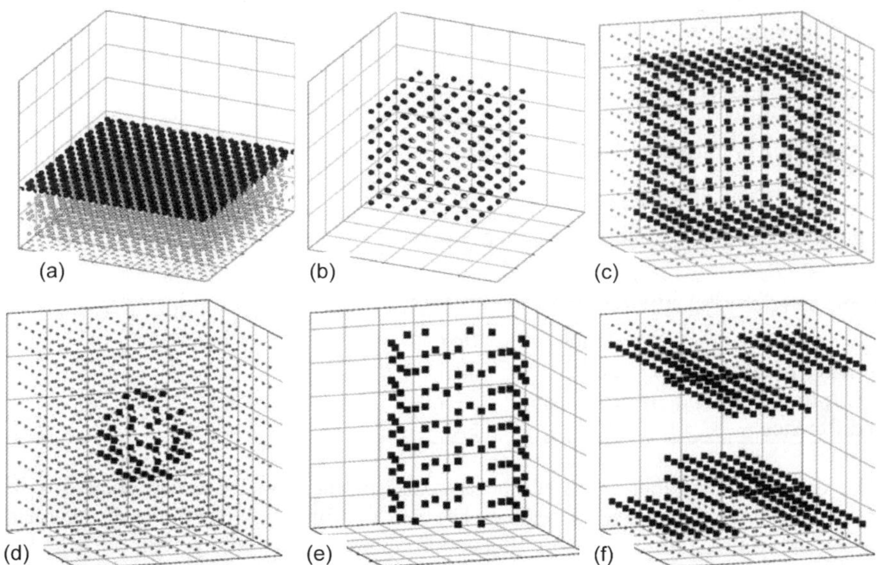

Figure 3 *Configurations used to test the PSD (a) Graphite, (b) Box (c) Hollow cube (d) Hollow sphere, (e) Nanotube (f) Constricted Pore. (●) Solid atoms, (●) Solid atoms in the surface layer.*

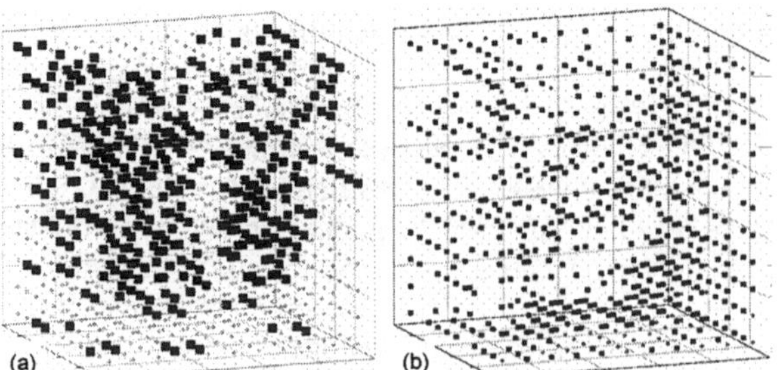

Figure 4 *Porous solids used to test the PSD (a) solid with 10 % porosity (b) solid with 50% porosity. (•) Solid atoms, (•) Solid atoms in the surface layer.*

3 RESULTS

Before results obtained are discussed, the assumptions used to calculate the expected accessible volume and surface area of each idealized configuration in Figure 3 are defined. The first assumption is that all the surfaces of the configurations are flat meaning that the roughness of the surface is not accounted for. The second assumption is that the minimum distance between a probe particle and a surface atom $\sigma_{S,F}$, which neglects that this distance can be smaller due to the interaction of the probe particle with the other solid atoms. Estimations of accessible volume and surface area for the various configurations, using these assumptions, will be referred to as the initial estimation here in. Argon, with the LJ parameters of $\sigma = 0.3405$ nm and $\varepsilon/k_b = 128$ K, will be used as the probe particle for all calculations unless otherwise noted.

3.1 Accessible Volume and Surface Area

The results obtained for the accessible volume of the configurations in Figures 3a and 3b are presented in Table 1. Note that variance of MCI results is < 0.01%.

Table 1 Accessible volume of the configuration proposed.

	Graphite	Box
Initial estimation (nm³)	28.8	32.1
Result from MCI (nm³)	29.5	34.7
% difference	2.5	8.0

The data indicates that the assumptions used in the initial estimation give a lower accessible volume than the MCI method. The Box configuration has the largest percentage of error with 8.0%, followed by the Bed and Graphite configurations with 2.6% and 2.5% respectively. This is partly because a flat surface does not include the irregularities of the surface. In addition, the second assumption also underestimates the accessible volume since the argon particle can be closer to the surface than $\sigma_{S,F}$ because the argon particle interacts with all the solid atoms. So even for these simple configurations a significant difference can exist between the accessible volume estimated by MCI and by using an assumption of a flat surface and fixed minimum approach of $\sigma_{S,F}$. One may expect to see a similar trend when calculating the surface area. However, it will be shown that depending on the solid configuration, the initial estimate using flat surface assumption can be higher or lower than the result obtained with MCI.

Table 2 presents the results obtained with the method outlined 3.1.2. As expected, the surface area of Bed and Graphite configurations are larger than the initial estimation due to the roughness of the surface. In addition, the results indicate that the Graphite configuration presents the smallest percentage of error due to close packing of carbon. In contrast, the initial estimation of surface area for the Box configuration is larger than the value calculated with the method proposed. This is because the edges and corners have geometry closer to a sphere, which present less surface area than the corners of a cube.

Table 2 Surface areas values for each configuration.

	Graphite	Box
Initial estimation (nm^2)	16.2	44.4
Area value from MCI (nm^2)	16.6	42.3
% difference	2.5	-4.7

3.2 Pore Size Distribution

The first thing that is considered for the PSD is the PSD of the hollow cube. This is important because it shows the importance of the choice of the parameter L_{max}. The plot in Figure 5a shows that the CPD for this configuration decreases by nearly 50% up to the diameter of 2.1 nm when $L_{max} = r$. The reason for this is that a ghost particle inserted in the corner cannot move out of the corner and reach the centre of the pore when $L_{max} = r$. To solve this problem, the value of L_{max} is set to $3^{1/2}r$. This allows particles inserted in the corners of the cube to reach the centre of the cube. This is a choice that has to be made when one is trying to define a pore size and is part of the ambiguous nature of pore sizes. We consider a pore in this idealised case to be the cube and propose that the PSD in Figure 5b is more closely aligned to our perception of a pore than the PSD in Figure 5b. Since a cube represents the extreme case, this should be sufficient for all cases and is used in all calculations presented in the other results.

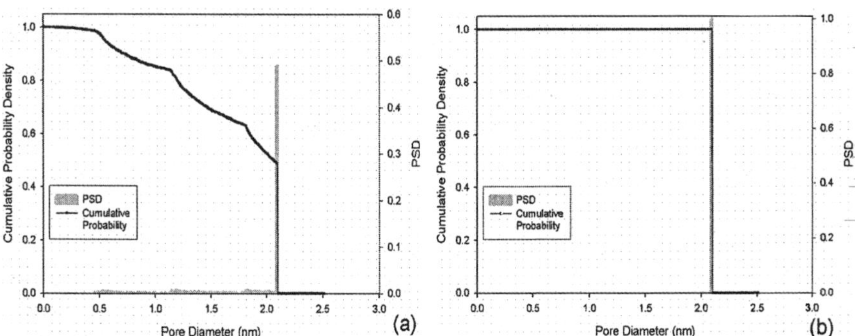

Figure 5 *Comparison between two PSD obtained by using different L_{max} values. (a) $L_{max} = r$ (b) $L_{max} = 3^{1/2}r$, where r is the radius of the ghost particle.*

The next solid configuration considered is the hollow sphere configuration. The hollow sphere was constructed from a grid of atoms with all atoms within a certain radius removed. This means that the sphere has irregularities at its surface and is not a perfect sphere. The plot in Figure 6a shows that it is possible to "find" pores smaller than the actual size of the hollow region. This is due to the restriction of the maximum displacement of the balloon particle. Figure 6b and c shows where some of the smaller pores are found in the irregularities in the surface of the sphere. This shows the balance between identifying unique pores and the characteristic size. It is not possible to completely remove ambiguity but, $L_{max} = 3^{1/2}r$ seems to be the best compromise.

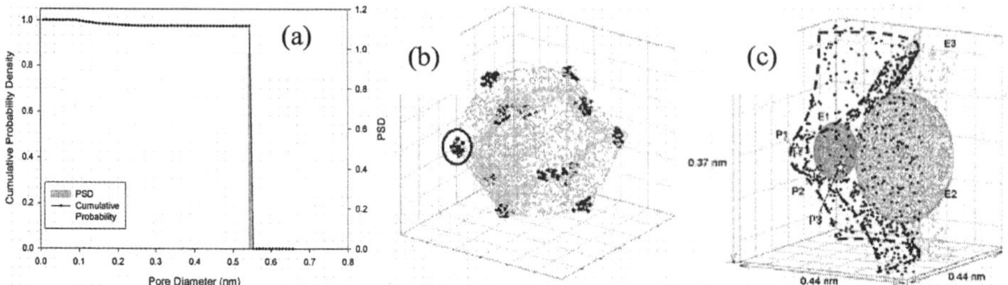

Figure 6 *a) PSD of hollow sphere, b) location of smaller pores, c) final balloon particles for smaller pores*

The other regular configurations considered were those of the nanotube and the constricted pore. The PSD for the nanotube in Figure 7a is a single peak, as we expect, and shows that the technique used here is valid for cylindrical configurations. The constricted pore is a step closer to a real system than the previous configurations considered. The PSD is Figure 7b has two clear and isolated pore sizes corresponding to the constricted and larger section of the pore. This illustrates the utility of the technique to find individual pore sizes for connected systems and not just find the largest possible pore size.

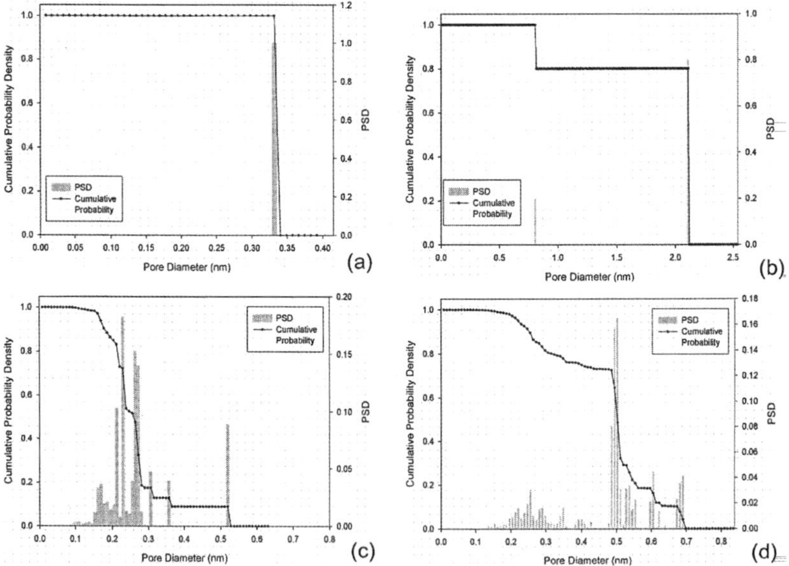

Figure 7 *Two PSDs obtained for a) nanotube, b) constricted pore, c) 10 % porous solid and d) 50% porous solid.*

The most complex PSDs studied were those for porous solids with 10% and 50% porosity (Figures 7c and d respectively). Naturally the PSDs of these solids show a much greater range of pore sizes than the configurations previously considered. The solid with 10% porosity has a pore size range from 0.085 nm to 0.52 nm and the solid with 50% porosity has a pore size range from 0.085 nm to 0.7 nm. The smallest pores result from the removal of a single solid atom which leaves a 0.085 nm space for the ghost particle to fit into. For the solid with 10% porosity, the finite size of the system and the lower porosity leads to more distinct and isolated peaks in the PSD. Whilst it is not possible to know in advance the PSD of a disordered solid such as these, the PSD method presented follows the expected trends. The reproducibility of the PSDs of the disordered solid was tested with several runs using different sequences of random numbers. It was found that the PSDs were very reproducible and varied only a negligible amount from one to the next. This means that the PSDs obtained are unique for the method proposed using a given solid and probe particle.

4 CONCLUSIONS

Different methods for estimating the accessible volume, the surface area and the pore size distribution using Monte Carlo integration have been presented. The last two methodologies are new and they were tested on a wide range of idealized configurations showing these methodologies produce the expected results with slit, cubic, spherical, cylindrical and constricted pore configurations. From the study of the accessible volume, it is clear that the results obtained from the flat surface assumption produce an underestimation of this value. In contrast, the results for the accessible surface area from the method proposed are variable compared to expected values depending on the geometry of the system. This is due to the fact that complex geometries have corners and edges whose areas are over estimated when the assumption of a flat surface is applied. The method of calculating PSDs was applied to a range of solid configurations and was shown to give consistent results for cylindrical, spherical, and cubic geometries. PSDs were calculated for disordered solids and seemed to be accordance with expected trends since the PSD of the solid with higher porosity had a wider ranged and smoother PSD.

Acknowledgements: This research was supported by the Australian Research Council.

References

1 Do, D. D.; Do, H. D. Modelling of Adsorption on Non-graphitized Carbon Surface: GCMC Simulation Studies and Comparison with Experimental Data. *J. Phys. Chem. B* **2006,** 110(35), 17531-8.

2 Pikunic, J.; et. Al. Structural Modelling of Porous Carbons: Constrained Reverse Monte Carlo Method. *Langmuir* **2003,** 19(20), 8565-82.

3 Gelb, L. D.; Gubbins, K. E. Pore Size Distribution in Porous Glasses: A Computer Simulation Study. *Langmuir* **1999,** 15(2), 305-8.

4 Kendall, T. T.; Gubbins, K. E. Modelling Structural Morphology of Microporous Carbons by Reverse Monte Carlo. *Langmuir* **2000,** 16(13), 5761-73.

5 Bhattacharya, S.; Gubbins, K. E. Fast Method for Computing Pore Size Distributions of Model Materials. *Langmuir* **2006,** 22(18), 7726-31.

COMPUTER SIMULATION OF ENANTIOSELECTIVE ADSORPTION IN A SIMPLE MODEL OF MOLECULARLY IMPRINTED POLYMERS

R. Spörl and L. Sarkisov

Institute for Materials and Processes, University of Edinburgh, Edinburgh, EH9 3JL, United Kingdom

1. INTRODUCTION

Two-thirds of the currently available prescription drugs are chiral and the majority of new chiral drugs are formulated as a single enantiomer. Formulation of a drug as a pure enantiomer has benefits associated with both efficacy and safety, and it is expected that the proportion of single enantiomer drugs will further grow. This stimulated a renewed recent interest in efficient synthetic routes to single enantiomer drug precursors, chiral separations and chiral sensing[1]. In general, chiral separation is a difficult task since the excess properties of the racemic mixtures are very small and the conventional separation methods such as distillation can not be applied directly. Chiral separation methods include crystallization, favoured more on the industrial scale, and various derivatization methods, where the chemical structure of the racemic compounds is modified to make them suitable for distillation or other standard methods. Gas or liquid chromatography is another commonly used method. This method requires so called chiral selectors, additional species that induce chiral separation through stereoselective recognition and binding of one of the enantiomers. This species can be a part of the stationary chromatographic phase (making it chiral) or an additive to the mobile phase. Chiral selectors can be of a biological origin (proteins, polysaccharides, etc.), or semi-synthetic and synthetic origin, such as helical polymers. Although a number of chiral selectors have been proposed over the years, their stability and cost is often an issue and as a result these methods have been limited to relatively small scale operations.

Recently, another approach to chiral applications has been proposed based on molecularly imprinted polymers (MIPs). MIPs are a family of polymeric materials, where polymerization takes place in the presence of additional structure directing agents (templates). Removal of the templates after polymerization leaves imprints in the structure that are able to recognize and bind original template molecules. This unique biomimetic functionality makes these materials very promising for a variety of applications from separations and sensing to catalysis, drug delivery and artificial immunoassays. Not surprisingly, molecularly imprinted polymers have been a bourgeoning area of research over the last 20 years and we refer the reader to an excellent recent review by Alexander and co-workers[2].

The first application of a MIP to the chiral drug separation was reported by Fischer and co-workers in 1991, when they prepared a polymer imprinted with S-timolol (a drug from the β-blockers family, used for several conditions such as high blood pressure and heart attack)[3]. Since then, this approach has been extended to a number of systems, including α- and β-agonists, non-steroidal anti-inflammatory drugs and other classes of chiral therapeutic molecules. A review of recent advances in chiral applications of MIPs has been provided by Maier and Lindner[4]; specific applications of MIPs to chiral drugs have been recently reviewed by Ansell[5].

Despite a number of obvious successes, rational design of new materials for chiral applications seems to be hindered by our limited understanding of the actual process of molecular recognition in these imprinted structures. Chiral selectivity in MIPs results from several contributions, including steric compatibility, van der Waals and electrostatic interactions, hydrogen bonds and so on. Commonly adopted strategies for the development of efficient chiral MIPs are based on forming strong associations between the functional monomer in the precursor mixture and a functional group on the template molecule. These associations are typically hydrogen bonds; for example in the aforementioned study by Fischer and co-workers these associations form between the carboxylic group of metacrylic acid (functional monomer) and the amine group of *S*-timolol (template)[3]. Relative contributions of this and other interactions to the specificity of the formed material are not clear, although some attempts to distinguish and assess these contributions have been made[6]. It has been widely recognized that molecular modelling and theoretical approaches may play a vital role in understanding the nature of molecular recognition in MIPs. Recently, a number of computational studies have emerged in an effort to provide more efficient, streamlined design strategies to novel polymeric structures[7-13].

In this article we consider a simple model of a chiral imprinted polymer, where selectivity is a result of purely steric contributions. The importance of this study is twofold. On one hand we would like to provide a general estimate of this contribution to the overall molecular recognition capability of a MIP. On the other hand, this can be considered as a model of an imprinted material that does not involve strong associations between matrix and template species, such as a MIP composed solely of the cross-linker (no functional monomers) or imprinted silica materials. The benefit of these materials compared to the more conventional MIPs, is that they do not involve design decisions associated with the choice of functional monomers, thus significantly simplifying the design cycle. We would like to understand the scope and applicability of these imprinted materials to enantioselective separations and sensing.

2. MODEL AND SIMULATION DETAILS

Here we adopt a model of chiral species previously studied in the context of racemic phase equilibria[14]. A chiral molecule is formed by five hard sphere sites, with the first site of size σ located in the centre of the molecule and four other sites of size *0.6σ*, *0.8σ*, *1.0σ* and *1.2σ* respectively in tetrahedral arrangement at the distance of *0.4σ* from the centre of the first site. This is the simplest model of a chiral centre, however other similar models have been suggested in previous studies[15, 16]. Two mirror images *R* and *S* of this molecule are depicted in Figure 1.

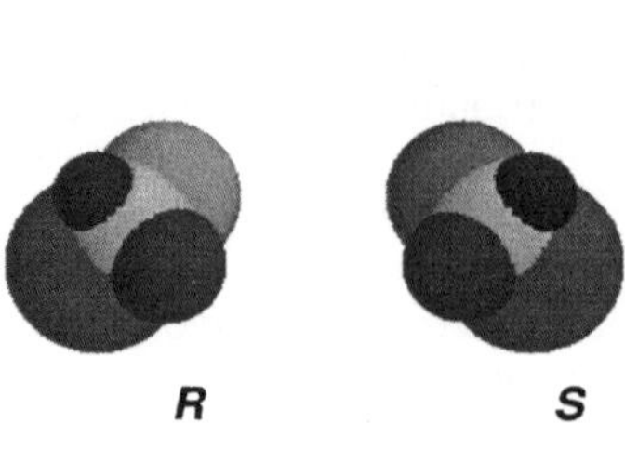
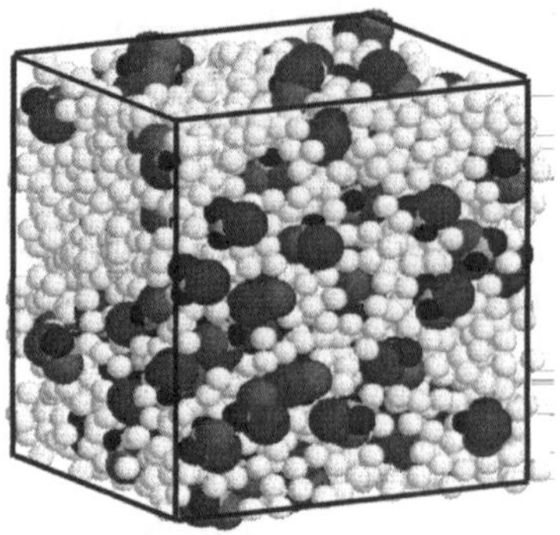

Figure 1: Two enantiomers, *R* and *S*, of the chiral model employed in this work are shown on the left. A typical configuration for the binary matrix-template mixture is shown on the right. This configuration corresponds to the matrix density $\rho_m \sigma_m^3 = (N_m/V)\sigma_m^3 = 0.540$ with the matrix particles (shown in light grey) and the template density $\rho_t \sigma^3 = (N_t/V)\sigma^3 = 0.138$.

Polymer is modelled as a fluid of hard spheres of size $\sigma_m = 0.6\sigma$. The simulation protocol employed in this study has been explored in a series of works by Van Tassel and co-workers and mimics the actual process of the imprinted polymer formation[17-23]. We start with a binary mixture of the matrix and template species. The system is a cubic simulation cell with edges of length 10σ and periodic boundary conditions applied in all directions. The binary mixture is equilibrated for ca. 10^8 canonical Monte Carlo moves. After equilibration, the template species are removed leaving the complementary enantioselective cavities in the structure. For each system, a total of 5 different matrix realizations are generated in this fashion. This is followed by adsorption studies. For each structure realization we perform grand canonical Monte Carlo simulation of adsorption of individual enantiomers. The total number of moves for each run is of order 10^8. In order to improve the efficiency of sampling, we implement the excluded volume map protocol for particle insertions and deletions, discussed in application to zeolites by Snurr and co-workers[24]. In this method the system is divided into small cubelets. A probe hard sphere particle is placed in the centre of each cubelet and tested for overlaps with the particles of the structure. If no overlaps are registered, this cubelet is saved in a list of accessible cubelets. Insertions are then performed by random selection of a cubelet from the list. A molecule of adsorbate is randomly placed within the selected cubelet, with this move accepted or rejected based on the following biased probability criterion:

$$P_{insertion} = \min\left(1, \frac{q_1^{rot} e^{\beta\mu_1} V_C}{(N_1 + 1)\Lambda_1^3}\right) \tag{1}$$

where N_1 is the number of adsorbate molecules in the system, V_C is the total volume of all accessible cubelets, μ_1 is the chemical potential of the adsorbate and $\beta = 1/kT$ is the reciprocal temperature. Note that the de Broglie wavelength, Λ_1, and the ideal gas rotational partition function, q_1^{rot}, are implicitly set to σ and 1 respectively. In order to preserve the microscopic reversibility, the acceptance criterion for particle deletions also has to be biased:

$$P_{deletion} = \min\left(1, \frac{N_1 \Lambda_1^3}{q_1^{rot} e^{\beta \mu_1} V_C}\right) \tag{2}$$

In this work, each cubelet has edges of length 0.1σ and the probe particle is a hard sphere of diameter 0.6σ. Translations and rotations of molecules are implemented in a standard way for rigid bodies. Attempts to translate, rotate (in the case of molecular species), insert or delete a particle are weighted equally. Adsorption isotherms from five matrix realizations are used to calculate the average isotherm $\rho\sigma^3 = \dfrac{1}{n_{sim}} \sum_i^{n_{sim}} \rho_i \sigma^3$ where $\rho_i = \dfrac{N_1}{V}$ is the adsorbed particle number density for realization i, V is the volume of the system, $\rho\sigma^3$ is the average reduced adsorbed density, n_{sim} is the number of matrix realizations.

3. RESULTS AND DISCUSSION

We have explored a range of systems with the matrix density varying from $\rho_m \sigma_m^3 = (N_m/V)\sigma_m^3 = 0.216$ to 0.540 (N_m is the number of matrix particles). Molecular recognition in these structures is a result of unique geometric compatibility between a binding cavity and a guest molecule. This compatibility manifests itself only if the material is sufficiently dense to form tight cavities. As a result, in these studies only the system with the highest matrix density exhibited appreciable molecular recognition effects. Therefore, here we focus on this latter system. To create a system with the reduced density of 0.540, 2500 matrix particles are placed in the simulation box. As a convention, isomer R is the template species throughout the article. The first system we consider is imprinted with $N_t = 46$ template particles ($\rho_t \sigma^3 = (N_t/V)\sigma^3 = 0.046$). Thus, the ratio of the geometric volumes of the template and matrix phases is $N_t v_t / N_m v_m = 0.25$, where v_t and v_m are geometrical volumes of template and matrix particles, respectively. Figure 2 shows adsorption isotherms calculated for this system, where closed circles correspond to the adsorbed density of the original template species R and open circles correspond to the adsorption of the mirror image S of the template molecule. At the highest loading, there are about 30 adsorbed molecules in the system. The matrix also clearly exhibits appreciable molecular recognition as the adsorption of the R species (template) is systematically enhanced compared to the S species (the mirror image analogue). This effect is further enhanced by higher concentration of the template species. In Figure 3, adsorption isotherms are shown for the system templated with 138 molecules ($\rho_t \sigma^3 = (N_t/V)\sigma^3 = 0.138$), while the reduced density of the matrix remains the same. This corresponds to the ratio of the geometric volumes of the template and matrix components $N_t v_t / N_m v_m = 0.75$. Presence of a larger number of template molecules leads to a higher overall density of the system at the stage of the structure formation and as a result to lower loadings as can be seen from the figure.

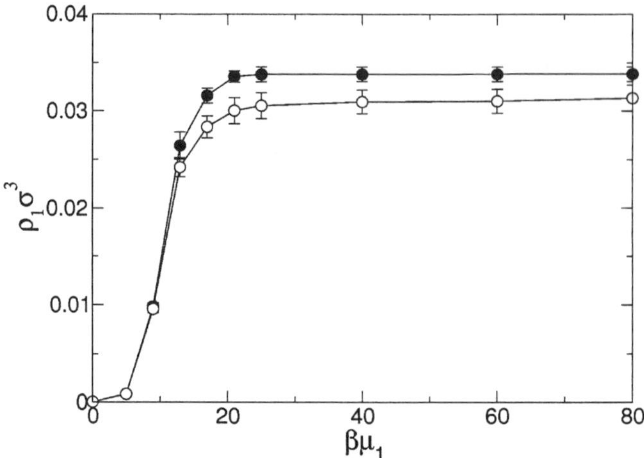

Figure 2: The adsorption isotherms (adsorbate density $\rho_1\sigma^3=(N_1/V)\sigma^3$ as a function of the adsorbate chemical potential $\beta\mu_1$). The matrix density is $\rho_m\sigma_m^3=(N_m/V)\sigma_m^3=0.540$, the template density is $\rho_t\sigma^3=(N_t/V)\sigma^3=0.046$. Adsorption of the original template species R is shown in black circles, whereas open circles correspond to the mirror image species S.

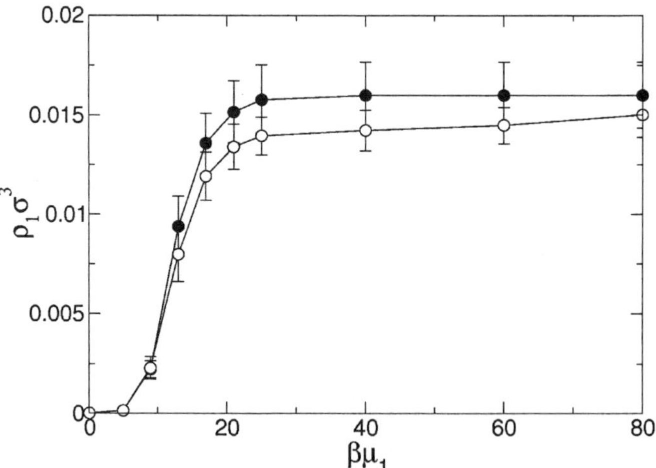

Figure 3: The adsorption isotherms (adsorbate density $\rho_1\sigma^3=(N_1/V)\sigma^3$ as a function of the adsorbate chemical potential $\beta\mu_1$). The matrix density is $\rho_m\sigma_m^3=(N_m/V)\sigma_m^3=0.540$, the template density is $\rho_t\sigma^3=(N_t/V)\sigma^3=0.138$. Adsorption of the original template species R is shown in black circles, whereas open circles correspond to the mirror image species S.

In Figure 4, we plot separation factors for two considered systems for a range of chemical potentials. The separation factor is defined as a ratio of the adsorbed density of the original template to its chiral counterpart at the same chemical potential $\Phi_1 = \dfrac{\rho_1^R(\beta\mu_1)}{\rho_1^S(\beta\mu_1)}$. The first few points on the adsorption isotherms are not included in the analysis because calculation of the separation factors in that region involves a ratio of two small values of the density and thus is a subject to significant statistical errors. Enantioselectivity of the model is somewhat enhanced by higher overall density of the matrix-template mixture; however it is within the statistical error of the calculations. As the difference is not significant we did not study other systems with the number of template particle varying between 46 and 138. We however did consider adsorption from binary racemic mixtures (as opposed to pure component adsorption) and observed similar trends. The observed separation factors do not exceed values of 1.2, and we believe this is a useful estimate of the possible role of steric factors in enantioselectivity in imprinted structures. Furthermore, the separation factor somewhat diminishes with the loading. At the same time, these studies reveal a rather limited potential for the chiral imprinted materials based on purely steric compatibility between the binding site and the guest molecule. As it has been already discussed, molecular recognition effects in these systems manifest themselves only at relatively high densities of the matrix. Our analysis of the morphology of these structures shows that at these densities the porous space is not fully accessible with respect to the adsorbate molecules; in other words different binding sites and porous regions are separated from each other by channels and windows too narrow for the adsorbate molecules to pass. Under conditions where the porous network is percolated and fully accessible with respect to the adsorbate species, steric molecular recognition effects are negligible. It seems, molecular recognition effects in these systems remain visible even at relatively high adsorbate densities. However, this is not a typical regime of operation for sensing or chromatographic applications. It is worthwhile to note, that several systems with lower matrix densities exhibit some signs of confined fluid crystallization. Interestingly, these features are more pronounced when probing adsorption from binary mixture of racemates. However, these effects must be explored in more detail to achieve conclusive results.

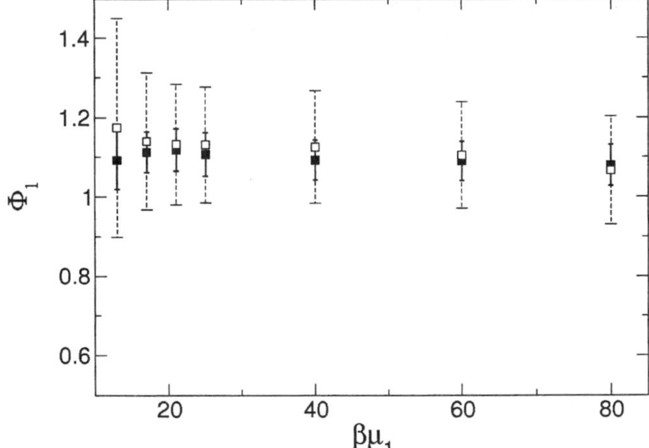

Figure 4: Chiral separation factors $\Phi_1 = \dfrac{\rho_1^R(\beta\mu_1)}{\rho_1^S(\beta\mu_1)}$ **as a function of the adsorbate chemical potential**

$\beta\mu_1$ **for systems with 46 template particles (closed squares, solid error bars) and 138 template particles (open squares, dashed error bars).**

The proposed models can be further extended to systematically incorporate and assess other effects participating in chiral recognition, such as van der Waals, electrostatic and hydrogen bond interactions. A particular attraction of these simplified models comes from a possibility of a theoretical treatment within the framework of integral equation theories. Although this path implies a number of well-recognized challenges, theory may provide more general, fundamental insights with less computational effort and therefore, we believe, it is an interesting and viable avenue of research.

Acknowledgements

We would like to thank the Engineering and Physical Sciences Research Council for financial support through grant no. EP/D074762/01.

References

1. N. M. Maier, P. Franco and W. Lindner, *Journal of Chromatography A*, 2001, **906**, 3-33.
2. C. Alexander, H. S. Andersson, L. I. Andersson, R. J. Ansell, N. Kirsch, I. A. Nicholls, J. O'Mahony and M. J. Whitcombe, *J. Mol. Recognit.*, 2006, **19**, 106-180.
3. L. Fischer, R. Muller, B. Ekberg and K. Mosbach, *Journal of the American Chemical Society*, 1991, **113**, 9358-9360.
4. N. M. Maier and W. Lindner, *Analytical and Bioanalytical Chemistry*, 2007, **389**, 377-397.
5. R. J. Ansell, *Advanced Drug Delivery Reviews*, 2005, **57**, 1809-1835.
6. I. A. Nicholls, *Chemistry Letters*, 1995, 1035-1036.
7. S. A. Piletsky, K. Karim, E. V. Piletska, C. J. Day, K. W. Freebairn, C. Legge and A. P. F. Turner, *Analyst*, 2001, **126**, 1826-1830.
8. S. Subrahmanyam, S. A. Piletsky, E. V. Piletska, B. N. Chen, K. Karim and A. P. F. Turner, *Biosens. Bioelectron.*, 2001, **16**, 631-637.
9. I. Chianella, M. Lotierzo, S. A. Piletsky, I. E. Tothill, B. N. Chen, K. Karim and A. P. F. Turner, *Anal. Chem.*, 2002, **74**, 1288-1293.
10. D. Pavel and J. Lagowski, *Polymer*, 2005, **46**, 7528-7542.
11. E. V. Piletska, N. W. Turner, A. P. F. Turner and S. A. Piletsky, *J. Control. Release*, 2005, **108**, 132-139.
12. I. Chianella, K. Karim, E. V. Piletska, C. Preston and S. A. Piletsky, *Anal. Chim. Acta*, 2006, **559**, 73-78.
13. S. Monti, C. Cappelli, S. Bronco, P. Giusti and G. Ciardelli, *Biosens. Bioelectron.*, 2006, **22**, 153-163.
14. M. Cao and P. A. Monson, *Journal of Chemical Physics*, 2005, **122**.
15. J. Vatamanu and N. M. Cann, *Journal of Chemical Physics*, 2001, **114**, 7993-8007.
16. J. Largo, C. Vega, L. G. MacDowell and J. R. Solana, *Molecular Physics*, 2002, **100**, 2397-2415.
17. P. R. Van Tassel, *Phys. Rev. E*, 1999, **60**, R25-R28.
18. L. H. Zhang and P. R. Van Tassel, *Molecular Physics*, 2000, **98**, 1521-1527.
19. L. H. Zhang and P. R. Van Tassel, *Journal of Chemical Physics*, 2000, **112**, 3006-3013.
20. S. Cheng and P. R. Van Tassel, *Journal of Chemical Physics*, 2001, **114**, 4974-4981.
21. L. H. Zhang, S. Y. Cheng and P. R. Van Tassel, *Phys. Rev. E*, 2001, **6404**, 4.
22. L. Sarkisov and P. R. Van Tassel, *Journal of Chemical Physics*, 2005, **123**, 10.
23. L. Sarkisov and P. R. Van Tassel, *Journal of Physical Chemistry C*, 2007, **111**, 15726-15735.
24. R. Q. Snurr, A. T. Bell and D. N. Theodorou, *Journal of Physical Chemistry*, 1993, **97**, 13742-13752.

MIXED GEOMETRY CHARACTERIZATION OF ACTIVATED CARBONS PSD

D. C. S. Azevedo[1], C. L. Cavalcante[1], R. H. López[2], A. E. B. Torres[1], J. P. Toso[2], G. Zgrablich[1,2]

[1] Universidade Federal do Ceará, Departamento de Engenharia Química, Fortaleza, Ceará, Brasil
[2] Universidad Nacional de San Luis-CONICET, Instituto de Física Aplicada (INFAP), San Luis, Argentina

1 INTRODUCTION

The slit pore geometry is usually assumed for the characterization of activated carbons, especially in determining the Pore Size Distribution (PSD). However, observed high values of the heat of adsorption in activated carbons at very low pressure cannot be explained by slit micropores, even for ultra-small sizes. On the other hand, it is reasonable to assume that, at least in part of the material, carbon plates may accommodate in such a way as to form pores with other geometries. Steele *et al.* and Seaton *et al.* [1,2] have investigated, for example, the use of rectangular micropores. In this work, we propose micropores with a triangular section. This kind of geometry would provide adsorption regions of higher heats of adsorption, since adsorption in the center of a triangular pore is affected by 3 graphitic plates instead of 2.

In the present work, pores with slit and triangular geometry are assumed to co-exist in undetermined proportions in activated carbons, and a method is proposed to obtain the corresponding PSD of such a material. By using a Grand Canonical Monte Carlo (GCMC) simulation method in the continuum space, families of N_2 adsorption isotherms are generated both for slit and for triangular geometry corresponding to different pore sizes. These families of isotherms are then used to fit experimental adsorption data corresponding to a controlled set of samples of activated carbons, allowing the determination of the micropore volume, the proportion of slit and triangular pores and the PSD corresponding to the combined geometry. Experimental N_2 isotherms for a number of activated carbon samples from different sources were used in the fitting procedure. The same experimental data were fit using each of the "pure" geometries and combinations thereof so that conclusions were drawn about the reliability and convenience of the proposed characterization method.

2. DESCRIPTION OF THE MODEL

Slit and triangular geometry pores are represented in Figure 1, only equilateral triangles are considered for the latter in order to keep the number of parameters to a minimum. The gas-solid potential for the slit geometry is given, as usual, by the superposition of two Steele potentials, one per each infinite plate:

$$U_{gs-STEELE}(z) = 2\pi\, \varepsilon_{gs}\rho_C\, \sigma^2{}_{gs}\Delta\left\{ \frac{2}{5}\left(\frac{\sigma_{gs}}{z}\right)^{10} + \left(\frac{\sigma_{gs}}{z}\right)^4 - \frac{\sigma^4{}_{gs}}{3\Delta(z+0.61\Delta)^3} \right\}$$

For the triangular geometry, the gas-solid potential is obtained by summing the contributions of three semi-infinite plates. The potential of each semi-infinite plate is given by [1]:

$$U_{gs}(z,y_e) = 4\,\varepsilon_{gs}\rho_C\left\{ -\sigma^6{}_{gs}I_3(z,y_e) + \sigma^{12}{}_{gs}I_6(z,y_e) \right\}$$

$$I_n(z,y_e) = \int_{-\infty}^{+\infty}dx \int_{ye}^{\infty}dy\, \frac{1}{\left(x^2 + y^2 + z^2\right)^n}$$

where y_e is the distance from the truncated edge. The gas-gas potential is taken as the usual Lennard-Jones potential.

Figure 1. *Slit (left) and triangular (right) geometry pore with pore size S.*

For N$_2$ adsorption at 77 K, with the values of the parameters for the potentials given in Table 1, the minimum gas-solid potential (corresponding to the maximum differential heat of adsorption at very low adsorbed quantity) is represented in Figure 2 as a function of the pore size S. From the figure, it can be seen that only the triangular geometry can provide a

Parameter	Value	Ref.
ρ_C	0.114 Å^{-3}	(6)
Δ	3.35 Å	(6)
ε_{gs}/k_B	53.22	(7)
σ_{gs}	3.494 Å	(7)
ε_{gg}/k_B	101.5	(7)
σ_{gg}	3.615 Å	(7)

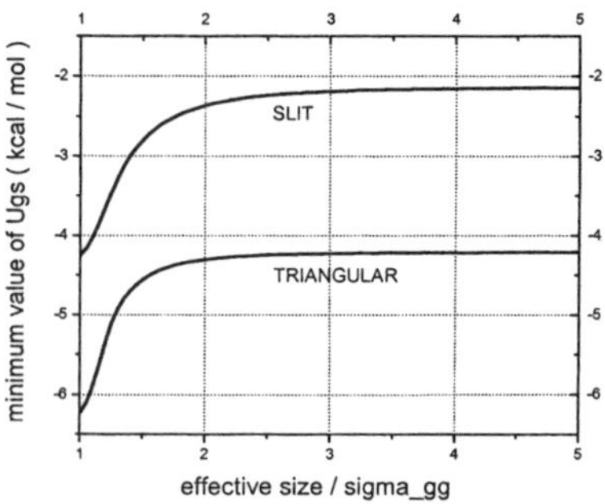

Table 1. *Values of potential parameters.*   **Figure 2.** *Gas-solid potential minima.*

high value of $\approx$ 6 kcal/mol for the differential heat of adsorption as observed experimentally [3].

3. EXPERIMENTAL

Five selected samples of activated carbon were used to obtain N_2 adsorption isotherms at 77 K.

- Samples 1 and 2, were chosen from a series of activated carbons prepared from coconut shells, using phosphoric acid as the activation agent, under different carbonization conditions, as described in Ref. [4]. These two samples have different characteristics: Sample 1 is almost exclusively microporous, while Sample 2 presents an important contribution from the mesoporous region.
- Sample 3 is a commercial microporous activated carbon supplied by Westvaco, USA (WV1050) and recommended for gas storage.
- Samples 4 and 5 correspond to activated carbons prepared from petroleum coke, using KOH as the activation agent, under different carbonization conditions, as described in Ref. [8].

Experimental adsorption isotherms are shown in Figure 3 (symbols) together with fits obtained with the mixed pore geometry model (lines).

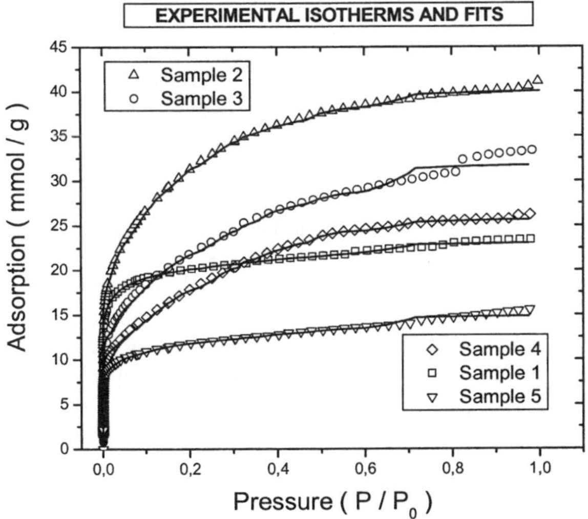

Figure 3. *Experimental N_2 isotherms at 77 K (symbols) and theoretical fits (lines).*

4. GCMC SIMULATION METHOD

A collection of adsorption isotherms (the local isotherms, θ_L) was obtained through the Grand Canonical Monte Carlo method, following the algorithm outlined in Ref. [9], both for the slit and the triangular geometries with the values used for the interaction potential parameters as given in Table 1. Transition probabilities for each Monte Carlo attempt, displacement, adsorption and desorption of molecules, are given by the usual Metropolis rules. The lateral dimensions of the cell for the slit geometry, and for the longitudinal dimension for triangular geometry, were taken as 10.3 nm and periodic boundary conditions were used in those directions. Equilibrium was achieved generally after 10^7 MC attempts, after that mean values were taken over the following 10^6 MC attempts over configurations spaced by 10^3 MC attempts, in order to assure statistical independence.

A minimization method for the mean square error, with a regularization term, as described in Ref. [5], was used to fit each experimental isotherm with the theoretical isotherm given by:

$$\theta_i^{theor} = \sum_{j=1}^{m} \theta_L(S_j^*, P_i, T)\, f(S_j^*)\, \delta S_j$$

For each of the three models: a) pure slit geometry, b) pure triangular geometry and c) mixed slit-triangular geometry, the corresponding Pore Size Distributions (PSD) were obtained. In the above equation, θ_i^{theor} stands for the theoretical value of adsorbed amount at pressure P_i and S_j^* for the value of the pore size in the middle of the jth interval. The number of fitting size parameters, m, was kept fixed for the three models.

5. RESULTS AND DISCUSSION

All results for the PSDs are shown in Figure 4. The first column shows the Pore Size Distribution obtained with the mixed geometry model; dark gray bars denote the contribution of the slit pores, while light gray ones stand for the total PSD. In this way, the contribution of each pore geometry can be readily appreciated. The resulting theoretical isotherms are shown by the continuous lines in Figure 3. The second and third columns correspond to the slit geometry and to the triangular geometry model, respectively. Fitting errors are given in the inset for each model. These fitting errors are small and of approximately comparable magnitude, so that it can be said that the three models provide satisfactory fits to experimental isotherms.

By examining the experimental adsorption isotherms, one can easily see that samples 1 and 5 are essentially microporous, while samples 2, 3 and 4 seem to present an important contribution to the PSD in the mesoporous range. PSDs obtained through simulations clearly reflect this property. On the other hand, following the line of reasoning given in Refs. [1,2], a real activated carbon is very likely to present a mixture of pores of different geometries. Pores with rectangular cross section have been studied in the literature [1,2]. However, in order to account for the large differential heat of adsorption observed at very low adsorbed amounts [3], a triangular geometry seems to be mandatory. Therefore, on the basis of the observed maximum adsorption energies, we propose that the PSD should be calculated through a model where the porous space is represented by a mixture of slit and triangular pores. PSDs obtained by considering the material as formed by a unique geometry results to be biased toward either smaller or larger pores. In fact, the pure slit geometry requires an excess of ultra-small pores in order to account for the strong adsorption region, thus shifting the PSD toward smaller pores. On the other hand, a pure triangular geometry, due to the effect of corners and the joint interaction with three graphitic   plates,   can   account   for   the   strong   adsorption   energies   even

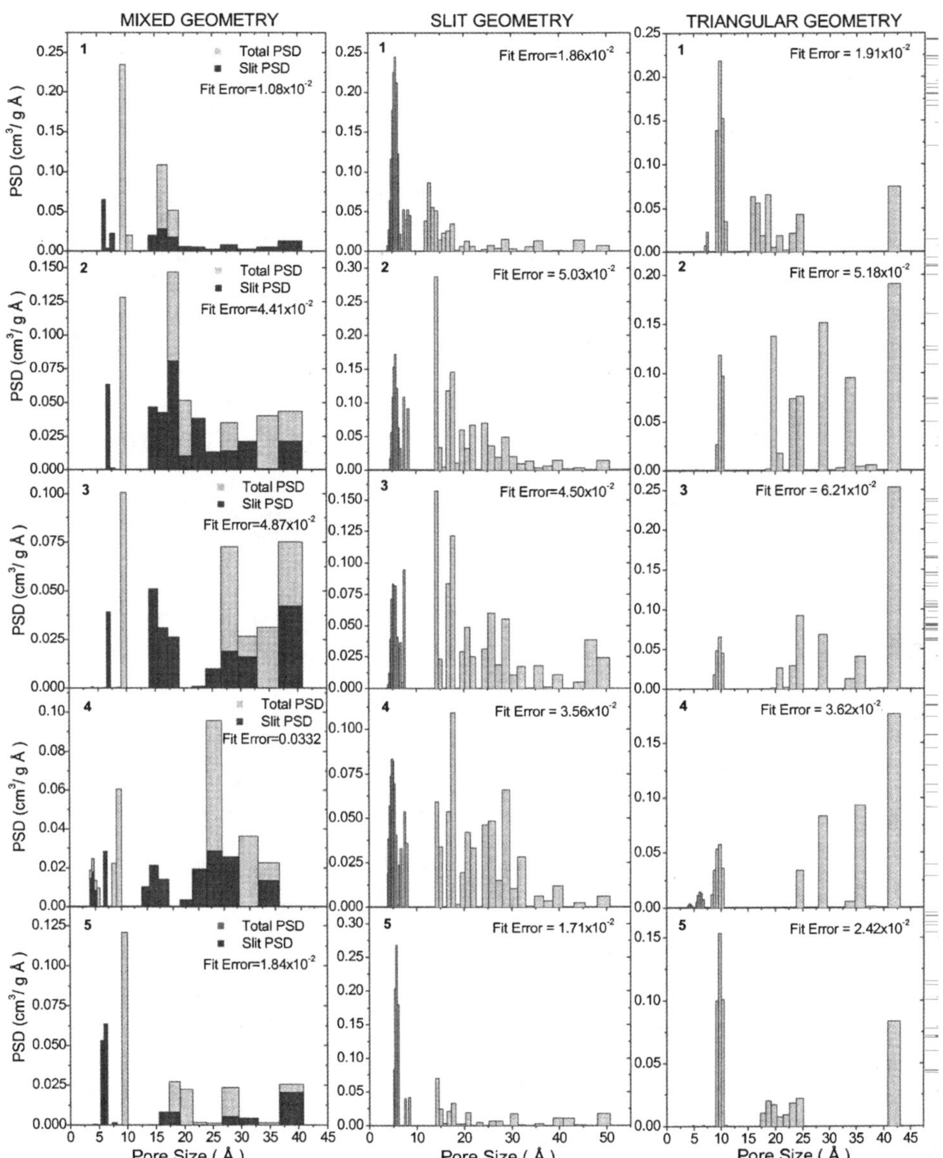

Figure 4. *PSD for the different models tested. The fitting error for each model is shown.*

with larger pores, thus shifting the PSD toward larger pores. Hence, a mixed-geometry PSD may probably better capture the energetics of the adsorption system. However, a more accurate analysis should be carried out by simultaneously fitting the adsorption isotherm and the differential heat of adsorption as a function of adsorbed amount, both obtained experimentally for the same sample. So far, we have not been able to find this kind of data in the literature.

Finally, in order to reinforce the reliability of the proposed model, we should briefly discuss the problem of the unicity of the solution for the mixed geometry. In fact, it could be argued that if the adsorption isotherm of a slit pore of size S_i is similar to that of a triangular pore of size S_j, then any fraction x of one and $1 - x$ of the other will give the same fit for all $0 \leq x \leq 1$. To this end, we have calculated a matrix $D(i, j)$, for all slit geometry sizes i and all triangular geometry sizes j, where each element represents the percentual relative absolute error between the isotherms corresponding to the two geometries. The smallest matrix element found is of 23%, the second smallest value found was 30%. Figure 5 shows a comparison between adsorption isotherms for the two geometries in both cases. The mean matrix element value turned out to be 71%, wile the maximum value found was 87%. We therefore conclude that the solution for the mixed geometry is unique.

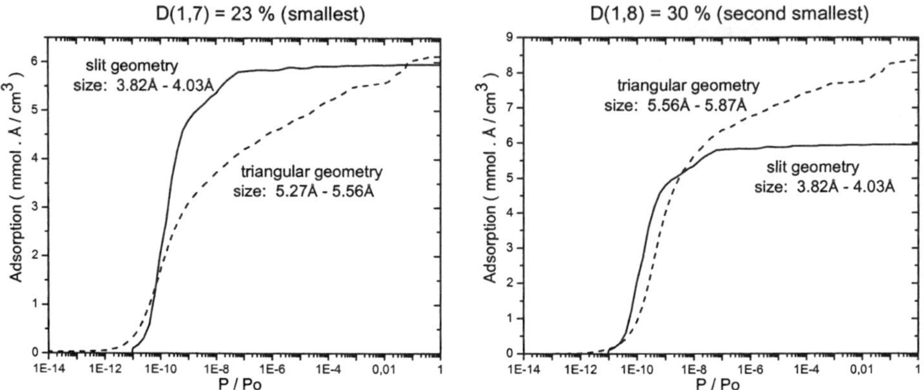

Figure 5. *Comparison between adsorption isotherms for the two geometries for* $D(1,7) =$ 23% *and* $D(1,8) = 30\%$.

6. CONCLUSIONS

Real activated carbons very likely present a mixture of pore geometries, as a result of different packings of graphitic plates. The maximum adsorption energy provided by each graphitic plate is approximately 2 kcal/mol. Therefore, in order to account for experimentally observed values around 6 kcal/mol, the interaction of a nitrogen molecule with three graphitic plates is necessary. This interaction can only be provided by pores with a triangular section.

A mixed geometry model, combining slit and triangular pores, is proposed and PSDs are obtained by fitting experimental N2 isotherms at 77K for three cases: a) pure slit geometry; b) pure triangular geometry; c) mixed geometry.

Slit geometry PSDs are likely to be shifted toward smaller pores, while triangular geometry PSD are shifted toward larger pores. The PSD which perhaps better captures the energetics of the adsorption system should be in between, and is probably close to the mixed geometry PSD.

Simultaneous adsorption and calorimetric experiments performed on the same sample should be extremely useful in order to carry out a more rigorous evaluation of the mixed geometry model.

Acknowledgements

The authors gratefully acknowledge CNPq (Brazil) and CONICET (Argentina) for financial support to carry out this research, and to CAPES (Brazilian Ministry for Education) for a Visiting Professor grant to Giorgio Zgrablich.

References

1. M.J. Bojan and W.A. Steele, *Carbon*, 1998, **36**, 1417.
2. G.M. Davies and N.A. Seaton, *Carbon*, 1998, **36**, 1473.
3. M. Prasad *et al.*, *Carbon*, 1999, **37**, 1641.
4. M. Bastos-Neto, D.V. Canabrava, A.E.B. Torres, E. Rodriguez-Castellón, A. Jiménez-López, D.C.S. Azevedo and C.L. Cavalcante Jr., *Appl. Surf. Sci.*, 2007, **253**, 5721.
5. G.M. Davies, N.A. Seaton and V.S. Vassiliadis, *Langmuir*, 1999, **15**, 8235.
6. W.A. Steele, *The Interaction of Gases with Solid Surfaces*; Pergamon, Oxford, 1974
7. P.I. Ravikovitch, A. Vishnyakov, R. Russo and A.V. Neimark, *Langmuir*, 2000, **16**, 2311.
8. M. O. A. Mendez, A. R. Coutinho and A. C. Lisboa. *Journal of the Brazilian Chemical Society*, 2006, **17**, 1144.
9. D.L.Valladares, F. Rodríguez-Reinoso and G. Zgrablich, *Carbon*, 1998, **36**, 1491.

MODIFIED BET EQUATION FOR DETERMINATION OF MICROPORE PORE-VOLUME AND MESOPORE SURFACE AREA IN MICROPOROUS-MESOPOROUS SOLIDS

O. Šolcová[1], L. Matějová[1], P. Hudec[2] and P. Schneider[1]

[1]Institute of Chemical Process Fundamentals ASCR, v.v.i., 16502 Prague 6, Rozvojová 135, Czech Republic
[2]Department of Petroleum Technology and Petrochemistry, Faculty of Chemical and Food Technology, Slovak University of Technology, 81237 Bratislava, Slovak Republic

1 INTRODUCTION

The importance of obtaining information on textural properties for microporous-mesoporous solids has enlarged tremendously with the huge increase of synthesized new materials possessing micro-mesoporous structure. Usually, the physical adsorption of inert gases is routinely applies. The presence of micropores makes the textural analysis based on transformation of adsorption isotherms much less straightforward than in the case of porous solids with mesopores only. The simple BET analysis [1] is, nevertheless, usually (and incorrectly) performed. The reason is that the simple BET isotherm was developed explicitly for non-microporous (i.e. purely mesoporous) solids and in the relative pressure range, x_{BET}, which guarantees no capillary condensation in pores (e.g. $0.25 > x_{BET} > 0.05$). The correct way is to use the comparative plots (t-plots, α-plots) which can supply the micropore volume, V_μ, as well as the mesopore surface area, S_m. The complication with comparative plots arises from the requirement to know the adsorption isotherm for a non-porous solid with the same chemical character as the analyzed porous sample (standard isotherm). The generalized standard isotherms of Lecloux and Pirard [2], $a = a(x, C)$ (see Appendix), are also of no use because the required C parameter obtained by application of the BET equation is invalid when micropores are present.

The other possibility is using the modified BET equation (1) (three-parameter BET [3]), which when applied to the analyzed sample gives the adsorbate volume completely filling the micropores, a_μ, the monolayer capacity, a_m, and the parameter C.

$$a = a_\mu + \frac{a_m C x}{(1+x)\left[1+(C-1)x\right]} \tag{1}$$

Here a is the adsorbed amount at relative pressure x ($x = p/p_o$ with p the adsorbate pressure and p_o the saturation adsorbate pressure at the measurement temperature) and C is a constant proportional to adsorption equilibrium constant in the first adsorbed layer. a, a_m and a_μ can be expressed either as adsorbate gas volume at standard conditions or as condensed liquid adsorbate volume. Transformation between these quantities can be performed with the use of data in Table 1.

Table 1. *Quantities for analysis of adsorption isotherm*

	N$_2$ at 77 K
Liquid molar volume, V$_m$, cm$^3{}_{liq}$/mol	34.679
Conversion factor, mm$^3{}_{liq}$/ cm$^3{}_{STP}$	1.5468
Area covered by one adsorbate molecule, σ, nm^2	0.162

The mesopore surface area, S$_m$, is obtained from the monolayer capacity a$_m$ as

$$S_m = \frac{a_m}{V_m} A\sigma \qquad (2)$$

with molar volume, V$_m$, Avogadro constant, A, and the area covered by one adsorbate atom/molecule, σ. Micropore volume, V$_\mu$, is simply V$_\mu$ = a$_\mu$/$\mathcal{V}_m$. As can be seen the modified (three-parameter) BET equation differs from the classic (two-parameter) BET equation by the presence of the a$_\mu$ term.

2 EXPERIMENTAL

A series of porous samples with variable micropore volume and nearly the same mesoporous structure was prepared as the mechanical mixtures of microporous zeolite, NaY (batch Nr. NaY-091087V3 with Si/Al=2.24), and mesoporous γ-alumina (laboratory prepared from boehmite) with γ-alumina weight fractions, f, varying between 0 and 1. Textural properties of both samples are summarized in Table 2. Nitrogen adsorption isotherms for pure NAY and γ-alumina are shown in Figure 1.

Adsorption measurements were performed with the volumetric instrument ASAP2020 (Micromeritics, USA). Prior to the measurement all samples were degassed overnight at 350 °C and 2 Pa.

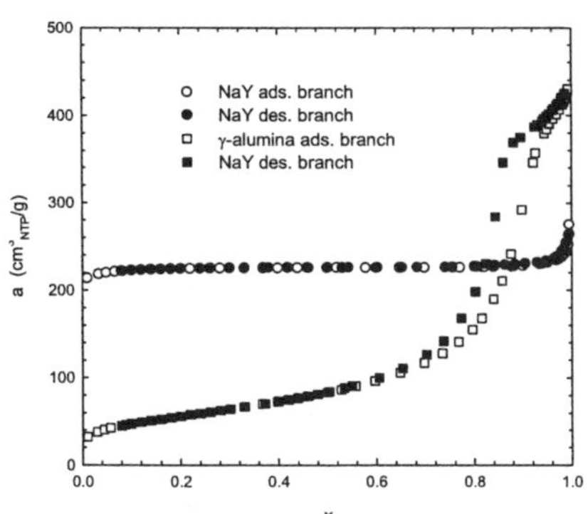

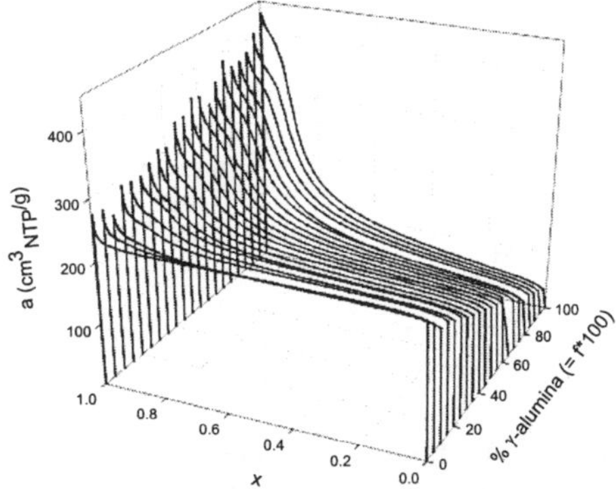

Figure 2 Adsorption branches of isotherms (77K) of pure NaY and γ−alumina

Figure 1 Nitrogen adsorption isotherms of NaY/γ-alumina mixtures

Table 2. *Textural properties of pure NaY and γ-alumina*

	NaY	**γ–Alumina**
Monolayer capacity, a_m (mm^3$_{liq}$/g) [*)	11.7	71.2
Mesopore surface area, S_m (m^2/g) [*)	32.9	199.7
Micropore volume, a_μ (mm^3/g) [*)	335	0
C [*)	22.6	95.2
Most frequent pore radius (nm) [**)	-	5.1[+), 6.9[++)

[*) from Eq.(1), for recalculation of adsorbed amounts in cm^3$_{NTP}$/g to mm^3$_{liq}$/g multiplication by the factor 1.4568 was used;

[**) from BJH [4] pore-size distribution with α–alumina standard isotherm [5]
[+) from adsorption isotherm branch; [++) from desorption isotherm branch

3 RESULTS AND DISCUSSION

3.1 Three-parameter BET equation

To show the correct determination of micropore pore-volume and mesopore surface area in microporous-mesoporous samples via the three-parameter BET equation nitrogen adsorption isotherms (77K) were determined for a series mixed NaY+γ-alumina samples. The adsorption branches for all isotherms are shown in Figure 2.

A thorough analysis of mixed sample isotherms with the modified BET equation was performed with i) C parameter fixed at the value for pure γ-alumina (C=95.2 see Table 1) and ii) without fixing the C parameter. Figure 3 summarizes the micropore volumes, V_μ, and mesopore surface areas, S_m, obtained with the fixed C parameter. The non-linear data fitting was performed with the use of the simplex algorithm.

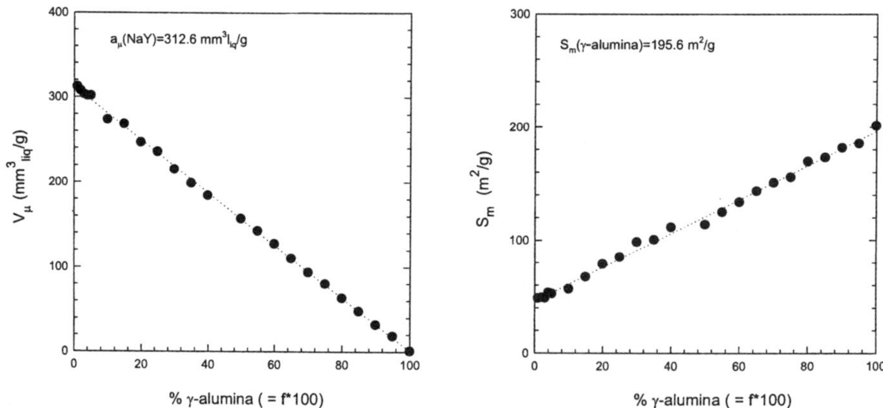

Figure 3 Micropore volumes, V_μ, and mesopore surface areas, S_m, for mixed samples determined by the modified BET equation with C fixed at C = 95.2 (i.e. value for pure γ-alumina).

As can be seen, micropore volumes as well as mesopore surface areas depend linearly on the amount of NaY in samples. Similar results were obtained by fitting the experimental mixed sample data to Equation (1) without fixing the C parameter. For both cases the micropore volume of NaY and the mesopore surface area of γ-alumina were determined by extrapolation of the linear dependences $V_\mu(f \to 0)$ for micropore volumes and $S_m(f \to 1)$ for mesopore surface areas. The data are summarized in Table 2.

3.2 t-plots

To obtain comparative textural properties (micropore volumes and mesopore surface areas) for all mixed samples, t-plots were constructed with application two kinds of standard isotherms. As the source of film thicknesses [5] the isotherm of nonporous α-alumina and the standard isotherm of Lecloux-Pirard [2] were used. Lecloux-Pirard present an interpolation formula for standard isotherms which depend on the BET parameter C. Figure 4 shows the t-plots for individual mixed samples and the resulting micropore volumes, V_μ, for C = 95.2 (i.e. value for pure γ-alumina sample). As can be seen V_μ decreases linearly with the fraction of γ-alumina in the sample, f. Nearly identical results were obtained when the nonporous α-alumina master isotherm was used.

For both standard isotherms the micropore volume of NaY and the mesopore surface area of γ-alumina were determined by the same way as for the three-parameter BET equation - extrapolation of the linear dependences $a_\mu(f \to 0)$ and $S_m(f \to 1)$. The data are also shown in Table 3.

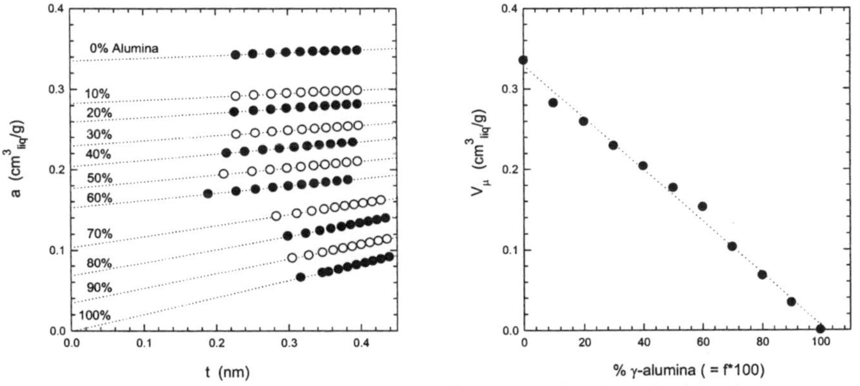

Figure 4 t-plots and micropore volumes for mixed samples (standard isotherms according to Lecloux-Pirard [2] with C from pure γ-alumina sample)

Table 3 *Extrapolated a_μ and S_m for pure NaY and γ-alumina*

Obtained by	V_μ (mm³/g)	S_m(m²/g)
Modified BET equation with fixed parameter C 95.2	313	196
Modified BET equation with free parameter C	337	200
t-plots with the standard isotherm of nonporous α-alumina	312	180
t-plots with the standard isotherm of Lecloux and Pirard	321	192

It can be seen that the extrapolated V_μ and S_m for pure NaY and γ-alumina, evaluated by independent methods, differ only slightly and also the results from the modified BET equation with free parameter C are nearly equal to values for pure NaY (335 mm^3/g) and γ-alumina (200 m^2/g). This proves the applicability of the modified BET equation for analysis of microporous-mesoporous samples.

3.3 Classic BET equation

The application of the classic and modified BET equations to microporous-mesoporous mixed samples is compared in Figure 5. For demonstration, adsorption points from the BET region (0.05<x<0.25), for the sample with 50% γ-alumina and 50% NaY, were chosen. It can be seen that the isotherm calculated with use of parameters from the classic BET equation (solid line) deviates significantly from experimental points. Quite the opposite is true when parameters from the modified BET equation were used (dashed line). Also the physically incorrect value of the C constant obtained from the classic BET equation (C = -71) points to the inappropriateness of application of the classic BET equation for isotherms analyses of microporous-mesoporous samples. It is worth noticing that the specific surface area from the classic BET equation is several-fold higher than

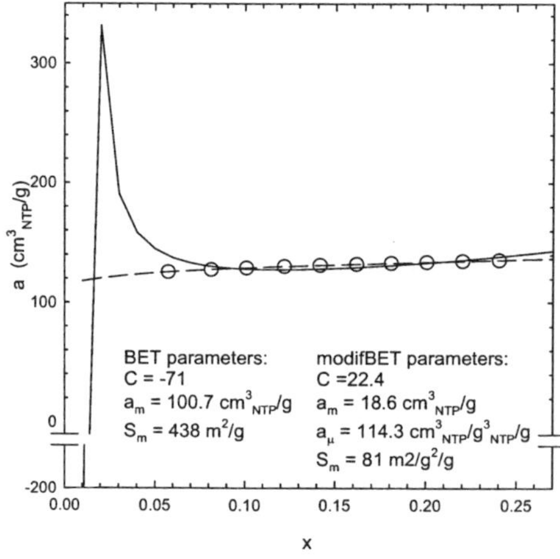

Figure 5 Experimental adsorption points and isotherms calculated with parameters of the classic (solid line) and modified BET (dashed lines) equations for mixed sample with 50% NaY/50% γ-alumina

from the modified BET equation. This is because in the classic BET equation the adsorbate volume condensed in micropores is considered as covering large mesopore surface.

To demonstrate the failure of classic BET equation for the other type of materials with various amounts of micropores the analysis of active carbon, alumina Alcoa and magnesia isotherms by the use of the modified BET equation, t-plot with Lecloux and

Pirard master isotherm and classic BET equation were performed. Data are summarized in Table 4.

Table 4 *Comparison of texture properties of materials with various amount of micropores*

	Three-parameter BET			t - plot		Classic BET plot	
	V_μ (mm^3/g)	S_m (m^2/g)	C (-)	V_μ (mm^3/g)	S_m (m^2/g)	S_{BET} (m^2/g)	C (-)
Active carbon	283	325	25.3	282	336	896.4	-157
Alumina Alcoa	71.9	182.8	4.2	71.7	193.5	267.8	81.2
Magnesia	0.37	10.6	120.3	0.14	11.4	12.4	181.1

It is evident that micropore volumes and mesopore surface areas evaluated with the three-parameter BET equation and t-plots agree very well for all three samples. The failure of the classic BET equation is directly noticeable for active carbon owing to the negative value of constant C. In this case calculated BET surface area is three times higher than the real mesopore surface area. As far as alumina Alcoa and magnesia is concerned the error in the surface area determination is not easily visible. For both samples the C constants look reasonably (81 and 181), but the overestimation of the surface area by the classic BET equation is evident even for magnesia with very low micropore volume (about 0.2 mm^3/g).

3.4. Accuracy of the three parameter BET equation

The accuracy of micropore volume, mesopore surface area and parameter C obtained by modified BET analysis was studied for a SBA-15 material. The optimum values were V_μ = 73 mm^3/g, S_m = 727 m^2/g and C = 54.9. The accuracy of evaluated texture properties is demonstrated in the Figure 6 for parameter C, where the sum of squared deviations (SSD) for varied C is shown. It is noticeable that the curve has a deep minimum for optimum parameter C.

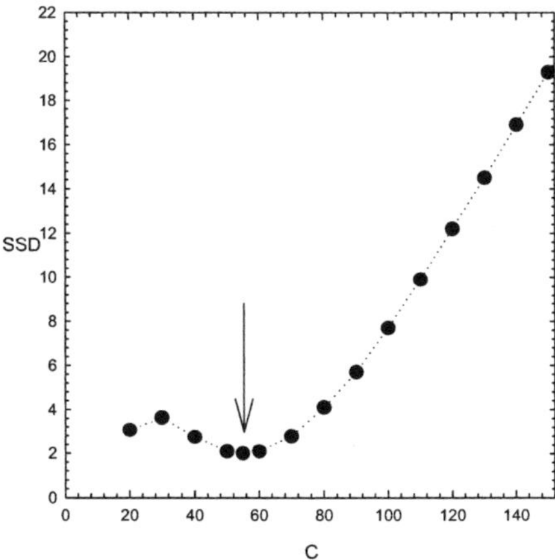

Figure 6 Sum of square deviations for varied C and corresponding optimum S_m and V_μ

4 CONCLUSIONS

To show the applicability of modified (three-parameter) BET equation for evaluation of micropore-volume and mesopore surface area in microporous-mesoporous samples a series of mechanical mixtures of microporous zeolite, NaY, and mesoporous γ-alumina was prepared. Samples possessed variable micropore volume and nearly the same mesopore volume. Textural characteristics of mixed samples isotherms obtained directly from the modified BET equation agreed very closely with independent results based on t-plots. This proves the suitability of the modified BET equation for analysis of adsorption isotherms of microporous-mesoporous samples. For illustration, the failure of classic (two-parameter) BET equation usually (and incorrectly) performed for samples of this type was clearly demonstrated. Unfortunately, the present frequent literature on attempts to synthesize new samples with prescribed pore structures, is filled with information obtained from the classic BET equation even if the studied samples contained non-negligible amount of micropores.

Acknowledgement

The financial support of the Grant Agency of Academy of Sciences of the Czech Republic (A 4072404), Slovak Grant Agency VEGA (1/3575/06) and the Czech Academy of Sciences, Program Nanotechnology for Society (KAN400720701) are gratefully acknowledged.

References

1   S. Brunauer, P.H. Emmett, E. Teller, Journal of the American Chemical Society 60 (1938) 309.
2   A. Lecloux, J. P. Pirard, Journal of Colloid and Interface Science 70 (1979) 265.
3   P. Schneider, Appllied Catalysis A: General 129 (1995) 157-165.
4   E.P. Barrett, L.G. Joyner, P.P. Halenda, Journal of the American Chemical Society 73 (1951) 373.
5   L. Matějová, O. Šolcová, P. Schneider, Microporous and Mesoporous Materials 107 (2008) 227.
6   W.H. Press, B.P. Flannery, S.A. Teukolsky, W.T. Vetterling, Numerical Recipes, Cambridge University Press, Cambridge, 1986.

CHARACTERISATION OF NANOSTRUCTURED MACRO-MESOPOROUS TiO_2-ZrO_2 IMPREGNATED BY NOBLE METALS FOR VOC OXIDATION

M. Hosseini [1], S. Siffert [1]*, H. L. Tidahy [1], R. Cousin [1], A. Aboukaïs [1], G. De Weireld [2], X. Canet [2], Z. Hadj-Sadok [3], B.-L. Su [3]

[1] Laboratoire de Catalyse et Environnement, E.A. 2598, Université du Littoral Côte d'Opale, 145 avenue Schumann, 59140 Dunkerque, France (siffert@univ-littoral.fr)
[2] Laboratoire de Thermodynamique, Faculté Polytechnique de Mons, 31 boulevard Dolez, 7000 Mons, Belgium
[3] Laboratoire de Chimie des Matériaux Inorganiques, Université de Namur, 61 Rue de Bruxelles, 5000 Namur, Belgium

1   INTRODUCTION

It is of great interest to study bimetallic catalysts due to their different properties from either of the constituent metals, and enhance catalytic stabilities, activities and/or selectivities [1, 2]. Pd-Au bimetals are studied as catalyst phases for several applications including VOC oxidation to carbon dioxide and water, at lower temperature and atmospheric pressure [3-5]. The addition of Au to Pd can significantly improve its catalytic properties [6]. Moreover, it was found that the support is one of the more important factors influencing the catalyst behaviour. The support can contribute to the catalyst stability and participate also in the improvement of catalytic efficiency. Many oxides as Al_2O_3, SiO_2, TiO_2 and ZrO_2 have been studied as support for catalysts based on noble metals for use in hydrocarbons oxidations [7-11]. Titania and zirconia supported by noble metals have been reported as active catalysts in the reaction of complete VOC oxidation [12-14]. Recently, nanostructured macro-mesoporous oxides attracted much attention for their specific properties of the high surface area and their influence on the catalytic activity [15]. Therefore, in this work, we studied new nanostructured macro-mesoporous TiO_2-ZrO_2 mixed oxides samples impregnated by Pd and/or Au for VOC oxidation.

2   EXPERIMENTAL

Nanostructured macro-mesoporous TiO_2-ZrO_2 mixed oxides were prepared using a micellar aqueous acidic solution of surfactant (pH=2) as templating agent. The inorganic sources were added drop by drop to the solution. The precursors employed were Ti $(OC_3H_7)_4$ (98%, Aldrich) and Zr $(OC_3H_7)_4$ (70%, Chempur) and CTMABr was used as surfactant. The molar ratio of Ti/Zr were 80/20, 50/50 and 20/80 and the surfactant/precursors molar ratio was 0.33. The obtained gel after stirring for 1h at room temperature was sealed into teflon autoclaves and heated. The hydrothermal treatment was performed during 24h at 80°C. The template was completely removed after 48h by ethanol extraction. The mesoporous titania-zirconia mixed oxide was dried at 40°C and calcined in air at 400°C for 4h. The synthesised supports were denoted TZ 80/20, TZ 50/50 and TZ 20/80 (titania-zirconia is denoted as TZ).

The new supports are impregnated by 0.5 or 1.5wt% of palladium and 1wt% of gold. The catalysts were prepared in accordance to the order of deposition of the promoters (gold and palladium):

- 0.5 and 1.5wt% of palladium on mesoporous supports denoted respectively as Pd/TZ and 1.5Pd/TZ.
- 1wt% of gold on mesoporous supports denoted as Au/TZ.
- 0.5wt% of palladium on mesoporous supports promoted by 1wt% of gold, denoted as PdAu/TZ.
- 1wt% of gold on mesoporous supports promoted by 0.5wt% of palladium, denoted as AuPd/TZ.

The palladium and/or gold samples were prepared by a method, previously described by our group [6]. Gold was deposited using Deposition-Precipitation method with a solution of $HAuCl_4$. Palladium supported samples were prepared by aqueous impregnating method using palladium nitrate. The samples were dried at 80°C and calcined in air at 400°C for 4h.

The solids obtained were all characterised by transmission electron microscopy (TEM), thermal analysis, specific areas mesurements, X-ray diffraction (XRD), H_2 temperature programmed reduction (TPR), electron paramagnetic resonance (EPR) and diffuse reflectance UV-Visible spectroscopy (DR-UV-VIS).

TEM of synthesized macro-mesoporous material was carried out with a Phillips TECNAI-10 at 100 kV.

The structures of solids were analyzed by powder XRD technique at room temperature with a Bruker D8 Advance diffractometer using Cu-Kα radiation scanning 2θ angles ranging from 10 to 80°.

TPR experiments were carried out in an Altamira AMI-200 apparatus. The TPR profiles were obtained by passing a 5% H_2/Ar flow (30 ml min^{-1}) through the calcined sample (about 100 mg). The temperature was increased from -40 to 300°C at a rate of 5°C min^{-1}. The hydrogen concentration in the effluent was continuously monitored by a thermo conductivity detector (TCD).

BET surface area and N_2 adsorption-desorption isotherm of solids were determined by using Thermo Finnigan Sorptomatic 1990 apparatus and the gas adsorbed at -196°C is pure nitrogen.

EPR measurements were performed at -196°C and 25°C on an EMX Bruker spectrometer. A cavity operating with a frequency of 9.5 GHz (X band) was used. Precise g values were determined from simultaneous precise measuring of frequency and magnetic field values.

Thermal analysis measurements of the coked samples were performed using a Netzsch STA 409 equipped with a microbalance differential analysis (DTA) and a flow gas system. The dried catalyst was treated under air; the temperature was raised at a rate of 5°C.min^{-1} from room temperature to 1000°C.

The DR-UV-VIS spectra of the calcined samples were recorded in a Varian UV-VIS spectrophotometer in diffuse reflectance (cary 5000).

The mono and bimetallic catalysts were tested in the oxidation of toluene. Toluene oxidation was carried out in a conventional fixed bed micro reactor and studied between 25 to 400°C (1°C.min^{-1}) and with "operando" Diffuse Reflectance Infrared Fourier Transform (DRIFT) spectroscopy. The reactive flow (100mL.min^{-1}) was composed of air and 1000 ppm of gaseous toluene. Before the catalytic test, the solid (100 mg) was calcined under a flow of air (2L.h^{-1}) at 400°C (1°C.min^{-1}) and reduced under hydrogen flow (2 L.h^{-1}) at 200°C (1°C.min^{-1}).

Table 1: *BET surface areas, average pore diameter of the supports before and after the catalytic test*

Sample	BET surface area (m^2/g)		Pore diameter (nm)
	Before calcination	Calcined at 400°C	
Ti-Zr 80/20	601	553	3.3
Ti-Zr 50/50	410	353	2.2
Ti-Zr 20/80	372	323	2.0

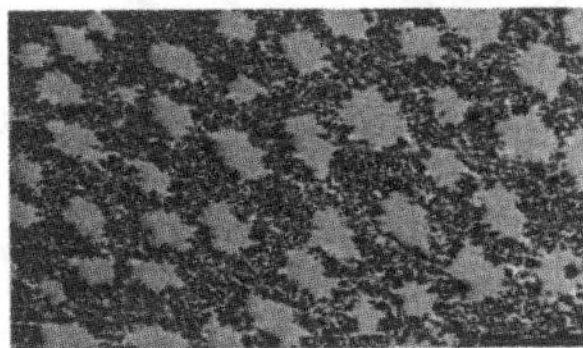

Figure 1: *TEM micrograph of macro- mesoporous TiO$_2$-ZrO$_2$ 50/50 mixed oxides*

3   RESULTS AND DISCUSSION

The TEM image of the synthesized TiO$_2$-ZrO$_2$ 50/50 mixed oxides (Fig. 1) reveals the presence of the macroporosity (300-700 nm) and a disorderd wormlike assembly of mesopores into the macroporosity walls. Morover, the synthesized mixed oxides have a homogeneous distribution of the components.

The UV-visible spectra of supports calcined at 400°C have been presented in Fig. 2. Macro-mesoporous TiO$_2$ and ZrO$_2$ previously studied [16] were added to compare the spectra of the new macro-mesoporous TiO$_2$-ZrO$_2$ mixed oxides samples. The absorption bands at 212, 304 and 329 nm in case of TiO$_2$ are characteristic of the anatase phase [17] and for ZrO$_2$ at 209 nm, characteristic of ligand to metal charge transfer band of an isolated Zr atom in a tetrahedral geometry [18]. These bands are not found in the samples TiO$_2$-ZrO$_2$ mixed oxides. However, the spectra of the mixed oxides are in accordance with their titanium-zirconium ratios. The TiO$_2$-ZrO$_2$ mixed oxides keep their amorphous phase after calcination in the contrary of pure TiO$_2$ and ZrO$_2$ oxides. This is supported by a XRD study (patterns not given).

The BET areas and the average pore diameter obtained by nitrogen adsorption-desorption isotherms (not shown) of the supports are presented in Table 1. The BET supports areas changes according to the titanium/zirconium molar ratio but not the average pore diameter. However, the low differences between the surface area of the untreated and calcined sample can be explained by the amorphous character of the titania-zirconia mixed oxides.

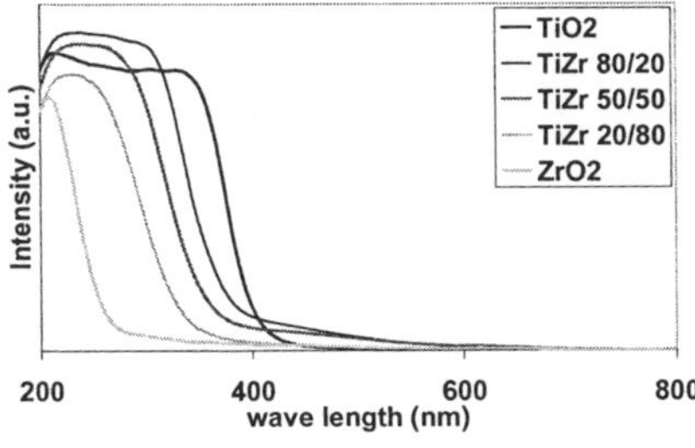

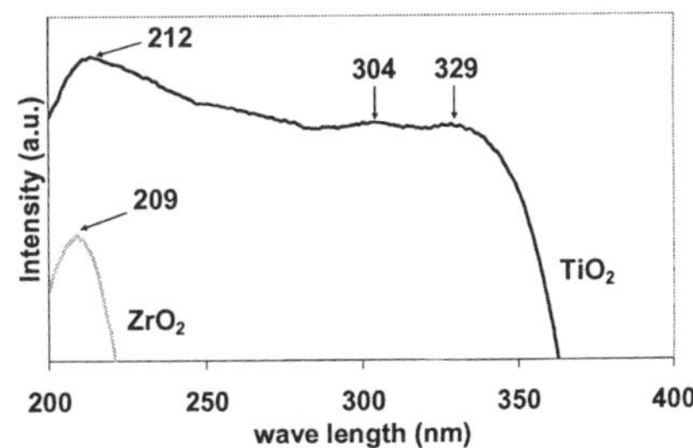

Figure 2: *DR-UV-Visible spectra of various TiO$_2$-ZrO$_2$ supports calcined at 400°C*

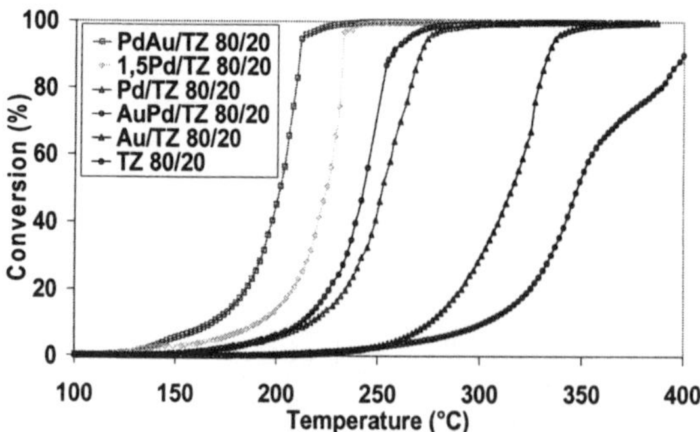

Figure 3: *Conversion of toluene on the various PdAu-TiO$_2$-ZrO$_2$ samples*

The catalytic activity of the supports was tested for toluene total oxidation in which H$_2$O and CO$_2$ were found to be the only products at total conversion. Results for toluene conversion show that with an increase in the titanium amounts and the surface areas, catalytic activity of the samples at the temperature for 50% of toluene conversion (T$_{50}$) increases in the following order : TZ 80/20 > TZ 50/50 > TZ 20/80. Among these supports the catalyst TZ 80/20 sample is the most active. However, these solids are not enough active for industrial applications. Catalytic tests were performed on macro-mesoporous TiO$_2$-ZrO$_2$ with various Ti/Zr molar ratio supported by Pd-Au (Fig. 3). The macro-mesoporous TiO$_2$-ZrO$_2$ supported by Pd-Au presented then the same activity order before and after deposition of noble metals.

The results of the catalytic activity measurements obtained for the catalysts on the best mesoporous titania-zirconia solid supported by Pd and/or Au are shown in Fig. 4. It is important to note that, whatever the support used, the catalytic activity, given by the T50 values (Table 2), follows the same order: PdAu/TZ > 1.5Pd/TZ > AuPd/TZ > Pd/TZ > Au/TZ > TZ. The PdAu/TZ sample exhibits the highest catalytic activity in comparison with the other samples. For this catalyst, the complete toluene conversion is reached at low temperature (220°C) compared to the same sample without precious metals, total oxidation is not observed under 400°C. Depending on the order of the deposition of the palladium and gold, the PdAu/TZ sample shows significantly a higher conversion in comparison with the AuPd/TZ sample. In the case of all types of the TiO$_2$-ZrO$_2$, when palladium is loaded before gold the catalytic activities are lower in comparison with the samples with gold firstly loaded. The impregnation of palladium on Au/TZ is therefore, very interesting for VOC oxidation and there must be a synergetic effect between Pd and Au on the catalyst obtained.

Table 2: *T$_{50}$ values for the different samples*

	T$_{50\%}$ of toluene conversion (°C)	
Sample	TZ 80/20	TZ 20/80
PdAu/TZ	202	218
AuPd/TZ	242	250
Pd/TZ	251	280
Au/TZ	340	309
1.5Pd/TZ	225	231
TZ	351	371

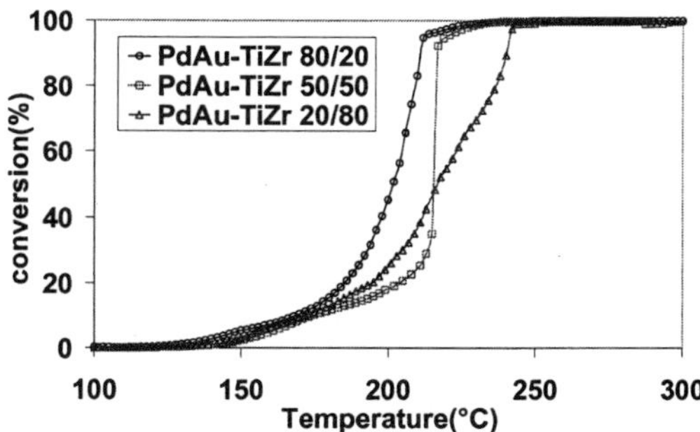

Figure 4: *Conversion of toluene on the various catalysts TiO_2-ZrO_2 80/20*

The bimetallic core-shell structures has been studied for complete benzene oxidation over gold-vanadia catalysts supported on mesoporous TiO_2 and ZrO_2 by Idakiev et al. [19] and for the both series of the supports the high activity was observed when the gold is loaded firstly. Moreover, Enache et al. [20] have reported that the gold can electronically influence the catalytic properties of Pd. Thus, Hutching and co-workers [21] had correlated the high activity for H_2 oxidation on AuPd/TiO_2 catalysts to a core-shell morphology with a gold-rich core and a palladium-rich surface. The best activity for PdAu/TZ can be explained by this morphology and the lower activity of AuPd/TZ by an inverse core-shell. This interesting result was also observed in our previous results using pure TiO_2 as support [6].

Fig. 5 shows "operando" DRIFT spectra recorded at 100°C under toluene flow on PdAu supported on TiO_2-ZrO_2 80/20, 50/50 and 20/80. For PdAu/TZ 80/20, adsorption bands at 1526 and 1498 cm^{-1} corresponding to the aromatic C-C band of toluene are observed. These bands are shifted to lower wavenumbers on the samples with less Ti amounts. The electronic density of the aromatic cycle increases with the observed catalytic activity. Therefore, the less interaction between toluene and the catalyst the best is the activity. The possible explanation of these properties might be based on the various electron transfers from support to noble metals.

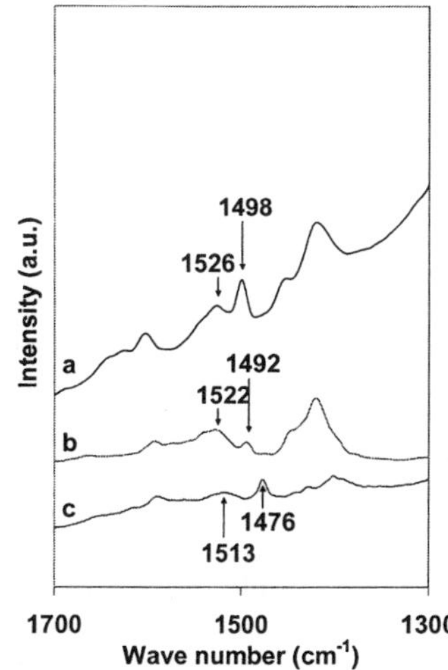

Figure 5: *DRIFT spectra of toluene adsorption at 100°C on: (a) PdAu/TZ 80/20, (b) PdAu/TZ 50/50 and (c) PdAu/TZ 20/80*

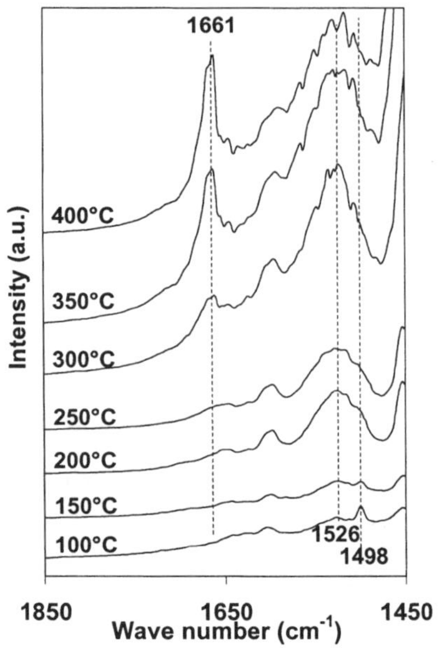

Figure 6: *"Operando" DRIFT spectra of toluene on PdAu/TZ 80/20*

Figure 7: *Correlation of the light off curve and the IR study*

The electronic interaction between titania and supported noble metal and the easy electron transfer from support to the supported metals have been demonstrated in some catalytic systems [22, 23]. Therefore, it seems that the catalyst present a strong support-metal interaction which increases in the same order than the activity for oxidation reaction.

For the "operando" DRIFT carried out for toluene reaction on the samples TiO$_2$-ZrO$_2$ 80/20 supported by Pd and/or Au, the band appearing at 1498 cm^{-1} (Fig. 6) (due to the toluene C-C absorption) was shifted to lower wavenumbers in case of PdAu supported on TiO$_2$-ZrO$_2$ 50/50 and 20/80 (not shown) and follow also the activity order. Moreover, the spectra of PdAu/TZ 80/20 sample displayed a broad band at 1661 cm^{-1} corresponding to the formation of hydrocarbons (aromatic substitution) or cokes during the reaction. As operando DRIFT carried out for toluene oxidation in the same order as classical test, there is a correlation with the light-off curve and the IR study. A decrease of the bands corresponding to toluene from 150°C and also an increase of coke formation were observed during toluene conversion (Fig. 7).

The hydrocarbon residues oxidation was then followed under air flow. Study of oxidation on the solids after catalytic test was performed. The intensity of the broad peaks corresponding to coke formed during toluene oxidation, decreases between 300 and 400°C. The presence of coke on the catalyst was confirmed by the EPR g factor at 2.003 [24] (Fig. 8).

Moreover, the both DTA exothermic signals at about 328 and 608°C observed for the studied samples after the toluene tests (Fig. 9) should correspond to two types of hydrocarbons molecules, respectively light coke corresponding to (polysubstituted) monoaromatic compounds and more heavy coke to polyaromatics [25]. The lowest corresponding weight loss between 300 and 700°C was observed for our best PdAu/TZ catalyst (Table 3).

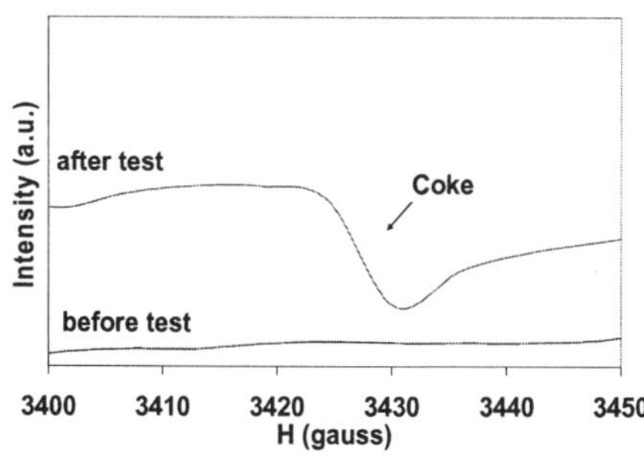

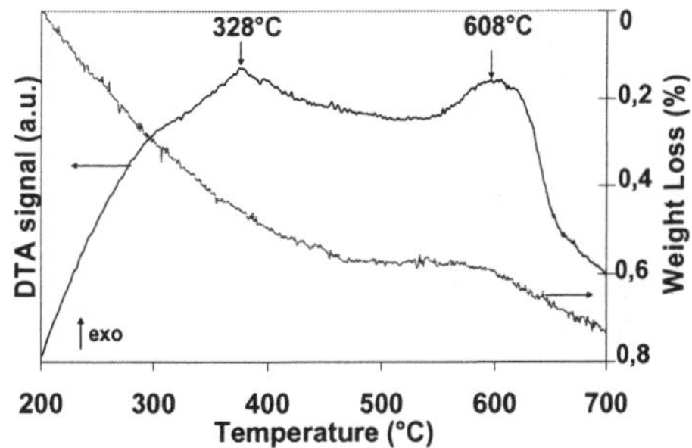

Figure 8: *EPR spectra of PdAu/TZ 80/20 at -196°C*

Figure 9: *DTA-TGA profile of PdAu/TZ 80/20 under air*

Table 3: *Mass losses for the used samples between 300 and 700°C under air flow*

Sample	Mass loss (%)
PdAu/TZ 80/20	0.8
Pd/TZ 80/20	1.0
Au/TZ 80/20	1.8

4   CONCLUSION

Nanostructured macro-mesoporous TiO_2-ZrO_2 mixed oxides with various Ti/Zr mole ratios of 80/20, 50/50 and 20/80, were synthesised using a mixture of zirconium propoxide and titanium isoporopoxide as Zr and Ti source and also CTMABr as surfactant. The new supports stay amorphous and keep their high specific surfaces after calcination at 400°C. They are impregnated by 0.5 or 1.5wt% of palladium and 1wt% of gold. The gold addition before palladium has then several advantages: the activity is well increased and less coke deposit is observed. Whatever the impregnated active phase, the more the Ti/Zr ratio the more the activity for toluene total oxidation. Moreover, whatever the support used, the catalytic activity for toluene total oxidation follows the same order: PdAu/TZ > 1.5Pd/TZ > AuPd/TZ > Pd/TZ > Au/TZ > TZ. The promotional effect of gold added to palladium was then observed for the PdAu/TZ sample. "Operando" DRIFT allowed following the VOC oxidation but also suggesting a metallic electrodonor effect to the adsorbed molecule which increases in the same order than the activity for oxidation reaction. The coke formation is also observed. The presence of coke after test was shown by DTA-TGA by exothermic signals between 300 and 700°C, by EPR with the signal at a factor g = 2.003.

References

1   M.A.C. Nascimento, Theoretical Aspects of Heterogeneous Catalysis, vol. 8, Kluwer Academic Publishers, Dordrecht, Boston, MA, 2001.
2   L. Guczi, A. Beck, A. Horvath, Zs. Koppany, G. Stefler, K. Frey, I. Sajo, O. Geszti, D. Bazin, J. Lynch, J. Mol. Catal. A: Chem., 2003, 204-205, 545.

3   M. S. Chen, K. Luo, T. Wei, Z. Yan, D. Kumar, C.-W. Yi, D. W. Goodman, Catal. Today, 2006, 117, 37.
4   J. Storm, R. M. Lambert, N. Memmel, J. Onsgaard, E. Taglauer, Sur. Sci., 1999, 436, 259.
5   D. Andreeva, T. Tabakova. L. Ilieva, A. Naydenov. D. Mehanjiev, M. V. Abrashev, Appl. Catal. A: Gen., 2001, 209, 291.
6   M. Hosseini, S. Siffert, H. L. Tidahy, R. Cousin, J.-F. Lamonier, A. Aboukaïs, A. Vantomme, B.-L. Su, Catal. Today, 2007, 122, 391.
7   E. M. cordi, J. L. falconer J. Catal., 1996, 162, 104.
8   P. Papaefthimiou, T. Ioannides, X. E. Verykios, Appl. Catal. B: Env., 1997, 13, 175.
9   M. Paulis, L. M. Gandia, A. Gil, J. Sambeth, J. A. Odriozola, M. Montes, Appl. Catal. B: Env., 2000, 26, 37.
10 K. Okumura, T. Kobayashi, H. Tanaka, M. Niwa, Appl. Catal. B: Env., 2003, 44, 325.
11 M. Lamallem, H. El Ayadi, C. Gennequin, R. Cousin, S. Siffert, F. Aïssi, A. Aboukaïs, Catal. Today, 2008, in press.
12 C. Zhang, H. He, K.-I. Tanaka App. Catal. B: Env., 2006, 65, 37.
13 M.P. Kapoor, Y. Ichihashi, K. Kuraoka, Y. Matsumura, J. Mol. Catal. A: Chem., 2003, 198, 303.
14 C. Gennequin, M. Lamallem, R. Cousin, S. Siffert, F. Aïssi, A. Aboukaïs, Catal. Today, 2007, 122, 301.
15 V. Idakiev, T. Tabakova, Z.Y. Yuan, B.L. Su, Appl. Catal. A: Gen., 2004, 270, 135.
16 H. L. Tidahy, S. Siffert, J.-F. Lamonier, R. Cousin, E.A. Zhilinskaya, A.Aboukaïs, Z.-Y. Yuan, A. Vantomme, B.-L. Su, X. Canet, G. De Weireld, M. Frere, B. N'Guyen, J.-M. Giraudon, G. Leclercq, Appl. Catal. A: Gen., 2006, 310, 61.
17 S. Zheng, L. Gao, Q. Zhang, J. Sun, J. Solide State Chem., 2001, 162, 138.
18 D. Das, H. K. Mishra, K. M. Parida, A. K. Dalai, J. Mol. Catal. A: Chem., 2002, 189, 271.
19 V. Idakiev, L. Ilieva, D. Andreeva, J.-L. Blin, L. Gigot, B.-L. Su, Appl. Catal. A: Gen., 2003, 243, 25.
20 D. I. Enache, J. K. Edwards, P. Landon, B. E. Solsona-Espriu, A. F. Carley, A. Herzing, M. Watanabe, C. J. Kiely, D. W. Knight, G. J. Hutchings, Science, 2006, 311, 362.
21 J. K. Edwards, B. E. Solsona, P. Landon, A. F. Carley, A. Herzing, C. J. Kiely, G. J. Hutchings, J. Catal., 2005, 236, 69.
22 J. Rasko, T. Kecskes, J. Kiss J. Catal., 2004, 224, 261.
23 J. Rasko, J. Catal., 2003, 217, 478.
24 C.L. Li, O. Novaro, E. Munoz, J.L. Boldu, X. Bokhimi, J.A. Wang, T. Lopez, R. Gomez, Appl. Catal. A: Gen., 2000, 199, 211.
25 M. Guinet, P. Dege, P. Magnoux, Appl. Catal. B: Env., 1999, 20, 1.

VISUALISATION OF DYNAMIC ADSORPTION BY 2D X-RAY RADIOGRAPHY AND 3D X-RAY μ-TOMOGRAPHY

M.C. Almazán-Almazán[a], A. Léonard[a], D. Toye[a], P. Lodewyckx[b], J.P. Pirard[a], F.J. López-Garzón[c], S. Blacher[a]

[a]Dep. Chemical Engineering, University of Liège, Sart-Tilman, B-4000, Liège, Belgium
[b]Royal Military Academy, Dep. Chemistry, Renaissancelaan 30, B-1000 Brussels, Belgium
[c]Dep. Química Inorgánica, Universidad de Granada, Fuentenueva s/n, 18071 Granada, Spain

1 INTRODUCTION

Activated carbon beds are largely used to adsorb organic compounds from contaminated gas streams[1]. Typically, a carbon filter is a random packing of active carbon grains which can be regarded as a porous structure, with pores formed between grains at the millimetric scale and micropores within the grains, at the nanometric one.

It is well known that, in addition to the quality of the adsorbent, the efficiency of a filter depends mainly on the bed and adsorbent geometries and on the ratio between the bed diameter and the adsorbent particle size[2,3]. These two characteristics determine the state of dispersion of the adsorbent in the bed from which depend crucially the transport and adsorption properties. Then, to predict and improve the efficiency of a filter, it is essential to characterise the internal structure of the bed.

X-ray visualization techniques are very effective non-destructive methods for visualising the interior details of this kind of package. These techniques are based on the principle that X-rays intensity is attenuated when passes through a material according to the Beer–Lambert law:

$$I = I_0 e^{-\mu x} \tag{1}$$

where I_0 and I are the incident and the transmitted X-ray intensities, μ is the linear attenuation coefficient which depends on the atomic number, the incident beam energy and the density of the studied material and x is the path length through the sample. When submitted to an X-ray radiation, materials with large attenuation coefficients block the beams, whereas the rays pass more easily through materials with lower μ.

A radiograph is the 2D projected image of an object, produced by the transmitted X-rays through it, under a given angle. On this radiograph, each grey level is determined by the cumulative value of the μ located along the X-ray path: light colours correspond either to low attenuation values or to short X-ray path, e.g. at the periphery of a cylindrical object, whereas dark colours correspond to high attenuation values or long X-ray path, e.g.

near the vertical axis of a cylindrical object. Using an appropriate device, a radiograph can be generated easily in a few seconds. This results in a set of real-time images from which, using image analysis techniques, it is possible to quantify the various parameters which allows describing the dynamic behaviour of the system. This methodology has been applied with success to characterise hygroscopic water distribution in adsorbent beds[4] and the porosity and permeability of rocks[5,6] using neutron radiographs. However, as the object is examined under only one angle, projections of all points of the object along the ray path are superimposed on to another. This feature usually complicates the visualisation and the quantification of the image.

If radiographs are recorded for several angular positions by rotating the sample, a back-projection algorithm can be used to reconstruct a stack of 2D cross sections of the object. This is the principle of the X-ray μ-tomography technique which provides a 3D image of the object. This technique enables to obtain an accurate image of the internal structure of the object and it enables to follow in situ the motion of a gas in the package. However, it must be taken into account that to have a good resolution, ~500 2D cross sections must be reconstructed and then, the resulting 3D image must be processed and quantified. These processes require good computer resources and in most cases are long time consuming.

In 2D tomographic images of cylindrical object cross sections, the grey level value of each pixel correspond to the local attenuation coefficient value. Light colours correspond to low attenuating regions, while dark colours correspond to high attenuating regions

In a previous work[7], some of us reported the ability of X-ray μ-tomography to quantify the dynamic flow of a gas through a carbon bed. The evolution of the propagation front was represented by the axial profile defined as the total grey level intensity of the successive 2D sections, from the inlet of gas flow in function of the depth of the bed.

In the present study, information obtained from 3D μ-tomography and 2D radiography is compared. To achieve this goal, line and axial profiles are determined and evaluated based on physical measurements. Then, the influence of both inlet gas concentration and the flow rate on the propagation front is studied.

2 EXPERIMENTAL

2.1. Experimental set-up.

The adsorption experiments were carried out in a classical breakthrough measurement system. The cylindrical canister (diameter = 15 mm and height =33 mm) was filled with 2g of a polydispersed commercial granular activated carbon (grains of ~1 mm mean diameter), Chemviron Carbon BPL, which is an essentially microporous adsorbent (Specific surface area (BET), S_{sp} = 1010 m^2 g^{-1}, Dubinin-Radushkevich micropore volume, V_0 = 0.36 cm^3 g^{-1} and characteristic energy of adsorption, E_0 = 22.7 J mol^{-1}). The experiments were performed exposing the carbon to a flow of dry air (flow rate = 5 and 10 L min^{-1}) containing a small amount of organic vapour ([CH$_3$I] = 5 and 10 g m^{-3}), at room temperature.

The vapour flowed through the bed for different periods of time, each a multiple of 3 min, until the maximum adsorption on the bed was reached. For each test, fresh carbon was used. The samples were weighed before and after each experiment in order to know the exact amount of CH$_3$I adsorbed. Microtomographic images were obtained after each adsorption experiment and compared with the images corresponding to the virgin carbon; i.e. without any exposure to the vapour.

2.2. X-ray radiography and tomography

The X-Ray microtomographic device was a "Skyscan-1074 X-ray scanner" (Skyscan, Belgium). The cone beam source operated at 40 kV and 1 mA. The detector was a 2D, 768 pixels × 576 pixels, 8-bit X-ray camera with a spatial resolution of 41 µm. The rotation step was fixed at the minimum, in order to improve image quality, giving total acquisition times close to 10 min.

Image analysis measurements were performed on both the radiographs and the tomographic images by determining line and axial profiles, respectively.

3 RESULTS AND DISCUSSION

3.1. Line and axial profiles

Figure 1 shows the radiographs for a $[CH_3I]$ of 5 g m^{-3} and a flow rate of 5 L min^{-1}, using false colours, in tones from orange (virgin) to dark brown (maximum uptake), in order to facilitate the observation of the gradient of adsorption. Yellow corresponds to the voids between grains.

It is seen that for virgin carbon (Figure 1a) grey level intensities have almost the same values on all the bed. When the carbon bed is submitted to the gas flow (Figure 1b-h), radiographs become darker and darker with time, from the top to the bottom. These dark pixels indicate the carbon bed zones in which the attenuation coefficient has increased, blocking the pass of the X-rays. The dark progression in radiographs is related to the progression of the adsorption front with time. This evolution can be represented by a line profile of the radiograph defined as the total grey level intensity on successive lines perpendicular to the inlet gas flow in function of the depth of the bed[5,6].

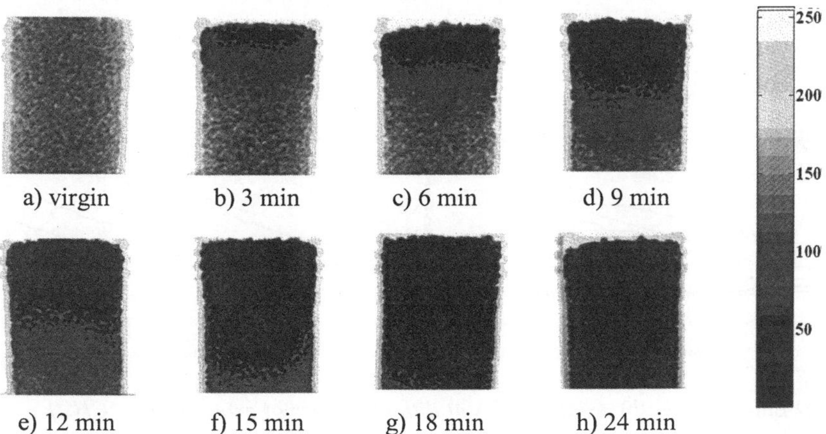

Figure 1 *Radiographs for $[CH_3I]$ = 5 g m^{-3} and flow rate = 5 L min^{-1}, at different time of exposure*

It must be noticed that all the radiographs are darker in the centre and lighter when going to the canister walls. This is due to the evolution of the thickness of the cylinder but it is not related to the adsorption process under study.

The axial profiles were obtained by the sum of the grey-level intensity of all the pixels in successive 2D cross sections as a function of the depth of the bed. Figure 2 shows, as an example, a typical cross section obtained from the virgin carbon bed, for an intermediate time of exposure and for the saturated bed. Two phases can be distinguished according to the different grey levels: the porous regions, represented by white pixels, and the carbon grains represented by darker pixels. When adsorption occurs, the pixels belonging to the carbon grains darken according to the adsorbed mass.

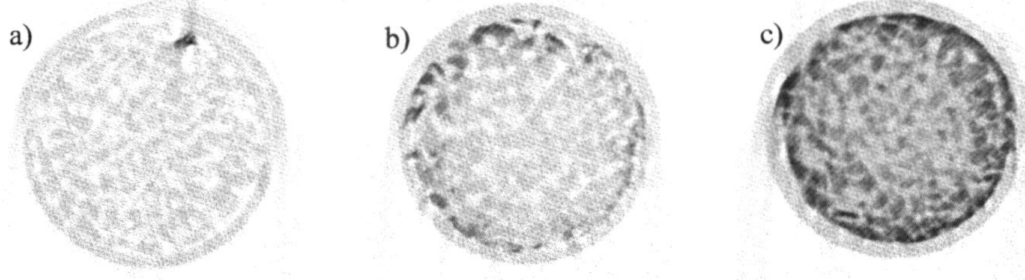

Figure 2 *Cross sections from: a) virgin bed, b) exposure time = 15 min, c) exposure time = 24 min*

Line profiles obtained from the 2D radiographs and axial profiles from all the 2D reconstructed images of the beds, for a $[CH_3I] = 5$ g m^{-3} and a flow rate = 5 L min^{-1}, are presented in Figure 3a and b, respectively. These profiles indicate how the adsorption front moves through the carbon bed. In the two cases, the profile for the pure virgin carbon, i.e. without any exposure to the vapour flow, has been included to facilitate the comparison (highest intensity in the graph). Intensity values can not be compared because on Figure 3a it corresponds to a cumulative value of μ along the X-ray path while it corresponds to the mean μ value within a cross section.

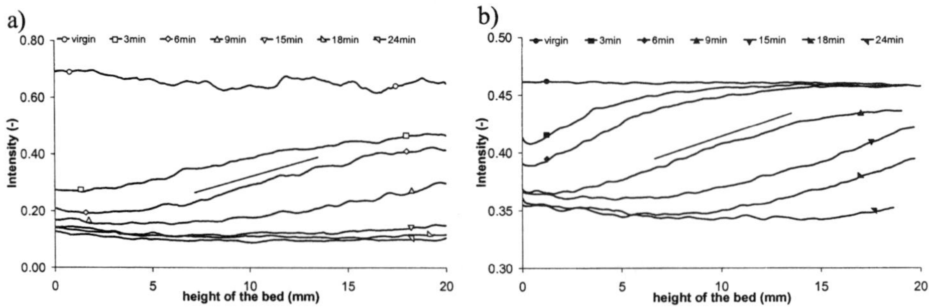

Figure 3 *a) Line profiles; b) Axial profiles at different times of exposure for $[CH_3I] = 5$ g m^{-3} and flow rate = 5 L min^{-1}*

Three zones can be distinguished in both line and axial profiles (Figure 2a and b). In the first one, starting from the inlet of the bed, the grey level is constant and takes the lowest value of the profile. The extent of this region corresponds to the depth of maximum uptake, i.e. an apparent saturation of the carbon. In the second one, grey level intensity increases with depth and it corresponds to the zone in which the mass transport is occurring (adsorption zone), in which there is a gradient of the amount of adsorbed vapour. Finally, the grey level intensity is constant and takes the highest value. This is the part of the bed not yet reached by the adsorbate. It is observed that the thickness of the maximum uptake layer increases with the time of exposure. After the maximum uptake layer, the adsorption zone is characterised by a linear decrease of the adsorption with the depth of the bed. The thickness of this adsorption zone decreases when the time is increased and it is negligible at the longest time. Only for the shortest times, it can be observed the third zone corresponding to the virgin layers of the bed.

It must be noticed that, the line profiles are almost superimposed for times of exposure longer than 15 min whereas the axial profiles are clearly differentiated. The difference of adsorbed CH_3I (first line of Table 1) confirms that at the longest times of exposure, the bed continues adsorbing vapour. This indicates that the sensibility of line profiles is not high enough to permit a good discrimination between darkness intensity of neighbour pixels above a given X-ray attenuation level. This fact is much more pronounced for higher inlet gas concentrations and flow rates (Figures 4a and b). Nevertheless, under those conditions, both kinds of profiles show the same trend: the differences between the line and axial profiles at different times of exposure are less pronounced than for low concentrations and vapour flow rate (Figures 3a and b). When the adsorption process is fast, it is thus difficult to resolve clearly the advance of the adsorption front from the radiographs.

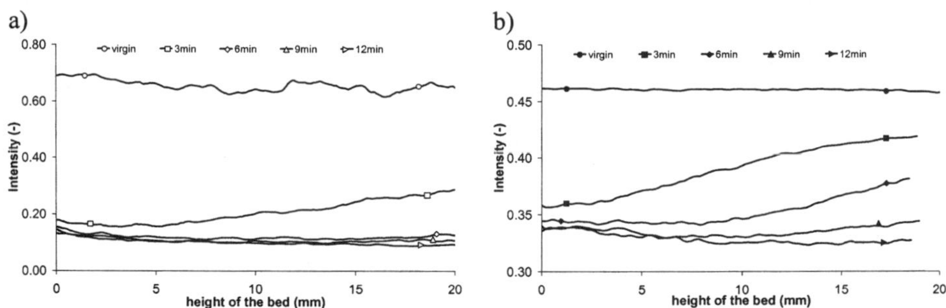

Figure 4 *a) Line profiles, b) Axial profiles, at different times of exposure for $[CH_3I]$ = 10 g m^{-3} and flow rate = 10 L min^{-1}*

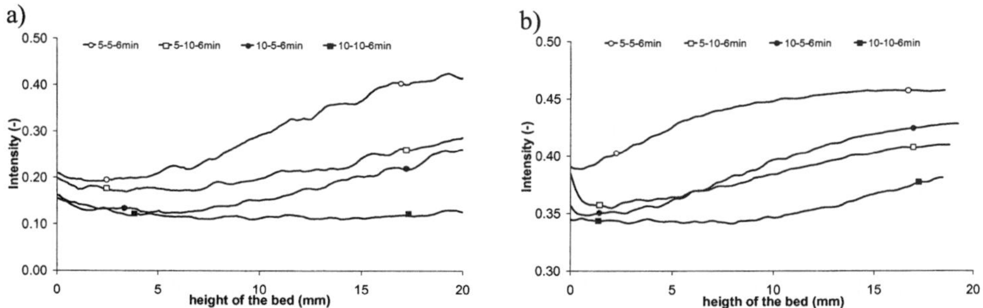

Figure 5 *Influence of the variation of the experimental conditions: a) line profiles, b) axial profile; Legend: x-y-z min : x = concentration, y = flow rate, z = exposure time.*

Figures 5a and b compare the line and axial profiles obtained after 6 min when the flow rate or the concentration is changed, keeping constant the other experimental factors. In both kinds of profiles, it can be observed that when the flow rate is increased and the concentration of vapour is kept constant, the adsorption front moves faster and the maximum uptake, i.e. the apparent saturation, is reached earlier. If the flow rate is constant and the concentration is raised, the difference between the profiles at a given time of exposure is more pronounced; that is, the velocity of advance of the adsorption front increases in a larger extension than when the flow rate is varied.

Table 1 collects the amounts of CH_3I adsorbed determined by physical measurements (see section 2.1) for the different experimental conditions. It can be observed that as the concentration and flow rate are increased, the maximum uptake is reached earlier, in agreement to the axial profiles (Figures 3a and b).

For the same time of exposure, n_{ads} increases when the flow rate is raised from 5 to 10 L min^{-1}. This observation is even more pronounced for a CH_3I concentration of 10 g m^{-3}. This can be related to the enhancement of the external mass transfer coefficient[8-10]. Table 1 also shows clearly, that, for the small flow rate and the same exposure times, the adsorbed amount of CH_3I increases with the concentration of vapour. This can be explained by an increase of the mass transfer driving force, due to the increase of the adsorption capacity at the equilibrium[8-10]. Moreover, it must be noticed that, for a CH_3I concentration of 5 g m^{-3}, the adsorbed amount at the longest time of exposure is larger when the flow rate is smaller. Indeed, at low flow rates, the progression of the adsorption front that corresponds to the mass transfer zone, is slowed down so that the carbon can adsorb a higher amount of vapour. An opposite situation is observed at a concentration of 10 g m^{-3}: for the longest exposure time, the adsorption is more important at the highest flow rate. This would suggest that, in this case, the increase of the external mass transfer coefficient together with a high mass transfer driving force counterbalances the faster progression of the adsorption front within the bed[8-10]. However, all these explanations should be confirmed by modelling the adsorption process. This will be performed in a further work.

Table 1 *Amount of CH_3I adsorbed, n_{ads}, (mmol g^{-1}) at the different experimental conditions*

[CH$_3$I] inlet (g m^{-3})	Flow rate (L min^{-1})	exposure time (min)						
		3	6	9	12	15	18	24
5	5	0.106*	0.229	0.428	0.491	0.827	0.946	1.308
5	10	0.108	0.328	0.614	0.975	1.205		
10	5	0.226	0.489	0.816	1.068	1.280	1.493	
10	10	0.360	0.917	1.345	1.625			

*The error values are 0.001 in all cases

4 CONCLUSIONS

2D X-ray radiographs and 3D μ-tomography coupled with image analysis were used to analyse dynamic adsorption of a gas in a carbon bed. Line and axial profiles were determined for various experimental conditions, i.e. varying flow rate and vapour concentration. The results obtained by these techniques were compared to those determined by physical measurements. The comparison shows that 2D analysis fails to resolve the evolution of the adsorption front when the carbon beds are close to saturation. This is more pronounced for larger flow rates and concentrations. On the contrary, 3D analysis allows obtaining an accurate description of the process, independently of the experimental conditions. However, for all the experimental conditions, both line and axial profiles present similar patterns. Then, in our opinion, 2D analysis keeps a particular interest because it provides a simple and rapid tool to characterise the global aspect of the dynamic adsorption.

References

1. P.N. Cheremisinoff and F. Ellerbush, *Carbon Adsorption Handbook*, Ann Arbor Science Publishers, Michigan, 1978.
2. M. Winterberg and E. Tsotsas, *AIChE Journal,* 2000, **46**, 1084.
3. M. Suzuki, T. Shimura, K. Iimura and M. Hirota, Proc 5th World Congress on Industrial Process Tomography, Bergen, Norway, 2007; 304.
4. H. Asano, N. Takehiko, T. Nobuyuki and F. Terushige, *Nuclear Instruments and Methods in Physics Research A*, 2005, **542**, 241.
5. F.C. de Beer, M.F. Middleton and J. Hilson, *Applied Radiation and Isotopes*, 2004, **61**, 487.
6. F.C. de Beer and M.F. Middleton, *South African Journal of Geology*, 2006, **109**, 541.
7. P. Lodewyckx, S. Blacher and A. Leonard, *Adsorption,* 2006, **12**, 19.
8. D.M. Ruthven, *Principles of adsorption and adsorption processes*, Wiley, New York, 1984.
9. R.T. Yang, *Gas separation by adsorption processes*, Imperial College Press, 1997.
10. J. D. Seader and J.E. Henley, *Separation Process Principles,* Wiley, New York, 1998.

SUPERCRITICAL METHANE ADSORPTION – A TOOL FOR CHARACTERIZATION OF CARBONACEOUS ADSORBENTS

M.R. Bałys[1], L. Czepirski[1], S. Furmaniak[2], A.P. Terzyk[2], P.A. Gauden[2]

[1] AGH – University of Science and Technology, Faculty of Fuels and Energy, Department of Coal Chemistry and Adsorption Engineering, al. Mickiewicza 30, 30-59 Cracow, Poland
[2] Nicolaus Copernicus University, Physicochemistry of Carbon Materials Research Group, 7 Gagarin St., 87-100 Toruń, Poland

1 INTRODUCTION

Studies on the physical adsorption of methane on carbonaceous adsorbents at supercritical conditions are stimulated both by theoretical and practical interest. For example, an analytical representation of adsorption equilibrium data is needed in connection with natural gas storage, PSA separation and purification processes, as well as methane recovering from coal seams. Modelling methane isotherms provides an effective tool for studying adsorption mechanism, characterization of porous solids (i.e. pore volume, pore size distribution) and collecting useful information on adsorption systems.[1-10] From a theoretical point of view, this system seems to be interesting, since the intermolecular interactions are relatively simple to modelling for the assumed pore geometry.

In this work high-pressure adsorption isotherms of methane for different carbonaceous adsorbents were analyzed by the potential theory.

Grand Canonical Monte Carlo (GCMC) simulation was also used for calculation of the local adsorption isotherms. The Global Adsorption Isotherm Equation was solved taking the local isotherms (simulated for different pore diameters) as the kernel. Obtained pore size distributions were compared with those calculated from the low – temperature nitrogen adsorption data.

2 EXPERIMENTAL

Two types of carbon materials have been used in this study: a commercial granulated active carbon R2 (Norit, Holland) and monolithic form of active carbon from MAST Carbon Ltd. (Guilford, Great Britain). Porous structure analysis of the samples has been carried out by subatmospheric N_2 adsorption at 77K in a Fisson's Sorptomatic'1900 apparatus.

The measurements of methane adsorption isotherms were made using a modified version of volumetric system described before.[11] The experimental data were measured in the range 276-308K and the pressure range about 0-10 MPa. The pressure was taken by a transducers with the 0,1% accuracy in the whole range. Corrections for the non-ideality of the gas phase were made using the Benedict - Webb - Rubin equation of state. The mean relative uncertainties of the experiments were about 2% by brief error calculations.

3 RESULTS AND DISCUSSION

3.1 Potential theory in interpretation of supercritical adsorption data

Adsorption potential theory, including the Dubinin - Radushkevich (DR) and the Dubinin - Astakhov (DA) equations, has been widely applied to investigating gas adsorption equilibria on microporous adsorbents.[12-14] Although this theory is considered a semiempirical approach, it has achieved great success in many practical applications. The potential theory of adsorption is simply expressed by:

$$W = F(A/\beta) \qquad A = RT \ln(P_s / p) \tag{1}$$

where: W, F, A and β denote the volume of adsorbed gas, the universal adsorption function, the adsorption potential and the affinity coefficient, respectively. Since there exist a number of ways of representing the saturation vapor pressure, adsorbed volume, and the form of adsorption potential.[15-19]

Methane adsorption isotherms were measured at temperatures above the critical temperature, where the concept of saturation vapor pressure is meaningless. Therefore, above the critical temperatures of adsorbates, one must estimate the "vapor" pressures at the adsorption temperatures. This difficulty of correlating data using the potential theory was overcome in the proposed method as follows. First, a hypothetical saturation pressure was calculated using the reduced Kirchhoff equation.

According to Dubinin the purpose of using an affinity coefficient is that all characteristic curves for various adsorbates on a given adsorbent at different temperatures are superimposable on single reduced characteristic curve. In the work, the molar volume of methane in adsorbed state was chosen as this correlating divisor. We propose to introduce so called reduced adsorption potential:

$$A_{red} = A/v^a \tag{2}$$

where: v^a - the molar volume of methane in adsorbed state. The last one parameter was approximated assuming that gas in adsorbed state can be regarded as a sort of superheated liquid by the method proposed by Ozawa from p-V-T relations reported in literature.[20-21]

Adsorption isotherms of methane at supercritical temperatures and nitrogen at 77K were correlated by plotting the logarithm of the volume adsorbed *vs.* the square of the reduced adsorption potential (Figure 1).

It can be seen that methane characteristic curve is temperature independent and at low values of reduced adsorption potential superimposes on nitrogen characteristic curve. This allows us to conclude that in this region methane adsorption at supercritical temperatures provides similar information as low temperature nitrogen adsorption.

Experimental methane isotherms were also interpreted by the use of the Dubinin - Radushkevich equation. Table 1 contains parameters of microporous structure of active carbons under study obtained from supercritical methane isotherms and low temperature nitrogen isotherms.

The results demonstrate the possibility of application of DR equation for analysis the microporous structure of carbonaceous adsorbents because methane kinetic diameter is enough to be accessible to the micropores of smaller size.

Similar good correlation was also confirmed for experimental data published in literature on supercritical gas adsorption.[22]

Table 1   *Parameters of microporous structure of active carbons calculated from characteristic curve*

| | Methane (supercritical) | | Nitrogen (77K) | |
	W_o [cm g^{-1}]	E [kJ/mol^{-1}]	W_o [cm g^{-1}]	E [kJ/mol^{-1}]
Norit R2	0.543	9.48	0.539	15.49
Monolith	0.345	11.79	0.337	17.28

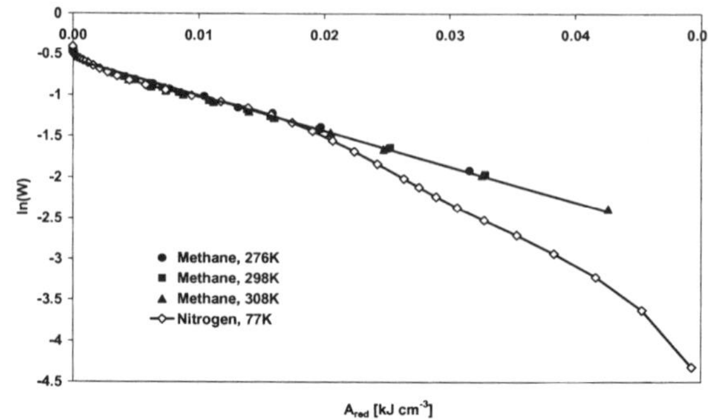

Figure 1   *Generalized characteristic curve for active carbon Norit R2*

The success of this correlation justifies the selection of the molar volume of adsorbed phase as a correlating divisor. Generalized correlation curve was next used for the calculation adsorption equilibrium points for hydrogen on the same carbon at temperatures 196, 243 and 298K. Values of experimental data are compared in Table 2 with those calculated from characteristic curve.

A generalized characteristic curve gives a possibility for succesful prediction of adsorption equilibria for different gases. Only one adsorption isotherm is practically necessary to obtain the characteristic curve and this is sufficient to describe the adsorption at all other temperatures and pressures. The results can be used for the design of adsorption systems and for predicting adsorption equilibrium behaviour of binary and/or multicomponent gaseous mixtures on active carbon under wide range of conditions, without time consuming and expensive experimental determination.

Table 2   *Values of experimental and predicted adsorption for hydrogen on active carbon Norit R2*

| $T = 196K$ | | | $T = 243K$ | | | $T = 298K$ | | |
| p | n_{exp} | n_{pred} | p | n_{exp} | n_{pred} | p | n_{exp} | n_{pred} |
MPa	mmol g^{-1}		MPa	mmol g^{-1}		MPa	mmol g^{-1}	
0.127	0.135	0.137	0.585	0.256	0.248	0.502	0.023	0.026
0.552	1.231	1.228	1.106	0.641	0.629	1.025	0.108	0.113
1.117	2.190	2.187	1.514	1.060	1.048	1.563	0.240	0.244
1.594	2.999	3.009	2.039	1.479	1.488	2.063	0.403	0.410
1.951	3.693	3.702	2.516	1.887	1.902	2.523	0.567	0.571
2.578	4.297	4.288	3.050	2.279	2.268	3.166	0.787	0.792
3.153	4.830	4.815	3.461	2.653	2.641	3.587	0.992	0.996
3.680	5.305	5.330	4.073	3.008	3.020	4.104	1.201	1.210
4.157	5.733	5.740	4.470	3.345	3.336	4.482	1.408	1.395
4.655	6.120	6.115	5.076	3.554	3.544	5.132	1.645	1.652

An important aspect of the design of adsorption process equipment for gas storage and separation is a proper understanding of possible thermal effects. Isosteric heat of adsorption is one of the thermodynamic properties that are of special relevance to gas-phase adsorption systems.

It can be simply expressed analytically from the Clausius - Clapeyron equation and using the DR equation for the equilibrium adsorption isotherm:[23]

$$\Delta H_S = \frac{RT^2}{P_s}\frac{dP_s}{dT} + A - \alpha T\left(\frac{\partial A}{\partial \ln n}\right)_T \qquad n = \frac{W}{v^a} \qquad (3)$$

Figure 2 presents the dependence of isosteric heat of methane adsorption on the volume adsorbed for active carbon Norit R2. The run of adsorption heat shows heterogeneity associated with active carbon texture. For the system under study the isosteric heat of adsorption was higher than the heat of condensation and of the sufficient magnitude to be characterized as physical adsorption.

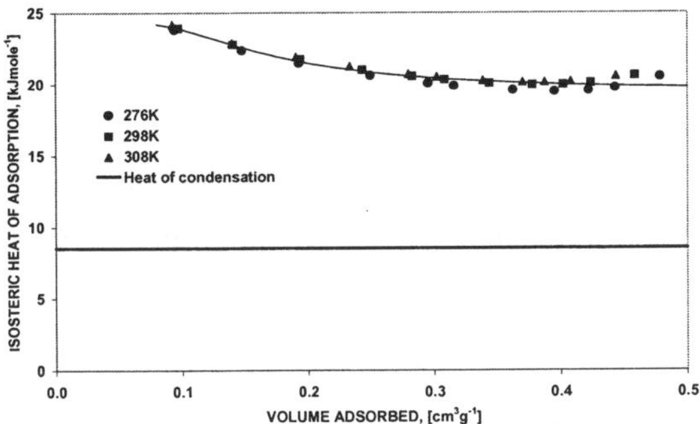

Figure 2   *Isosteric heat of methane adsorption vs. volume adsorbed for active carbon Norit R2*

It was considered that eq. (3) would be sufficient for numerical calculations of thermodynamic characteristics for a given sample as a method facilitating and speeding up calculation for various operative conditions.

3.2   Determination of PSD from low temperature N_2 adsorption data and simulation of methane adsorption in slit-like pores

The PSDs of studied carbons from low temperature nitrogen adsorption isotherms was determined using the method proposed by Nguyen and Do (ND)[24] with the basis of 82 ND local isotherms generated for the same effective width range (from 0.465 to 233.9 nm, i.e. micro-, meso-, and macropores), as in the DFT software (DFT PLUS, ASAP 2010, Micromeritics, USA). The ND method is a relatively simple approach, and leads to exactly the same PSDs as calculated from DFT method (as was shown for many porous materials having different structures and origin - see the results published by Kowalczyk and coworkers.[25] To invert the global adsorption isotherm equation recently proposed *"Karolina"* algorithm was applied.[26]

The Grand Canonical Monte Carlo simulation method of CH_4 adsorption in slit-like pores was applied. For each adsorption point $25 \cdot 10^6$ iterations were performed during the equilibration, and next $25 \cdot 10^6$ equilibrium ones, applied for the calculation of the averages (one iteration = an attempt to change the state of the system by displacement, creation or annihilation, all three attempts with the same probability).

The energy of fluid-fluid (*ff*) intermolecular interactions between methane molecules was modeled by the five-center potential (each atom is represented by one LJ and electrostatic centre (Table 3), assuming the length of CH bond equal to 0.109 nm):[27]

$$U_{ff}(r) = \sum_{i=1}^{5} \sum_{j=1}^{5} \left[U_{LJ}^{(ij)}(r_{ij}) + U_{C}^{(ij)}(r_{ij}) \right] \tag{4}$$

where r is the distance between the mass centers of the interacting molecules, r_{ij} is the distance between the pair of the centers, $U_{LJ}^{(ij)}(r_{ij})$ is the energy of dispersion interactions between the pair of centers modeled using the Lennard – Jones potential:

$$U_{LJ}^{(ij)}(r_{ij}) = \begin{cases} 4\varepsilon^{(ij)} \left[\left(\dfrac{\sigma^{(ij)}}{r_{ij}} \right)^{12} - \left(\dfrac{\sigma^{(ij)}}{r_{ij}} \right)^{6} \right] & r_{ij} < r_{cut}^{(ij)} \\[2ex] 0 & r_{ij} \geq r_{cut}^{(ij)} \end{cases} \tag{5}$$

where $r_{cut} = 5\,\sigma$. The energy of electrostatic interactions between the pair of centers, $U_{C}^{(ij)}(r_{ij})$ was modeled using the approach proposed recently by Fennel and Gezelter[28] (the alternative method to the well-known Ewald summation one):

$$U_{C}^{(ij)}(r_{ij}) = \begin{cases} \dfrac{q_i q_j}{4\pi\varepsilon_0} \left[\dfrac{erfc(\alpha r_{ij})}{r_{ij}} + \dfrac{erfc(\alpha r_{cut,C})}{r_{cut,C}} + \right. \\[2ex] \left. + \left(\dfrac{erfc(\alpha r_{cut,C})}{r_{cut,C}^2} + \dfrac{2\alpha \exp(-\alpha^2 r_{ij}^2)}{r_{ij}\sqrt{\pi}} \right)(r_{ij} - r_{cut,C}) \right] & r_{ij} \leq r_{cut,C} \\[3ex] 0 & r_{ij} > r_{cut,C} \end{cases} \tag{6}$$

where ε_0 is the permittivity of free space (8.8543 $C^2J^{-1}m^{-1}$), and α is the damping parameter. We assumed: $\alpha = 2.5$ nm^{-1} and $r_{cut,C} = 1.5$ nm.[28]

The energy of solid-fluid was modeled by the 10-4-3 Steele's potential, where the interaction between methane molecule and carbon wall is given by:

$$U_{sf}(z) = 2\pi\rho\Delta \sum_{i=1}^{5} \varepsilon^{(si)} \left(\sigma^{(si)} \right)^2 \left[\frac{2}{5} \left(\frac{\sigma^{(si)}}{z_i} \right)^{10} - \left(\frac{\sigma^{(si)}}{z_i} \right)^{4} - \frac{\left(\sigma^{(si)} \right)^4}{3\Delta(z_i + 0.61\Delta)^3} \right] \tag{7}$$

where z is a distance between the centre of the mass of CH_4 molecule and carbon wall, z_i is the distance between *i-th* center and the wall, ρ and Δ are the density of carbon atoms forming the wall, and the interlayer spacing (both values were assumed in simulations as the same as for graphite, i.e. $\rho = 114$ nm^{-3} and $\Delta = 0.3354$ nm).

Table 3 shows the collected values of the applied LJ parameters and the values of charges located on each atom forming methane molecule.

Table 3   *The values of the parameters of LJ potential.[29] The parameters between two different centers were calculated using the Lorentz - Berthelot mixing rules.*

Center	σ [nm]	ε/k_B [K]	q [e]
Fluid			
H	0.281	51.2	+ 0.143
C	0.335	8.6	- 0.572
Solid			
C	0.34	28.0	-

The simulation box was formed from the parallel walls forming a slit-like pore. The following geometric diameters of slits were considered: 0.75, 0.8, 0.85, 0.9, 0.95, 1.0, 1.05, 1.1, 1.15, 1.2, 1.25, 1.3, 1.35, 1.4, 1.5, 1.6, 1.7, 1.8, 1.9, 2.0, 2.1, 2.2 and 2.3 nm (for larger slits no differences between simulated local isotherms were recorded, therefore only micropores are considered). The length of the simulation box was equal to 4 nm, with periodic boundary conditions applied in two remaining directions.

Figure 3 shows examples of simulated methane excess adsorption isotherms in slit-like pores with different geometric diameters.

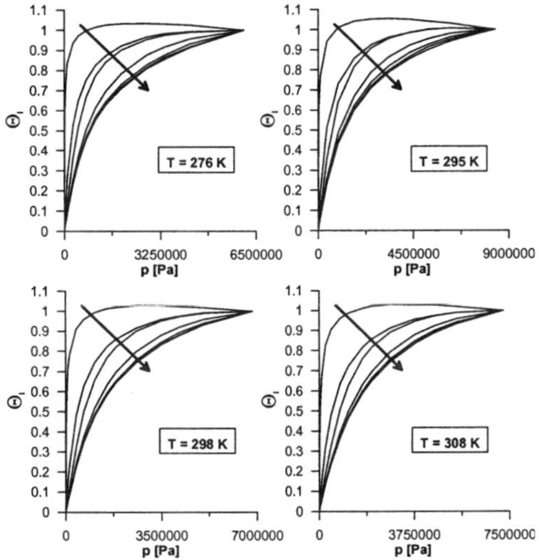

Figure 3   *Examples of simulated excess isotherms of methane adsorption in slit-like pores with geometric diameters: 0.75, 1.0, 1.25, 1.5, 1.8, 2.0, 2.3 nm (the arrow shows the rise in pore diameter).*

One can see that there are not differences recorded between methane adsorption isotherms simulated for pores larger than ca. 2 nm. It is well known fact and therefore one can expect that from methane adsorption data the PSD for larger pores than micropores will not be detected correctly.

Figure 4 shows that for all studied carbons the fit between GCMC results and experimental data is very good however, at larger pressures the deviation occurs. Obtained PSDs from nitrogen adsorption data and ND method show the typical bimodal structure for all studied carbons. This bimodal structure is also recovered from GCMC results of methane adsorption. However in this case, the first peak is usually shifted towards smaller diameters.

The second peak shows the presence of larger pores, and is usually intensive. This shows the weakness of the method. Since there are no remarkable differences observed in local isotherms generated for larger pores than micropores, *"Karolina"* algorithm tries to fit the experimental CH_4 data raising the weights of local isotherms simulated for adsorption in larger pores. Summing up, since all studied carbons contain (sometimes small number) pores larger than micropores the PSD in this range will not be recovered by simulated CH_4 isotherms.

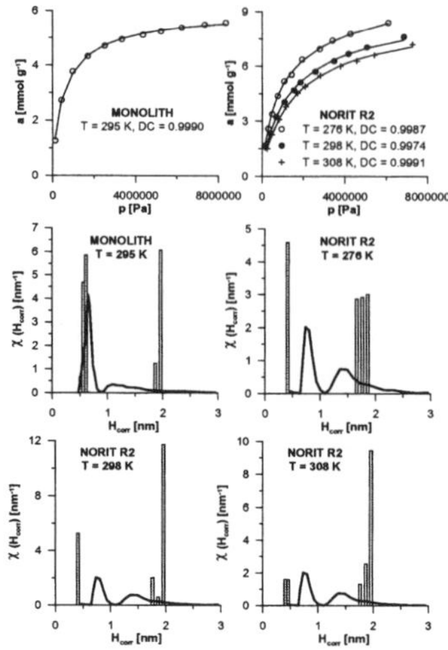

Figure 4   *The results of the fitting of experimental isotherm with "Karolina" algorithm and local isotherms from GCMC simulations. The averaged results from 3 runs are shown (in each run almost the same fit was obtained). Solid line – ND method (nitrogen adsorption at 77 K), bars – methane adsorption.*

The problem in the gap between peaks on PSDs obtained from the set of the local adsorption isotherms (N_2, T = 77K) was previously discussed by Gauden et al.[30] From the comparison of the PSDs collected in Figure 4 one can state that the bimodal shape of distributions is the result of the similarity of adsorption mechanisms and the local adsorption isotherms (for CH_4 more significant than for N_2) in the range of the pore widths for which the gap between peaks (related to the primary and secondary micropore filling mechanism) exists. Therefore, for methane data the first peak on the PSDs is always shifted to smaller pore sizes.

4 CONCLUSIONS

Adsorption potential theory successfully describes the adsorption equilibria of methane on active carbon at supercritical conditions. A generalized characteristic curve gives a possibility for prediction of adsorption equilibria. GCMC simulation was also used for calculation of the local adsorption isotherms and pore size distributions. Both approaches give the possibility to characterize the adsorbent in terms of structural characteristics and obtain the corresponding pore size distributions. In further studies we expect that procedures proposed will be especially useful for characterization of systems with diffusional limitations like carbon molecular sieves or natural coals.

Acknowledgements

The authors from AGH-UST wish to acknowledge to the Ministry of Science and Higher Education for its financial support of this work (PBZ-MEiN-2/2/2006).
The authors from Nicolaus Copernicus University acknowledge the use of the computer cluster at Poznań Supercomputing and Networking Center and the Information and Communication Technology Center of the Nicolaus Copernicus University.

References

1   M.J. Bojan, R. van Sloten, W. Steele, *Separ. Sci. Technol.*, 1992, **27**, 1837.
2   R.F. Cracknell, P. Gordon, K.E. Gubbins, *J.Phys.Chem.*, 1993, **97**, 494.
3   P.B. Balbuena, K.E .Gubbins, *Langmuir*, 1993, **9**, 1801.
4   C.R.C. Jansen, N.A. Seaton, *Langmuir*, 1996, **12**, 2866
5   P.N. Aukett, N. Quirke, S. Riddiford, S.R. Tennison, *Carbon*, 1992, **30**, 913.
6   K. Shethna, S.K. Bhatia, *Langmuir*, 1994, **10**, 870.
7   X.S. Chen, B. Mc Enaney, T.J. Mays, J. Alcaniz - Monge, D. Cazorla - Amoros, A. Linarez-Solano, *Carbon*, 1997, **35**, 1251.
8   J. Jagiełło, P. Sanghani, T.J. Bandosz, J.A. Schwarz, *Carbon*, 1992, **30**, 507.
9   Li Zhou, *Adsorp. Sci. Technol.*, 2005, **23**, 509.
10   Yaping Zhou, Li Zhou, Shupei Bai, Bin Yang, *Adsorp. Sci. Technol.*, 2001, **19**, 681.
11   L. Czepirski, S. Hołda, B. Łaciak, M. Wójcikowski, *Adsorp. Sci. Technol.*, 1996, **14**, 77.
12   D.D. Do, *Adsorption Analysis: Equilibria and Kinetics*, Imperial College Press, London, 1998.
13   R.C. Bansal, M. Goyal, *Activated Carbon Adsorption*, CRC Press, Taylor & Francis Group, Boca Raton FL, 2005.
14   J. Toth, *Adsorption Theory, Modeling and Analysis*, Marcel Decker Inc., New York, 2002.
15   R.K. Agarval and J.A. Schwarz, *Carbon*, 1998, **26**, 873.
16   K.A.G. Amankwah, J.A. Schwarz, 1995, *Carbon*, **33**, 1313.
17   L. Zhan, K.X. Li, R. Zhang, Q.F. Liu, CH.X. Lü, L.Ch. Ling, *J. Supercrit. Fluid*, 2004, **28**, 37.
18   M. Li and A.Z. Gu, *J. Colloid Interf. Sci.*, 2004, **273**, 356.
19   K. Kaneko, *Adsorption*, 1997, **3**, 197.
20   S.Ozawa, S. Kusumi, Y.Ogino, *J. Colloid Interf. Sci.*, 1981, **79**, 399.
21   Li Ming, Gu Anzhong, Lu Xuesheng, Wang Rongshun, *Carbon*, 2003, **41**, 579.
22   L. Szepesy and V. Illes, *Acta Chimica Acad.Sci.Hungaricae*, 1963, **35**, 37.
23   D. Ramirez, Shaoying Qi , M.J. Rood, *Environ. Sci. Technol.*, 2005, **39**, 5864.
24   C. Nguyen and D.D. Do, *Langmuir*, 1999, **15**, 3608.
25   P. Kowalczyk, A.P. Terzyk, P.A. Gauden, R. Leboda, E. Szmechtig-Gauden, G. Rychlicki, Z. Ryu and H. Rong, *Carbon*, 2003, **41**, 1113.
26   S. Furmaniak, A.P. Terzyk, P.A. Gauden, K. Lota, E. Frąckowiak, F. Béguin, P. Kowalczyk, *J. Colloid Interf. Sci.*, 2008, **317**, 442.
27   D.D. Do and H.D. Do, *J. Phys. Chem. B*, 2005, **109**, 19288.
28   Ch.J. Fennel and J.D. Gezelter, *J. Chem. Phys.*, 2006, 234104.
29   S.M. El-Sheikh, K. Barakat, N.M. Salem, *J. Chem. Phys.*, 2006, **124**, 124517.
30   P.A. Gauden, A.P. Terzyk, M. Jaroniec, P. Kowalczyk, *J. Colloid Interf. Sci.*, 2007, **310**, 205.

THE ADSORPTION OF TRIACETIN ON ACTIVATED CARBON

P.J.Branton[1], L.Song[2], Y.Hattori[3] and K.Kaneko[2]

[1] Group Research & Development, British American Tobacco, Regents Park Road,
Millbrook, Southampton SO15 8TL, UK
[2] Chemistry Department, Chiba University, Inage Ku, Chiba 263-8522, Japan
[3] Textile Science and Technology, Shinshu University, Tokida, Ueda 386-8567, Japan

1 INTRODUCTION

Glycerol triacetate (commonly referred to as triacetin, CAS number 102-76-1) is widely
used as a food additive and for its humectant function. It is widely used in the cigarette
industry as a plasticizer (*i.e.* a cross-linking agent) for cellulose acetate filters. Its structure
is shown in Figure 1. Available physical constants for triacetin are a density of $1.15g/cm^3$,
melting point of 3°C and boiling point of 258-260°C (at 25°C and 100kPa pressure) and a
vapour pressure of 1Pa at 37.6°C.[1]

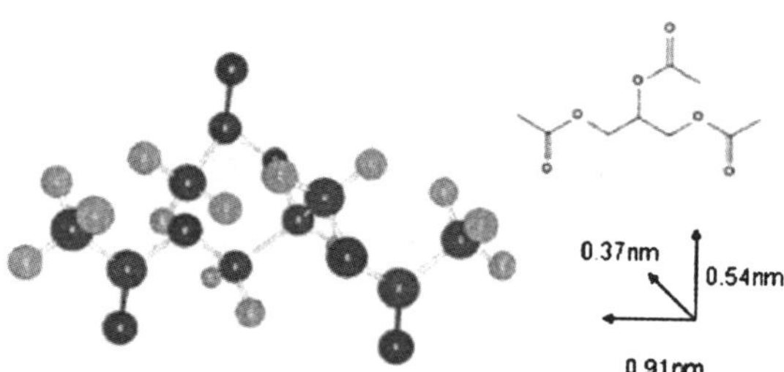

Figure 1 *Molecular Structure of Triacetin*

Tobacco smoke is an ever changing and extremely complex mixture of chemicals
dispersed as an aerosol. It is formed when tobacco, itself a complex mixture of over 2000
chemical constituents,[2] is burnt incompletely during the smoking of cigarettes, cigars, or
pipes. Inside the burning cigarette the tobacco is exposed to temperatures ranging from
ambient up to approximately 950°C in the presence of varying concentrations of oxygen.[3]
Over four thousand chemical substances are generated, many of which arise from several
distinct routes, as demonstrated by studies carried out by Baker.[4] The components are
distributed between the gas phase and condensed phase particles which constitute the
smoke aerosol.

Cellulose acetate is a good filtration media for particulate matter. It does not however show any affinity towards gas phase compounds. Activated carbon is widely used in cigarette filters where it can selectively filter, to varying degrees, certain volatile and semi-volatile (distributed between gas and condensed phases) compounds. The carbon is usually dispersed among cellulose acetate fibres but is, alternatively, sometimes placed in a cavity in the centre of the filter. During filter rod manufacture the carbon will interact with the triacetin through adsorption processes and these reduce the available surface area of carbon and hence the filtration efficiency of the carbon.

Described in this paper is a method to expose carbon to (and measure) triacetin vapour sorption under different environmental conditions. Carbons have been selected based on their pore size, surface area and surface chemistry. The aim of the work was to study the interactions between the different carbons and triacetin to allow the future design of carbons whereby the interactions with triacetin are minimized.

2 EXPERIMENTAL

Five carbons have been investigated and their parameters are shown in Table 1. Micropores are defined as pores up to 2nm in diameter, and mesopores as having pores between 2nm and 50nm in diameter.[5]

Carbon	BET surface area (m^2/g)	Micropore Volume (cm^3/g)	Mesopore Volume (cm^3/g)	Mean micropore width (nm)
A7	820	0.31	-	0.7
A20	1830	0.89	-	1.2
Meso A7	1180	0.38	0.28	0.7
GC	1070	0.38	-	0.7
FA20	730	0.35	-	1.1

Table 1 *Carbon Characteristics*

Surface area and porosity measurements were made on a Quantachrome Autosorb 1 instrument. The mean micropore width and micropore volume were determined using the subtracting pore effect (SPE) method for an α_s-plot.[6-8] The mesopore size was determined using the Dollimore and Heal (DH) method.

Samples A7 and A20 were pitch based activated carbon fibres. Meso A7 was prepared as follows: A7 was immersed at room temperature for 1 day in 50ml of a 0.3mol/l aqueous solution of $Ca(NO_3)_2$ after pre-evacuation at 10mPa and 383K for 2 h. The resulting impregnated material was then air dried at 383K for 1 day and then chemically activated at 1123K for 1h under an argon flow of 300ml/min. The activated sample was soaked in 1mol/litre hydrochloric acid, stirred for 1 day and then washed in deionised water to remove any residual chemical agent.

FA20 was prepared by the direct reaction of elemental fluorine (99.7%) with A20 at 303K. GC was a coconut based carbon activated in steam for 1-2 days.

30-40mg of carbon was placed in an open glass dish of area 3x3x1.5cm and held in a conditioning cabinet together with triacetin liquid (98%, Wako Pure Chemical Industries,

Ltd). The temperature of the system was controlled between 20 and 80°C to an accuracy of ± 2°C and the relative humidity (RH) controlled between 10% and 90% using appropriate salt solutions.

After selected time intervals, carbon was removed and the quantity of adsorbed triacetin determined using Thermo Gravimetric Analysis (EXSTAR 6000 thermoanalyzer (SII)).

The reduction in carbon porosity due to adsorbed triacetin was evaluated using nitrogen adsorption at 77K following outgassing at 100°C for 2 hours. (These conditions were sufficient to remove adsorbed water but not adsorbed triacetin).

Triacetin adsorption isotherms on the GC carbon were measured at Hiden Isochema (Warrington, UK) using an IGA-002 instrument. Prior to experiment, samples were outgassed at 300°C until a pressure lower than 1×10^{-7} mbar was achieved. Isotherms were measured at 60°C. Triacetin vapour was introduced into the system from a liquid reservoir also at 60°C where trapped air had been previously removed using a 'block and bleed' method. The saturation vapour pressure (P_0) was determined by saturating the instrument with triacetin vapour and recording the pressure. Equilibrium was assumed to be complete when the weight was 99% relaxed. Equilibrium was achieved within 2 hours for each data point.

3 RESULTS AND DISCUSSION

3.1 Triacetin adsorption mechanism

In order to confirm whether triacetin was chemically or physically adsorbed on carbon, triacetin was adsorbed at 25°C and 50-60% relative humidity for time intervals from 0 to 7 days. The amount of adsorbed triacetin was determined using TGA. Samples were analysed using nitrogen adsorption following outgassing at temperatures from room temperature to 300°C. Results are summarized in Table 2 using A7 as the carbon. It can be seen that the triacetin is very tightly held in the pores, but can be fully removed using heat and a vacuum to yield the original carbon.

Outgassing Temp (°C)	Surface area (m^2/g)	Total Pore volume (cm^3/g)
22	510	0.18
100	540	0.19
300	790	0.30

Table 2 *Triacetin removal from carbon*

The outgassing at 300°C gives a surface area and pore volume similar to the original A7 material and indicates that triacetin is physically adsorbed in the carbon pores.

3.2 Measurement of triacetin adsorption isotherms

Due to the extremely low vapour pressure of triacetin at ambient temperature, and its large molecular size relative to pore size, the adsorption kinetics was slow and isotherm measurements were difficult. Isotherms were measured on carbon GC (a microporous only carbon with $1070\text{m}^2/\text{g}$ surface area, $0.38\text{cm}^3/\text{g}$ micropore volume). For comparison, an isotherm is also shown for a micro/mesoporous carbon (not described in Table 1) with a $1370\text{m}^2/\text{g}$ surface area, $0.62\text{cm}^3/\text{g}$ micropore volume and $0.31\text{cm}^3/\text{g}$ micropore volume. Isotherms are shown in Figure 2. The amount adsorbed is shown as a percentage of the outgassed carbon weight.

(i) **(ii)**

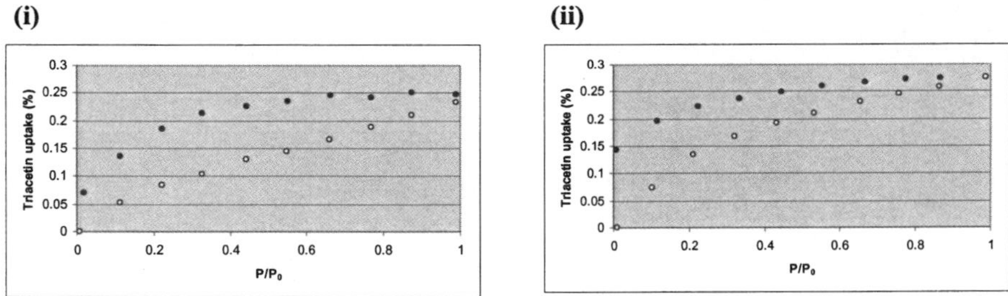

Open symbols and closed symbols denote adsorption and desorption respectively

Figure 2 *Triacetin sorption at 60°C on (i) microporous and (ii) micro/mesoporous carbon*

The long equilibrium times and low uptake are indicative of steric effects. It is interesting to note that no significant differences were seen in the triacetin adsorption capacity when larger mesopores were introduced into the structure.

3.3 Effect of temperature on triacetin adsorption

Figure 3 shows nitrogen adsorption isotherms of the carbons (following outgassing at 100°C) after triacetin adsorption at (i) 80°C and < 30% RH for 1day and (ii) 20°C and < 30% RH for 1week.

(i) **(ii)**

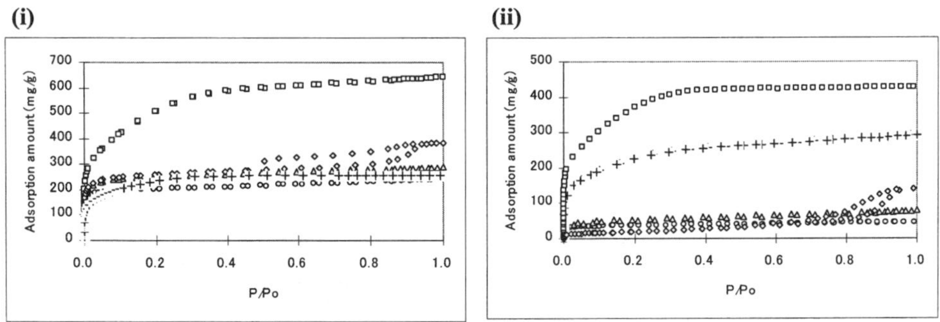

Figure 3 *Nitrogen sorption following triacetin adsorption at (i) 80°C and < 30% RH for 1day and (ii) 20°C and < 30% RH for 1week*
(o) A7, (□) A20, (◊) Meso A7, (Δ) GC and (+) FA20

The pore characteristics are shown in Table 3.

	BET surface area (m²/g)		Micropore volume (cm³/g)		Mesopore volume (cm³/g)		Adsorbed triacetin (g/g)	
Temperature	20°C	80°C	20°C	80°C	20°C	80°C	20°C	80°C
Carbon								
A7	630	120	0.22	0.04	-	-	0.034	0.137
A20	1260	690	0.68	0.51	-	-	0.042	0.175
Meso-A7	790	50	0.28	-	0.20	0.14	0.058	0.263
GC	670	150	0.29	0.05	-	-	0.033	0.122
FA20	660	580	0.31	0.27	-	-	0.011	0.091

Table 3 *Nitrogen sorption following triacetin adsorption at 80°C and < 30% RH for 1day, and 20°C and < 30% RH for 1week*

After exposure to triacetin at 20°C, the available surface areas of A7, A20, Meso-A7 and GC have decreased by *ca* 30% of their original values. After exposure to triacetin at 80°C, loss of available surface area was *ca* 80% of the original value (95% in the case of the mesoporous carbon). FA20 is the exception where the surface area has decreased by less than 20% even at 80°C.

The adsorption data suggests that triacetin adsorption is temperature-dependent and increases with increasing temperature (over the range 20 to 80°C). As the large triacetin molecule strongly interacts with micropores and the intrapore diffusion is slow, adsorption at a higher temperature is preferable to attain an equilibrium adsorption regardless of physical adsorption mechanism. Meso-A7 has the greatest triacetin adsorption capacity. FA20 shows the greatest resistance to triacetin adsorption. Although the slit pore width of FA20 is 1.1 nm, which is close to the 1.2 nm slit pore width of the A20 carbon, it has a lower triacetin adsorption capacity.

3.4 Effect of humidity on triacetin adsorption

Figure 5 shows water adsorption isotherms of the carbon materials at 30°C.

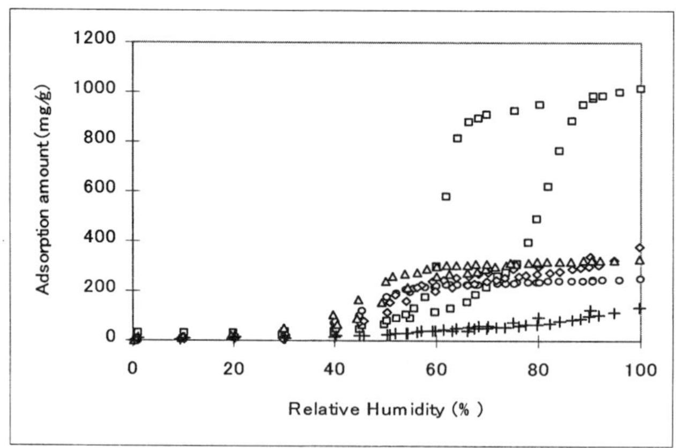

Figure 5 *Water adsorption isotherms of carbon materials at 30°C*
(○) A7,   (□) A20,   (◊) Meso A7,   (Δ) GC and   (+) FA20

With the exception of the extremely hydrophobic FA20, water sorption increases sharply from 40% RH. The water adsorption isotherm of FA20 has no sharp increase in sorption and the adsorption amount is much smaller then that of A7, which has similar pore size and surface area. Meso-A7 shows a greater water adsorption capacity than A7. Water adsorption behaviour on carbon fibre can be explained via a cluster-stabilized mechanism.[9,10]

Figure 6 shows nitrogen adsorption isotherms at 77 K for A7 after triacetin adsorption under different relative humidities at 20°C for 1week. Similar isotherms were obtained for the other carbons, although the mesoporous meso-A7 carbon displayed more extreme differences when humidities were changed. Figure 7 shows the nitrogen adsorption isotherms at 77 K for meso-A7 after triacetin adsorption under different relative humidities at 80°C for 1week.

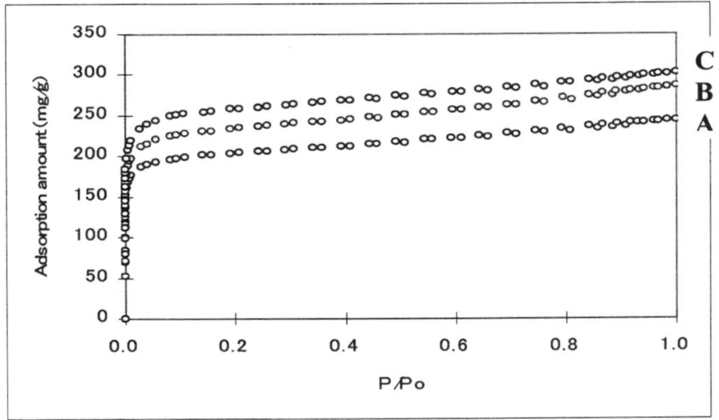

Figure 6 *Nitrogen sorption on A7 following triacetin adsorption at 20°C and various RH for 1 week*

   A 10% RH   **B** 60% RH   **C** 90% RH

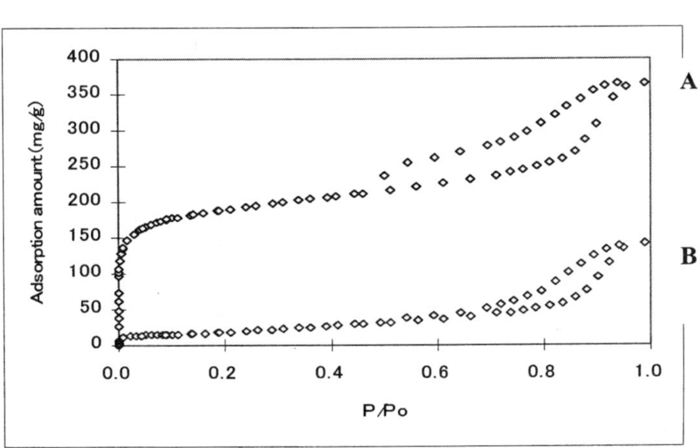

Figure 7 *Nitrogen sorption on meso-A7 following triacetin adsorption at 80°C and various RH for 1 week*

   A 60-70% RH   **B** <30% RH

In summary, the results showed that the lower the relative humidity, the greater the triacetin adsorption.

For the meso-A7 carbon it can be seen that the surface area and porosity is significantly affected by the triacetin adsorption in the presence of coadsorbed water. However, the fundamental shape of the adsorption hysteresis remains even after the remarkable depression in the extent of nitrogen adsorption. This is probably because almost all the triacetin molecules are adsorbed in the micropores.

The pore characteristics of all the carbons after triacetin adsorption at 20°C are shown in Table 5.

Carbon	Relative Humidity (%)	BET surface area (m^2/g)	Micropore volume (cm^3/g)	Mesopore volume (cm^3/g)	Adsorbed triacetin (g/g)
A7	10	630	0.22	-	0.034
	55	720	0.26	-	0.023
	90	800	0.30	-	LOQ
A20	10	1260	0.68	-	0.042
	55	1400	0.76	-	0.033
	90	1550	0.83	-	0.026
GC	10	670	0.29	-	0.033
	55	850	0.34	-	0.021
	90	970	0.37	-	LOQ

LOQ (Below the limit of quantification)

Table 5 *Nitrogen sorption following triacetin adsorption at 20°C and various RH for 1 week*

4 CONCLUSIONS

A method has been established to measure triacetin adsorption on carbon. Triacetin is physically adsorbed on carbon. Temperature, relative humidity, pore size and surface chemistry are all important factors for the adsorption of triacetin.

Triacetin sorption increases as temperature increases and decreases as the relative humidity increases (due to competing adsorption by water vapour).

Pore size plays an important role in triacetin adsorption; the introduction of mesopores was shown to increase the extent of triacetin adsoption by a factor of two.

Fluorinated activated carbon fibre showed the strongest resistance to triacetin adsorption, compared to carbon materials of similar surface area.

Triacetin adsorption on carbon can be minimized using a microporous carbon with surface fluorine groups and maintaining a low temperature and high humidity.

References

1. D.R.Lide ed-in-chief, CRC Handbook of Chemistry and Physics, 87th edition, CRC Press, Taylor & Francis Group, Florida, 2006.

2. D.Layten and M.T.Nielsen in Tobacco: Production, Chemistry and Technology, pp. 265–284, Blackwell Science, Oxford, 1999.

3. M.Muramatsu in Studies on the Transport Phenomena in Naturally Smouldering Cigarettes, Sci. Papers Central Res. Inst. Japan Tob. Salt Mon. Corp., 1981, **123**, 9-77.

4.   D.Layten and M.T.Nielsen in Tobacco: Production, Chemistry and Technology, pp. 398–404, Blackwell Science, Oxford, 1999.

5.   K.S.W.Sing, D.H.Everett, R.A.W.Haul, L.Moscou, R.A.Pierotti, J.Rouquerol and T.Siemieniewska, Pure Appl.Chem, 1985, **57**, 603

6.   M.Kenny, K.Sing and C.Theocharis, Fundamentals of Adsorption, 1992, 323-332.

7.   K.Kaneko and C.Ishii, Colloids Surf 1992;**67**, 203-212.

8.   N.Setoyama, T.Suzuki and K.Kaneko, Carbon, 1998; **36(10)**, 1459-1467.

9.   T.Ohba, H.Kanoh and K.Kaneko, J.Amer.Chem.Soc, 2004; **126**, 1560-1562.

10. T.Ohba, H.Kanoh and K.Kaneko, J. Phys. Chem. B, 2004, **108**, 14964-14969.

ADSORPTION OF CO_2 ON CARBONACEOUS ADSORBENTS

B.Buczek and J.Ziętkiewicz

Faculty of Fuels and Energy, AGH – University of Science and Technology, Cracow, Poland

1 INTRODUCTION

Carbon molecular sieves (CMSs) find today wide industrial application in gas mixture and gas separation processes.[1,2]
CMSs belong to the group of microporous materials characterized by a relatively high adsorption capacity and selectivity towards a wide spectrum of gases. They possess amorphous structures with well developed surface area and pore size comparable to the effective diameter of small molecules.[3]
The mechanism of gas mixture separation depends on the type of adsorbent. Two different mechanisms are distinguished. In the first mechanism, the pore system is sufficiently wide to enable fast diffusion, while the separation is caused by the selective adsorption dependent on different adsorption energies of gas components. The second mechanism is based on the kinetically controlled gas diffusion, caused by constrictions of the pore apertures. Here the mouth of pores are in the same range as those of gas molecules.
The porous structure (pore volume, pore size distribution) is a feature that controls their applicability in the separation processes.

2 METHOD AND RESULTS

Two types of adsorbents produced on an industrial scale were investigated. One was originally used in equilibrium processes and the second in kinetically controlled separation processes.

2.1 Porous structure of equilibrium CMSs.

Porous structure of carbon molecular sieves BF-H_2 and D55/2 produced by Carbotech was charecterized by nitrogen adsorption isotherms obtained at 77K using volumetric apparatus SORPTOMATIC 1900 (CE Instruments).
Volume of micropores (W_0) and characteristic energy of adsorption (E_0) were determined according to Dubinin-Radushkevich equation.[4] Mesopores surface (S_{me}) was calculated by Dollimore-Heal method[5] and specific surface area from the BET equation. Pore size distribution in micropore range was calculated by the method of Horvath-Kawazoe.[6] which maximum value (D_{HKmax}) is shown in Table 1.

Average micropores width was calculated from relationship given by Stoeckli, Rebstein and Ballerini (D_{St_av})[7] and for comparison result from McEnaney (D_{ME_av})[8] equation. Results of measurements and calculations are presented in Table 1 and Figure 1 and micropore distribution in Figure 2.

Table 1 *Porous structure parameters*

Adsorbent	W_0 cm^3/g	E_0 KJ/mol	D_{HKmax} nm	D_{St_av} nm	D_{ME_av} nm	S_{me} m^2/g	S_{BET} m^2/g
BF-H$_2$	0.271	23.2	0.71	0.92	0.99	16	607
D-55/2	0.366	18.9	0.58	1.44	1.33	21.9	825

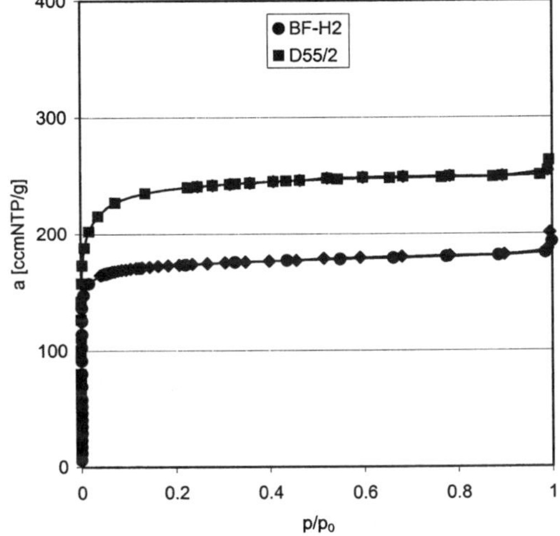

Figure 1 *Isotherms of nitrogen adsorption at 77K*

Both active carbons with molecular sieves properties have isotherms of I-type according to IUPAC classification.[9]

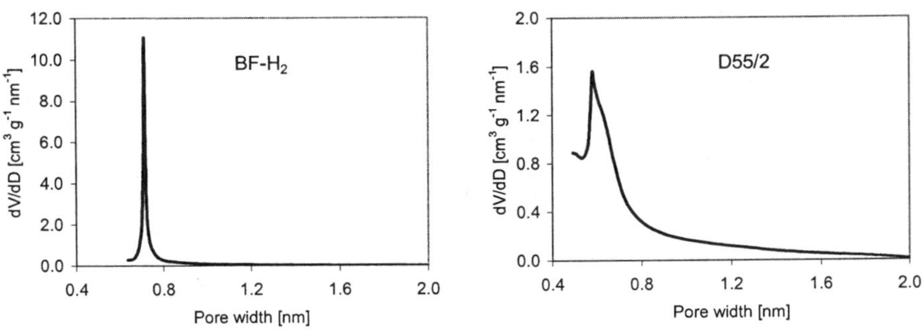

Figure 2 *Micropore size distribution of BF-H2 and D55/2 adsorbents*

2.2 Adsorption of CO_2 on equilibrium CMSs.

The isotherms of CO_2 sorption were measured in the pressure range up to 1000 Torr at temperature 273K using the SORPTOMATIC 1900 apparatus. Isotherms are shown in Figure 3.

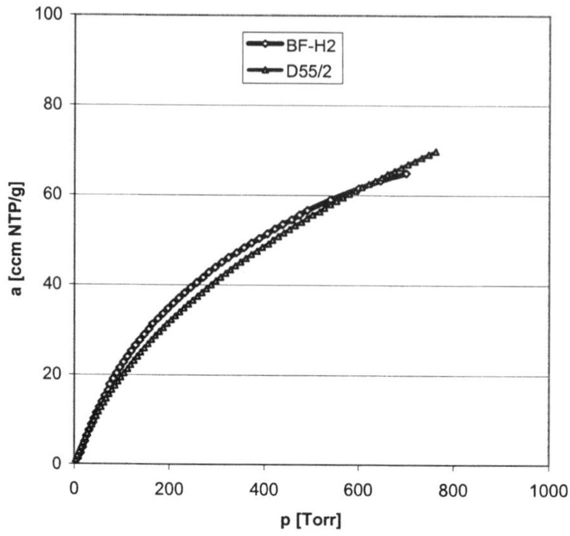

Figure 3 *Isotherms of carbon dioxide adsorption at 273K*

Approach based on Polanyi's potential theory of adsorption and Dubinin's theory of volume filling of micropores was used for the interpretation of adsorption isotherms.[10] The amount of CO_2 adsorbed in micropores was calculated.

In this case extrapolated value in co-ordinate of DR equation corresponds to adsorption in micropores at saturation pressure equal to 34.4 atm (from equilibrium line liquid-vapour). Above data was taken from NIST (National Institute of Standards and Technology).[11]

Figure 4 presents dependence of density of liquid CO_2 on temperature and predicted maximal density value of adsorbed phase in the range from critical temperature 304.1 to 220 K.[10] At the region close to critical point the significant differences between densities are observed.

The average density of adsorbed phase of CO_2 in micropores was calculated. It is higher than liquid density on saturation line at 273K (0.928 g/cm^3) ; these are respectively for BF-H$_2$; 1.08 g/cm^3 and D55/2; 0.97 g/cm^3.

Differences in this value can be connected with higher characteristic energy (see Table 1) and narrow micropore size distribution (Figure. 1) for BF-H$_2$ carbon molecular sieve. In the case of the first analysed adsorbents, the value of density corresponds to the maximum value of density of adsorbed phase at that temperature.[10]

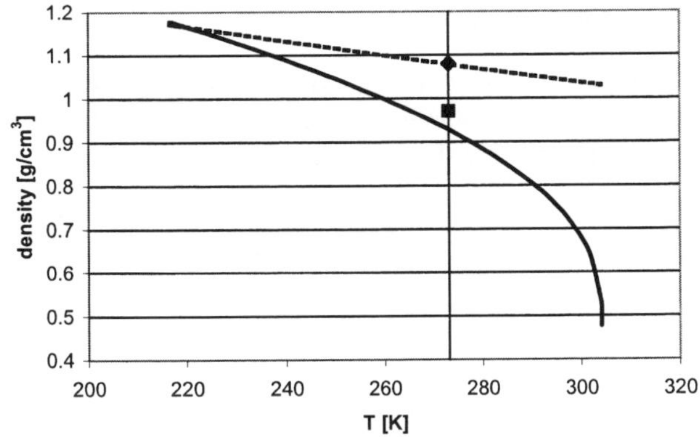

Figure 4. *Density of CO_2 on saturation line (solid line); in adsorbed phase (doted line);* ◆ *CMS-H_2;* ■ *D55/2*

2.3 Adsorption of CO_2 on kinetic CMSs.

The isotherms of adsorption CO_2 were also measured for two adsorbents used originally for separation gaseous mixtures e.g. CO_2/CH_4, N_2/O_2 in processes controlled by by kinetics of adsorption:

D55/2-3A (Carbotech) and WSC/K (Norit).

These adsorbents cannot be characterized by nitrogen adsorption at 77K, due to diffusion limitation. Carbon dioxide isotherms are presented in Figure 4.

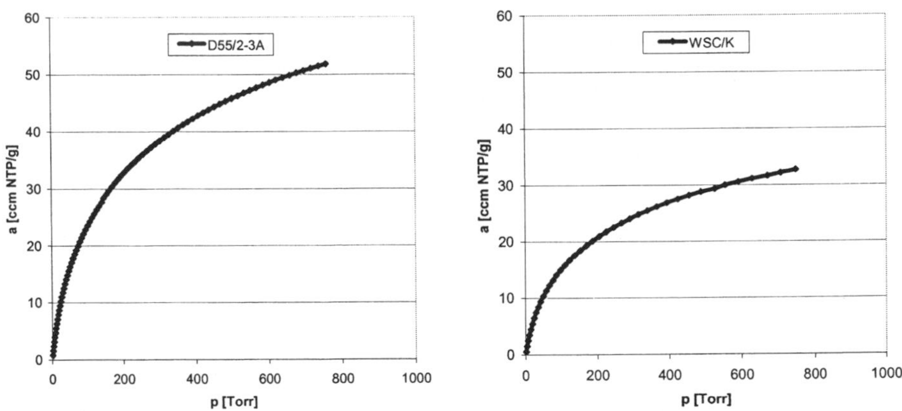

Figure 4 *Isotherms of carbon dioxide adsorption at 273K*

Volume of micropores is one of the parameters which affect the separation processes. Thus evaluation this parameter of micropore structure is of great importance. From the Dubinin-Radushkevich equation, the amount of CO_2 adsorbed in micropores (a) was calculated, and next their volume (W_0): D55/2-3A; 0.177 cm³/g and for WSC/K; 0.111 cm³/g. For calculation density of adsorbed phase of CO_2 equal 1.08 g/cm³ was taken.

2.4 Kinetics of adsorption CO_2 and CH_4.

For adsorbents (D55/2-3 and WSC/K) measurements of kinetics of CO_2 and CH_4 sorption were carried out.[13,14] in volumetric apparatus at temperature 293 K and roughly constant pressure 1 bar. Results are shown in Figure 5.

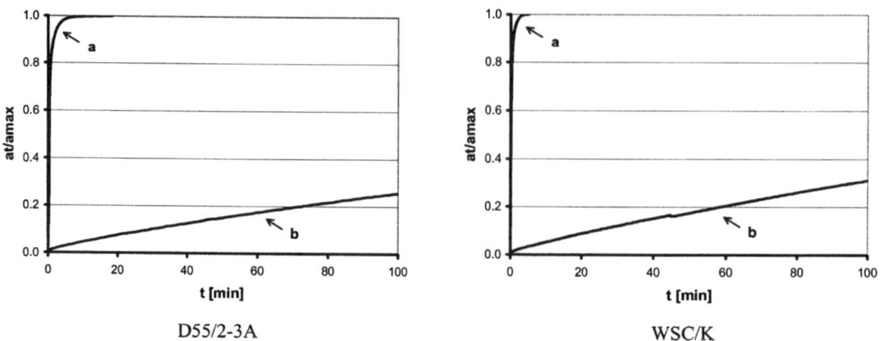

Figure 5 *Kinetics of adsorption of CO_2 (a) and CH_4 (b)*

From kinetics parameter of diffusion De/R^2 was calculated using equation given by Crank[12] and parameters ratio for adsorbates (Table 2).

Table 2 Diffusion parameters

Adsorbent	Methane $De/R^2 * 10^6$ s^{-1}	Carbon dioxide $De/R^2 * 10^3$ s^{-1}	Parameters ratio CO_2/CH_4
D55/2-3A	2,56	2,31	902
WSC/K	2,01	2,60	1290

Kinetically controlled separation processes are influenced by volume of micropores and their aperture. For carbon molecular sieve WSC/K sorption of carbon dioxide is faster and parameters ratio is greater.

2.5 Results of test in dynamic conditions

Separation abilities of kinetics CMSs were checked in a test column under dynamic conditions.[9,10] The separated mixture had a concentration of 50% CO_2 and 50% CH_4.
The measurement consisted of three steps:
    i. sample degassing
    ii. pressurization with feed to 1.5 bar
    iii. cocurrent flow connected with methane analysis in outflow gas.
The results of test are shown on Figure 6.

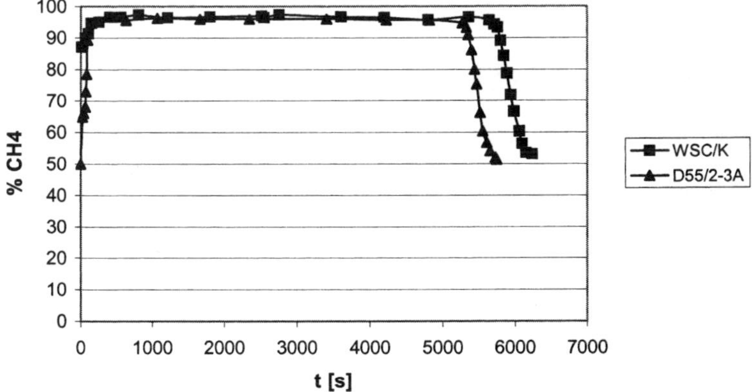

Figure 6 *Dependency of methane concentration in outflow gas in time*

In dynamic experiment gas with content 95% of methane was obtained. Time in which concentration of methane in outlet rapidly decreases can be used as the parameter which differentiated adsorbents usage for separation processes.

CONCLUSIONS

Carbon dioxide as the adsorbate is useful for characterization microporous adsorbents especially when standard methods (recommended by IUPAC) cannot be applied.[15,16] Isotherms of CO$_2$ sorption at low pressure give information relating to the structure of micropores, thereby allow control the process of preparation of kinetic CMSs adsorbents.

Tests in dynamic conditions can determine influence of two factors: differences in micropores volume and in kinetics of sorption on separation processes in which CO$_2$ is one of components.

The authors are grateful to MNiSzW (Project 11.11.210.125) for its financial support of this work

References

1 N.O. Lemcoff, *Studies in Surface Sciences and Catalysis,* 1998, **120**, 347

2 P. Samaras, X. Dabou, G.P. Sakellaropoulos, *Studies in Surface Sciences and Catalysis,* 1998, **120**, 425

3 M. Bałys, B. Buczek, E. Vogt, *Studies in Surface Sciences and Catalysis (COPS VI)*, Elsevier, 2002, **144**, 225.

4 M. M. Dubinin, *Carbon,* 1989, **27**, 457.

5 D. Dollimore, G. R. L. Heal, *J. Appl. Chem.*, 1964, **14**, 109.

6 G. Horvath, K. Kawazoe, *Chem. Engn. Jpn.*, 1983, **16**, 6, 470.

7 H. F. Stoeckli, P. Rebstein, L. Ballerini, *Carbon*, 1990, **28**, 6, 907.

8 B. Mc Enaney, *Carbon,* 1987, **25**, 69.

9 IUPAC Recommendations *Pure Appl. Chem.* **1994**, 66, 1739.

10 M. M. Dubinin, *Adsorption and Porosity*, WAT, Warszawa 1975.

11 http://webbook.nist.gov/chemistry/fluid/.

12 J. Crank, *The mathematics of diffusion*, Clarendon Press, Oxford, 1975.

13 M. Bałys, B. Buczek, J. Ziętkiewicz, *Inżynieria i Ochrona Środowiska*, 2002, **5**, 2, 117.
14 M. Bałys, B. Buczek, J. Ziętkiewicz, *Inżynieria i Ochrona Środowiska*, 2004, **7**, 3-4, 305.
15 D. Lozano-Castello, D. Cazorla-Amoros, A. Linares-Solano, *Carbon*, 2004, **42**. 1233.
16 M.B. Sweatman, N. Quirke, *Langmuir*, 2001, **17(16)**, 5011.

CHARACTERIZATION OF VARIOUS MICRO AND MESOPOROUS MATERIALS FOR TOLUENE ADSORPTION AT LOW AND HIGH CONCENTRATION

X. Canet,[1] A. Vantomme,[2] F. Gilles,[2] B.-L. Su,[2] and G. De Weireld[1*]

[1]Laboratoire de Thermodynamique et Physique mathématique, Faculté Polytechnique de Mons, boulevard Dolez 31, 7000, Mons, Belgium (guy.deweireld@fpms.ac.be)
[2]Laboratoire de Chimie des Matériaux Inorganiques, Faculté Notre Dame de la paix, rue de Bruxelles 61, 5000 Namur, Belgium

1 INTRODUCTION

Volatile Organic Compounds (VOCs) released into the atmosphere by industrial activities and automobiles lead to some important environmental problems such as greenhouse effect, ozone layer destruction and can often modify the Chapman cycle (photochemical reaction).[1] VOCs, also responsible for direct health effects, are emitted by a wide array of products numbering in thousands. All of these products can release organic compounds during their use and, to some degree, when they are stored. Concentrations of many VOCs are consistently higher indoors (up to ten times higher) than outdoors. Ever increasingly stringent legislation forces many companies to become tougher monitoring and controlling solvent emissions. Thus, in order to reduce these solvent emissions, industry needs efficient filters and processes at both high and low concentrations.

The determination of adsorption isotherms for adsorbents – VOCs systems is then of prime importance for industry and science. Adsorption on microporous solids is a well known technique for gas purification, often used after a cryogenic process to obtain high purity gases.

Faujasite type zeolites are well known for their adsorption properties especially at low concentration.[2-9] Microporous zeolites have a high selectivity and not important adsorption capacities.

Adsorption on granular activated carbons bed is a common process for VOC recovery because of its low cost. Activated carbons have important adsorption capacities at high concentration but they are non selective.[10-14] These materials are a main interest for industrial air treatment units.

What about mesoporous materials? Some of them have already been tested[15] only at low concentration, then it might be interesting to undertake a comparison of existing materials in a large range of concentration and temperature.

In this work, we characterized various adsorbent types (activated carbon, zeolites faujasite X and Y, highly ordered mesoporous carbon materials CMI8, mesoporous SBA-15 carbonated, and mesoporous ZrO_2-TiO_2) and we carried out toluene adsorption measurements: at low concentration (between 200 and 400°C) with pulse chromatography to obtain Henry constants and at high concentration (between 30 and 210°C) with gravimetric technique to measure maximum adsorption capacities.

These two sets of adsorption parameters obtained with two different techniques in a large temperature range cover a wide range of applications.

2 MATERIALS

2.1 Materials preparation

Adsorbents used in this work are of various types :
- commercial granular activated carbon: NC100 (PICA Company, France);
- commercial faujasite zeolites samples: LiLSX (Si/Al ratio = 1 provided by Tricat GmbH), NaX (Si/Al ratio = 1.2, provided by Zeolist), and NaY (Si/Al ratio = 2.4, provided by Union Carbide);
- highly ordered mesoporous carbon materials CMI-8;
- mesoporous SBA-15 carbonated;
- mesoporous ZrO_2-TiO_2

2.1.1 Synthesis of NC100. [16] This commercial activated carbon is based on coconut shell which is carbonised at 850°C and activated at the same temperature with steam.

2.1.2 Synthesis of LiLSX, NaX and NaY. These zeolitic materials are synthesized in a basic medium (NaOH and KOH) containing silica and alumina sources. This medium is then heated without stirring at appropriate temperature (depending on the desired zeolite).The final product is finally washed and dried.[17] LiLSX is obtained by ion-exchanging of NaLSX with a lithium chloride solution.

2.1.3 Synthesis of CMI 8. [18] The synthesis of the ordered mesoporous carbon material with high surface area nominally designated CMI-8, has been achieved using the hexagonally ordered mesoporous silica CMI-1 as an exotemplate.

Impregnation and carbonization experiments were performed with sucrose. The optimum nanocasting procedure for the mesoporous carbon replica CMI-8 was to dissolve 1.25 g of sucrose in a 5 ml acidic solution (pH 0.5; H_2SO_4). A defined amount of CMI-1 materials corresponding to a porous volume of 1 cm^3 was added to this solution and then stirred for 3 hours. The resultant mixture was dried at 100°C in a muffle oven for 6 hours, followed by 6 hours at 160°C. The CMI-8 containing the partially carbonizing masses was added into an aqueous acidic solution (pH 0.5; H_2SO_4) containing 0.8 g of sucrose. The resultant mixture was dried again as describe above. The color of the sample turned dark brown. This powder sample was heated to 950°C under N_2 atmosphere for 6 hours. The carbon-silica composite obtained was washed in 40% fluorhydric acid solution for 56 hours in order to dissolve completely the silica CMI-1 exotemplate.

2.1.4 Synthesis of SBA-15c [19]. The synthesis of SBA-15c was performed using SBA-15 silica as the template and sucrose as the carbon source. The calcined SBA-15 was impregnated with aqueous solution of sucrose containing sulfuric acid. Briefly, 0.38 g of SBA-15 was added to a solution obtained by dissolving 0.48 g of sucrose in 5 ml of H_2SO_4 solution (pH = 0.5) during 3 hours. The mixture was placed in a drying oven for 6 hours at 100°C, and subsequently the oven temperature was increased to 160°C and maintained there for 6 h. The sample turned dark brown or black during the treatment in the oven.

The silica sample, containing partially polymerized and carbonized sucrose at the present step was treated again at 100°C and 160°C using the same drying oven after the addition of 0.31 g of sucrose in 5 ml of H_2SO_4 solution (pH = 0.5). The carbonization was completed by pyrolysis with heating to typically 950°C under N_2 atmosphere for 6 hours.

The carbon-silica composite obtained after pyrolysis has been washed with 40% fluorhydric acid at room temperature for 24 hours, to remove the silica template.

2.1.5 Synthesis of ZrO₂-TiO₂.[20] A 15 %wt micellar solution of cetyltrimethylammonium bromide (CTMABr) is prepared by dissolving CTMABr in an aqueous acidic solution (pH=2) at 40°C under stirring for at least 3 h. An appropriate content of zirconium propoxide (Zr(OC₃H₇)₄) and titanium propoxide (Ti(OC₃H₇)₄) is added dropwise into the above solution with a surfactant/(Zr + Ti) molar ratio of 0.33. This solution has a 50/50 metal to metal molar ratio as Zr–Ti source. After further stirring for 1 h, the mixture is transferred into a Teflonlined autoclave, and heated at 80°C for 1 day. The product is filtered by Soxhlet extraction with ethanol for at least 30 h in order to remove the surfactant species, and then dried at 40°C in vacuum.

2.2 Materials characterization

Nitrogen adsorption-desorption isotherms were obtained at -196°C over a wide relative pressure range from 0.01 to 0.995 with a volumetric adsorption analyzer ASAP2010 or TRISTAR 3000 manufactured by Micromeritics. Mean pore size and pore size distributions were determined by the BJH method from the analysis of the adsorption branch of the isotherms. The BET surface areas, pore volumes and pore size obtained from these measurements are summarised in Table 1.

Table 1 *Specific surface area, pore volume and mean pore size of adsorbents*

	Specific surface area (m²/g)	Pore volume (cm³/g)	Mean pore size (nm)
CMI 8 [18]	1999	1.2	20.4
SBA-15c [19]	1103	0.87	2.5
LiLSX 1.0	704	0.395	0.52
NaY 2.4	621	0.323	
NaX 1.2	627	0.333	
NC 100 [16]	1493	0.667	
ZrO₂-TiO₂ [20]	746	0.6	2.3

3 ADSORPTION STUDY

3.1 Pulse Chromatography Technique

The liquid VOC is injected at the inlet of a column filled with the adsorbent in which circulates a carrier gas (nitrogen). The adsorbate concentration profile at the outlet of the column is then measured using an appropriate VOC detector. The Henry constant of adsorption is determined by a mathematical treatment of this concentration profile. [2-9, 21]

A column filled with the adsorbent is located in the oven of a chromatograph equipped with a Thermal Conductivity Detector (TCD). A purificator, filled with 3A zeolite, is placed in the system to remove water contained in the carrier gas. A mass flow controller (0-5 Nl/h) was used to regulate the flow of carrier gas which is measured at the TCD outlet with a soap film flowmeter in order to equilibrate the flow in each part of the TCD. Before each measurement, the temperature was raised up to 400°C with a rate of 1°C/min and maintained at this temperature during a 10-hour period under an inert gas flow in order to regenerate the adsorbent. After this regeneration step, temperature is set to experimental temperature. Liquid adsorbate is injected with a syringe in the heated injector, passes through the column where it is adsorbed-desorbed and goes to the TCD. The outlet concentration profile is then recorded as a function of time.

The first moment μ of the signal (retention time of the adsorptive in the column) is related to the dimensionless Henry constant K (Equation 1):

$$\mu = \frac{V_{column}}{Q_T}\left[\varepsilon + (1-\varepsilon)K\right] + \frac{V_d}{Q_T} \tag{1}$$

in which V_{column} is the column volume (m^3), V_d is the dead volume (m^3), Q_T is the total volume flow rate (m^3.s^{-1}), ε is the total bed porosity.

The determination of K is only valid if the column is working in the Henry's domain. This hypothesis is checked by performing several experimental runs with different injected volumes of adsorbate for which retention times are calculated. When the retention time does not depend on the injected volume, the hypothesis is valid. Typically, the injected volume is between 0.02 and 0.2µl.

The temperature dependence of the dimensionless Henry constant shows that ln K as a function of $1/T$ is a linear plot which allows the determination of both energetic and entropic parameters ΔU and K_0 (Equation 2).

$$K = K_0 \exp\left(\frac{-\Delta U}{RT}\right) \tag{2}$$

in which ΔU is the adsorption potential energy (J.mol^{-1}) and K_0 is a pre-exponential constant which takes into account the reduction of the freedom degree of the molecule when passing from the gas phase to the adsorbed phase.

At low concentration in the Henry domain, the adsorption is more important on LiLSX than on mesoporous carbonated CMI8 and SBA-15c despite a specific area strongly lower (Figure 1).

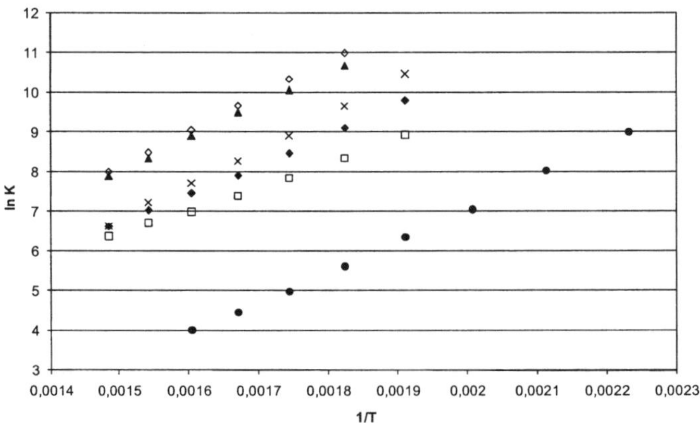

Figure 1 *Plot of Henry constant logarithm versus 1/T (in K^{-1}) for Toluene adsorption on (◊) LiLSX 1.0, (▲) NaX 1.2, (♦) NaY 2.4, (□) CMI8, (x) SBA-15c and (●) ZrO$_2$-TiO$_2$*

For other materials, Faujasite and oxide, the Henry constant follows the trend the higher is the specific area, the higher is the Henry constant. The activated carbon Pica

NC100 has not been studied by the pulse chromatography technique because it can not be heated higher than 150°C (it is not compatible with the temperature range used for other adsorbents).

If we consider the potential adsorption energy ΔU for the studied materials, the lowest one is for CMI8 with 49.7 kJ.mol^{-1} and the highest is 74.3 kJ.mol^{-1} for LiLSX (close to SBA-15c at 74.0 kJ.mol^{-1}). NaX 1.2, NaY 2.4, and ZrO_2-TiO_2 have intermediate values, 69.0, 61.8 and 66.8 kJ.mol^{-1}, respectively.

3.2 Gravimetric Technique

The dynamic gravimetric technique consists in exposing a clean adsorbent maintained at fixed temperature to a constant carrier gas flow rate containing a VOC. A thermobalance allows the direct measurement of the uptaken mass of the sample: it is related to the VOC partial pressure in the VOC–inert gaseous mixture. A full description of the gravimetric apparatus as well as the experimental procedure is completely presented elsewhere.[21-22] The experimental device is a thermobalance (model B111) which is part of a TG 111 thermoanalyser (Setaram) allowing the measurement of massic (TG signal) adsorption data.

The experimental setup also contains a gas generation system producing adsorbate–inert mixtures. It consists in saturating the inert gas flow by bubbling it through liquid VOC contained in a saturator maintained at constant temperature.

The adsorbate partial pressure (P_{voc}) is calculated in Equation 3:

$$P_{VOC} = y \cdot P_{atm} \tag{3}$$

in which y is the adsorbate molar ratio (Equation 4) and P_{atm} is the atmospheric pressure (the operating pressure of the TG 111 thermoanalyser).

$$y = \frac{P_{sat}(T_{sat})}{P_{out}} \tag{4}$$

where $P_{sat}(T_{sat})$ is the VOC vapour pressure at the saturation temperature T_{sat} and P_{out} is the pressure at the outlet of the saturators measured using a pressure transducer. The VOC molar ratio of the mixture is close to 18300 ppm which is given by the vapour pressure law of toluene at 12°C (which is the saturation temperature T_{sat}) with a flow of 2 Nl.h^{-1}.

The excess specific adsorbed amount θ (g.g^{-1}) is given by Equation 5:

$$\theta = \frac{m_a}{m_{os}} \tag{5}$$

where m_a is the adsorbed mass (g), m_{os} is the outgassed sample mass (g) and the calculation of m_a and m_{os} is based on Equation 6 and Equation 7:

$$m_a = TG_{upt-eq} - TG_{upt-outg} \tag{6}$$

$$m_{os} = TG_{upt-outg} - TG_{empty} \tag{7}$$

The indices 'outg' and 'eq' refer to the values of the TG_{upt} signal just before adsorption (outgassed adsorbent) on the one hand and at the adsorption equilibrium on the other hand. The complete treatment of this signal requires the determination of the TG_{empty} signal measured with both empty crucibles. This measurement is performed under the same experimental conditions (temperature and gas flow rate) as during the adsorption experiment.

CMI-8 adsorbed amount presents an important temperature dependence between 30°C and 90°C, this amount is divided by more than 3.3 (decrease of 70.1% between 30°C and 90°C). At the same time, the commercial activated carbon NC100 exhibits a decrease of 26.8% and the mesoporous SBA-15c a decrease of 45.7%. The CMI-8 has the highest adsorption capacity at 30°C and it can be desorbed quite easily with increasing temperature. This CMI8 temperature behaviour can be explained by the ΔU far lower from the other materials; which is the result of low interactions between toluene and the material structure. The activated carbon is difficult to regenerate with increasing temperature; moreover, it cannot be heated higher than 150°C. The temperature influence on adsorption capacities can be used to adapt the regeneration mode for many adsorbents like the activated carbon or faujasite zeolites which are less temperature dependent than CMI-8.

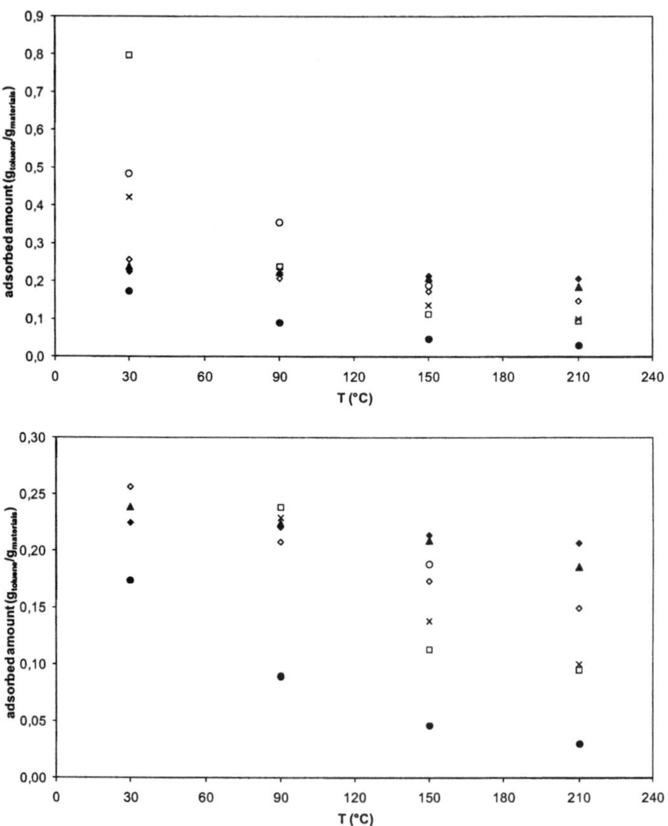

Figure 2 *Excess adsorbed amount versus temperature for (◊) LiLSX 1.0, (▲) NaX 1.2, (♦) NaY 2.4, (□) CMI8, (x) SBA-15c, (○) NC100 and (●) ZrO₂-TiO₂*

In order to compare structural differences such as specific area, the surface excess adsorbed amount is calculated (Equation 8).

$$\theta_S = \frac{\theta}{S_{BET}} \tag{8}$$

θ_S is expressed per unit of adsorbent surface area ($g_{toluene}.m^{-2}$).

The excess adsorbed amount per unit of surface area, more representative of the interactions between adsorbent and adsorbate, is plotted in Figure 3. For zeolites, the temperature effect is linear as for activated carbon NC100, whereas the temperature dependence is exponential for mesoporous materials (CMI8, SBA-15c, ZrO_2-TiO_2). At low temperature, CMI-8 has the highest surfacic excess adsorbed amount and the lowest one between 90 and 210°C. For high concentration, CMI-8 appears to be one of the most promising materials of our selection for adsorption capacities.

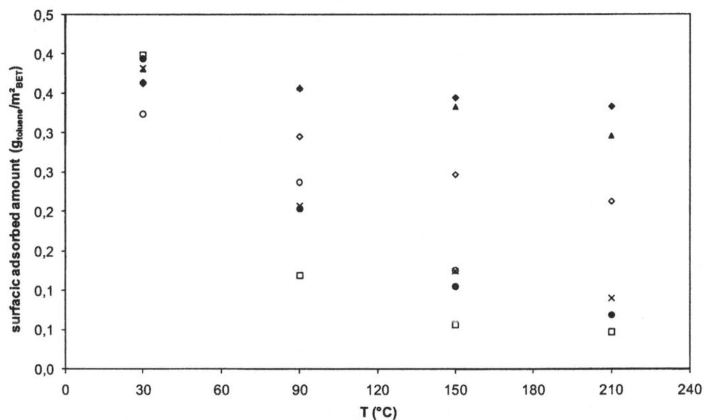

Figure 3 *Surface adsorbed amount versus temperature for (◊) LiLSX 1.0, (▲) NaX 1.2, (♦) NaY 2.4, (□) CMI8, (x) SBA-15c, (○) NC100 and (●) ZrO₂-TiO₂*

4 CONCLUSION

This behaviour comparison gives each material advantage at low or high concentration and is interesting to obtain a first screening of adsorbents. The influence of temperature on adsorption capacities is described and this can be useful to adapt the regeneration mode for certain adsorbents like the activated carbon or zeolites which are less temperature dependent than CMI-8. This study can eventually be used to find new ways to develop materials with higher performance either at high or low VOC concentration.

Acknowledgements

This work was supported by the European Program InterReg III (Programme France-Wallonie-Flandre, FW-2.1.5) and by the Ministry of Environment and Natural Resources of the Walloon region DGRNE (Belgium).

References

1 P. Le Cloirec, Les composés organiques volatils (COV) dans l'environnement, *Technique et Documentation Lavoisier*, 1998.

2 J.F. Denayer and G.V. Baron, *Adsorption*, 1997, **3**, 251.

3 J.F. Denayer and G.V. Baron, J.A. Martens, P.A. Jacobs, *J. Phys. Chem. B*, 1998, **102** (17), 3077.

4 E. Diaz, S. Ordonnez, A. Vega and J. Coca, *J. Chromatogr. A*, 2004, **1049**, 139.

5 O. Inel, D. Topaloglu, A. Askin and F. Tumsek, *Chem. Eng. J.*, 2002, **88**, 255.

6 D. M Ruthven, *Principles of adsorption and adsorption processes*, John Wiley and Sons, Canada, 1984.

7 F. Tumsek and O. Inel, *Chem. Eng. J.*, 2003, **94**, 57.

8 X. Canet, F. Gilles, B.-L. Su, G. De Weireld, M. Frere and P. Mougin, *J. Chem. Eng. Data*, 2007, **52**(6), 2117.

9 X. Canet, F. Gilles, B.-L. Su, G. De Weireld and M. Frere, *J. Chem. Eng. Data*, 2007, **52**(6), 2127.

10 P. Pre, F. Delage and P. Le Cloirec, *Environ. Sci. Technol.*, 2002, **36**, 4681.

11 F. Delage, P. Pre and P. Le Cloirec, *Environ. Sci. Technol.*, 2000, **34**, 4816.

12 P. Pre, F. Delage and P. Le Cloirec, *Fundamentals of Adsorption 7* , K Haneto, Kanoh H, Hanzawa Y, editors, IK International, 2001, 700.

13 S. Giraudet, P. Pre, H. Tezel and P. Le Cloirec, Carbon, 2006, **44**, 2413

14 K.-J. Kim, C.-S. Kang, Y.-J. You, M.-C. Chung, M.-W. Woo, W.-J. Jeong, N.-C. Park and H.-G. Ahn, *Catalysis Today*, 2006, **111**, 223–228

15 H.L. Tidahy, S. Siffert, J.-F. Lamonier, E.A. Zhilinskaya, A. Aboukaïs, Z.-Y. Yuan, A. Vantomme, B.-L. Su, X. Canet, G. De Weireld and M. Frere, *Studies in surface science and catalysis*, 2007, **160**, 201.

16 G. Finqueneisel, C. Vagner, T. Zimny and J. V. Weber, *Colloids and Surfaces A: Physicochem. Eng. Aspects*, 2004, **232** (2-3), 175.

17 H. Robson, *Verified Syntheses of Zeolitic Materials*, Second Revised Edition, Elsevier Netherlands, 2001.

18 A. Vantomme, L. Surahy and B.L. Su, *Colloids and Surfaces A: Physicochem. Eng. Aspects*, 2007, **300**, 65.

19 T.-Z. Ren, Z.-Y. Yuan and B.-L. Su, *Colloids and Surfaces A: Physicochem. Eng. Aspects*, 2007, **300**, 79.

20 A. Vantomme, A. Leonard, Z.-Y. Yuan and B.-L. Su, *Colloids and Surfaces A: Physicochem. Eng. Aspects*, 2007, **300**, 70.

21 J. Nokerman, X. Canet, P. Mougin, S. Limborg-Noetinger and M. Frere, *Meas. Sci. Technol.*, 2005, **16**, 1802.

22 S. Dutour, J. Nokerman, S. Limborg-Noetinger and M. Frere, *Meas. Sci. Technol.*, 2004, **15**, 185.

COMPUTER ASSISTED DESIGN OF POROUS SOLIDS FOR VOC'S EMISSION CONTROL BY ADSORPTION IN VEHICLE COLD START

M.V. Navarro, A. Rodriguez, T. García, R. Murillo, J. M. López and A. M. Mastral

Instituto de Carboquímica, CSIC, M Luesma Castán 4, 50018 Zaragoza, Spain

1 INTRODUCTION

Catalysts for vehicle emission control usually start working at around 473 K. The catalyst requires a period of 60 to 120 s to reach this temperature; during that time up to 80% of hydrocarbons in a drive cycle are emitted[1]. The emissions control during this "cold start" period is essential to reduce the environmental impact of gasoline engines. Among all the solutions studied up to now, the use of HC adsorbents, "hydrocarbon traps", before the three-way catalyst seems to be the most relevant from a scientific-technological point of view. The critical factors for any emission trap are: i) high adsorption capacity of hydrocarbons at low temperatures, ii) desorption starting at temperatures higher than 473 K, iii) a reversible adsorption process and iv) solid material resistant at temperatures higher than 1000 K. The maximum values of these factors are limited by properties of the solids like pore volume, pore structure, chemical nature of the solid, extra-framework atoms, etc. Zeolites have been found to be the preferred adsorbents for these applications, mainly due to their stability under severe process conditions.

In addition to experiments, molecular simulations have been proved to be a very useful tool to gain qualitative and quantitative insight on fluid diffusion and equilibrium loading of porous materials at a molecular level. Molecular simulations have some advantages over experiments like the calculation convenience and reducing costs. A variety of computational studies have been reported on alkane adsorption in zeolites but works concerning the simulation of adsorption of alkenes in zeolites are scarce[2-8] mainly due to the absence of a suitable force field.

Grand Canonical Monte Carlo (GCMC) simulations have been carried out to gain further insight into the propene adsorption process at the microscopic scale in ZSM-5. Data obtained for model crystalline solids with propane and propene are compared with experimental isotherms in order to optimise the parameters that reproduce the isotherm shapes and adsorption capacity ratio for these two adsorbates. The influence of specific types of extra-framework cations interacting with the adsorbate molecule is studied with this tool in order to suggest an optimum solid for process conditions of cold start emission control. This way, it would be possible to optimise the synthesis process of porous solids.

2   METHOD AND RESULTS

ZSM-5 zeolite is a siliceous micropore material. It comprises two different types of interconnected micropore channels[9]. One channel is a straight with dimensions 0.53 nm x 0.56 nm along the XZ plane; the other is a sinusoidal channel of dimensions 0.51 nm x 0.55 nm along the YZ plane ordered in the MFI framework.

Adsorption isotherms were computing using the GCMC algorithm implemented in Materials Studio 4.0 Software provided by Accelrys Inc. To generate the isotherms, the structures optimisation of adsorbate and adsorbent with the same force field is previously needed. Selection of a proper force field is a crucial factor for the simulation, because it is related to energy calculations of all molecular interactions directly. To get a reasonable force field, we used the force field recommended by MS software for zeolites and hydrocarbons, PCFF.

Propene was chosen as model molecule of hydrocarbon emissions from cold start emissions since it is one of the simplest alkenes found, with alkanes and aromatics, in the engine exhaust. Propane was also used in this study to test the force field comparing the simulated results with the ones presented in literature. To simulate the different ZSM-5 zeolites, several Si atoms were changed by Al atoms and Na^+, Ca^{2+} and Cu^{2+} atoms were introduced in the MFI structure as extra-framework cations to compensate the charge loss generated.

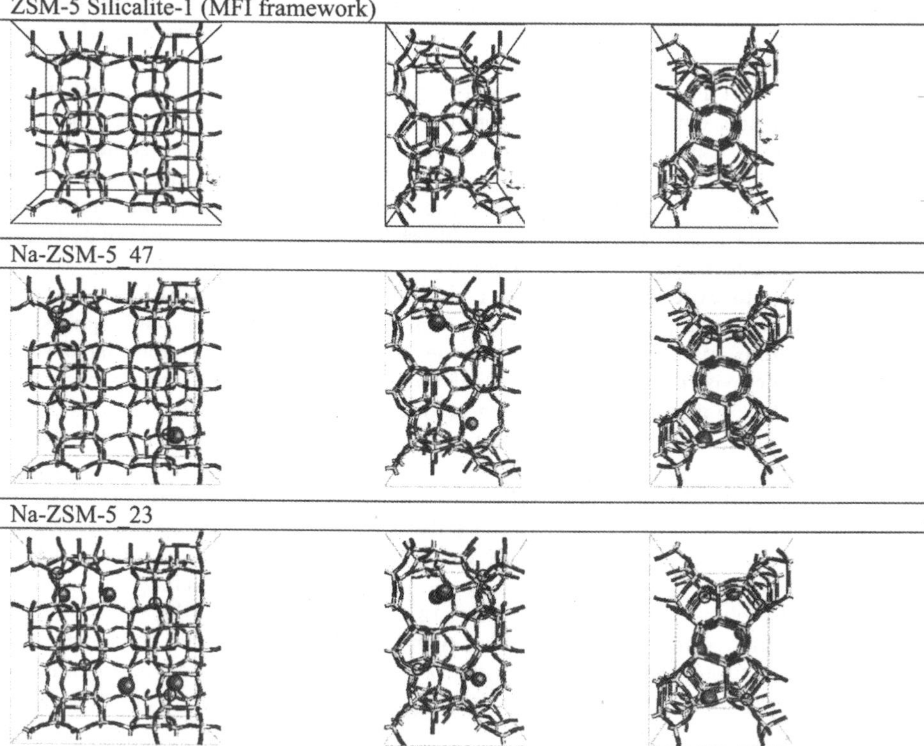

Figure 1   *Optimised unit cell representation of Silicalite-1 and solids exchanged with sodium. Dark atoms are O, light are Si, surrounded are Al and balls are Na.*

Ca-ZSM-5_47

Ca-ZSM-5_23

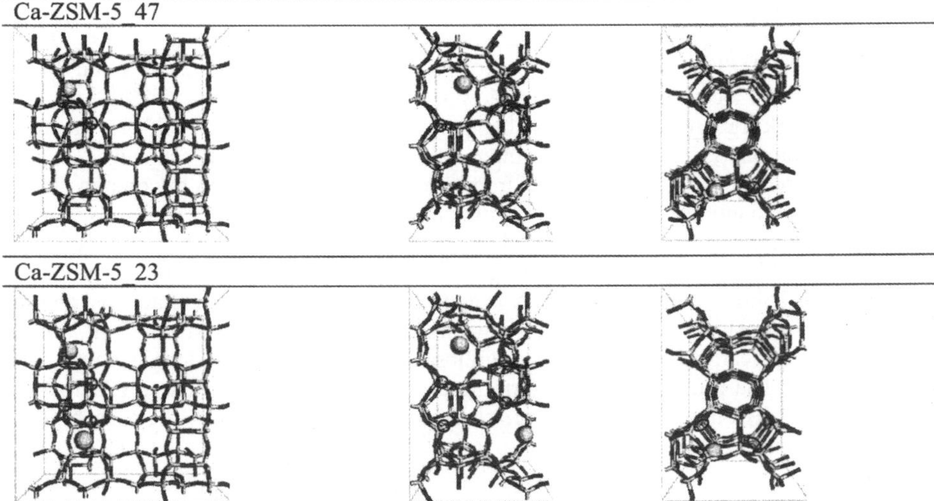

Figure 2   *Optimised unit cell representation of solids exchanged with calcium. Dark atoms are O, light are Si, surrounded are Al and calls are Ca.*

Cu-ZSM5_47

Cu-ZSM5_23

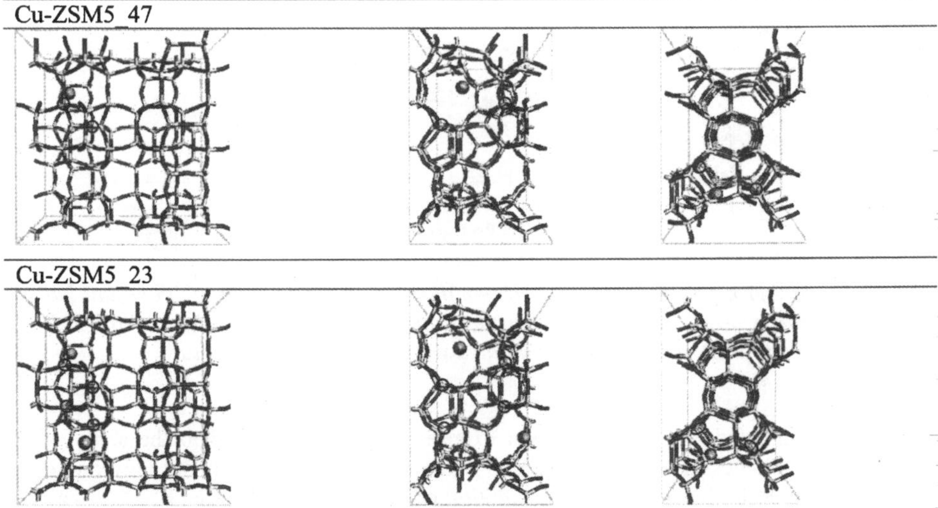

Figure 3   *Optimised unit cell representation of solids exchanged with copper. Dark atoms are O, light are Si, surrounded are Al and balls are Cu*

Figures 1, 2 and 3 compile the unit cells of the optimised solids. Solids are named M-ZSM-5_r, where M is the extra-framework cation incorporated to the ZSM-5 structure and r the Si/Al molar ratio. In addition, the base ZSM-5 silicalite unit cell is also shown for comparative purposes. The positions in the unit cell of both extra-framework cations and aluminium atoms were elucidated by the so-called Löwenstein's rule[10] and following studies[11-13] of energetic viability and statistically preferential location of atoms. These studies describe MFI structure with 12 nonequivalent crystallographic T_sites and T12 is the most probable location of Al atoms followed by T6. The scarce information related to extra-framework cation location place them close to T2 and T3, which are the most accessible positions in the edge of the intersection and the sinusoidal channel [14,15]. The relative positions of Al atoms and extra-framework cations allow them to compensate the strong Coulombic interactions of cations with the zeolite framework.

Eight unit cells of the zeolite are used in the simulation to construct the simulation box (2x2x2 cells) and periodic boundary conditions are adopted in three directions to produce a homogeneous solid. A cut-off of 18.5 Å is applied to Lennard-Jones interactions, and the long-range electrostatic interactions are calculated using the Ewald sum. $5 \cdot 10^6$ Monte Carlo steps were carried out to achieve equilibration and $2 \cdot 10^6$ more steps were used for data analysis. The probability of performing a molecule displacement was 0.2, rotation around the centre of mass was 0.2, change conformer was 0.2 and finally the probability of exchange with the reservoir was 0.4.

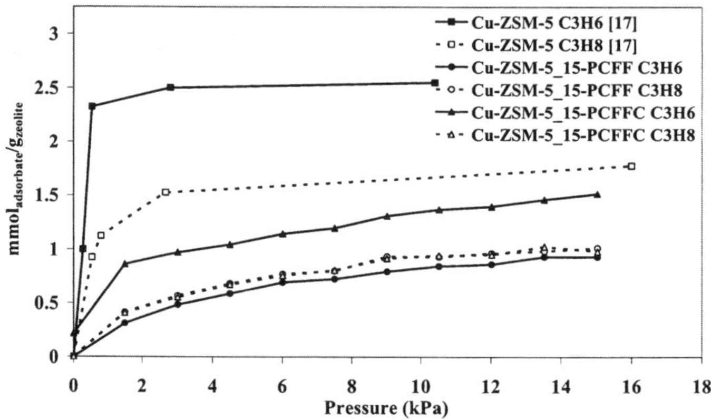

Figure 4 *Isotherm comparison of experimental and two different force fields simulation results; experimental isotherms are squares, isotherms simulated with PCFF are circles and isotherms simulated with PCFFC are triangles.*

To our knowledge there are several experimental works related to the study of the adsorption of alkanes in exchanged ZSM-5 and alkenes in silicatile-1, but only one paper has been devoted to the study of the adsorption of propene in exchanged ZSM-5, which is carried out with Cu-ZSM-5[16]. By comparing the results obtained in the simulations with the PCFF force field implemented in MS to the experimental results available in the literature[16], it is observed that the simulation predicts a lower adsorption capacity for both propane and propene, see figure 4. In addition, there is also a change in the relative order of propane and propene adsorption capacity. Cu-ZSM-5 experimentally adsorbs more propene than propane while simulation predicts more propane than propene adsorption capacity. In order to overtake these differences between simulated and experimental results we modified the force field changing the charges applied (reference in figure 2 PCFFC). Within the set of charges found in the literature E. Beerdsen et al.[17] (Table 1) provided the ones which produce the highest increase in the adsorption capacity of propene above the adsorption capacity of propane. This trend is shown in different works not only for ZSM-5[5,16,18] but also for other zeolites[7,19].

Table 1		*Static atomic charges relation*				
Si	O_{Si}	O_{Al}	*Al*	*Na*	*Ca*	*Cu*
+2.05	-1.025	-1.2	1.75	+1	+2	+2

Differences between simulated and experimental results are probably due to the substantially strong interaction between cooper and propene caused by a process of chemisorption added to the physisorption[16]. Simulations only take into account the physisorption but indicate qualitative insights on trends on adsorption capacity depending on type and density of cations. Computed adsorption isotherms of propene in the seven different solids of this study carried out at 303, 323, 343 and 473 K and pressures from $1 \cdot 10^{-3}$ to 0.5 kPa have been compared in figure 5. These temperatures and pressures were chosen within the range of the real application in cold start engines, from ambient temperature to the temperature at which three-way catalysts start working, 473 K.

The adsorption isotherms of propene in MFI-type zeolites look very similar in shape to previous reported data. The shape of the isotherms change by increasing the temperature, within the range of pressures studied, from a linear behaviour of the adsorption capacity at 473 K to a type I isotherm at 303 K. This behaviour is more evident for copper zeolites, which have the highest adsorption capacity of propene.

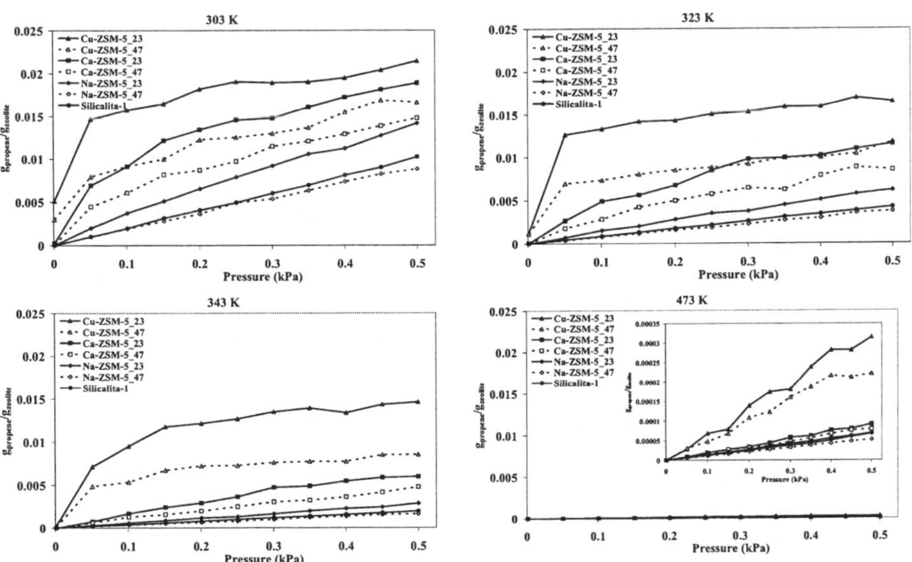

Figure 5 *Simulated adsorption isotherms of propane in MFI-type structures with Na, Ca and Cu cations, exchanged 0, 1 or 2 atoms per unit cell, at 303 K to 473 K*

Our results show that density, size and charge of nonframework cations influence on the propene adsorption inside MFI-type structures. In the case of Ca and Cu, adsorption capacity increases from 0 to 2 cations per unit cell with density of extra-framework cations in the structure. In the case of Na, although adsorption capacity also increases with density of extra-framework cations in the structure (from 2 to 4 cations), the structure free of Na extra-framework cations shows a higher adsorption capacity than Na-ZSM-5_47 at all the temperatures. Comparing isotherms of Silicalite-1 and Na-ZSM-5_23, silicalite-1 retains more propene than the Na zeolite at 473 K. However, Na-ZSM-5_23 already adsorbs more alkene than the MFI structure at 343 K. It has been observed that the more the temperature the more the difference between solids. All three extra-framework cations, Na, Ca, and Cu, balance in the solid structures Si/Al ratios of 47 and 23 but they show different trends in adsorption capacity. Therefore, the influence of the properties of the extra-framework cations in the exchanged zeolites is very important in propene adsorption.

The structure considered in this work, MFI, comprises two different channels, straight and sinusoidal. The preferential positions of Al atoms and cations are in the intersections of these channels partially blocking the space available to place propene in those positions. This seems to be the case when we compare isotherms obtained for solids Ca-ZSM5 and Cu-ZSM5. For the same number of extra-framework cations in the unit cell, the MFI structure exchanged with the biggest atom of Ca shows the lowest adsorption capacity of propene.

To the same Si/Al ratio, we can observe in Figure 5 that the adsorption capacity of Ca and Cu exchanged zeolites is higher than the adsorption capacity of Na-ZSM5, for all the temperatures and the whole range of pressures studied. This fact can be explained because of the smallest space occupied by divalent cations; while Na atoms only compensate one Al atom, the introduction of one divalent cation compensates the negative charge created by two Al atoms.

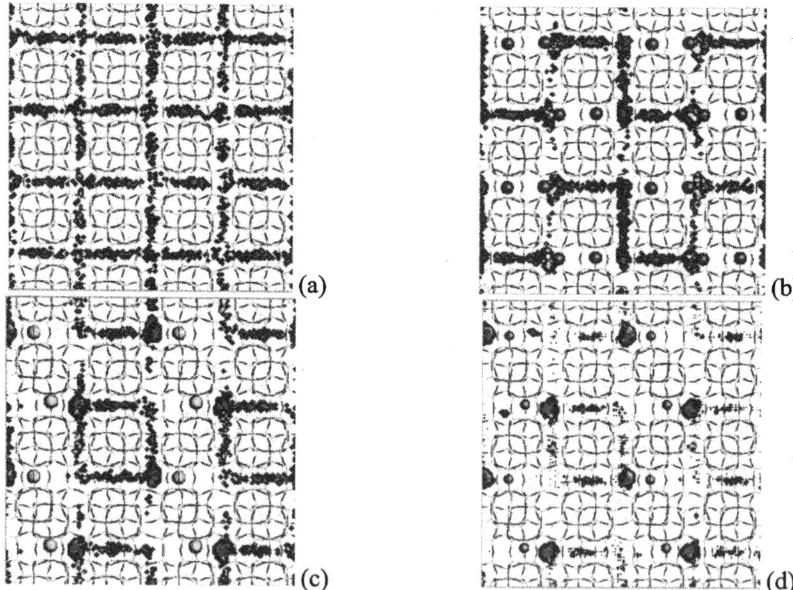

Figure 6   *Minimum energy structures at 303 K and 0.001 kPa for a) Silicalite-1, b) Na-ZSM-5_23, c) Ca-ZSM-5_23, d) Cu-ZSM-5_23.*

Summarizing, Cu-ZSM5-23 shows the highest adsorption capacity of all the solids studied at the range of temperature and pressures studied because it optimises charge and size as well as density of extra-framework cations.

Figure 6 compiles representations of minimum energy at 0.001 kPa of propene adsorbed in the different structures studied. This is the lowest pressure studied that allows to visualize the probability of density distribution of propene molecules. We also chose 303 K, the lowest temperature in the range studied, because its higher amount of molecules adsorbed allows a better understanding of the placement of molecules in the structure. For solids exchanged with extra-framework cations only results obtained with the highest density are compiled in Figure 6, because trends are the same with both densities.

In the case of silicalite-1 (Figure 6.a) the distribution of molecules is mostly homogeneous within the channels with a slight preference for sinusoidal channels. For solids exchanged with Na atoms (Figure 6.b), the molecules of adsorbate try to avoid the extra-framework cations, being located along the sinusoidal channels with a higher probability than along the straight channels. For calcium (Figure 6.c), the molecules of propene show a preference to be place firstly in the intersections of channels close to the locations of extra-framework cations and, secondly, in the sinusoidal channels intersecting with cation places. Copper divalent atom (Figure 6.d) produces a remarkable disproportion between the higher probability of finding the molecules adsorbed near to the extra-framework cation or in any other location. These locations of propene molecules indicate a higher interaction with Cu atoms than Na or Ca and it could represent another reason for the highest adsorption capacity shown by the solid Cu-ZSM-5_23.

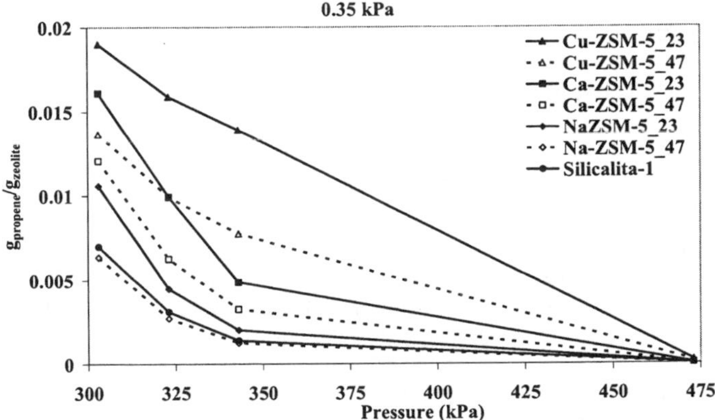

Figure 7 *Simulated adsorption isobars of propene in MFI-type structures with sodium, calcium and copper cations, exchanged 0, 1 or 2 atoms per unite cell, at 0.35 kPa*

Figure 7 compares the performance of the different solids studied. Adsorption isotherms were simulated at a fixed pressure of 0.35 kPa in a range of temperatures from 303 K to 473 K. It can be observed here a better performance of Cu-ZSM-5_23 than the other solids. First, it shows higher adsorption capacity (see Figure 5) and second, this solid is able to adsorb propene at higher temperatures. These are two key properties for solids intended to be used to retain hydrocarbons produced during cold start of gasoline engines. At these conditions, the most suitable performance would be to adsorb the highest amount of hydrocarbons at ambient temperature and to retain it until the three-way catalysts start working at around 473 K.

3   CONCLUSION

At the studied conditions, adsorption capacity of propene in exchanged MFI structures is positively influenced by, on one hand, a higher density and charge and, on the other hand, a smaller extra-framework cation size.   In addition, Cu atoms seems to have a stronger interaction with propene molecules than other extra-framework cations, which produces a remarkable disproportion between the probability of finding the propene molecules adsorbed near to the extra-framework cations or in any other location.   Finally, the solid Cu-ZSM-5_23 presents, not only the highest adsorption capacity of propene at room temperature, but also, at higher temperatures.

References

1   J. Park, S.J. Park, I. Nam, G.K. Yeo, J.K. Kil, Y.K. Youn, *Micropor. Mesopor. Mat.*, 2007, **101**, 264.
2   W. Zhu, F. Kapteijn, J. A. Moulijn, M.C. den Exter, J.C. Jansen, *Langmuir*, 2000, **16**, 3322.
3   J.H. ter Horst, S.T. Bromley, G.M. van Rosmalen, J.C. Jansen, *Micropor Mesopor. Mat.*, 2002, 53, 45.
4   P. Pascual, P. Ungerer, B. Tavitian, A. Boutin, *J. Phys. Chem. B*, 2004, **108**, 393.
5   S. Jakobtorweihen, N, Hansen, F.J. Keil, *Mol. Phys.*, 2005, **103**, 471.
6   X. Yang, B.H. Toby, M.A. Clambor, Y. Lee, D.H. J. Olson, *J. Phys. Chem. B*, 2005, **109**, 7894.
7   M.A. Granato, T.J.H. Vlugt, A.E. Rodrigues, *Ind. Eng. Chem. Res.*, 2007, **46**, 321.
8   B. Liu, B. Smit, F. Rey, S. Valencia, S. Calero, *J. Phys. Chem. C*, 2008, **112**, 2492.
9   C. Bacerlocher, W.M. Meier, D.H. Olson, Atlas of Zeolita Framework Types, Elsevier Press, Ámsterdam, 2001.
10   W. Löwenstein, *Am. Miner.*, 1954, **39**, 92.
11   J.G. Fripiat, F. Berger-Andre, J.M. Andre, E.G. Derouane, *Zeolites*, 1983, **3**, 306.
12   S.R. Lonsinger, A.K. Chakraborty, D.N. Theodorou, A. Bell, *Catal. Lett.*, 1991, **11**, 209.
13   D.H. Olson, N. Khosrovani, A.W. Peters, B.H. Toby, *J. Phys. Chem. B*, 2000, **104**, 4844.
14   E. de Vos Burchart, H. van Bekkum, B. van der Graaf, *Collect. Czech. Chem. Commun.*, 1992, **57**, 681.
15   B.F. Mentzen, G. Bergeret, H. Emerich, H.P. Weber, *J. Phys. Chem. B*, 2006, **110**, 97.
16   H.W. Jen, K. Otto, Catal. Lett. 1994, 26, 217.
17   E. Beerdsen, D. Dubbeldam, B. Smit, T.J.H. Vlugt, S. Calero, *J. Phys. Chem. B*, 2003, **107**, 12088.
18   A.V. Ivanov, G.W. Graham, M Shelef, *App. Cat. B: Env.*, 1999, **21**, 243.
19   M.A. Granato, T.J.H. Vlugt, A.E. Rodrigues, *Ind. Eng. Chem. Res.*, 2007, **46**, 7239.

THE USE OF SILICA-ALUMINA MIXED OXIDES WITH WELL DEFINED SURFACE CHARACTERISTICS AND POROSITIES FOR THE TRANSFORMATION OF RENEWABLE FEEDSTOCKS INTO VALUE ADDED CHEMICALS.

M. Yates[1], M.A. Martin-Luengo[2] and M.J. Martinez Domingo[1,2] M. Esteban Ramos[1]

[1] Instituto de Catálisis y Petroleoquímica (CSIC), c/Marie Curie 2, 28049 Madrid, Spain.
[2] Instituto de Ciencia de Materiales de Madrid (CSIC), c/ Sor Juana Inés de la Cruz s/n, 28049 Madrid, Spain.

1 INTRODUCTION

Aluminas possess good mechanical and thermal stabilities, high surface areas and resistance to sintering. Their Al-OH and Al^{3+} groups play a crucial role in their activity as acid catalysts. When silicon is present Si^{4+} ions can substitute Al^{3+} ions in tetrahedral positions, giving rise to a negative charge that is neutralized with positive ions that increase the acidity with respect to alumina but maintain the thermal and mechanical stability. The major advantage for the employment of silica-aluminas is the ease of preparation and tailoring of their textural and surface properties, making them versatile solids for many catalytic reactions.

At present industrial production of *p*-cymene is carried out by Friedel-Crafts alkylation of benzene or toluene with alkyl halides and $AlCl_3$,[1,2] or by alkylation of toluene with isopropylic alcohol.[3] However, both these procedures have low selectivities and employ highly toxic substances. The model reaction chosen for this work was the transformation of limonene, a renewable biodegradable subproduct of very low toxicity and cost, extracted from citrus peels, to *p*-cymene under microwave irradiation. Microwave irradiation was chosen as an environmentally friendly means to increase the reaction rate and thus greatly reduce the overall time required to achieve high conversions and selectivities to the desired product. These results were related to the textural characteristics: surface area and porosity, and the acidities of the silica-alumina mixed oxides.

2 EXPERIMENTAL

2.1 Raw Materials

The catalysts used in this investigation were a series of silica-aluminas produced by Sasol. These materials are prepared by an acid hydrolysis of aluminium hexalate dissolved in hexanol and ortosilicic acid mixtures. By varying the amount of orthosilicic acid different amounts of silica may be incorporated. In this study four materials with silica contents between 1 to 40 wt.% were employed and designated as Siral 1, 10, 20 and 40, accordingly. The D-limonene employed in the reactions supplied by Sigma Aldrich (99.9%) was used without any further purification.

2.2 Characterisation techniques

The texture of the materials: specific surface area and mesopore volume were analysed by adsorption/desorption of nitrogen at -196°C, in a Tristar apparatus from Micromeritics. The solids were outgassed overnight at 300°C to a vacuum of less than 10^{-2} Pa to ensure that they were clean, dry and free from any loosely adsorbed species. All subsequent calculations were made subject to the outgassed weight of the sample. The BET method was used to determine the specific surface areas (S_{BET}) from the adsorption data in the relative pressure range of 0.05 to 0.30 $p/p°$, typical for this type of sample [4]. The mesopore size distributions were calculated from the desorption branch of the corresponding nitrogen isotherm using the Kelvin equation and the BJH method with the parameters for the thickness of the adsorbed layers from the Harkins-Jura equation since this employed a metal oxide as the non-porous standard.[5]

Mercury intrusion porosimetry (MIP) analyses were employed to determine the particle and pore size distributions and pore volume, utilising CE Instruments Pascal 140/240 apparatus, on samples previously dried overnight at 150°C. Approximately 0.2g of sample was accurately weighed into the sample holder that was subsequently outgassed at room temperature for five minutes to a vacuum of 0.1 kPa before filling with mercury and starting the analysis. The pressure/volume data were analysed by use of the Washburn Equation assuming a cylindrical nonintersecting pore model, taking the mercury contact angle as 141° and surface tension as 484 mN m^{-1}.[6] Combination of the results from nitrogen isotherms and MIP leads to the characterisation of the total pore volume. However, from the shapes of the curves all pores above 1 μm were assumed to be related with the interparticulate pore filling of the powders. The cumulative pore volume curves in pore diameters greater than 1 μm were used to determine the primary particle sizes of the samples assuming a spherical particle geometry.[7]

The acidities of the solids were analysed from their ammonia adsorption capacities, determined on a Micromeritics ASAP 2010 device. The samples were first outgassed overnight at 300°C to a vacuum of about 10^{-2} Pa. Ammonia adsorption isotherms at 30°C were then determined up to a pressure of about 350 torr, thus obtaining the chemisorption plus physisorption capacities. The samples were subsequently outgassed at 30°C to remove the physisorbed species and a second adsorption isotherm at 30°C determined to measure the physisorption capacity of the sample. The chemisorption capacity was calculated from the differences in the uptakes between the first and second isotherms. The number of acid centres were calculated assuming that each molecule of ammonia reacted with one acid site.[8]

Pyridine adsorption coupled with infrared analysis was used to determine qualitatively the Lewis and Brønsted acid sites of the mixed oxide supports. The solids were made into self supporting discs and then outgassed at 300°C for 5 hours under high vacuum to clean the surface. A first spectrum was then measured and subsequently the samples were exposed to pyridine vapour at room temperature and a second spectrum determined. From the differences in the two spectra, the Lewis and Brønsted acid sites were analysed from their typical adsorption peaks located at 1455 and 1545 cm^{-1}, respectively.[9]

2.3. Catalytic reaction

The focalised monomodal microwave apparatus employed in this study, a Synthewave 402 from Prolabo, allows the programming and measurement of the sample temperature by infrared detection. Microwave induced reactions in dry media are considered a clean technology as they avoid the use of solvents, thus being an efficient and economic method for fast screening of catalysts. For the reactions 50 µl of limonene were physically mixed with 200 mg of solid. These mixtures were then placed in a glass reactor and irradiated at maximum power output for fixed periods of time: 5, 10 or 20 minutes. The reaction mixtures were allowed to cool and the reactants and products then extracted by dissolution in ethanol. These mixtures were subsequently analysed by GC-MS (Hewlett Packard 5890 series II GC with a 25 m methyl silicone capillary column heated in a helium flow from 50°C to 170°C at 6°Cmin^{-1}, coupled to a Hewlett Packard series 5971 mass spectrometer. To ensure the reproducibility of the results and avoid condensation of the mixtures the injector and detector were heated to 180°C and 250°C, respectively. Following the extraction of the reaction products the catalytic activities of the samples were measured once more using the same protocols as described above to determine whether the materials had suffered any loss in activities or selectivities.

3 RESULTS AND DISCUSSION

3.1 Textural properties of the raw materials

For this reaction the available surface area coupled with the accessibility to the active acid sites are important in controlling the catalytic process.[10] The results from the textural characterisation and acidity of the solids used are summarised in Table 1. The particle size distributions of the materials determined by analysis of the MIP curves in pore diameters of greater than 1 µm gave similar results for all the materials, with an average particle size of about 30 µm.

Table 1 *Textural properties of the raw materials*

Sample	Area S_{BET} (m^2/g)	Mesopore Volume (cm^3/g)	Macropore Volume (cm^3/g)	Total Pore Volume (cm^3/g)	Mesopore Diameters (nm)		Ammonia Adsorption (mmol/g)	Acidity (centres/m^2)
Siral 1	321	0.45	0.01	0.46	4.5	5.8	0.236	$4.44 \cdot 10^{17}$
Siral 10	370	0.58	0.05	0.63	4.8	6.8	0.402	$6.54 \cdot 10^{17}$
Siral 20	432	0.67	0.12	0.79	5.4	9.4	0.491	$6.84 \cdot 10^{17}$
Siral 40	506	0.91	0.25	1.16	5.8	12.7	0.669	$7.97 \cdot 10^{17}$

The nitrogen ad/desorption isotherm for Siral 1 was of Type IV with a well defined plateau at high relative pressures, characteristic of mesoporous solids, with a H1 hysteresis loop according to the IUPAC classification, corresponding to porous materials consisting of agglomerates of regular shape with a narrow mesopore size distribution.[6] However, as the silica content was increased there was a progressive increase in the specific surface area, mesopore volume and average mesopore diameters of the materials. The increase in the width of the pores as the silica content was increased may be clearly observed from the change in shape of the hysteresis loops, shown in Figure 1.

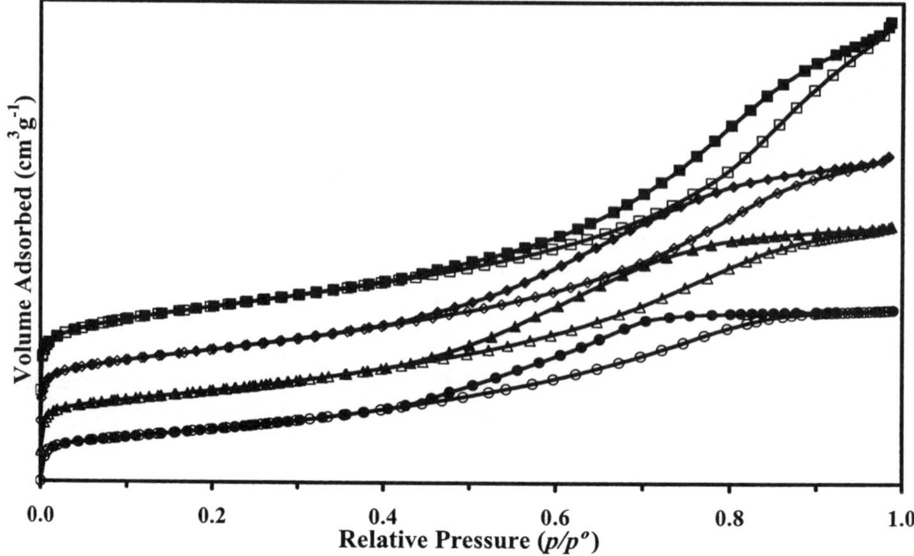

Figure 1 *Nitrogen Adsorption/Desorption isotherms for Siral 1 (○●), Siral 10 (△▲), Siral 20 (◇◆), and Siral 40 (□■). Isotherms are successively off set 50 cm^{3}g^{-1} for clarity.*

From the pyridine adsorption FTIR spectra, shown in Figure 2, it may clearly be observed that as the silica content was augmented there was a general increase in both Lewis and Brønsted acid sites.

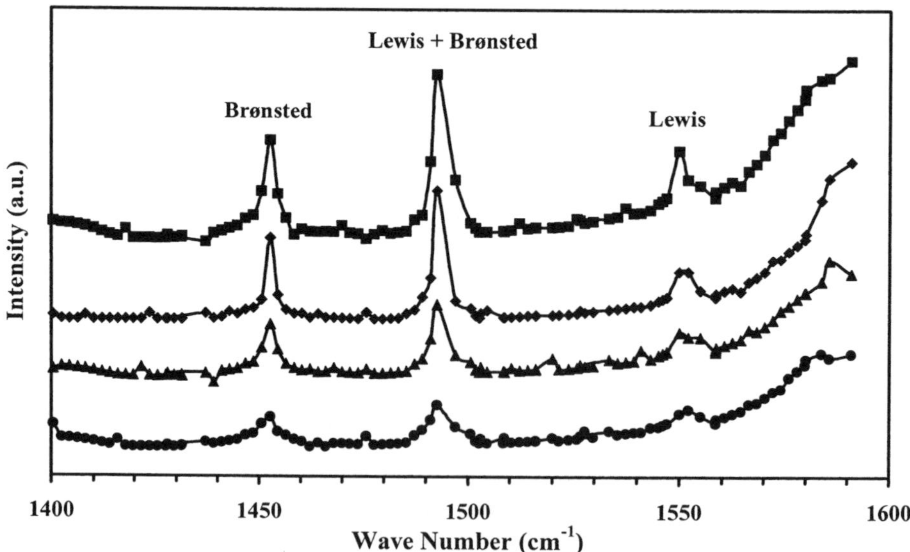

Figure 2 FTIR spectra for *Siral 1 (●), Siral 10 (▲), Siral 20 (◆), and Siral 40 (■).*

From the ammonia adsorption results presented in Table 1 it may be observed that the total acidities of the solids rose as the silica content in these mixed oxides was increased, in agreement with the results shown from the adsorption of pyridine. However, it should be noted that the observed increases in acidity were not directly proportional in either per gram or per square metre basis of chemisorbed ammonia versus silica content. This was probably due to the reported encapsulation process suffered by the solids where the surface silica composition was much higher than the bulk composition as analysed by XPS.[11]

Microwave irradiation was chosen for its characteristics of rapid and homogeneous heating. Previous studies with this heating source have demonstrated surprising conversions and selectivities in various reactions.[12] The transformation of limonene and other terpenes has been previously attributed to their reaction with acid sites that protonate the unsaturated molecules, along with the possibility of dehydrogenation or polymerisation. The reaction mechanism of limonene and other terpenes in the presence of solid acids has been related to an initial isomerisation on acid sites, followed by dehydrogenation of these intermediates to give *p*-cymene, disproportionation (slow reaction, compared to isomerisation or dehydrogenation, due to the need for two molecule intermediates) and in some cases polymerisation.[13] The mechanism involves adsorption of the extracyclic double bond of limonene on the acid site, forming a primary carbonium ion, followed by displacement of the proton to form the more stable tertiary carbonium ion from which terpinolenes, or terpinenes are formed,[14] followed by dehydrogenation of the intermediates produced after isomerisation. The results obtained with microwave irradiation of limonene are presented in Table 2.

Table 2 *Catalytic activity results after different periods of microwave heating.*

Sample	Time (min)	Conversion (%)	α-terpinene (%)	γ-terpinene (%)	α-terpinolene (%)	p-cymene (%)
	5	15	85	0	15	0
Siral 1	10	56	52	14	0	34
	20	100	0	0	0	100
	5	65	48	11	31	10
Siral 10	10	69	15	0	14	71
	20	100	0	0	0	100
	5	88	31	6	15	48
Siral 20	10	100	0	0	0	100
	20	100	0	0	0	100
	5	93	0	0	3	97
Siral 40	10	100	0	0	0	100
	20	100	0	0	0	100

Since the reaction is catalysed by the acid sites the increase in conversion activity as the number and accessibility of the surface acid sites rose with increased silica content was to be expected. Although higher activities were found as the total acidity increased, since there was an increase in the number of both Lewis and Brønsted acid sites with higher silica contents, Figure 2, no correlation could be made with only one type of site. The reactants and products found with microwave irradiation of limonene are presented in Figure 3.

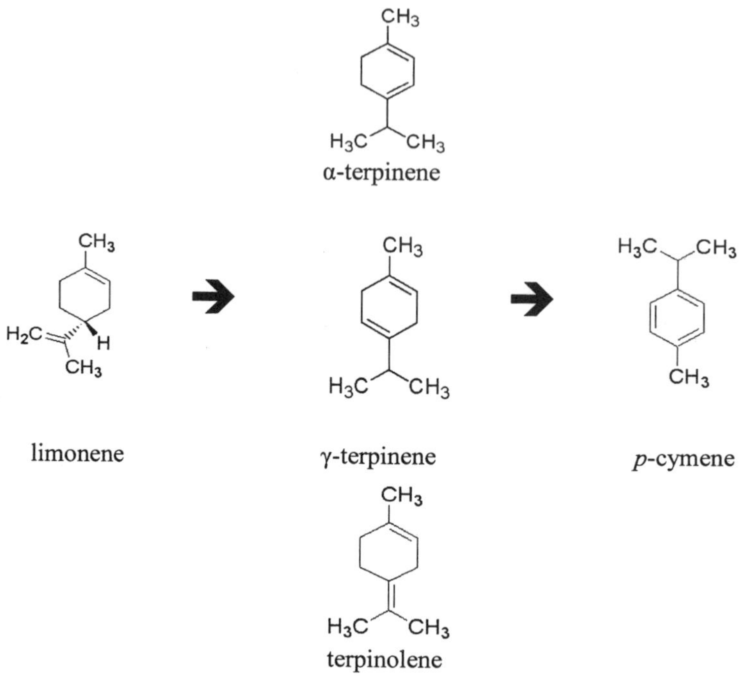

CH3
H3C   CH3
α-terpinene

CH3
H2C   H
CH3
limonene

CH3
H3C   CH3
γ-terpinene

H3C   CH3
CH3
p-cymene

CH3
H3C   CH3
terpinolene

Figure 3 *Reactants and products found with microwave irradiation of limonene over silica-alumina catalysts.*

From the results presented in Table 2 it should be noted that only α- and γ-terpinene, γ-terpinolene and *p*-cymene were found as products of the reaction. No undesirable compounds such as menthanes or menthenes, produced by disproportionation or polymerisation, were detected. From these results it may be observed that with longer irradiation times both the limonene conversion and the selectivity to *p*-cymene increased. The conversions and selectivities were also greater as the silica content of the mixed oxides increased. For the sample with the highest silica content after only five minutes a conversion of greater than 90 % was reached with almost 100 % selectivity to the desired product. Since this sample had more acid centres, calculated on both per gram or per m^2g^{-1} basis, it would appear that the selectivity to *p*-cymene was due to the rapid aromatisation of the intermediates produced by isomerisation over the acid centres. Thus, although conversion of limonene to *p*-cymene over Lewis acid sites with conventional heating has been previously observed, the much longer reaction times necessary to increase the overall conversion (3 hours) leads to reduced selectivities due to the formation of undesirable mentanes etc..[15]

It should be remembered that the industrial production of *p*-cymene is carried out by Friedel-Crafts alkylation's of benzene or toluene. Thus, although recent research has been carried out to make this process less toxic by the use of solid acids instead of the conventional homogeneously catalysed reaction the selectivity was still well below 100 % and thus expensive separation techniques would still be required.[16] Furthermore, under gas-solid conditions to alkylate toluene with isopropylalcohol temperatures in excess of 300°C are required and even then toluene conversions and selectivities to *p*-cymene of little more than 50 % were achieved.[17] In contradistinction the process presented here with microwave irradiation achieved selectivities of 100 % in very short reaction times with the additional advantages of using limonene as reactant, whose toxicity is orders of magnitude lower than that of toluene or benzene and has the added advantage of being a natural product of low cost that is extracted during the processing of citrus fruits.

4 CONCLUSIONS

These results illustrate that the use of microwave irradiation is particularly favourable for the production of *p*-cymene from limonene. This production method is attractive since it avoids the use of highly toxic feedstocks presently used for its manufacture such as: benzene, toluene and aluminium trichloride. The high conversions and excellent selectivity towards the desired product (*p*-cymene), avoiding undesirable by-products were believed to be due to the accelerated heating and reaction rates achieved with microwave irradiation. As the reaction is governed by the number and accessibility of the acid sites there was a clear relationship between the greater reactivities achieved with increased silica contents of the silica-alumina mixed oxides. Higher silica contents led to increases in the specific surface area, pore volume and average pore size in addition to increases in the number of acid sites. These effects were complimentary since they caused an increase in both the number and accessibility of the active acid centres necessary for this reaction.

Acknowledgements
The authors would like to acknowledge the kind provision of solids by SASOL.

References

1. J.M. Derfer and M.M. Derfer, Kirth–Orthmer Encyclopedia Chem. Technol. 22 (1978) 709.
2. J. Du, H. Xu, J. Shen, J. Huang, W. Shen and D. Zhao, Appl. Catal. A, 296 (2) (2005) 186.
3. BASF, US Patent: 3,555,103 (1967).
4. S. Brunauer, P.H. Emmett and E. Teller, J. Amer. Chem. Soc., 60 (1938) 309.
5. W.D. Harkins and G. Jura, J. Amer. Chem. Soc., 66 (1944) 1362.
6. J. Rouquerol, D Avnir, C.W. Fairbridge, D.H. Everett, J.H. Haynes, N. Pericone, J.D.F. Ramsay, K.S.W. Sing and K.K. Unger, Pure and Appl. Chem., 66, 8 (1994) 173.
7. R.P. Mayer and R.A. Stowe, J. Colloid Interf. Sci., 20 (1965) 893.
8. M. Gómez-Cazalilla, J.M. Mérida-Robles, A. Gurbani, E. Rodríguez-Castellón and A. Jiménez-López, J. Solid State Chem., 180 (3) (2007) 1130.
9. C.A. Emeis, J. Catal. 141 (1993) 347.
10. C. Fernandes, C. Catrinescua, P. Castilho, P.A. Russo, M.R. Carrott and C. Breen Appl. Catal. A: General; 318 (2007) 108.

11. W. Daniell, U. Schubert, R. Glöckler, A. Meyer, K. Noweck and H. Knözinger, Appl. Catal. A: General; 196 (2) (2000) 247.
12. R. Salvador, B. Casal, M. Yates, E. Ruiz-Hitzky and M.A. Martín-Luengo. Appl. Clay Sci.; 22 (2002) 103.
13. C. Fernandes, C. Catrinescua, P. Castilho, P.A. Russo, M.R. Carrott and C. Breen Appl. Catal. A: General; 318 (2007) 108.
14. R.J. Grau, P. D. Zgolicz, C. Gutierrez and H. A. Taher, J. Molec. Catal.; 148 (1999) 203.
15. C. Catrinescu, C. Fernandes, P. Castilho and C. Breen, Appl. Catal. A: General; 311 (1) (2006) 172.
16. N.N. Binitha and S. Sugunan, Catal. Commun.; 8 (2007) 1793.
17. D. Yadav Ganapati and A. Purandare Suraj, Micro. Meso. Mat.; 103 (1-3) (2008) 363.

DESIGNING (AL-)SBA-15 CATALYST PELLETS WITH UNIQUE PROPERTIES

Pavel Topka[1], Yuriy L. Zub[2], Jindřich Karban[1] and Olga Šolcová[1]

[1]Institute of Chemical Process Fundamentals of the ASCR, v. v. i., Rozvojová 135, CZ-165 02 Praha 6, Czech Republic, e-mail: p.topka@seznam.cz
[2]O. O. Chuiko Institute of Surface Chemistry, National Academy of Sciences of Ukraine, 17 General Naumov Str., UA-03 164 Kyiv, Ukraine

1 INTRODUCTION

In 1992, the discovery of mesoporous molecular sieves opened new possibilities in many areas of chemistry and material science [1]. Thanks to high surface areas and large pores with narrow pore size distribution, mesoporous molecular sieves attracted much attention as sorption materials, catalysts and catalyst supports [2]. However, only little attention has been paid to industrial aspects of possible commercial applications. Owing to that the powder catalyst cannot be used directly for industrial applications [3], the development of methods for the preparation of extrudates or pellets from mesoporous molecular sieves is desirable.

One of the most promising materials is SBA-15, which is obtained by template method using triblock copolymer surfactant [4]. This highly ordered hexagonal mesoporous silica is similar to MCM-41 materials, but possesses recognizable advantages. The most important of them is the thickness of the hexagonal walls (3.1 - 6.4 nm), which results in high thermal and mechanical stability of SBA-15 materials [5, 6].

The first paper dealing with the preparation of extrudates from SBA-15 materials appeared in 2007 [3]. Here, small amounts of bentonite as a binder and methylcellulose as a plasticizer were used for the preparation of extrudates from powder Al-SBA-15. Consequently, a reduction of pore volume and surface area in comparison with powder sample was observed (e.g. decrease in surface area from ~ 1000 m^2/g to ~ 600 m^2/g). Moreover, a decrease of surface acidity of extrudates compared to powder Al-SBA-15 was revealed using temperature-programmed desorption of pyridine. Similarly, 50 % of alumina binder was used for extrudate preparation of hydrocracking catalyst based on MCM-41 and SBA-15 [7]. Recently, extrudated Fe_2O_3/SBA-15 catalyst was used for the wet hydrogen peroxide oxidation of phenolic aqueous solutions [8]. In this case, the catalyst was agglomerated by extrusion using bentonite and methylcellulose.

To the best of our knowledge, only one paper dealing with the preparation of (Al-)SBA-15 pellets without the use of additives has been reported so far. In 2004, Chiu et al. [9] described the preparation of Al-SBA-15 meso/macroporous monoliths and their activity in Friedel-Crafts alkylation reactions. In order to compare the properties of monoliths with mesoporous catalyst, powder Al-SBA-15 was formed into pellets under ~ 120 MPa pressure and employed in the investigated reaction. However, the detailed characterization of pelleted Al-SBA-15 lacks in this paper. Moreover, it can be expected that the pellets formed under such high pressure did not retain the mesoporous structure of powder Al-SBA-15.

Friedel-Crafts alkylations are industrially important reactions that are used for production of numerous aromatic compounds. Such processes often rely on liquid acid catalysts, such as $AlCl_3$, HF, and BF_3 [10], however, the separation and handling of the acid waste present undesirable economic, environmental, as well as health and safety issues. Recently, much effort has been devoted to investigating the feasibility of using solid acids, such as zeolites and mesoporous aluminosilicas, as new means of catalyzing the alkylation reactions [11].

The aim of this preliminary study was to compare the texture and structure properties of self-supporting SBA-15 pellets prepared without the addition of binders or plasticizers with the powder catalyst. Furthermore, Al-SBA-15 pellets prepared by the same way were employed in the Friedel-Crafts alkylation of toluene with benzylalcohol under moderate conditions.

2 EXPERIMENTAL

2.1 Preparation of (Al-)SBA-15

Siliceous SBA-15 was synthesized by the method described in details in ref. [4]. 8.00 g of Pluronic P123 (Aldrich) was dissolved in 232 ml of 1.9 M HCl and 60 ml of distilled water under vigorous stirring, resulting in a clear solution. Afterwards 18.2 ml of tetraethyl orthosilicate (Aldrich) was added. After stirring for 12 h at 40 °C the reaction mixture was allowed to stand at 100 °C for 24 h in a polypropylene bottle. The solid product was collected by filtration, thoroughly washed with distilled water and dried overnight in air. The template was removed by calcination in air at 500 °C for 8 h with a temperature ramp of 1 °C/min.

Synthesis of Al-SBA-15 was carried out in the following way: at first, siliceous SBA-15 was prepared according to the above-described procedure. After that, parent SBA-15 material was impregnated with ethanolic solution of aluminum chloride according to the procedure described in ref. [12]. Aluminum(III) chloride (Aldrich) was used as the aluminum source and dry ethanol as a solvent. A typical synthetic procedure was as follows: required amount of $AlCl_3$ was combined with 200 ml of dry ethanol and 2 g of SBA-15. This mixture was kept with magnetic stirring at room temperature for 6 h. The solid material was then filtered, washed vigorously with dry ethanol and then water to eliminate chloride anions, dried at room temperature in air and calcined in air at 500 °C for 8 h with a temperature ramp of 1 °C/min.

For the preparation of self-supporting pellets, powder (Al-)SBA-15 without any binders or plasticizers was pressed into approximately 0.5-cm-thick × 0.5-cm-diameter pellets using different pressures.

2.2 Characterization

Powder (Al-)SBA-15 and the pellets were characterized by X-ray diffraction, nitrogen adsorption measurement and chemical analysis. Powder X-ray diffraction data were obtained on Bruker AXS D8 diffractometer in the Bragg-Brentano geometry arrangement using Cu Kα radiation with a graphite monochromator and a scintillation detector. Adsorption isotherms of nitrogen at -196 °C were measured with a Micromeritics ASAP 2020M instrument. Prior to the measurement, the samples were degassed overnight at

120 °C and 0.1 Pa. To guarantee the precision of the obtained data the purity of the used nitrogen (Technoplyn, Linde) was 99.9995 %. Silicon-to-aluminum molar ratio in prepared Al-SBA-15 was determined by chemical analysis using an XP sequential WD-XRF spectrometer Thermo ARL 9400 (Institute of Chemical Technology Prague, Czech Republic) with the standard deviation lower than 5 %.

2.3 Catalytic reaction

Friedel-Crafts alkylation of toluene with benzyl alcohol was carried out in a glass batch reactor stirred with inert gas under reflux conditions. In a typical experiment, weighted amount of Al-SBA-15 catalyst in the form of powder or pellets (about 100 mg) was activated 0.5 h at 500 °C in air. The activated catalyst was placed into a three-neck 100-ml round-bottom flask equipped with a condenser, a nitrogen inlet, and an outlet for product withdrawal. After that, calculated amount of toluene (Lach-Ner, Czech Republic) was injected into the reactor filled with nitrogen and placed into thermostated bath. The reaction was initiated by adding the calculated amount of benzyl alcohol (Aldrich) to boiling reaction mixture. The reaction temperature was 110 °C, initial Al to benzyl alcohol molar ratio was 0.01, and initial toluene to benzyl alcohol volume ratio was 24. At given reaction times, 20 µl of liquid phase were sampled for GC analysis.

The samples of reaction mixture were analyzed on an Agilent 6890 gas chromatograph equipped with a DB-5 column (30 m × 0.25 mm × 1 µm), using He as a carrier gas. The products were identified by GC-MS analyses performed on an Agilent 6890 gas chromatograph coupled to an Agilent 5973 mass spectrometer operating in 70 eV ionization mode (m/z = 35 − 400). DB-5MS column (30 m × 0.25 mm × 0.25 µm) was used with He as a carrier gas. Conversion of benzyl alcohol K and selectivity to o-benzyl toluene and p-benzyl toluene S were calculated based on the material balance using Eq. 1 and 2:

$$K = (n_P + n_O + n_D)/(n_S + n_P + n_O + n_D) \tag{1}$$

$$S = n_P/(n_P + n_O + n_D) \tag{2}$$

where n_S, n_P, n_C, n_D are the molar amounts of benzyl alcohol (n_S), o-benzyl toluene and p-benzyl toluene (n_P), m-benzyl toluene (n_O), and dibenzyl ether (n_D), respectively.

3 RESULTS AND DISCUSSION

3.1 Self-supporting pellets prepared from siliceous SBA-15

Nitrogen adsorption isotherms on powder SBA-15 and self-supporting pellets prepared from this material using the forming pressure 2.7 MPa are shown in Fig. 1a. Both isotherms are of types I and IV according to the IUPAC classification and are typical for microporous-mesoporous materials [13]. The isotherm of parent SBA-15 exhibits steep increase in the adsorbed amount at p/p_0 0.7 − 0.8. This indicates the narrow pore-size distribution of synthesized material (Fig. 1b). On the other hand, the isotherm of self-supporting pellets formed with 2.7 MPa pressure possesses slightly broader hysteresis loop in comparison with parent SBA-15 (Fig. 1a), and the pore-size distribution is also broader (Fig. 1b).

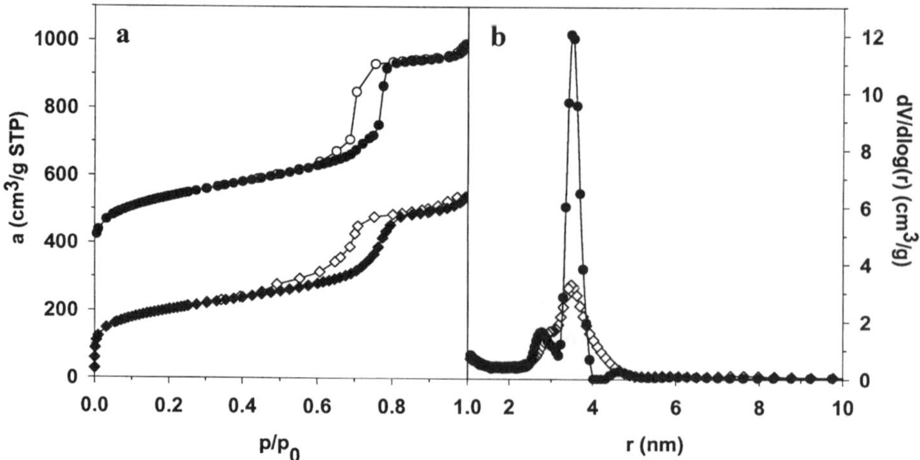

Figure 1 *a) Adsorption isotherms of nitrogen at -196 °C on parent SBA-15 (●), and self-supporting pellets formed with 2.7 MPa pressure (◆); desorption branches are depicted using open symbols. For clarity, 300 cm³/g (STP) was added to the adsorption isotherm of parent SBA-15. b) Pore size distributions determined from adsorption branches of adsorption isotherms for parent SBA-15 (solid symbols), and self-supporting pellets formed with 2.7 MPa pressure (open symbols).*

Texture properties of the self-supporting pellets prepared by using four various pressures are summarized in Tab. 1 together with texture properties of parent siliceous SBA-15. Owing to that the use of classic (two-parameter) BET equation for analysis of adsorption isotherms of microporous-mesoporous samples is not correct [14], a modified (three-parameter) BET equation [15] was employed together with t-plot analysis. The C parameter obtained from modified BET equation was used for the construction of t-plot according to Lecloux and Pirard [16]. From the t-plot, the micropore volume V_{micro} and the mesopore surface area S_{meso} were revealed. Mesopore volume V_{meso} was calculated as the difference between total pore volume (measured at $p/p_0 = 0.97$) and V_{micro}. The maximum of pore diameter d_{max} was determined from mesopore-size distribution evaluated according to advanced Barrett, Joyner and Halenda (BJH) approach [17, 18] from the adsorption branch of adsorption isotherm.

Table 1 *Texture properties of parent SBA-15 and self-supporting pellets formed under different pressures.*

Pressure (MPa)	V_{micro} (cm³/g)	V_{meso} (cm³/g)	S_{meso} (m²/g)	d_{max} (nm)	C	S_{BET} (m²/g)
0	0.200	0.835	475	7.0	17.5	827
0.7	0.203	0.786	439	7.0	16.5	799
2.0	0.194	0.708	404	7.2	16.5	750
2.7	0.185	0.630	373	7.1	16.4	703
17.7	0.174	0.481	324	6.7	15.3	637

For comparison with the literature, where the classic BET analysis [19] is, nevertheless, usually (and incorrectly) performed, surface area S_{BET} is also included into Tab. 1.

It can be seen that in studied range of pressures (i) micropore volume decreased only slowly with increasing pressure applied for pellet forming, (ii) a little bit higher decrease can be recognized for mesopore volume with increasing pressure, (iii) a decrease of surface area with increasing pressure is not significant, and (iv) a substantial decrease in pore size appeared only for the highest forming pressure, 17.7 MPa. For lower pressures, only unimportant decrease of surface area less than 20 % can be noticed. Moreover, with further pressure increase the loss is negligible. It is also observed that the maxima of pore-size distributions are nearly the same for the powder sample and pellets prepared applying low pressures. It should be noted that both 2.7 MPa and 17.7 MPa samples possessed bimodal pore size distribution with second maximum at 3.8 nm and 3.9 nm, respectively. Thus, it can be inferred that noticeable changes of porous structure appeared for forming pressures above 2 MPa. However, the reduction in mesopore surface area was not extreme even in the case of 17.7 MPa sample (70 % of S_{meso} in comparison with parent material). This is interesting especially if we take into account that the use of additives led to considerable diminishment of surface area of pellets (e.g. 60 % of S_{BET} in comparison with parent material in ref. [3]).

3.2 Self-supporting pellets prepared from Al-SBA-15

Adsorption isotherms and pore-size distributions of parent Al-SBA-15 and self-supporting pellets formed with 4.7 MPa pressure are shown in Fig. 2. Similarly as for SBA-15, both isotherms are of types I and IV according to the IUPAC classification [13] and are typical for microporous-mesoporous materials.

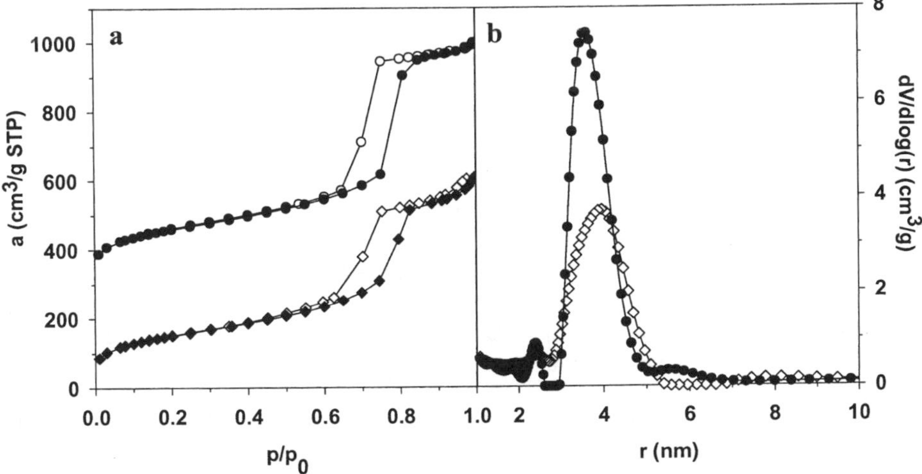

Figure 2 *a) Adsorption isotherms of nitrogen at -196 °C on parent Al-SBA-15 (●), and self-supporting pellets formed with 4.7 MPa pressure (◆); desorption branches are depicted using open symbols. For clarity, 300 cm³/g (STP) was added to the adsorption isotherm of parent Al-SBA-15. b) Pore size distributions determined from adsorption branches of adsorption isotherms for parent Al-SBA-15 (solid symbols), and self-supporting pellets formed with 4.7 MPa pressure (open symbols).*

It is seen that the shape of isotherms and pore size distribution is similar for both samples, which confirms the fact that the texture of parent Al-SBA-15 was not significantly affected during the preparation of the pellets.

The texture data were obtained from the adsorption isotherms using the same approach as described above. Parent Al-SBA-15 had the micropore volume 0.117 cm^3/g, mesopore volume 0.934 cm^3/g, mesopore surface area 365 m^2/g, and pore diameter 7.2 nm. The self-supporting pellets prepared from powder Al-SBA-15 using the forming pressure 4.7 MPa exhibited micropore volume 0.115 cm^3/g, mesopore volume 0.770 cm^3/g, mesopore surface area 334 m^2/g, and pore diameter 7.4 nm. Surface area S_{BET} decreased from 576 m^2/g to 523 m^2/g, while the C parameter was 18 in both cases. Thus, the decrease in mesopore surface area was only 8 %, while the pore diameter was slightly higher.

The structure of parent Al-SBA-15 and prepared pellets was also checked by X-ray diffraction (Fig. 3). The diffractogram of parent powder showed the properly developed hexagonal structure of Al-SBA-15 with three discernible diffraction lines. From the diffractogram of the pellets it is seen that the intensity of individual diffraction lines slightly decreased, the character of the diffractogram, however, remained unchanged. Therefore, the mesoporous structure of Al-SBA-15 was preserved during the preparation of self-supporting Al-SBA-15 pellets.

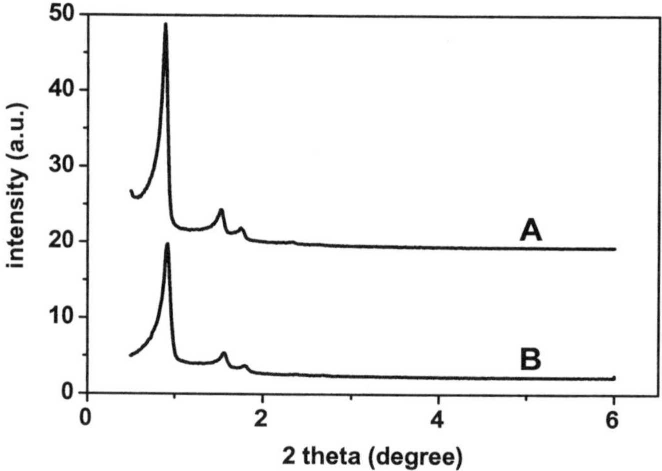

Figure 3 *X-ray diffraction patterns of Al-SBA-15: powder catalyst (A), and self-supporting pellets formed with 4.7 MPa pressure (B).*

The conversion and selectivity achieved with parent Al-SBA-15 and self-supporting pellets prepared using the forming pressure 4.7 MPa in batch alkylation of toluene with benzylalcohol are shown in Fig. 4. In the experiment with powder Al-SBA-15, the reaction mixture was stirred with a magnetic stirrer in order to exclude the effect of external diffusion. With the powder catalyst, 85 % conversion of benzyl alcohol was accomplished after 60 min of the reaction with selectivity to *o*- and *p*-benzyl toluene $S = 50$ % (Fig. 4).

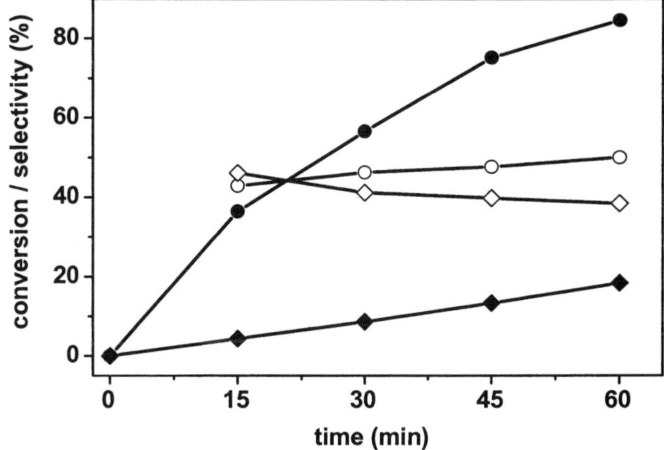

Figure 4 *Conversion of benzyl alcohol (solid symbols) and selectivity to o- + p-benzyl toluene (open symbols) dependence on time: parent Al-SBA-15 (●), and self-supporting pellets formed with 4.7 MPa pressure (◆); Si/Al = 38.*

On the other hand, the pelleted catalyst allowed achieving only 18 % conversion of benzyl alcohol with 38 % selectivity to *o*- and *p*-benzyl toluene at the same time. Thus, the pelleted Al-SBA-15 catalyst showed approximately 20 % activity in comparison with a powder one. This decrease in activity was probably caused by diffusion limitations inside the catalyst pellets, as described in the literature [9]. In ref. [9], even larger difference (about 8 times) was observed between the activity of powder and pelleted Al-SBA-15 catalyst in the same reaction. This can be explained by much higher pressure used for the preparation of the pellets in ref. [9] (~ 120 MPa) than in our case (4.7 MPa). Moreover, under such high pressure the collapse of mesoporous structure is expected.

To sum up, the preparation of self-supporting pellets using low forming pressures enables to preserve the mesoporous structure of the catalyst, which results in lower diffusion limitations inside the catalyst pellets. Consequently, enhanced catalytic activity can be reached.

4 CONCLUSIONS

Self-supporting pellets were successfully formed from SBA-15 and Al-SBA-15 powder without the addition of binder or plasticizer. Prepared self-supporting pellets maintained the porous structure of original powder. In comparison with parent sample, a decrease of mesopore surface area and mesopore volume up to 40 % was noticed in the case of pellets formed with the highest 17.7 MPa pressure.

The Al-SBA-15 pellets prepared using 4.7 MPa forming pressure were enough mechanically stable to retain the structural integrity during batch alkylation of toluene with benzyl alcohol at 110 °C. The conversion reached with Al-SBA-15 pellets after 60 min of the reaction was about five times lower than that accomplished with powder Al-SBA-15.

Acknowledgement

The financial support of Grant Agency of the Academy of Sciences of the Czech Republic (A 4072404) and AS CR Program Nanotechnology for Society (KAN 400720701) is gratefully acknowledged.

References

[1]   J. S. Beck, J. C. Vartuli, W. J. Roth, M. E. Leonowicz, C. T. Kresge, K. D. Schmitt, C. T. W. Chu, D. H. Olson, E. W. Sheppard, S. B. McCullen, J. B. Higgins and J. L. Schlenker, *J. Am. Chem. Soc.,* 1992, **114**, 10834.
[2]   A. Taguchi and F. Schüth, *Microporous Mesoporous Mater.,* 2005, **77**, 1.
[3]   G. Chandrasekar, M. Hartmann, M. Palanichamy and V. Murugesan, *Catal. Commun.,* 2007, **8**, 457.
[4]   D. Y. Zhao, J. L. Feng, Q. S. Huo, N. Melosh, G. H. Fredrickson, B. F. Chmelka and G. D. Stucky, *Science,* 1998, **279**, 548.
[5]   D. Y. Zhao, Q. S. Huo, J. L. Feng, B. F. Chmelka and G. D. Stucky, *J. Am. Chem. Soc.,* 1998, **120**, 6024.
[6]   M. Kruk, M. Jaroniec, S. H. Joo and R. Ryoo, *J. Phys. Chem. B,* 2003, **107**, 2205.
[7]   S. Ahmed, *React. Kinet. Catal. Lett.,* 2007, **90**, 285.
[8]   F. Martinez, M. I. Pariente, J. A. Melero and J. A. Botas, *J. Adv. Oxid. Technol.,* 2008, **11**, 65.
[9]   J. J. Chiu, D. J. Pine, S. T. Bishop and B. F. Chmelka, *J. Catal.,* 2004, **221**, 400.
[10] G. A. Olah, *Friedel–Crafts Chemistry*, Wiley, New York, 1973.
[11] A. Vinu, D. P. Sawant, K. Ariga, M. Hartmann and S. B. Halligudi, *Microporous Mesoporous Mater.,* 2005, **80**, 195.
[12] Z. H. Luan, M. Hartmann, D. Y. Zhao, W. Z. Zhou and L. Kevan, *Chem. Mater.,* 1999, **11**, 1621.
[13] S. J. Gregg and K. S. W. Sing, *Adsorption, Surface Area and Porosity*, Academic Press, New York, 1982.
[14] P. Schneider, P. Hudec and O. Solcova, *Microporous Mesoporous Mater.,* in press.
[15] P. Schneider, *Appl. Catal. A,* 1995, **129**, 157.
[16] A. Lecloux and J. P. Pirard, *J. Colloid Interf. Sci.,* 1979, **70**, 265.
[17] K. S. W. Sing and R. T. Williams, *Adsorpt. Sci. Technol.,* 2004, **22**, 773.
[18] L. Matejova, O. Solcova and P. Schneider, *Microporous Mesoporous Mater.,* 2008, **107**, 227.
[19] S. Brunauer, P. H. Emmett and E. Teller, *J. Am. Chem. Soc.,* 1938, **60**, 309.

CAPILLARY CONDENSATION AND SORPTION HYSTERESIS OF TOLUENE, METHYLCYCLOHEXANE AND NEOPENTANE ON MCM-41 AND MCM-48

P. A. Russo, M. M. L. Ribeiro Carrott, F. M. L. Conceição, P. J. M. Carrott

Centro de Química de Évora and Departamento de Química, Universidade de Évora, Colégio Luís António Verney, 7000-671 Évora, Portugal

1 INTRODUCTION

Gas adsorption is a technique widely used to characterise the textural properties of porous materials, *i.e.*, to evaluate specific surface areas, pore volumes and pore sizes, and normally involves the determination of adsorption isotherms. The calculation of the textural parameters from the adsorption data requires the application of analysis methods to the isotherms, which are based on models and theories for the mechanisms of adsorption. Since adsorption on open surfaces, micropores or mesopores occurs by different mechanisms, the methods of isotherm analysis applied depend on the type of porosity present in the solid. In the case of mesoporous materials, the adsorption first proceeds through the formation of a mono-multilayer on the pore surfaces until a critical thickness is reached and the pores are filled with condensate. Capillary condensation is reflected on the adsorption isotherm by the presence of a step, to which a hysteresis cycle is frequently associated. The size of the mesopores is assessed by analysis of this region of the isotherms. However, the origin of hysteresis and the nature of the two branches of the hysteresis cycle are aspects not completely understood, which makes the interpretation of the isotherms of mesoporous materials difficult.[1]

Until recently, what was known about the mechanisms involved in the adsorption of gases on mesoporous solids was based on studies carried out on materials with disordered pore structures.[1] However, the disclosure in 1992 of ordered mesoporous silicate materials[2] allowed the investigation of these processes to be performed on materials having highly regular and uniform pore structures. Consequently, since then a significant number of theoretical and experimental studies on the adsorption of gases by ordered mesoporous materials have been published.[3-12] From this research, dealing mostly with adsorption of nitrogen and argon on MCM-41, some new information about the adsorption/desorption processes in mesopores and adsorption hysteresis has been obtained.

In this work, the adsorption of toluene, methylcyclohexane and neopentane was measured at several temperatures on MCM-41 and MCM-48 silica of similar pore size. The MCM-41 has a hexagonal array of uniform cylindrical pores while MCM-48 has a three-dimensional pore system consisting of two interwoven channels separated by a continuous pore wall. The aim of this work is to investigate the effect of the pore structure of MCM-48 on the capillary condensation and adsorption hysteresis of the hydrocarbons.

2 MATERIALS AND METHODS

2.1. Materials and characterisation

The MCM-41 sample was synthesised as previously described, using hexadecyltrimethylammonium bromide as surfactant, tetraethoxysilane (TEOS) and ammonia.[13] The MCM-48 material was prepared hydrothermally at 383 K following the procedure previously described,[14] using hexadecyltrimethylammonium bromide as template, sodium hydroxide, and TEOS as silica source. The samples were calcined at 823K for 8 h using a heating rate of 3 K min^{-1}.

The X-ray diffraction (XRD) measurements were carried out on a Bruker AXS-D8 Advance powder diffractometer, using CuKα radiation (40kV, 30mA), with a step size of 0.01° (2θ) and 5s per step. The unit cell parameters of MCM-41 and MCM-48 were calculated from the XRD data by indexing the diffraction peaks to 2-D P6mm and 3-D cubic Ia3d lattices, respectively. Nitrogen adsorption isotherms at 77K were determined on a CE Instruments Sorptomatic 1990, using helium (for dead space calibration) and nitrogen of 99.999% purity. Prior to the adsorption measurements all the samples were outgassed for 8 h at 453 K. The specific pore volumes and surface areas were obtained by the α_s method using standard data for nitrogen adsorption on non-porous partially hydroxylated silica.[15]

2.2. Hydrocarbon adsorption experiments

Toluene (>99%, Riedel-de-Haën), methylcyclohexane (>99%, Aldrich) and neopentane (>99%, Linde) were used as adsorptives. The liquid adsorptives were outgassed in vacuum by repeated freeze-thaw cycles prior to the determination of the isotherms. The saturation pressures and density of the hydrocarbons at the working temperatures were calculated with equations given in the literature.[16]

The adsorption of toluene, methylcyclohexane and neopentane was determined gravimetrically in an apparatus equipped with a CI Electronics vacuum microbalance and a Disbal control unit. Pressure was measured with Edwards Barocel 600 and 622 capacitance manometers. The temperature of the circulating liquid jacket around the balance tubes was controlled within ±0.1 K using a Grant LTD thermostat and a Masterflex peristaltic pump. Prior to the adsorption experiments the samples were outgassed at 453K for 8 hours. Blank measurements indicated that buoyancy corrections were negligible.

3 RESULTS AND DISCUSSION

Figure 1 shows the X-ray diffraction patterns and nitrogen adsorption isotherms at 77 K of the MCM-41 and MCM-48 silica materials. The XRD patterns of the MCM-41 and MCM-48 samples are typical of well ordered 2-D hexagonal P6mm and 3-D cubic Ia3d mesostructures, respectively. The nitrogen isotherms of both samples exhibit capillary condensation steps in a very narrow range of p/p°, which indicates high uniformity of the pore sizes. While the isotherm of the hexagonal silica is completely reversible that of the MCM-48 shows hysteresis at high pressures (p/p°>0.45), associated with secondary mesoporosity formed between the particles of the solid, but the adsorption on the primary mesopores is reversible. The structural characteristics of the silicas calculated from the XRD and N$_2$ adsorption data are listed in Table 1. It can be seen from the results in Table 1 that the MCM-48 and MCM-41 materials have mesopores of similar dimension but, due to

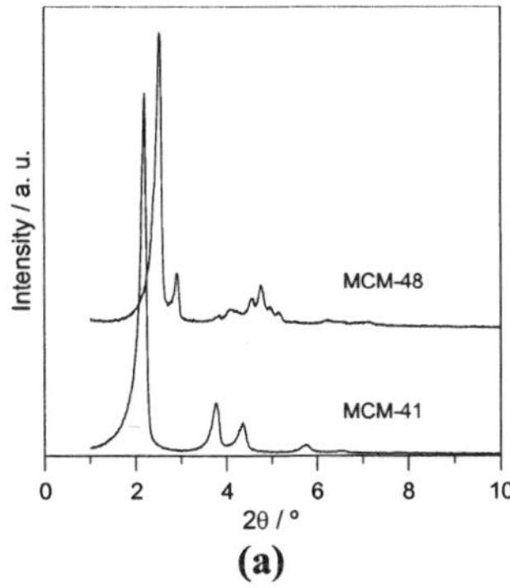
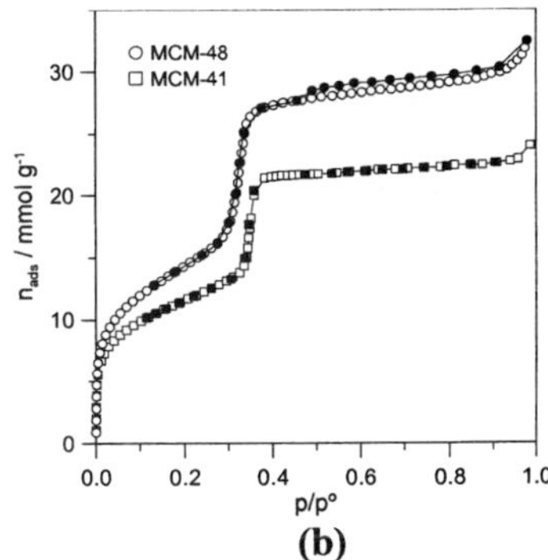

(a)   (b)

Figure 1 *(a) X-ray diffraction patterns and (b) nitrogen adsorption isotherms at 77 K (closed symbols-desorption) of the MCM-41 and MCM-48 samples.*

Table 1 *Structural parameters of the MCM-41 and MCM-48 samples determined by nitrogen adsorption and XRD.*

Sample	a_0/ nm	A_s/m^2 g^{-1}	A_{ext}/m^2 g^{-1}	V_p/cm^3 g^{-1}	D_p/ nm
MCM-41	4.71	925	23	0.75	3.9
MCM-48	8.66	1106	83	0.94	3.8

the differences in the porous structures, the pore volume of the cubic mesostructure is considerably higher as well as the pore volume/pore area ratio as previously noticed.[17]

The adsorption of toluene and methylcyclohexane at 298 K and neopentane at 273 K on the MCM-41 and MCM-48 are compared in Figure 2. The hydrocarbon isotherms have distinct shapes in the low pressure region, prior to the capillary condensation step, as previously noted.[12] While the amount of methylcyclohexane adsorbed at low pressures varies linearly with p/p°, the toluene and neopentane isotherms show relatively well defined knees. These features indicate that the gas-solid interactions are weaker in the methylcyclohexane-silica system in comparison with the other two systems. The isotherms of each organic adsorptive on the two silicas have similar shape at low p/p° and so, apparently, the gas-solid interactions are comparable. It is interesting to note that, despite the higher surface area and pore volume, the amounts of toluene and methylcyclohexane adsorbed prior to capillary condensation by the MCM-48 silica are similar to those adsorbed by the MCM-41. This is not observed with neopentane and the percentage of the MCM-48 total pore volume occupied at the beginning of the capillary condensation step is higher for this hydrocarbon. The total volume adsorbed by the cubic material is identical for the three adsorptives, as it is observed with MCM-41, meaning that when the pores are filled with condensate the volume occupied by the organics is the same. Also, the difference between the total hydrocarbon uptakes and that of nitrogen for the MCM-48 silica are in agreement with those usually found for MCM-41 materials.[18]

In order to compare the adsorption behaviour of the substances studied here on the MCM-41 and MCM-48 pore structures we have calculated the ratios $\Delta V_{cond}(48/41)/\Delta V_{total}(48/41)$, where $\Delta V_{cond}(48/41)$ is the difference between the volumes adsorbed at the beginning of the condensation step for a given adsorptive on the MCM-41 and MCM-48 samples and $\Delta V_{total}(48/41)$ is the difference between the total pore volumes

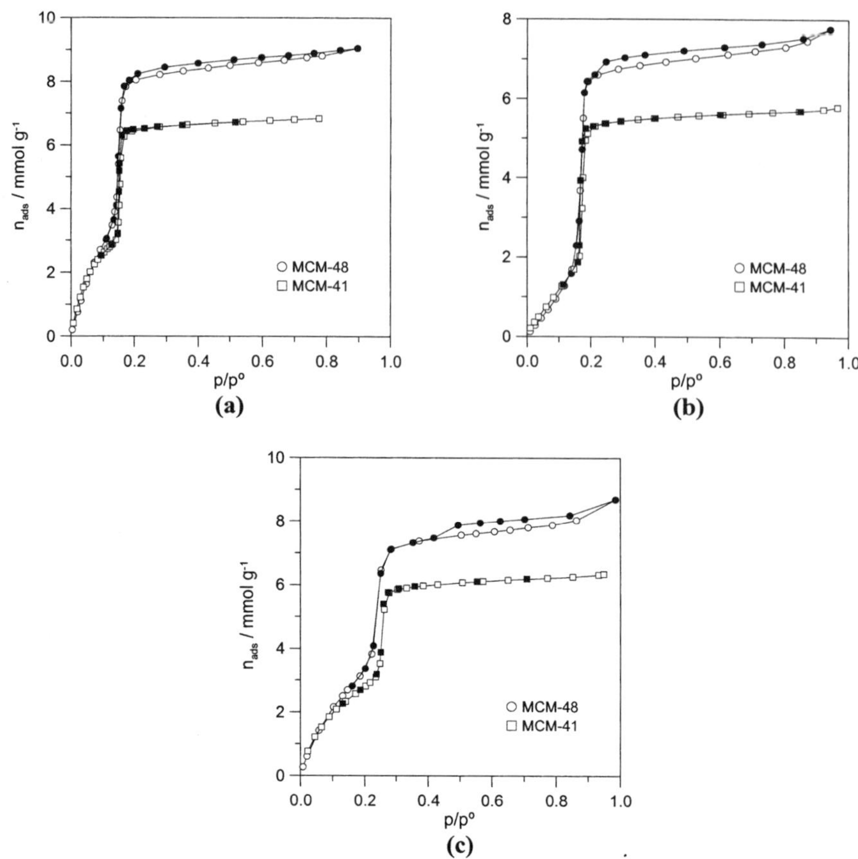

Figure 2 *Adsorption/desorption isotherms of (a) toluene at 298 K, (b) methylcyclohexane at 298 K and (c) neopentane at 273 K on MCM-41(D_p=3.9nm) and MCM-48(D_p=3.8 nm) materials (closed symbols-desorption).*

of the materials. Ratios of 0.52 for nitrogen and neopentane, 0.25 for toluene and 0.15 for methylcyclohexane were obtained. The results suggest that the adsorption behaviour of neopentane on MCM-48 with respect to MCM-41 is similar to that of nitrogen but quite different from that of toluene and even more distinct from that of methylcyclohexane. The molecules of the three organic compounds have similar kinetic diameters and, as already seen, the interactions of toluene and neopentane with the silica surface are stronger than those of methylcyclohexane. One possible explanation for the differences observed is associated with the inflexibility of the cyclic molecules. It is possible that the less rigid neopentane molecules are able to occupy the pore space of MCM-48 more efficiently. The cubic pore structure seems to affect to a greater extent the adsorption of methylcyclohexane than that of toluene. In this case, the slightly larger size and weaker gas-solid interactions of the former hydrocarbon may contribute to the results obtained.

By comparing the capillary condensation/evaporation pressures of the hydrocarbons on the MCM-48 with those corresponding to MCM-41 silicas with different pore sizes,[12] it is found that the capillary condensation/evaporation pressures of the organics observed for MCM-48 are consistent with the sample's pore size and, consequently, the differences found in the adsorption before condensation occurs do not affect it. The adsorption isotherms of the three organics on the MCM-48 sample have reversible steps, which means that in all cases the size of the primary mesopores is below the hysteresis critical diameter (the pore size below which adsorption occurs without hysteresis at a given temperature), D_{ch}.[4] On MCM-41, this is only observed for neopentane while for toluene and methylcyclohexane a very narrow hysteresis loop is seen.

Figures 3 and 4 show the adsorption-desorption isotherms of toluene and methylcyclohexane measured at several temperatures on the MCM-41 and MCM-48 materials.

It can be seen from Figure 3(a) that the toluene isotherm determined at 308 K on the MCM-41 silica is completely reversible while at 303 K the isotherm already exhibits a narrow hysteresis cycle, which broadens as the adsorption temperature decreases so that at 288 K a clear H1 type hysteresis loop is present. The methylcyclohexane isotherms determined on the same material [Figure 4(a)] start having irreversible steps at higher temperature than those of toluene. The sorption of methylcyclohexane is reversible at 318 K and the isotherms at temperatures equal or lower than 308 K show hysteresis. Therefore, the hysteresis critical temperature (T_{ch}), *i. e.*, the temperature above which adsorption hysteresis is not observed, of toluene and methylcyclohexane on MCM-41 with pore size 3.9 nm are 303 K and 308 K, respectively.

The toluene adsorption/desorption isotherm measured at 298 K on the MCM-48 has a reversible capillary condensation/evaporation step and the H4 hysteresis cycle closes at the beginning of the step. At 288 K a very narrow hysteresis cycle associated with adsorption/desorption on the primary mesoporosity is observed, combined with the hysteresis cycle at high pressures. The lower closure point of the H4 hysteresis loop on the methylcyclohexane isotherms is progressively displaced to lower p/p° as the temperature decreases and at 278 K it closes at the beginning of the step. In order for the methylcyclohexane isotherms on MCM-48 to exhibit hysteresis associated to the filling and emptying of the primary mesopores it is necessary to reduce the adsorption temperature down to 268 K. At this temperature a narrow hysteresis loop associated to the step is present. Therefore, the hysteresis critical temperatures of toluene and methylcyclohexane on MCM-48 with D_p=3.8 nm are 288 K and 268 K, respectively. In Table 2 the T_{ch} of the hydrocarbons on the MCM-41 with D_p=3.9 nm and MCM-48 with D_p=3.8 nm are compared with those previously found for a MCM-41 sample with D_p=3.4 nm.[12]

As expected, the hysteresis critical temperatures of both organic compounds are higher for the cylindrical pores with higher diameter. Sorption hysteresis occurs at the same temperature on the isotherms of toluene and methylcyclohexane of MCM-41 with pores of diameter 3.4 nm. The results obtained with the MCM-41 sample with D_p=3.9 nm are not very different, with the T_{ch} of methylcyclohexane being only 5 K superior, which indicates that the size of the cylindrical pores may affect differently the hysteresis behaviour of the two hydrocarbons. However, 5 K is not a significant difference and it would be necessary to investigate the sorption of both organics on MCM-41 silicas with pores of various sizes, in order to confirm this feature.

The hysteresis behaviour of toluene and methylcyclohexane on the MCM-41 and on the MCM-48 is clearly different. On the one hand, hysteresis associated to the filling and emptying of the primary mesopores of the cubic mesostructure occurs at higher temperature on the toluene isotherms, in contrast with what is observed for the hexagonal pore structures. It can be seen from the values of T_{ch} in Table 2 that the difference between

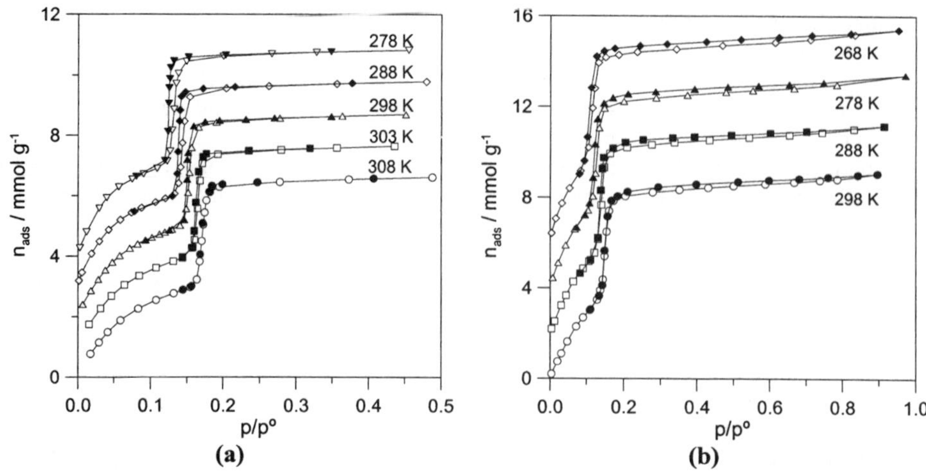

Figure 3 *Adsorption/desorption isotherms of toluene at several temperatures on (a) MCM-41(D_p=3.9nm) and (b) MCM-48(D_p=3.8 nm) materials (closed symbols-desorption). The amounts adsorbed are shifted for each temperature by 1 mmol g^{-1} for MCM-41 and 2 mmol g^{-1} for MCM-48.*

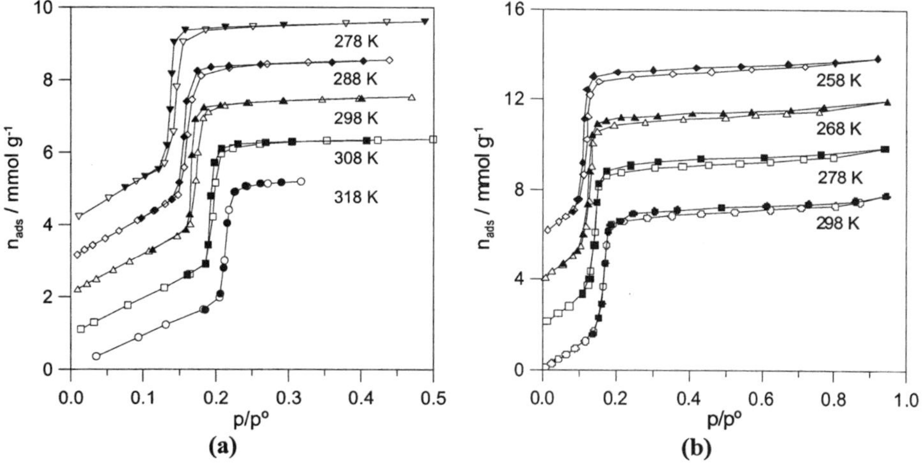

Figure 4 *Adsorption/desorption isotherms of methylcyclohexane at several temperatures on (a) MCM-41(D_p=3.9nm) and (b) MCM-48(D_p=3.8 nm) materials (closed symbols-desorption). The amounts adsorbed are shifted for each temperature by 1 mmol g^{-1} for MCM-41 and 2 mmol g^{-1} for MCM-48.*

the T_{ch} of the two organics on MCM-48 is 20 K. On the other hand, although it was expected that the T_{ch} of the two hydrocarbons would be lower on the MCM-48 in comparison with the MCM-41 with D_p=3.9 nm, due to the smaller pore size of the former,

Table 2 *Hysteresis critical temperatures of toluene and methylcyclohexane on MCM-41 and MCM-48 silica materials.*

Adsorptive	T_{ch} / K		
	MCM-41		MCM-48
	D_p=3.4 nm[12]	D_p=3.9 nm	D_p=3.8 nm
Toluene	268	303	288
Methylcyclohexane	268	308	268

the T_{ch} found are considerably smaller. This is particularly notorious with methylcyclohexane, for which the T_{ch} on the MCM-48 is similar to the T_{ch} on a MCM-41 with much narrower pores (3.4 nm).

It was shown that the relative shift of the hysteresis critical temperature from the bulk critical temperature, $(T_c-T_{ch})/T_c$, is linearly correlated with the reciprocal of the pore size for several gases, such as nitrogen, argon and oxygen, on MCM-41 and SBA-15.[4,7] From the variation of the $(T_c-T_{ch})/T_c$ of toluene and methylcyclohexane with the pore size on MCM-41 silicas, we have estimated the pore diameter that should correspond to the T_{ch} found experimentally for the MCM-48. Values of pore diameter of 3.7 nm and 3.4 nm were obtained for toluene and methylcyclohexane, respectively. This means that the hysteresis behaviour observed for toluene and methylcyclohexane on the MCM-48 sample is the same as the one that would be observed on MCM-41 materials with pore size 3.7 and 3.4 nm, respectively. The D_p estimated for toluene is not very different from the actual size of the MCM-48 channels but that of methylcyclohexane is considerably lower. Hence, the hysteresis behaviour of methylcyclohexane is more affected by the cubic pore structure, as was found for the adsorption behaviour at low p/p°.

All the neopentane isotherms determined at temperatures between 273 K and 258 K on the MCM-48 and MCM-41 materials with similar pore size (not shown) have completely reversible capillary condensation/evaporation steps, as was expected on the basis of the results obtained for MCM-41 silicas with pores of smaller and larger dimension.[12]

Morishige *et al.*[8] have found that the hysteresis behaviour of nitrogen on MCM-41 and MCM-48 was similar and have concluded that the interconnected pore structure has no significant effect on the adsorption hysteresis. Qualitatively similar pore condensation and hysteresis behaviour of argon was also observed between MCM-41 and MCM-48 silicas.[6] Considering the similarities encountered in the adsorption of nitrogen and neopentane on the MCM-48 compared to the MCM-41, it seems possible that the pore structure also does not affect the hysteresis behaviour of neopentane.

Therefore, the MCM-48 pore structure seems to have no influence on the hysteresis behaviour of some substances such as nitrogen, argon and perhaps neopentane, but it affects to a smaller or higher extent the hysteresis behaviour of other substances such as toluene or methylcyclohexane.

4 CONCLUSIONS

It was found that the MCM-41 and MCM-48 silicas adsorb similar amounts of toluene and methylcyclohexane at pressures below capillary condensation, despite the latter having a higher pore volume. This was not observed for the adsorption of nitrogen and neopentane. The adsorption behaviour of nitrogen and neopentane on MCM-48 compared to that on MCM-41 is identical and distinct from that of toluene and methylcyclohexane.

The hysteresis behaviour of toluene and methylcyclohexane was found to be significantly affected by the pore structure of the MCM-48 material, with that of methylcyclohexane being the most affected. While adsorption hysteresis is observed at identical or slightly higher temperature on the methylcyclohexane isotherms of MCM-41 materials, the T_{ch} of toluene adsorbed on MCM-48 is 20 K higher.

Acknowledgements

This work was financially supported by the Fundação para a Ciência e a Tecnologia (FCT, Portugal) (project no. PTDC/CTM/67314/2006). P.A.R. acknowledges FCT also for PhD grant no. SFRH\BD\17713\2004.

References

1. F. Rouquerol, J. Rouquerol and K.S.W. Sing, *Adsorption by Powders and Porous Solids*, Academic Press, London, 1999.
2. C. T. Kresge, M. E. Leonowicz, W. J. Roth, J. C. Vartulli and J. S. Beck, *Nature*, 1992, **359**, 710.
3. P. I. Ravikovitch, S. C. O'Domhnaill, A. V. Neimark, F. Schüth and K. K. Unger, *Langmuir*, 1995, **11**, 4765.
4. K. Morishige and M. Shikimi, *J. Chem. Phys.*, 1998, **18**, 7821.
5. S. Inoue, Y. Hanzawa and K. Kaneko, *Langmuir*, 1998, **14**, 3079.
6. M. Thommes, R. Köhn and M. Fröba, *J. Phys. Chem. B*, 2000, **104**, 7932.
7. K. Morishige and M. Ito, *J. Chem. Phys.*, 2002, **117**, 8036.
8. K. Morishige, N. Tateishi and S. Fukuma, *J. Phys. Chem. B*, 2003, **107**, 5177.
9. S. Z. Qiao, S. K. Bhatia and D. Nicholson, *Langmuir*, 2004, **20**, 389.
10. N. Tanchoux, P. Trens, D. Maldonado, F. Di Renzo and F. Fajula, *Colloids Surf. A: Physicochem. Eng. Aspects*, 2004, **246**, 1.
11. Y. He and N. A. Seaton, *Langmuir*, 2006, **22**, 1150.
12. P. A. Russo, M. M. L. Ribeiro Carrott and P. J. M. Carrott, *Adsorption*, 2008, **14**, 367.
13. M. Grün, K. K. Unger, A. Matsumoto and K. Tsutsumi in *Characterization of Porous Solids IV*, ed. B. McEnaney, T. J. Mays, J. Rouquérol, F. Rodriguez-Reinoso, K. S. W. Sing and K. K. Unger, Royal Society of Chemistry, London, 1997, p 81.
14. K. Schumacher, M. Grün and K. K. Unger, *Micropor. Mesopor. Mater.*, 1999, **27**, 21.
15. S. J. Gregg and K. S. W. Sing, *Adsorption Surface Area and Porosity*, 2nd ed., Academic Press, London, 1982.
16. R. C. Reid, J. M. Prausnitz and B. E. Poling, *The Properties of Gases and Liquids*, McGraw-Hill, New York, 1986.
17. M. Kruk, M. Jaroniec, R. Ryoo and J. M. Kim, *Chem. Mater.*, 1999, **11**, 2568.
18. M. M. L. Ribeiro Carrott, A. J. Candeias, P. J. M. Carrott, P. I. Ravikovitch, A. V. Neimark and A. D. Sequeira, *Micropor. Mesopor. Mater.*, 2001, **47**, 323.

PORE STRUCTURE AND MORPHOLOGY OF MESOPOROUS SILICATE AND ALUMINOSILICATE MOLECULAR SIEVES BY NITROGEN ADSORPTION, AFM AND PALS

J. Goworek [a], R. Zaleski [b], A. Borówka [a], R. Kusak [a] and A. Kierys [a]

[a] Department of Adsorption, Faculty of Chemistry, Maria Curie-Sklodowska University, 20 031 Lublin, Poland
[b] Department of Nuclear Methods, Institute of Physics, Maria Curie-Sklodowska University, 20 031 Lublin, Poland

1 INTRODUCTION

The synthesis of ordered silica materials of different porosity and controlling of the morphology of the particles is an intriguing research field in material science. MCM-41 family of highly uniform mesoporous silicates and aluminosilicates which consists of a hexagonal arrangement of cylindrical pores is in the centre of attention of many research works [1-5]. Many efforts are devoted to tailoring of the pore dimensions and modification of the framework surface. Pore dimensions are tailored using different synthesis route and post-synthesis restructuring procedures. It is well known that any additives in reacting mixture causes, from the aspect of colloidal chemistry, the changes of surface energy and influences the formation of primary particles constituting the framework of as-synthesized sample. The materials obtained on the templating route exhibit two types of pore structure. Dominant porosity represents the primary pores, usually hexagonally arranged as in the case of MCM-41 silica or mesoporous silica fibres MSF. The another group of pores of irregular character represents interparticle space and is determined, among others, by morphology of silica particles. The extent of the secondary porosity is influenced by temperature of synthesis as well as postsynthesis hydrothermal treatment of silica in mother liquor. Usually long lasting hydrothermal treatment causes the decrease of interparticle volume. For thermally restructured samples one can observe that hysteresis of adsorption at $p/p_0 \approx 0.9$ disappears [6].

Positron Annihilation Lifetime Spectroscopy (PALS) allows investigation of microporous and mesoporous materials [7-8]. PALS porosimetry uses as a probe long-lived positronium triplet state (bound state of electron and positron with parallel spins called ortho-positronium, o-Ps). Ortho-posirtonium forms in non-conducting porous solids, where is trapped in electron free volume (i.e. pore). Lifetime of trapped o-Ps is shortened from 142 ns in vacuum to the value dependent on pore size due to pick-off process (positron annihilation with electron which has anti-parallel spin). Determination of pore sizes basing on o-Ps lifetime pore radii over 1 nm became possible in the last decade due to extension of commonly used the Tao-Eldrup model made by Goworek at all [9]. Relation between o-Ps lifetime ($\tau_{o\text{-}Ps}$) and free volume radius (R) is described by the extended version of the Tao-Eldrup model (ETE). The ETE model interhits from the Tao-Eldrup model some approximations and needs experimental parameter Δ, which was calibrated for porous silica (Δ = 0.18 nm). Approximation of the shape of free volume by a sphere, as it was done in the Tao-Eldrup model, is not always correct for pores in solids. Pores are often described as infinitely long cylinders, what should be taken into account by assuming a proper geometry in model calculations. Unlike most porosimetric methods PALS is sensitive to both open and closed pores. Positrons penetrate investigated material freely and positronium can be formed in any free volume within positron range.

In the present paper the pore size distributions obtained by two different techniques i.e. low temperature adsorption of nitrogen (standard technique) and positronium annihilation lifetime spectroscopy (PALS) are compared for aluminosilicate samples of different ratio of primary and secondary mesopores. The aim of this paper is testing of the consistency of the pore structure characteristic determined from both these methods.

2 METHOD AND RESULTS

2.1 Materials

Pure siliceous (MCM-41) and aluminosiliceous (Al-MCM-41) mesoporous molecular sieves were prepared following the synthesis procedures reported previously – Refs [10,11], respectively. The source of silica was tetraethyl orthosilicate (TEOS, Aldrich, 98%) and alumina - aluminum sulfate octadecahydrate $Al_2(SO_4)_3*18H_2O$, (Aldrich, >98%). The cationic surfactant cetyltrimetylmammonium bromide (CTAB, Aldrich, >96%) was used as micellar template. The mesoporous silica fibres (MSF) were synthesized as described in Ref.[12]. The silica source was tetrabutyl orthosilicate (TBOS, Aldrich, 97%) and CTAB was a template. The spontaneous growth of the silica fibres was conducted in two-phase system. After 20 days the material was evacuated from solution. All as-synthesized surfactant-silica mesophases were filtered, washed with distilled water and dried in air at 90°C. Next the samples were calcined by heating up to 550°C (1°/min) for 8 hours. Then, all samples were additionally heated at 550°C in oxygen atmosphere to remove carbon traces which remain after calcination.

2.2 Methods

Nitrogen adsorption/desorption isotherms were measured at 77 K on ASAP 2405 volumetric adsorption analyzer Micromeritics (Norcross, GA). The pore size distributions were obtained from the adsorption branch of the isotherm using the BJH method [13]. The specific surface areas of the investigated samples were evaluated using the standard BET method for the adsorption data in a relative pressure p/p_0 range from 0.04 to 0.25. The total pore volume was estimated from a single point adsorption at the relative pressure of 0.985.

X-ray diffraction measurements were performed using HZG-4 diffractometer (Carl Zeiss Jena) using monochromatic Cu K_α radiation. The patterns were recorded in 2θ range of 1° to 10° with a step 0.02°.

The AFM measurements were performed with Digital Instruments Nanoscope III atomic force microscope in tapping mode with the standard silicon probe tips. AFM images were acquired in air at room temperature.

Positronium annihilation measurements (PALS method) were done using a conventional fast-slow delayed coincidence spectrometer with BaF_2 detector. The sample in the form of two 2 mm thick of powder, with the ^{22}Na positron source in a Kapton envelope between them, was placed inside the vacuum chamber. The sample chamber was kept at a low pressure of about 0.3 Pa, that is sufficient to avoid the *ortho-para* conversion of positronium on paramagnetic atmospheric oxygen. The coincidence counting rate was about 2×10^6 per hour; usually about 1.7×10^7 counts per spectrum were collected. Such number of counts is required by MELT program [14], that was used for numerical analysis of the spectra.

SEM studies were performed on a Tesla BS-301 microscope operating at 15.0 keV.

2.3 Results and discussion

2.3.1 AFM and SEM characterization. The atomic force microscopy was applied for characterization of morphology of investigated materials. Figure 1 illustrates roughness of the surface for these samples.

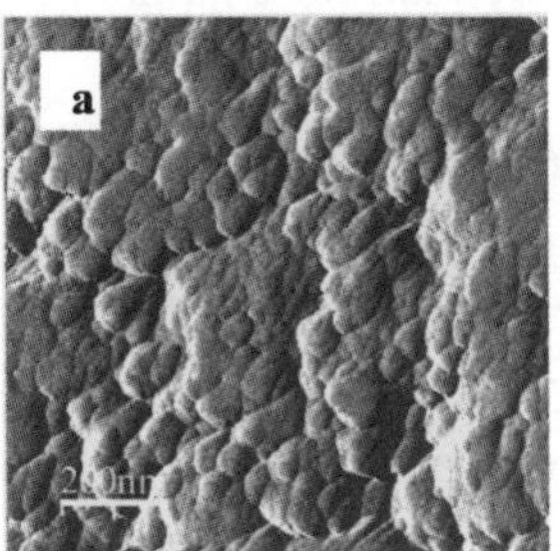
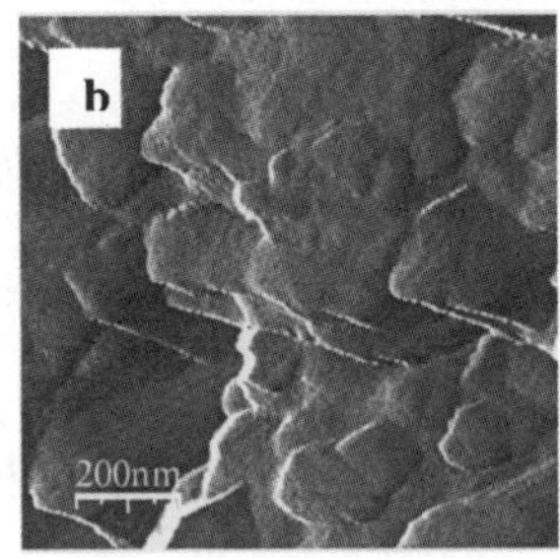
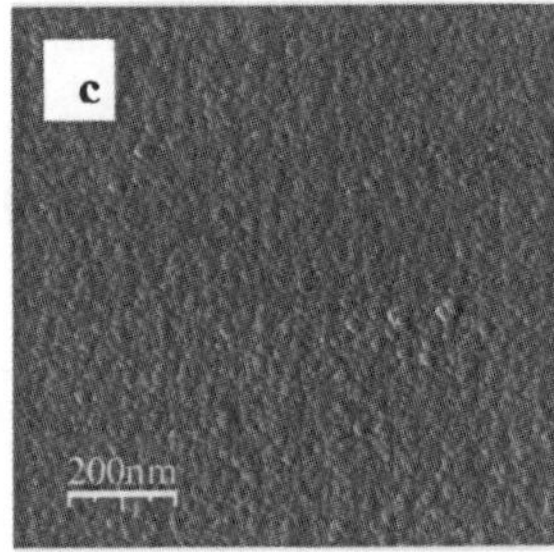

Figure 1 *Basic features of the surface morphology for samples under investigations: a – MCM-41; b – Al-MCM-41; c – MSF; as determined by AFM*

Comparing the results presented in Figure 1 a-c one can say that roughness of the surface, the shape and dimensions of the basic subparticles which make whole samples skeletal depend on the synthesis route and type of used components. The surface image is exceptionally different in the case of MSF. The shape of silica fibres is differentiated and this material consists of various silica species cylindrically shaped. The several examples of these species are presented in Figure 2a. Figure 2b,c shows the silica structure more precisely. The surface morphology has been determined by atomic force microscopy.

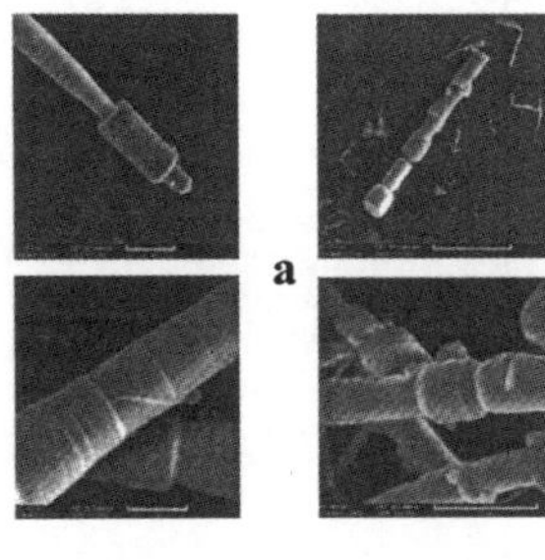
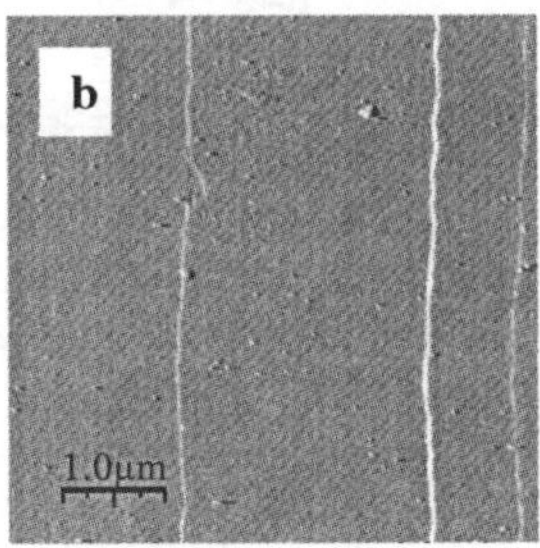
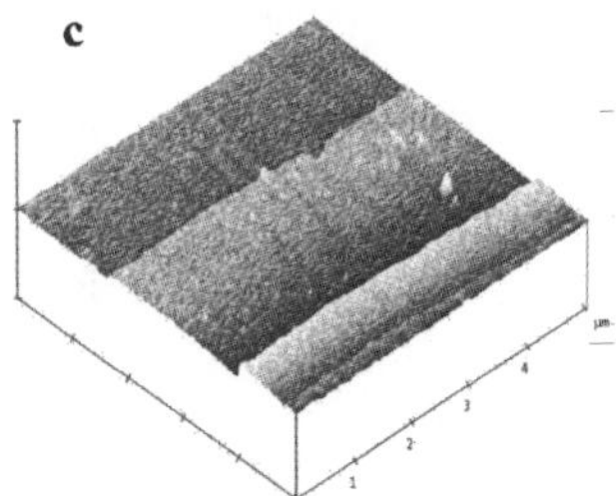

Figure 2 *Low-magnification images of the nanofibres synthesized with CTAB: a – SEM (scale bar: 20 nm),   b, c – AFM images (5 μm x 5 μm)*

Long lasting synthesis of MSF (about 3 weeks) causes that sub-particles of silica are tightly packed and interparticle space is extremely small. Very slow growing of silica fibres may be assumed as an equivalent to hydrothermal treatment. Generally, the cylindrically shaped silica fibres have diameter around 10 μm.

2.3.2 Nitrogen adsorption characterization. The absence of larger interparticle pores in MSF is confirmed by the shape of nitrogen adsorption isotherm (see Figure 3). The adsorption isotherms for MCM-41 and Al-MCM-41 are presented in the same figure. All isotherms are typical of well ordered hexagonally silica, being characterized by a steep increase at relative pressure $p/p_0 \approx 0.35$. For MCM-41 and Al-MCM-41 a narrow hysteresis loop is observed at $p/p_0 > 0.9$, which is caused by larger irregular mesopores. A flat segment at higher p/p_0 for MSF sample suggests that this sample does not contain larger pores. The pore size distributions derived from first step on adsorption isotherms using BJH procedure are presented in Figure 4 for samples under investigation. The specific surface area (BET method), total pore volume and pore radii for both groups of pores are collected in Table 1.

Table 1 *Parameters characterizing porosity of investigated samples obtained from nitrogen adsorption, PALS and XRD data*

Sample	N₂ adsorption			PALS		XRD
	S_{BET} (m²/g)	V_p (cm³/g)	D_{meso} (nm)	D_{meso} (nm)	D_{int} (nm)	w_d (nm)
MCM-41	1125	1.07	2.62	1.88	14.27	3.45
Al-MCM-41	1100	0.86	2.59	1.86	14.02	3.35
MSF	985	0.59	2.60	2.46	12.91	3.60

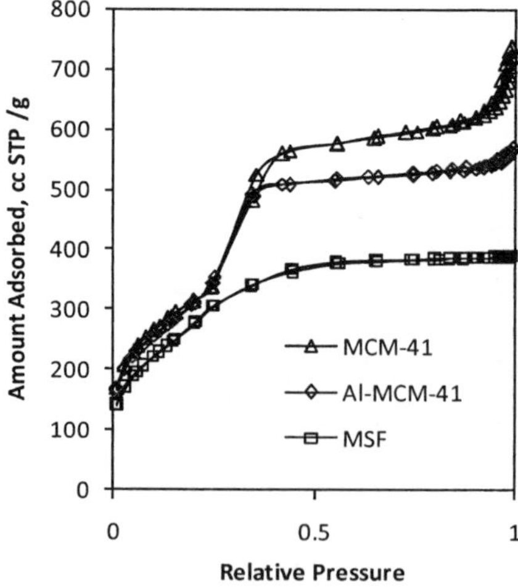

Figure 3 *Nitrogen adsorption/desorption isotherms of MCM-41, Al-MCM-41 and FSM*

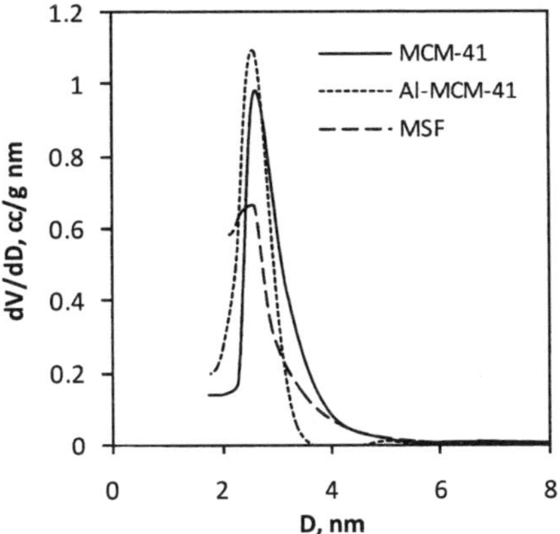

Figure 4 *Pore size distributions of primary pores for MCM-41, Al-MCM-41 and FSM*

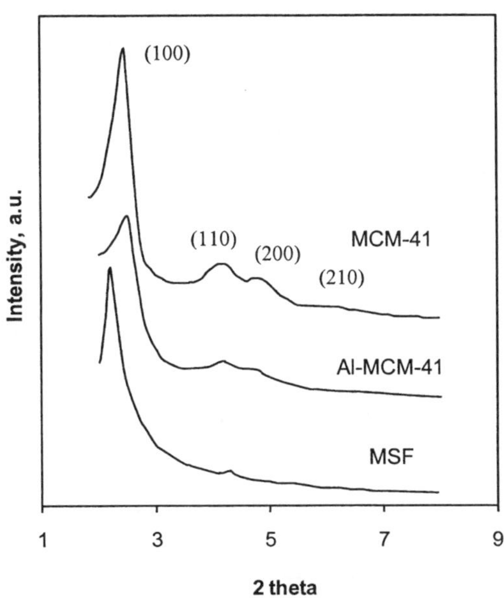

Figure 5 *X-ray diffraction pattern for samples under investigations*

2.3.3 XRD experiment. The X-ray diffraction patterns of samples under investigation are presented in Figure 5. It can be seen that the sample MCM-41 features the best ordered mesoporous structure. X-ray pattern exhibit 4 peaks in the range $2\theta < 7°$ typical for hexagonal ordering of silica framework. These reflections can be indexed as (100), (110), (200) and (210) assuming hexagonally symmetry. For uniform cylindrical pores with (100) interplanar spacing d_{100}, the pore size of mesopores w_d can be calculated from the relation [15-16]:

$$w_d = c \cdot d_{100} \left(\frac{\rho V_{meso}}{1 + \rho V_{meso}} \right)^{1/2} \tag{1}$$

where: ρ is the density of sample walls (2.2 g/cm^3) [17], V_{meso} is the primary mesopore volume and c = 1.213 is a constant. Numerical w_d values are summarized in column VII of Table 1.

2.3.4 PALS characterization. Determination of pore radius distributions by PALS can be done using relation

$$\frac{dV}{dD} \propto I(\tau) \frac{d\tau}{dD} \tag{2}$$

where $I(\tau)$ is the o-Ps intensity distribution obtained from experiment, relation $\tau(D)$ is given by the ETE model and $d\tau/dD$ is the derivative of this function. In order to obtain absolute values, as in the case of nitrogen adsorption/desorption data, the results have to be normalized to the total pore volume. The total pore volume is hardly obtained from PALS data, mainly due to large uncertainty of intensity of very short-lived para-positronium component. Even though such attempts were made [18], we do not find them appropriate in this case. Also data from liquid nitrogen adsorption cannot be used as they concern open porosity only. So relative distributions of pore size estimated from PALS data with use of equation (2) are presented in Figure 6.

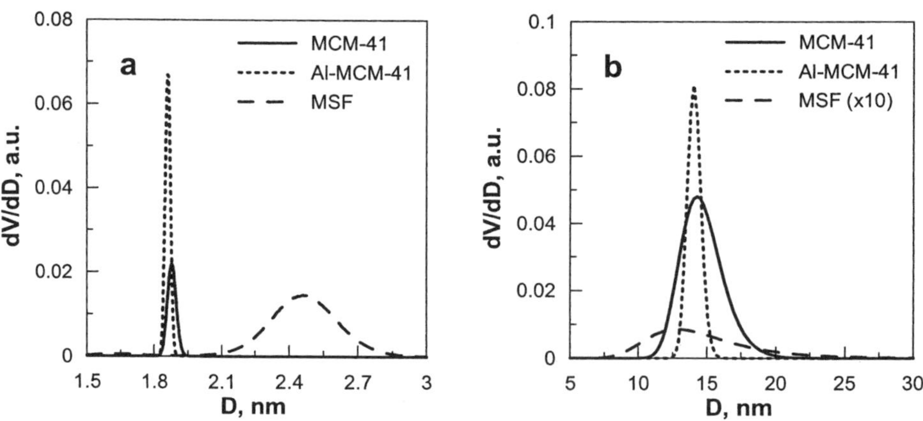

Figure 6 *Pore size distributions of MCM-41, Al-MCM-41 and MSF calculated from PALS spectra: a – primary pores; b – irregular pores*

Part a of Figure 6 illustrates pore size distributions for regular primary pores and part b shows pore radii distributions for larger mesopores. The presented results may be compared to PSDs derived from liquid nitrogen adsorption isotherms. Because in the range of micropores the BJH method is unreliable, only results for the R > 1 nm can be considered. Pore size distribution for MSF derived from adsorption data are in good agreement with the distribution obtained from PALS measurements. Circular channels of this sample are arranged around the fibre axis. According to growing mechanism and architecture of MSF fibres it is probable that part of pores is closed and openings of these pores should be assumed rather as a defects of the silica surface produced during calcination. Relatively long circular pores of MSF represent an ideal model system for PALS studies - the escape of positronium to the outer surface of fibre is in this case substantially restricted. Thus intensity and lifetimes of positronium represent real dimensions of circular channels. For MCM-41 and Al-MCM-41 the BJH method leads to larger radii of primary pores than ETE model. Discrepancies between PSD derived from nitrogen adsorption and PALS may be related to the detection of closed pores using PALS method which are inaccessible for nitrogen molecules. On the other hand, the difference between these PSDs of primary pores may be the result of positronium escape from the open pores to intergranular space in MCM-41. As a consequence observed lifetimes are shortened. In MSF most of the positronium annihilates inside the pores, while in MCM-41 large intensity of the intergranular component suggests escape of o-Ps from the pores. It is hard to compare PSDs for pores of large dimensions in which condensation takes place above $p/p_0 = 0.9$. PSDs for large mesopores derived from PALS experiment represent well formated peaks. Positronium annihilates inside the pores of mean diameter about 15 nm. Relatively small amount of pores of diameter above 10 nm is detected for MSF. These pores may be assumed as a defects in silica fiber produced in calcinations process.

It should be noted that PSDs for MCM-41 and Al-MCM-41 (not shown) are very extended along x-axis. Additionally calculated pore dimension are characterized by large experimental error. In the range of $p/p_0 = 0.95\text{–}0.98$ pore size dimensions change from ~15 nm to ~1000 nm and the maximum of the PSD peak is indefinable. Thus it may be assumed that PALS porosimetry is suitable for detection of large interparticle space (including closed pores).

3 CONCLUSIONS

The Extended Tao-Eldrup model gives the relation between ortho-positronium lifetime and the size of the free volume where o-Ps is trapped. Lifetime distributions obtained from PALS can be transformed into pore size distributions. Positron porosimetry gives reliable results for closed pores and large mesopores. In MCM-41 the annihilation takes place partially outside the microparticles of silica, while in MSF inside the pores. There exits quite good correlation between pore structure characteristics obtained by adsorption and PALS method. Pore dimensions derived from XRD experiment are larger than those determined by nitrogen adsorption and positron porosimetry.

References

1   C.T. Kresge, M.E. Leonowicz, W.J. Roth, J.C. Vartuli and J.S. Beck, *Nature*, 1992, **359**, 710.

2   U. Ciesla, and F. Schuth, *Micropor. Mesopor. Mater.*, 1999, **27**, 131.

3   A. Corma, *Chem. Rev.*, 1997, **97**, 2373.

4   *Nanoporous Materials IV*, A. Sayari, M. Jaroniec (eds), Studies in Surface Science and Catalysis, vol. 156, 2005.

5   G. Schultz-Ekloff, J. Ratousky and A. Zukal, *Micropor. Mesopor. Mater.*, 1999, **27**, 273.

6   A. Sayari, P. Liu, M. Kruk, M. Jaroniec, *Chem. Mater.*, 1997, **9**, 2499.
7   R. Zaleski, J. Goworek, A. Borówka, in: *Studies in Surface Science and Catalysis*, Elsevier, D.L. Llevellyn, F. Rodriguez-Reinoso, J. Rouquerol, N. Seaton (eds), vol. 160, 2007, p. 471.
8   R. Zaleski, A. Borówka, J. Wawryszczuk, J. Goworek, T. Goworek, *Chem. Phys. Lett.*, 2003, **372**, 794.
9   T. Goworek, K. Ciesielska, B. Jasińska and J. Wawryszczuk, *Chem. Phys. Lett.*, 1997, **272**, 91.
10   M. Grün, K.K. Unger, A. Matsumoto and K. Tsutsumi, in: *Characterization of Porous Solids IV*, The Royal Society of Chemistry, B. McEnaney, J.T. Mays, J. Rouquerol, F. Rodriguez-Reinoso, K.S.W. Sing and K.K. Unger (eds), 1997, p. 81.
11   M.M.L. Ribeiro Carrott, F.L. Conceição, J.M. Lopes, P.J.M. Carrott, C. Bernardes, J. Rocha and F. Ramoˆa Ribeiro, *Micropor. Mesopor. Mater.*, 2006, **92**, 270.
12   F. Marlow, M. McGehee, D. Zhao, B. Chmelka and G. Stucky, *Adv. Mater.*, 1999, **11**, 632.
13   E.P. Barrett, L.G. Joyner and P.H. Halenda, *J. Am. Chem. Soc.*, 1951, **73**, 373.
14   A. Shukla, M. Peter and L. Hoffman, *Nucl. Instr. and Meth. A*, 1993, **335**, 310.
15   T. Dabadie, A. Ayral, Ch. Guizard, L. Cot and P. Lacan, *J. Mater. Chem.*, 1996, **6**, 1789.
16   M. Kruk, M. Jaroniec and A. Sayari, *J. Phys. Chem. B*, 1997, **101**, 583.
17   *The Colloid Chemistry of Silica*; Bergna, H. E., (ed.); American Chemical Society, Washington, DC, 1994.
18   C.L. Wang, M.H. Weber and K.G. Lynn, *Appl. Phys. Lett.*, 2002, **81**, 4413.

TRANSPORT IN SIO$_2$-CATALYST SUPPORTS: A CONTRIBUTION TO PROCESS INTENSIFICATION

D. Enke, D. Stoltenberg, H. Preising, J. Kullmann, T. Hahn

Institute of Chemistry, University of Halle, Schlossberg 2, D-06108 Halle/Saale, Germany

1 INTRODUCTION

For a rational design of catalyst supports the knowledge of transport characteristics of the available porous materials is essential. Especially for the heterogeneous catalysis, the knowledge of the diffusion characteristics is important due to the great influence of the mass transport through the catalyst particle on the catalysts behaviour. To select the optimal catalyst, one has to consider not only the catalytic activity of the active component on the inner surface but also the mass transport properties. To estimate the pore structure often tabulated values are used to differentiate between the various supports. This results in oversimplified assumptions for the microstructure. The experimental determination of the transport properties proves to be very complicated in case of SiO$_2$ supports, because of their complex pore structure. Hence, there are just limited experimental data available.

In this study, the pore diffusion coefficients under reaction conditions were investigated for two commercial SiO$_2$ supports by combination of diffusion measurements on membranes with a catalytic test reaction. The obtained findings were used for a rational design of Ni/SiO$_2$ catalysts with hierarchical pore structure.

2 EXPERIMENTAL

Two mesoporous silica gels (pore size "SG - 1" 5.7 nm and "SG - 2" 8.4 nm respectively) were purchased from Acros Organics. The particles are 0.2 mm in diameter.

The two-phase-membranes were synthesized by the following procedure.[1] An initial glass of the composition 70 wt.-% SiO$_2$, 23 wt.-% B$_2$O$_3$ and 7 wt.-% Na$_2$O was cut into thin plates of the typical dimension 20 x 20 x 0.5 mm. The glass plates were heat treated at 720 °C for 72 h for phase separation. Afterwards an acid leaching treatment for 10 h in 3 M HCl at 90 °C followed. The membranes were washed in deionised water and dried at room temperature.

The catalyst supports with hierarchical pore structure were synthesized by a modified procedure described by Nakanishi et al.[2] Water and polyethylene oxide were mixed in a Teflon vessel. While stirring constantly sulphuric acid was added and the mixture was stirred until the polymer dissolved. Afterwards, tetraethoxysilane (TEOS) was added and the vessel was closed and held at 50 °C for 24 h. The synthesized gel was washed with water and dried for 72 h at 50 °C. Following, the sample was calcined at 500 °C for 4 h (heating rate 3 K min^{-1}).

To modify the surface of the materials with hierarchical pore structure with amino groups, the prepared supports were put in a 10 wt% solution of 3-aminopropyl-triethoxysilane (APTS), with 10 ml per g support. It was kept at 100 mbar for 24 h at room temperature and afterwards dried at 70 °C for 3 h.

The nickel supported catalysts (both silica gel and catalyst supports with hierarchical pore structure) were prepared by equilibrium adsorption in $Ni(NO_3)_2$ aqueous solution. First, the catalyst supports were activated at 210°C for 24 h. After that, the samples were covered with $Ni(NO_3)_2$ aqueous solution (2 mole $Ni(NO_3)_2 \cdot 6H_2O$ dissolved in 1000 ml water; 3 g support in 10 ml solution), kept in vacuum for 20 h and washed with deionised water. The products were dried overnight at 120 °C and then carefully calcined in air at 450 °C for 5 h (heating rate 3.25 K min^{-1}). The resulting samples were reduced in a hydrogen flow of 50 ml min^{-1} at 450 °C for 5 h (heating rate 7 K min^{-1}) to obtain the nickel supported catalysts and finally cooled down to the reaction temperature in a hydrogen flow. The metal loading of the nickel supported catalysts varied between 0.7 and 2.8 wt.-%.

Nitrogen sorption measurements were performed by using a Sorptomatic 1990 apparatus by ThermoFinnigan. All samples were degassed at 120°C before measurement for at least 24 h at 10^{-5} mbar. Adsorption and desorption isotherms were measured over a range of relative pressures (p/p^0) from 0 to 1.0. Surface areas were determined from the linear part of the Brunauer-Emmett-Teller (BET) equation in a relative pressure range (p/p^0) of the adsorption isotherms between 0.05 and 0.25.[3] A value of 0.162 nm^2 was used for the cross-sectional area per nitrogen molecule. The total pore volume was estimated from the amount of gas adsorbed at the relative pressure p/p^0 = 0.99 assuming that pores were filled subsequently with condensed adsorptive in the normal liquid state.

The two-phase porous glass membranes were examined for the permeability of helium at temperatures up to 140 °C and a pressure adjustable from 1.1 up to 1.5 bar. The measurements were performed as follows: The membrane was stuck on a brass plate containing a bore with 7 mm diameter. After that, the plate was fixed with a special dome on a turbo molecular pump (TMU 261, Pfeiffer). Following, the dome was evacuated and heated up to 100 °C by heating tape. The gas flow was determined by measuring the pressure below the membrane. The integral permeability was estimated using the pumping speed, the measured pressures above and below the membrane and the membrane thickness.

Additionally the membrane was examined using a classic Wicke-Kallenbach-cell. The membrane was fixed between two Teflon-rings in a special brass body, subdividing it into two cells. The cells were evacuated to remove adsorbed water and heated up to 100 °C. Afterwards the cells were flown through by helium and nitrogen respectively, ensuring isobaric conditions. The gas compositions were determined by conductivity measurements. The effective diffusion coefficients were calculated according to Raja Rao and Smith.[4]

The benzene hydrogenation on the Ni catalysts based on the commercial silica gels and the bimodal supports was carried out in a continuous-flow reactor at atmospheric pressure. A gaseous mixture of benzene and hydrogen flowed through the catalyst bed (0.1 or 0.2 g catalyst, 0.2 mm particle diameter) at temperatures between 80 and 140°C. The partial pressure of benzene was 0.141 atm. Only cyclohexane was detected as a product under the present conditions.

3 RESULTS AND DISCUSSION

The textural properties of the commercial SiO_2 supports and the prepared two-phase membranes were measured by nitrogen low temperature adsorption. The texture characteristics are shown in Table 1. The porosities are determined from the total pore volume assuming a frame density of 2.2 g cm^{-3}. The commercial silica gels show larger specific pore volumes and porosities compared to the membranes, what is understandable due to the different macroscopic structure of the two-phase membranes.

The membranes were built up by a macroporous glass body completely filled with colloidal silica species building the microstructure (Figure 1). The differences to ordinary controlled-pore glasses are explained in the following. The heat treatment of the glass at temperatures above 580 °C not only leads to a high degree of segregation (and thereby large pore diameters) but also to a higher solubility of SiO_2 in the sodium-rich borate phase. In contrast to the sodium-borate-phase the SiO_2-species are insoluble in acids. The following acid leaching treatment removes the sodium-borate phase, leaving the finely dispersed SiO_2 behind. The colloidal silica forms a microstructure comparable to silica gels.

Table 1 *Texture properties of the commercial silica gels and the two-phase membrane*

Sample	Spec. surface area [m² g⁻¹]	Spec. pore volume [cm³ g⁻¹]	Mean pore diameter [nm]	porosity
SG-1	598	0.690	5.7	0.59
SG-2	391	0.954	8.4	0.67
M	226	0.154	3.4	0.25

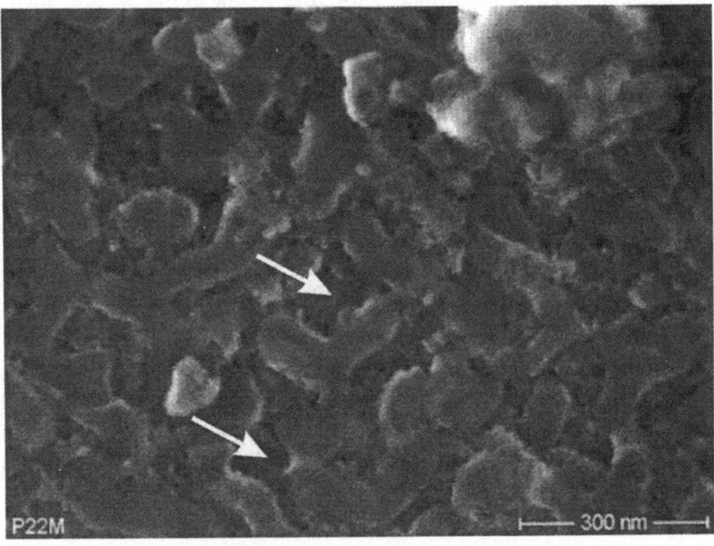

Figure 1 *SEM image of a mesoporous two-phase-membrane (arrows colloidal silica species)*

Table 2 *Transport properties of the two-phase membrane (He, 100 °C)*

Device	Mean pore diameter	D_{eff}	D_{K}	Tortuosity τ
	[nm]	[cm²/s]	[cm²/s]	
Permeability	3.4	$1.35 \cdot 10^{-3}$	$1.6 \cdot 10^{-2}$	11.8
WK-cell	3.4	$1.39 \cdot 10^{-3}$	$1.6 \cdot 10^{-2}$	11.5

The results of the permeability measurements and the measurements done with a Wicke-Kallenbach-cell of the two-phase membrane are shown in Table 2. Both were carried out at temperatures similar to those of the benzene hydrogenation (100 °C). Since porous glasses are known to be good adsorbers, the use of adsorbable gases would falsify the tortuosity factors due to surface diffusion as a second diffusion mechanism beside Knudsen diffusion. As a noble gas, Helium was taken as diffusing gas to avoid surface diffusion effects. The tortuosity factor of the membrane was calculated for both measurements according to equation (1).

$$\tau_{membrane} = \frac{\varepsilon \cdot D_K}{D_{eff}} \qquad (1)$$

Here, ε is the porosity of the membrane, D_{eff} is the measured diffusion coefficient and D_K is the calculated Knudsen diffusion coefficient for He at 100 °C assuming straight cylindrical pores with the determined pore diameter of 3.4 nm. The assumption of Knudsen diffusion as dominant mechanism was confirmed since the permeability of the membrane was independent of the pressure in the investigated range between 1.1 and 1.5 bar.

According to Park et al.[5] the permeability is equivalent to the effective diffusivity in case of Knudsen diffusion. This was confirmed under the experimental conditions by nearly identical results for the permeability measurements and the Wicke-Kallenbach-cell.

The effective diffusion coefficients of benzene diffusing through the commercial silica gels were estimated from the point of intersection between the kinetic and the mass transfer controlled region in the Arrhenius plots of the benzene hydrogenation.[6] Figure 2 shows the measured Arrhenius plots for both silica gels. The effective diffusivity of benzene was estimated using the intrinsic rate constant at the point of intersection, the particle radius and the Thiele diffusion modulus. Plug flow behaviour and first order kinetics for the benzene conversion were assumed. The pore diffusion coefficients of benzene under the present reaction conditions were determined by equation (2), using the porosity of the silica gels and the tortuosity factor measured for the two-phase membrane.

$$D_{P(benzene)} = \frac{\tau_{membrane} \cdot D_{eff(benzene)}}{\varepsilon_{gel}} \qquad (2)$$

The results of the kinetic investigations are given in Table 3.

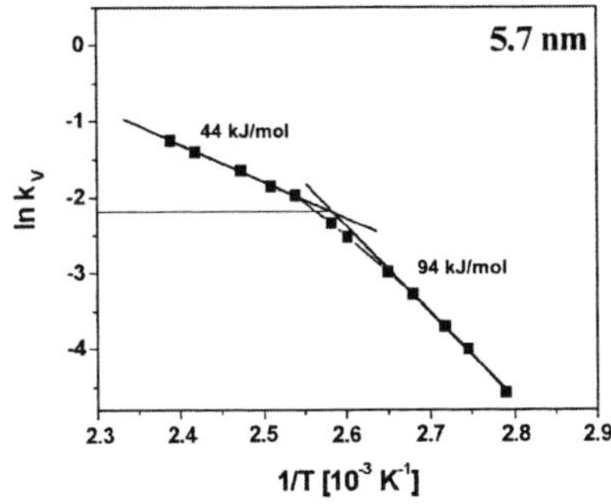

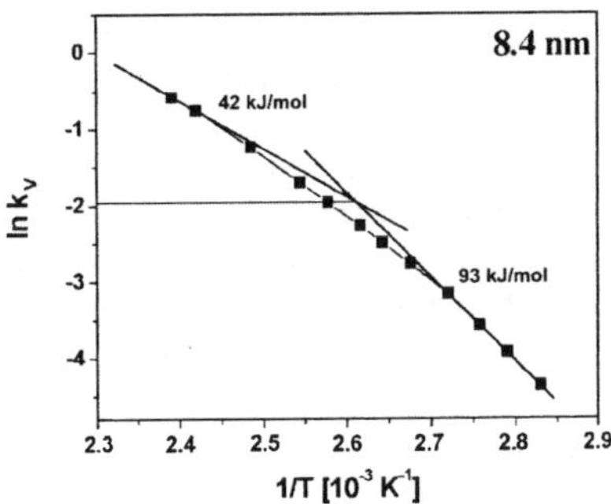

Figure 2 *Arrhenius plots of nickel supported catalysts based on commercial silica gels (apparent activation energies are included)*

Table 3 *Pore diffusivities from benzene hydrogenation for silica gel supported nickel catalysts*

Sample	Mean pore diameter [nm]	$D_{eff\,(benzene)}$ [cm^2 s^{-1}]	$\tau_{(membrane)}$	$D_{P\,(benzene)}$ [cm^2 s^{-1}]
SG - 1 - Ni	5.7	$1.2 \cdot 10^{-6}$	11.8	$2.4 \cdot 10^{-5}$
SG - 2 - Ni	8.4	$1.6 \cdot 10^{-6}$	11.8	$2.8 \cdot 10^{-5}$

The obtained pore diffusivities of benzene under the present conditions are low compared to theoretical Knudsen-Diffusivity in straight cylindrical pores. This can be explained by strong interactions between the benzene and the surface hydroxyl groups of the support.[7] The high tortuosities are due to the very different microstructure of the two-phase-membranes and the silica gels compared to the theoretical assumed straight cylindrical pores. The obtained results can now be used for a rational design of Ni/SiO$_2$ catalysts.

To overcome the compromise between either large surface areas with poor accessibility and small diffusivities or high flow rates with low surface areas, a modified procedure by Nakanishi et al.[2, 8] with a polymer induced phase separation was used to create silica supports with an optimised pore structure. This procedure leads to hierarchically structured supports with adjustable macropores and micro/mesopores - a so-called interconnected structure. Here, the size of the macropores is adjustable independently from the mesopore size between 100 and 8000 nm for the macropores and between 2 and 8 nm for the mesopores respectively. The synthesized silica based supports show excellent transport properties due to the large transport pores combined with high surface areas of 400 to 900 m^{2}g^{-1} caused by the primary (micro/meso-)pore system.

Furthermore the supports with hierarchical pore structure were amine-modified as described before to enhance the support-catalyst interactions, followed by the described Ni loading. This different strategy of loading with the precursor of the active component results in a stronger bonding and therefore in better catalyst performances. The so-prepared supported Ni catalysts were also used for benzene hydrogenation. Dependent on the pore size of the transport pores it is possible to leave the mass transfer controlled region in the

Arrhenius plot (Figure 3) under comparable reaction conditions (temperature, particle size). This gives an opportunity to optimize processes working in the region controlled by the diffusion through the catalyst particles.

By applying a novel pressure-solvent-exchange synthesis, developed by Preising [9], the optimised catalyst supports can be prepared as monoliths such as tubes (Figure 4), caps and discs. This technique is characterized by a gelation under a high-pressure, two solvent exchange steps and optimized drying and heat treatment for template removal. The monoliths are free of cracks and show a very high mechanical stability. This gives the opportunity to produce membrane reactors.

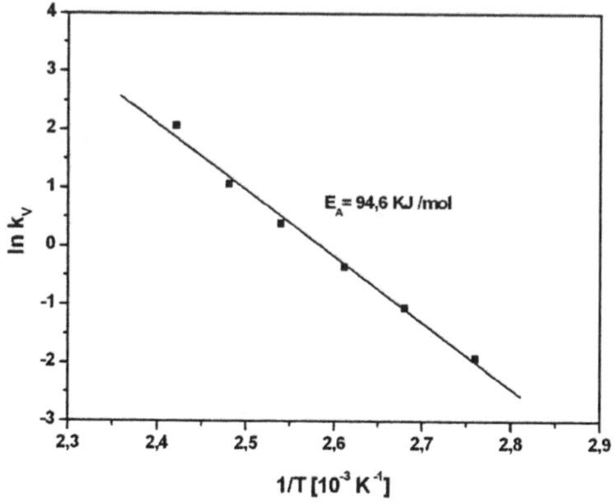

Figure 3 *Arrhenius plot of a Ni-catalyst based on a silica gel with hierarchical pore structure, Texture characteristics of the support: BET surface area 837 m^2/g, pore volume$_{micro/mesopores}$ (N$_2$) 0.37 cm^3/g, pore volume$_{macropores}$ (mercury intrusion) 1.2 cm^3/g, average pore diameter$_{macropores}$ (mercury intrusion) 2300 nm*

Figure 4 *Silica gel monolith with hierarchical pore structure prepared via a novel pressure-solvent-exchange synthesis*

4 CONCLUSIONS

The combination of diffusion measurements of membranes and a catalytic test reaction was successfully used to investigate the transport characteristics of commercial silica gels. This procedure can be used to estimate the transport properties of commercial catalysts under reaction conditions.

Furthermore silica supports with tailorable properties can be prepared by combination of phase-separation in sol-gel process with a novel pressure-solvent-exchange synthesis. The following parameters can be controlled in the range relevant for industrial applications:

- macroscopic geometry
- mechanical stability
- meso- and macropore size
- wall thickness (between adjacent macropores).

Additionally optimized preparation procedures of the nickel supported catalysts (i.e. surface modification of the support) can be used to realize a better catalyst performance.

Now all instruments for a rational design of nickel supported catalysts are available.

References

1   D. Enke, F. Janowski, W. Gille and W. Schwieger, *Colloids and Surfaces A: Physicochm. Eng. Aspects*, 2001, **187**, 131.
2   K. Nakanishi, R. Takahashi, T. Nagatane, K. Kitayama, N. Koheiya, H. Shikata and N. Soga, *Journal of Sol-Gel-Science and Technology*, 2000, **17**, 191.
3   S. Brunauer, P.H. Emmett and E. Teller, *J. Am. Chem. Soc.*, 1938, **60**, 309.
4   M. Raja Rao and J.M. Smith, *AIChE Journal*, 1964, **10**, 293.
5   I.-S. Park, D.D. Do and A.E. Rodrigues, *Catal. Rev.-Sci. Eng.*, 1996, **38 (2)**, 189.
6   M. Kotter, P. Lovera and L. Riekert, *Ber. Buns. Physik. Chemie*, 1976, **80**, 61.
7   J.A. Cusuamo and M.J.D. Low, *J. Phys. Chem.*, 1970, **74**, 792.
8   H. Preising and D. Enke, *Chemie Ingenieur Technik*, 2006, **78**, 1333.
9   H. Preising, PhD Thesis, Halle, 2006.

SURFACE, ACIDIC AND CATALYTIC PROPERTIES OF SULFATED TITANIUM OXIDES

N. Yacoub[1], C. Azer[1,2], S. Selim[2], M. Mekewi[2], A.R. Ramadan[1,3] and J. Ragai[1,3]

[1] The Youssef Jameel Science and Technology Research Center, American University in Cairo, 113 Kasr El Aini Street, P.O. Box 2511, Cairo 11511, Egypt
[2] The Department of Chemistry, Ain Shams University, Abbasiyya, Cairo, Egypt
[3] The Department of Chemistry, American University in Cairo, 113 Kasr El Aini Street, P.O. Box 2511, Cairo 11511, Egypt

1. INTRODUCTION

The so-called titania super acids have recently attracted a great deal of attention. Indeed sulfated oxides of titanium have been shown to act as superacids and to be active in a wide spectrum of reactions such as esterification, alkene polymerization, acylation and isomerization of hydrocarbons[1-9].

In a previous work on sulfated oxides[10,11] we investigated the surface, infrared, acidic and adsorption properties of two series of sulfated oxides prepared from a solution of titanium oxysulfate-sulfuric acid complex hydrate ($TiOSO_4.12H_2SO_4.12H_2O$) using ammonia and urea as two different precipitating agents. It was shown, that in both cases, the presence of sulfate ions allowed for the appearance of strong acidic sites on the surface of the sulfated oxides. It was also shown that the presence of sulfate ions affected the textural characteristics of the oxides and drastically reduced the extent of the surface area. The present work aims at complementing such work by investigating the effect of sulfate ions on the acidic and textural characteristics of pure commercial titanium oxide. Incorporation of the sulfate ions being achieved through impregnation in different increasing amounts of $(NH_4)_2SO_4$. The state of ligation of the generated surface sulfates as well as the catalytic activity of these oxides towards the polymerization of methylmethacrylate are also investigated.

2. EXPERIMENTAL

2.1 Materials

Four samples of TiO_2 (99.9% anatase, Aldrich) each of 100 grams were impregnated in four 250ml solutions of increasing amounts of $(NH_4)_2SO_4$ (99.999% Aldrich) and then dried to constant weight at 52°C. Sulfate contents of these oxides designated as T, T1, T2 and T3 were determined by elemental analysis and found to be 0.180%, 6.742%, 20.434% and 25.558% respectively.

Sample T* was prepared by impregnating 100g of the same initial TiO_2 in 250ml of distilled water followed by drying at 52°C. Sample T* as well as the sulfated oxides T, T1,

T2 and T3 were then heat-treated for two hours in air at different increasing temperatures of 335°C, 390°C, 450°C, 500°C and 580°C.

2.2 Techniques

The thermal studies were carried out on a Stanton Redcroft STA-780 thermal analyzer designed to give simultaneous thermogravimetric (TG), differential analysis (DTA) and differential thermogravimetric records (DTG). Determinations were carried out in an atmosphere of flowing N_2, the heating rate being kept at 10°C min^{-1}.

The infrared studies were carried out using a 337 Perkin-Elmer double beam grating spectrophotometer. Solid samples were prepared in the form of a KBr pellet. 2 mg of the compound were mixed with approximately 200 mg of Kbr spectroscopic grade. The mixture was then subjected to a pressure of about 1400 kPa in a hydraulic press. A disc of pure KBr was used in the reference cell.

The X-ray diffraction patterns were obtained by means of a compact x-ray diffraction analyzer system 1840 PHILIPS using Ni filtered Cu Kα radiation.

Adsorption measurements were carried out at 77K using a conventional volumetric technique. Typically, the sample was outgassed overnight at room temperature to residual pressures of ~25x10^{-4} torr. The gas pressures were measured on a mercury manometer. The time required for each point of the adsorption or desorption isotherm to attain equilibrium was between 15-20 minutes.

The number of acidic sites was determined through the adsorption of NaOH from solution. 0.2 g of the solid material were shaken overnight in 10 ml of the base of known normality with concentrations ranging from 0.005 N to 0.2 N. A blank titration was carried out on the base alone. The difference between the blank titration and that with the sample gave the amount adsorbed.

For the catalytic studies methylmethacrylate monomer stabilized by hydroquinone was used after being purified with 20% NaOH to remove the quinone inhibitor. To 125 ml of methylmethacrylate monomer, a 0.25 ml of 2.5N NaHSO$_3$ was added and then to each 1 ml of the solution mixture a 0.025 g of catalyst solid was added. The polymerization process was conducted as reported by Hussein et al[12]. The solution was filtered to remove the solid catalyst and the polymer was precipitated by methanol. Catalytic activity was expressed in terms of polymer percentage yield, calculated as the ratio of the weight of the polymer produced to the weight of the monomer used multiplied by 100.

3. RESULTS AND DISCUSSION

3.1 Thermal Studies

The thermogravimetric and differential thermal analyses show that sample T* and T have a negligible weight loss throughout the heating range from ambient to 900°C with no endothermic or exothermic peaks appearing in the DTA curves. Samples T1, T2 and T3 display endotherms with maxima at around ~280°C, ~360°C and ~420°C. In case of T1 an additional endotherm is observed at 320°C.

The endotherm observed at ~280°C is associated with water coordinatively bound to he titanium dioxide[13]. The endothermic peak appearing at 360°C is indicative of the elimination of chemisorbed ammonia from the titanium oxides[14]. The endothermic peaks appearing at 320°C in the case of T1 and at 420°C in the case of T1, T2 and T3 may be related to the evolution of SO_2 stemming from ionic and differently bound sulfate ligands[6].

TG curves for samples T*, T, T1, T2 and T3 indicate a total weight loss of 0.6%, 0.6%, 10%, 27% and 40% respectively. Representative DTA, DTG and TG curves are shown in Figure 1.

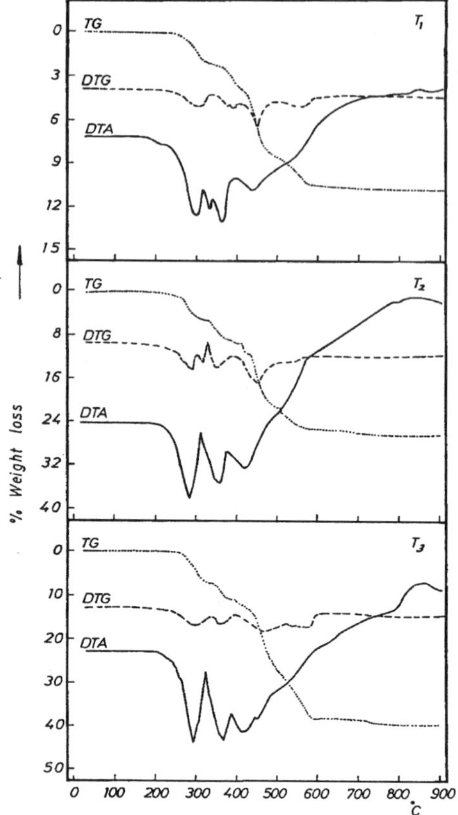

Figure 1 *Thermal analyses of samples T1,T2 and T3*

3.2 Infrared Studies

Infrared studies carried out on the original unheated T*,T, T1, T2 and T3 samples reveal in all cases three peaks at 1400 cm^{-1} ,1635 cm^{-1} and 3450 cm^{-1}. The latter two bands are assigned to the strongly H-bonded H-O-H species and may be due to residual hydroxyl groups or to ligated water in the sample[10]. The band at 1400 cm^{-1} is attributed to the NH_4^+ ion presumably originating from the ammonium sulfate in which the TiO_2 was impregnated [15].

Additional bands are observed at 680 cm^{-1}, 570 cm^{-1} and 350 cm^{-1} and may be attributed to Ti-O stretching modes, whereas the bands appearing at 2360-2320 cm^{-1} may be assumed to be overtone vibrations to the fundamental band appearing at 570 cm^{-1}. In all of the unheated samples T, T1, T2 and T3 additional bands appear at ~1110 cm^{-1}, ~980 cm^{-1} and ~615 cm^{-1} , these are respectively attributed to the υ_3 , υ_1 and υ_4 vibrational frequencies of the free sulfate anion in T_d symmetry [16].

Heat treatment of all samples T, T1, T2 and T3 to 335°C, 390°C, 450°C, 500°C and 580°C leads to a gradual decrease of the 1400 cm^{-1} band due to the elimination of the NH_4^+

ion in the form of ammonia. Representative IR bands for T3 are shown in Figure 2. In case of T1, T2 and T3, heat treatment to 335°C, 390°C and 450°C leads to the appearance of new bands at ~1240cm^{-1}, 1130 cm^{-1}, 1030 cm^{-1} and 980 cm^{-1} which persist to 450°C in the case of sample T1, to 500°C for T2 and to 580°C for sample T3. These bands are attributed to the bidentate bridging sulfate ion in C$_{2v}$ symmetry. The coexistence of monodentate sulfate is not to be precluded as the infrared spectra of the latter overlap with some of the bidentate bridging ligands [16].

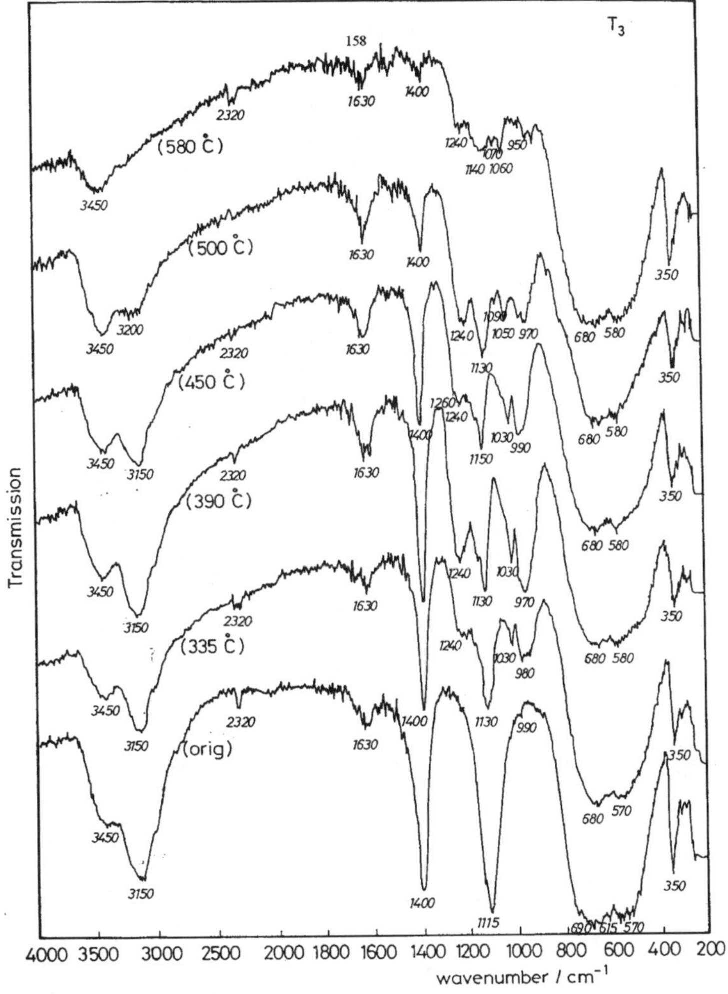

Figure 2 *Representative infrared spectra of T3 and its heat–treated products*

3.3 Adsorption Studies

Nitrogen adsorption isotherms for the various samples were of Type II displaying in some cases a weak hysteresis leading to the development of a Type IV character. The textural characteristics were analyzed by considering the various corresponding "α_s" plots. In the construction of each "α_s" plot the volume of nitrogen adsorbed was plotted as a function of the reduced adsorption "α_s" as measured on a nonporous reference oxide[17]. As explained previously[18], α_s is defined as the amount adsorbed at a particular value of relative pressure/amount adsorbed at a relative pressure p/p_0 equal to 0.4. The α_s plots displayed upward deviations which were indicative of mesoporosity whereas some of the plots suggested the presence of some microporosity [18].

Table 1 summarizes all of the adsorption results. It is seen that in general there is a lowering of the extent of the surface areas as a result of sulfatizing, suggesting some blocking of the pores by the sulfate anions in differently ligated forms.

Table 1 *Nitrogen adsorption and surface acidity data for the T series of oxides*

	Temp. (°C)	BET C-const	S_{BET} (m²g⁻¹)	S_s (m²g⁻¹)	Total pore volume ml g⁻¹	Micropore volume ml g⁻¹	Porosity	No. of acidic sites (μmol/g)
T*	Original	311	9.4	9.5	0.011	0.004	Meso+Micro	44
T	Original	112	9.8	10.0	0.012	0.003	Meso+Micro	101
T1	Original	155	7.0	7.1	0.007	0.003	Micro+Meso	312
T2	Original	328	5.3	5.6	0.006	0.002	Micro+Meso	336
T3	Original	73	5.5	5.2	0.005	0.002	Micro+Meso	581
T*	335	174	8.3	8.4	0.011	0.002	Meso+Micro	5
T	335	54	11.6	10.0	0.010	0.004	Meso+Micro	87
T1	335	116	4.7	4.2	0.004	0.002	Micro+Meso	1994
T2	335	31	4.4	3.6	0.003	0.001	Micro	3323
T3	335	55	2.9	2.6	0.003	0.001	Micro	3949
T*	390	113	9.2	9.1	,10.010	0.002	Meso+Micro	0
T	390	172	8.4	8.1	0.012	--	Nonporous	39
T1	390	36	4.7	3.7	0.004	0.001	Micro+Meso	2029
T2	390	28	6.6	3.3	0.003	0.001	Micro	3950
T3	390	18	2.9	2.1	0.001	0.001	Micro	4480
T*	450	130	8.4	8.2	0.013	0.001	Meso+Micro	20
T	450	83	10.5	9.4	0.012	0.003	Meso+Micro	35
T1	450	131	6.7	6.2	0.010	0.001	Meso+Micro	2196
T2	450	64	5.6	5.4	0.007	0.002	Micro+Meso	4102
T3	450	72	4.0	3.8	0.006	0.001	Micro+Meso	4193
T*	500	161	9.0	9.2	0.010	0.002	Meso+Micro	37
T	500	127	8.6	9.0	0.013	0.002	Meso+Micro	29
T1	500	191	7.6	8.0	0.009	0.002	Meso+Micro	1654
T2	500	118	6.7	6.2	0.009	0.002	Meso+Micro	3510
T3	500	62	6.4	6.2	0.009	0.002	Meso+Micro	4045
T*	580	118	8.8	8.6	0.012	0.002	Meso+Micro	16
T	580	142	8.8	8.4	0.012	0.001	Meso+Micro	21
T1	580	172	8.4	8.5	0.011	0.003	Meso+Micro	42
T2	580	122	8.9	8.7	0.013	0.002	Meso+Micro	139
T3	580	131	11.0	10.2	0.016	0.003	Meso+Micro	427

3.4 Surface Acidity

The total number of acidic sites was determined from the saturation values of the neutralization adsorption isotherms of NaOH [19, 20]. The latter were Langmuir-like in shape. The total number of basic sites was determined by phosphoric adsorption from solution.

Table 1 indicates that the total number of acidic sites expressed in μmol g^{-1} increases with higher sulfate content, and that the highest acidities are observed in the case of samples T2 and T3 heat–treated to 390°C and 450°C. In the latter case the infrared studies indicate that the sulfate ions are predominantly in a bidentate state of ligation.

3.5 X-Ray Diffraction Studies

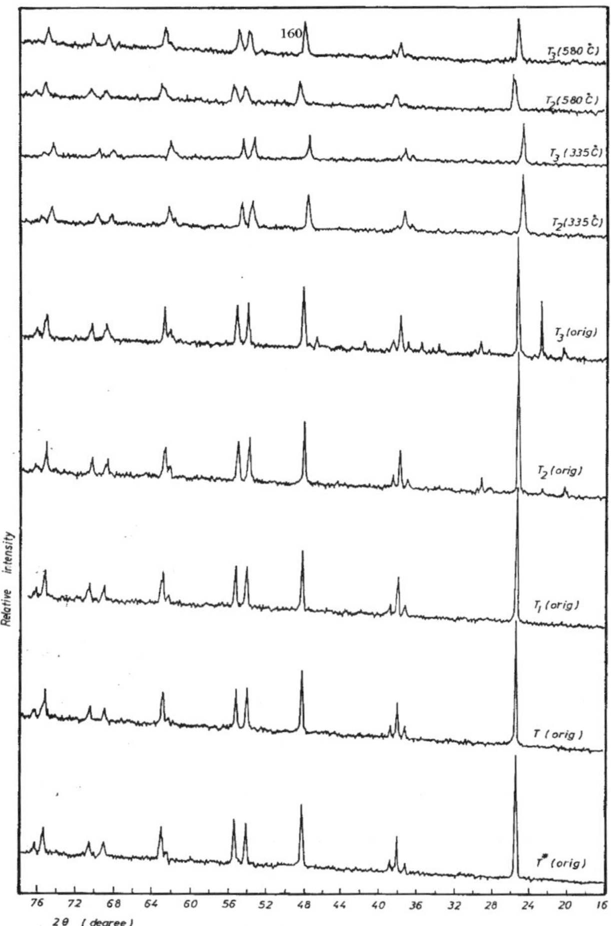

Figure 3 *X-ray diffraction of some samples of the T series*

The X-ray diffraction patterns of T*, T, T1, T2 and T3 and their heat-treated products display the anatase phase. The original samples T2 and T3 also exhibit diffraction peaks belonging to some remaining $(NH_4)_2SO_4$ phase used in the present preparations (most intense peaks are at 2θ = 20.46°, 20.24°, 29.52°, 22.42°). The latter peaks are seen to disappear at the higher heat treating temperatures i.e. 335°C, 390°C, 450°C, 500°C and 580°C. Representative X-ray diffraction patterns are shown in Figure 3.

3.6 Catalytic Activity

Table 2 *Data of catalytic polymerization of methylmethacrylate by the T series of samples*

	Temp. (°C)	No. of acidic sites (μmol/g)	Monomer to polymer % conversion	Polymer Average olecular wt
T*	Original	44	10.2	--
T	Original	101	23.8	36 X 10^4
T1	Original	312	26.0	37 X 10^4
T2	Original	336	51.0	39 X 10^4
T3	Original	581	61.3	41 X 10^4
T*	335	5	9.8	--
T	335	87	21.6	25 X 10^4
T1	335	1994	46.6	50 X 10^4
T2	335	3323	53.3	59 X 10^4
T3	335	3949	65.2	63 X 10^4
T*	390	0	6.7	--
T	390	39	17.5	--
T1	390	2029	29.2	52 X 10^4
T2	390	3950	38.8	65 X 10^4
T3	390	4480	69.9	71 X 10^4
T*	450	20	8.3	--
T	450	35	15.7	--
T1	450	2196	37.9	54 X 10^4
T2	450	4102	50.1	68 X 10^4
T3	450	4193	59.8	69 X 10^4
T*	500	37	12.5	--
T	500	29	13.7	--
T1	500	1654	57.9	47 X 10^4
T2	500	3510	13.5	--
T3	500	4045	48.4	66 X 10^4
T*	580	16	7.9	--
T	580	21	9.7	--
T1	580	42	9.5	--
T2	580	139	9.3	--
T3	580	427	10.7	--

The catalytic activity towards the polymerization of methylmethacrylate was measured for all of the original and heat-treated to 335°C, T, T1, T2 and T3 samples. For higher heating temperatures (i.e. 390°C, 450°C, 500°C and 580°C) the catalytic activity was only measured for samples T1,T2 and T3. Table 2 reports the percent conversion of the methylmethacrylate monomer to the polymer as well as the obtained average molecular weight of the latter. It is seen that both the average molecular weight as well as the percent conversion increase with the sulfate content in these oxides and are closely related to the sulfated ion in bidentate form of ligation.

4 CONCLUSIONS

Sulfated oxides of titanium exhibit an acidity which increases with increasing sulfate content. There is a lowering of the extent of the surface areas as a result of sulfatizing, suggesting some blocking of the pores by the sulfate anions in differently ligated forms. A bidentate state of ligation favors the generation of a higher acidity which in turn improves the catalytic properties of titanium oxides for the polymerization of methylmethacrylate. These results corroborate the mechanism put forward by Saur et al [21] who suggested that sulfated TiO_2 had the structure $(Ti_3O_3)S{=}O$ and that upon the addition of water the following would take place:

$$
\begin{array}{ccc}
\begin{array}{l} M{-}O \\ \ \ \ \ \ \ \searrow \\ M{-}O{-}S{=}O \ + \ H_2O \\ \ \ \ \ \ \ \nearrow \\ M{-}O \end{array}
&\longrightarrow&
\begin{array}{l} M{-}OH \\ M{-}O \ \ \ \ \ OH \\ \ \ \ \ \searrow \ \ \ \nearrow \\ \ \ \ \ \ \ S \\ \ \ \ \ \nearrow \ \ \ \searrow \\ M{-}O \ \ \ \ \ O \end{array}
&\longrightarrow
\begin{array}{l} M{-}O \ \ \ \ \ O \\ \ \ \ \ \searrow \ \ \ \nearrow \\ \ \ \ \ \ \ S \\ \ \ \ \ \nearrow \ \ \ \searrow \\ M{-}O \ \ \ \ \ O \end{array} \Bigg\} \, H
\end{array}
$$

Such a mechanism would account for the increase in acidity with increasing sulfate content either in a *bidentate state of ligation* or in *a free state*. The latter would also probably eventually become ligated to the surface when present in solution and would lead to the observed increase in acidity.

References

1. T. Jiang, Q. Zhao, M. Li, H. Yin, *J. Hazard. Mater.*, doi:10.1016/j.jhazmat.2008.02.008.
2. S.J. Yang, A.M. Bai, J.T. Sun, *J. Zhejiang Univ. Sci. B.*, 2006, **7**, 553.
3. C. Xie, Z. Xu, Q. Yiang, et al., *J. of Mol. Catal. A.*, 2004, **217**, 193.
4. P. Ciambelli, D. Sannino, V. Palma, V. Vaiano, *Stud. Surf. Sci. Catal.*, 2005, **155**, 179.
5. D. S. Muggli, L. Ding, *Appl. Catal. B*, 2001, **32**, 181.
6. A.K. Dalai, R. Sethuraman, S.P.R. Katikaneni, R.O. Idem, *Ind. Eng. Chem. Res.*, 1998, **37**, 3869.
7. S. Segunan, C.R.K. Seena, *Ind. J. Chem.*, 1999, **38A**, 947.
8. L. Rongsheng, C. Jingfeng, Z. Wuyang, Y. Hua, Z. Zhiming, W. Quan, *React. Kinet. Catal. Lett.*, 1992, **48**, 483.
9. S.K. Samantaray, T. Mishra, K.M. Parida, *J. Mol. Catal.*, 2000, **156**, 267.
10. C. Azer, S. Selim, N. Yacoub and J. Ragai, *Proceedings of COPS IV*, 1997, 306.
11. C. Azer, S. Selim, N. Yacoub and J. Ragai, *Proceedings of the Sixth International Conference on Fundamentals of Adsorption*, 1998, 306.
12. G. Hussein, N. Sheppard, M. Zaki, R. Fahim, *Chem. Soc. Farad. Trans.*, 1989, **85**, 1723.
13. J. Ragai, *J. Chem. Technol. Biotechnol.*, 1987, **40**, 75.
14. H. Nakabayashi, N. Kakuta, A. Ueno, *Bull .Chem. Soc. Jpn.*, 1991, **64**, 2428.
15. N.D. Parkyns, *Chemisorption and Catalysis*, Hepple Publication, 1970.
16. K. Nakamoto, *Infrared and Raman Spectra of Inorganic and Coordination Compounds*, Wiley, New York, 1986.
17. J. Ragai, K.S.W. Sing, R. Mikhail, *J. Chem. Technol. Biotechnol.*, 1980, **31**, 1.
18. S.J. Gregg, K.S.W. Sing, *Adsorption Surface Area and Porosity*, 2nd edition, Academic Press, 1982.
19. H.P. Boehm, *Discuss. Faraday Soc.*, 1971, **52**, 264.
20. A.R. Ramadan, N. Yacoub, S. Bahgat, J. Ragai, *Colloids Surf. A*, 2007, **302**, 36.
21. O. Saur, M. Bensitel, A.B.M. Saad, J.C. Lavalley, C.P. Tripp, B.A. Morrow, *J. Catal.*, 1986, **99**, 104.

A COMPARATIVE STUDY OF SULFATED AND PHOSPHATED TITANIUM OXIDES PREPARED BY HYDROLYSIS FROM TITANIUM ETHOXIDE

H. Amin[1], N. Yacoub[1], A. R. Ramadan[1,2] and J. Ragai[1,2]

[1]The Youssef Jameel Science and Technology Research Center, American University in Cairo, 113 Kasr El Aini Street, P.O. Box 2511, Cairo 11511, Egypt
[2]The Department of Chemistry, American University in Cairo, 113 Kasr El Aini Street, P.O. Box 2511, Cairo 11511, Egypt

1 INTRODUCTION

Titanium dioxide super-acids have been the subject of numerous investigations, attracting wide attention for their catalytic activity for a large number of processes such as isomerization of hydrocarbons, esterification and alkene polymerization[1-9]. They have mostly been prepared by sulfate enrichment of the oxide matrix. Sulfate groups, however, are lost when the oxide matrix is subjected to high temperatures (500°C and above), generally resulting in a loss of acidity[10]. Another possible method for obtaining titanium dioxide super-acids entails the enrichment of the metal oxide with phosphate ions. This has been the subject of a number of investigations[11-15]. Phosphate ions were found to remain within the oxide matrix up to high temperatures[11-12].

Within this context, this study entails a comparative investigation of the textural and acidic properties of sulfated and phosphated titanium dioxide. The metal oxide has been prepared by hydrolysis from titanium ethoxide, and impregnated with sulfate and with phosphate ions, then subjected to heat treatment at different temperatures till 800°C.

2 EXPERIMENTAL

Reagents, Apparatus and Techniques

The titanium ethoxide (in excess ethanol) and ammonium sulfate (99.999% pure) were obtained from Aldrich. Ammonium hydrogen phosphate (99.4% pure) was obtained from Fisher. Sample surface areas were determined using a Micromeretics ASAP 2020. A Thermolyne 48000 furnace was used for sample heating and a Shimadzu DTG-60 for thermal analysis. Infrared studies were carried out using a 337 Perkin-Elmer double beam grating spectrophotometer, with the solid samples prepared in the form of KBr pellets. 2 mg of the compound were mixed with approximately 200 mg of KBr (spectroscopic grade), and the mixture then subjected to a pressure of about 1400 kPa in a hydraulic press.

Surface acidity was determined volumetrically by the adsorption of sodium hydroxide (Aldrich 99.99% purity) from solutions of different concentrations. About 0.2 g of the solid oxide were shaken in 10 ml of the base for 6 hours, then left overnight. For each sample, a blank run was carried out first on the base alone and the difference between the blank run and that with the sample gave the amount adsorbed[16].

Sample Preparation

The titanium dioxide samples were prepared by dispersing the titanium ethoxide in ethanol (95%) and adding de-ionized water drop-wise with continuous stirring till a white precipitate formed. Excess de-ionized water was then added to ensure complete precipitation and facilitate filtration[17]. After filtration, the pH of the filtrate was determined to be 4.46. The precipitate was washed free of alcohol, then dried at $52^\circ C$ till constant weight.

The sulfated and phosphated samples, TS and TP respectively, were prepared by impregnating titanium dioxide with ammonium sulfate and ammonium hydrogen phosphate, respectively, through incipient wetting, followed by drying at $52^\circ C$ till constant weight. Amounts of titanium dioxide and concentrations of ammonium sulfate and ammonium hydrogen phosphate solutions used were determined to give 5% by weight sulfur and 5% by weight phosphorus contents respectively. A blank sample, TW, was also prepared by incipient wetting of titanium dioxide with de-ionized water.

The samples TW, TS and TP were then heat treated in air for 2 hours at temperatures of $80^\circ C$, $200^\circ C$, $300^\circ C$, $400^\circ C$, $600^\circ C$, $700^\circ C$ and $800^\circ C$ resulting in three sample series. These temperatures were chosen based on the thermal behaviour of the samples.

3 RESULTS AND DISCUSSION

Thermal studies

Samples TS and TP behave somewhat differently when heat treated. For the TS sample, a gradual weight loss of ~12% occurs till $400^\circ C$, followed by a further loss of ~10% occurring between $400^\circ C$ and $550^\circ C$. The latter is associated with the loss of the sulfate groups[10], possibly as SO_2, and is reflected in a broad endotherm from about $300^\circ C$ till about $700^\circ C$. The TP sample, on the other hand, shows a similar gradual weight loss of ~12% occurring till $400^\circ C$, after which sample weight remains virtually unchanged, demonstrating that the phosphate groups are not lost by heat treatment. This is corroborated by the infrared spectra, and also reflected in the surface acidity measurements carried out.

Infrared Results

Samples TW, TS and TP exhibit a peak at ~3400 cm^{-1} attributed to the combined symmetric and asymmetric stretching modes of chemisorbed water[18]. In addition, samples TS and TP exhibit a second band at ~3150 cm^{-1} associated with the symmetric and asymmetric stretching modes of adsorbed ammonia[19]. The original (unheated) TS and TP samples exhibit a peak at ~1400 cm^{-1} characteristic of the NH_4^+ group and associated with the bending vibration of H-N-H angle in the ammonium ions[19].

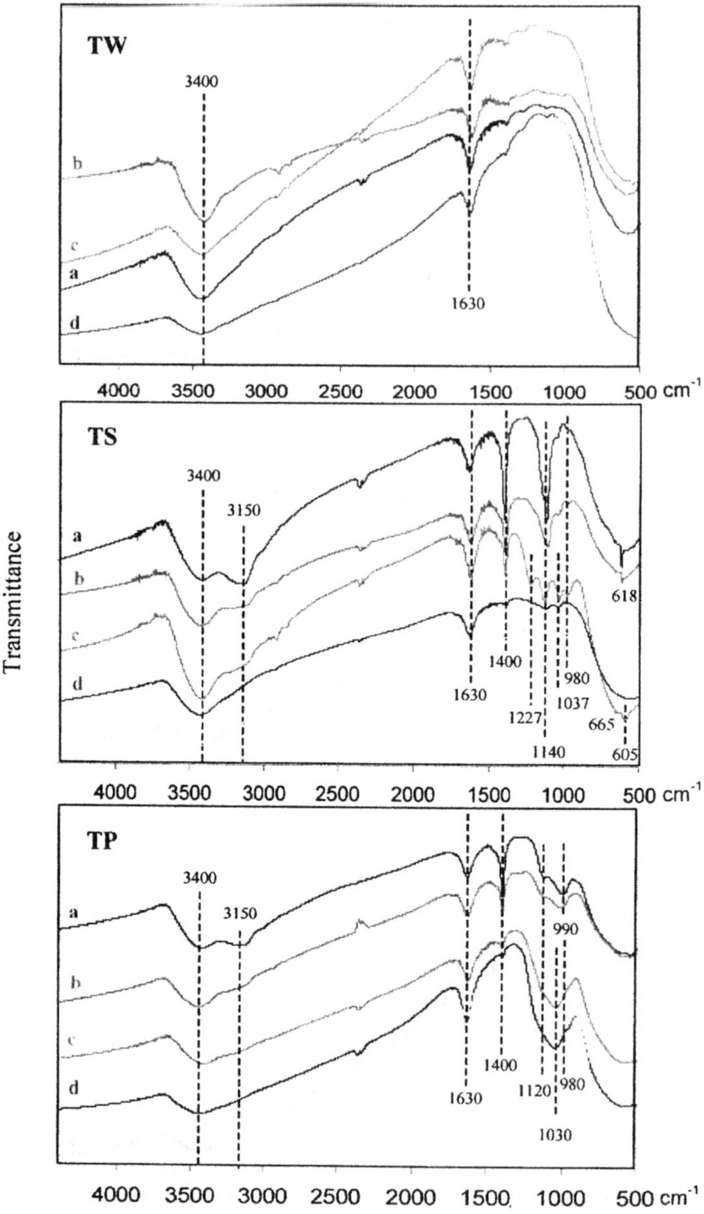

Figure 1 *Selected infrared spectra for TiO$_2$ samples: (a) Original temperature, (b) 200°C, (c) 400°C, (d) 600°C, (e) 700°C*

Both the ~3150 cm^{-1} and the ~1400 cm^{-1} peaks arise from the use of ammonium salts in the impregnation of the oxide samples with sulfate and phosphate ions. These peaks do not appear in the blank sample TW. All three samples also show a peak at ~1630 cm^{-1} associated with the bending mode of molecular water[20].

For the sample TS, three bands also appear at 1140 cm^{-1}, 980 cm^{-1} and 618 cm^{-1}. These are attributed to the v_3, v_1 and v_4 vibrational frequencies of the free sulfate ion in T_d symmetry[18]. As for the sample TP, two bands appear at 990 cm^{-1} and 1120 cm^{-1}. The former band is attributed to the v_1 vibrational frequency of the free phosphate ion in T_d symmetry. The latter band is possibly the 1080 cm^{-1} band, characteristic of the v_3 vibration of the phosphate ion, shifted to lower wavelength as a result of hydrogen bond interactions in the host metal oxide lattice[21].

Heat treatment led to the decrease in intensity of the 3400 cm^{-1} and the 1630 cm^{-1} bands for the TW, TS and TP samples, and of the 3150 cm^{-1} 1400 cm^{-1} for the TS and TP samples. These decreases are associated with the loss of water and ammonia, respectively, with increasing temperature. Heat treatment also led to the appearance of bands at 1227 cm^{-1}, 1138 cm^{-1} and 1037 cm^{-1} for TS sample, starting at 200°C. These bands are attributed to sulfate ion in a bidentate state of ligation, and correspond to the lowering of the T_d symmetry of the sulfate ion to C_{2v} by complex formation[18]. In addition, the splitting of the 618 cm^{-1} band, starting at 300°C, into two bands at ~665 cm^{-1} and ~605 cm^{-1} suggests the presence of a chelated sulfate ligand[18]. The decrease of all these bands with increasing temperature and their virtual disappearance at temperatures above 700°C is concomitant with the loss of the sulfate ions with increased temperatures.

For sample TP, heat treatment led to the gradual appearance, starting 400°C, of bands at ~1120 cm^{-1}, ~1030 cm^{-1} and ~980 cm^{-1}, attributable to the phosphate in a bidentate state of ligation[11], and corresponding to the lowering of the T_d symmetry of the phosphate ion to C_{2v} by complex formation[18]. These are persistent up to 800°C.

For the TW, TS and TP samples, two small peaks appear at ~2850 cm^{-1} and ~2930 cm^{-1}. These are attributed to the methylenic group[4,22], probably remnant from the starting material, titanium ethoxide. With heat treatment, these peaks appear to develop parallel with the weight loss of the samples, then virtually disappear at temperatures above 600°C.

Surface Acidity

Surface acidity values, expressed in μmol/g for samples TW, TS and TP heat treated at different temperatures, are presented in table 1. They indicate that for the TW series surface acidity decreases with increasing temperatures, probably due to the elimination of bridging hydroxo groups with heat treatment.

TS and TP series show interesting trends for surface acidity values. For the TS series, the increase in surface acidity is associated with the appearance of the bidentate sulfato group. Indeed, the highest acidity values of this series (3136 μmol/g) occurs for the sample heat treated at 400°C, with a predominance of the bidentate sulfato ligand. Heat treatment at higher temperatures is seen to result in a decrease of surface acidity for TS samples, related to the gradual loss of the sulfato groups. For the TP series, the increase in surface acidity is associated with the appearance of the bidentate phosphato group, and is seen to increase with increasing heat treatment till the temperature value of 700°C.

Onc noteworthy observation is the persistence of high surface acidity values for the phosphated samples with heat treatment at higher temperatures compared to the sulfated samples. Surface acidity for values for TP(600) and TP(700) are significantly higher than those of TS(600) and TS(700). Figure 2 is a schematic presentation of the variation of surface acidity values with temperature for the TW, TS and TP series.

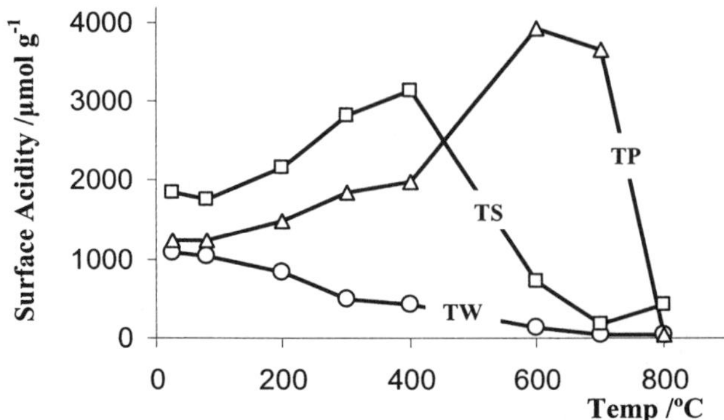

Figure 2 *Variation of surface acidity with heat treatment for the different TiO$_2$ samples*

Adsorption Measurements

Representative adsorption isotherms with the corresponding α_s plots are represented in Figure 3. α_s is defined as (amount adsorbed)/(amount adsorbed at p/p° = 0.4)[23]. Table 1 summarizes the analysis of the data of the nitrogen adsorption isotherms for the TW, TS and TP samples series. Different trends are observed for each of these series.

For the TW samples, the surface area is seen to initially increase with heat treatment, then slowly decrease till 400°C, after which the decrease is significant. This is associated with an initial loss of water at lower temperatures, together with a loss of the hydroxo groups of the metal oxide matrix which continues till higher temperatures. Beyond 400°C, the latter becomes significant, with an observed considerable decrease of surface area due to increased sintering. This trend is reflected in the variation of surface acidity of the samples. Associated with this variation of surface area, an increase in the total pore volume till 400°C is observed, followed by a decline at higher temperatures. A mixture of microporosity and mesoporosity is observed for these samples, with the reduction of the micropore volume and the appearance of macroporosity at higher temperatures.

For the TS samples, the trend in surface area variation with heat treatment is less clear. This is probably due to the presence of different processes occurring with increased temperature. These include the loss of water and of ammonium ions in the form of ammonia till about 400°C; the bidentate attachment of the sulfato groups to the surface, generally associated with a decrease in surface area due to pore blocking[10] and seen in the infrared spectra to be predominant at 300°C and 400°C; the loss of the sulfato groups starting 400°C; and the loss of the hydroxo groups of the metal oxide matrix. The surface area is, however, seen to decrease with increasing temperature beyond 400°C, with a predominance of mesoporosity. This can be associated with the loss of the sulfato groups and the hydroxo groups, as reflected in the infrared spectra and the significant decrease of surface acidity.

TP samples exhibit an initial small increase in surface area with heat treatment, followed by very limited changes between the temperatures of 200°C and 600°C. At 700°C a decrease is observed, which becomes significant at 800°C. The effect of the development of a bidentate attachment of the phosphato groups to the surface, observed in the infrared spectra starting 300°C, on the reduction of the surface area due to pore blocking is not observed. This, together with the limited variation of surface area values between 200°C and 600°C may be due to the different processes taking place with increasing temperature. These processes encompass the loss of water and ammonia, the loss of the hydroxo groups of the metal oxide matrix, in addition to the bidentate ligation of the phosphato group. The samples exhibit microporosity and mesoporosity, with a significant decrease in microporosity and the appearance of macroporosity at 800°C.

One interesting observation is that the presence of the phosphato groups in a bidentate state of ligation results in significantly higher values of surface area at temperatures of 600°C and 700°C for the TP samples compared to the TW and TS samples.

Table 1 Nitrogen adsorption and surface acidity data for the different TiO_2 samples

	V_m /ml g^{-1}	BET C-const	S_{BET} /m^2 g^{-1}	S_s /m^2 g^{-1}	Total pore volume /ml g^{-1}	Micropore volume /ml g^{-1}	Porosity	Surface acidity /μmol g^{-1}
TW	23.8	182	103.4	106.8	0.1113	0.0080	micro + meso	1074
TS	19.7	184	85.7	89.0	0.0941	0.0044	micro + meso	1853
TP	21.3	227	92.5	98.5	0.0955	0.0132	micro + meso	1248
TW(80)	49.2	202	214.2	231.5	0.2206	0.0198	micro + meso	1038
TS(80)	19.7	165	85.8	89.1	0.0954	--	meso	1754
TP(80)	22.3	228	97.1	103.6	0.1040	0.0127	micro + meso	1249
TW(200)	45.0	194	196.0	205.2	0.2306	0.0121	micro + meso	837
TS(200)	24.2	181	105.2	109.9	0.1072	0.0068	micro + meso	2171
TS(200)	22.9	189	99.9	104.0	0.1070	0.0115	micro + meso	1381
TW(300)	36.1	191	157.3	165.0	0.2420	0.0136	micro + meso	489
TS(300)	13.9	180	60.5	63.1	0.0769	0.0061	micro + meso	2826
TS(300)	24.1	177	104.9	109.2	0.1158	0.0100	micro + meso	1853
TW(400)	28.9	229	125.9	134.3	0.2211	0.0159	micro + meso	436
TS(400)	23.8	221	103.4	110.4	0.1475	0.0114	micro + meso	3136
TS(400)	24.8	138	108.1	108.6	0.1234	0.0073	micro + meso	1998
TW(600)	7.0	263	30.4	32.0	0.0876	--	meso	143
TS(600)	16.5	274	71.9	75.7	0.2165	--	meso	717
TS(600)	24.5	189	106.6	111.6	0.1462	0.0103	micro + meso	3940
TW(700)	0.8	253	3.3	3.5	0.0165	0.0005	micro + meso	50
TS(700)	7.0	266	30.3	31.9	0.1539	--	meso	179
TP(700)	18.5	201	80.4	84.5	0.1311	0.0066	micro + meso	3666
TW(800)	0.3	192	1.2	1.3	0.0028	0.0002	meso + macro	40
TS(800)	3.8	201	16.5	17.2	0.0531	--	meso	410
TP(800)	0.4	232	1.9	2.0	0.0060	0.0003	meso + macro	38

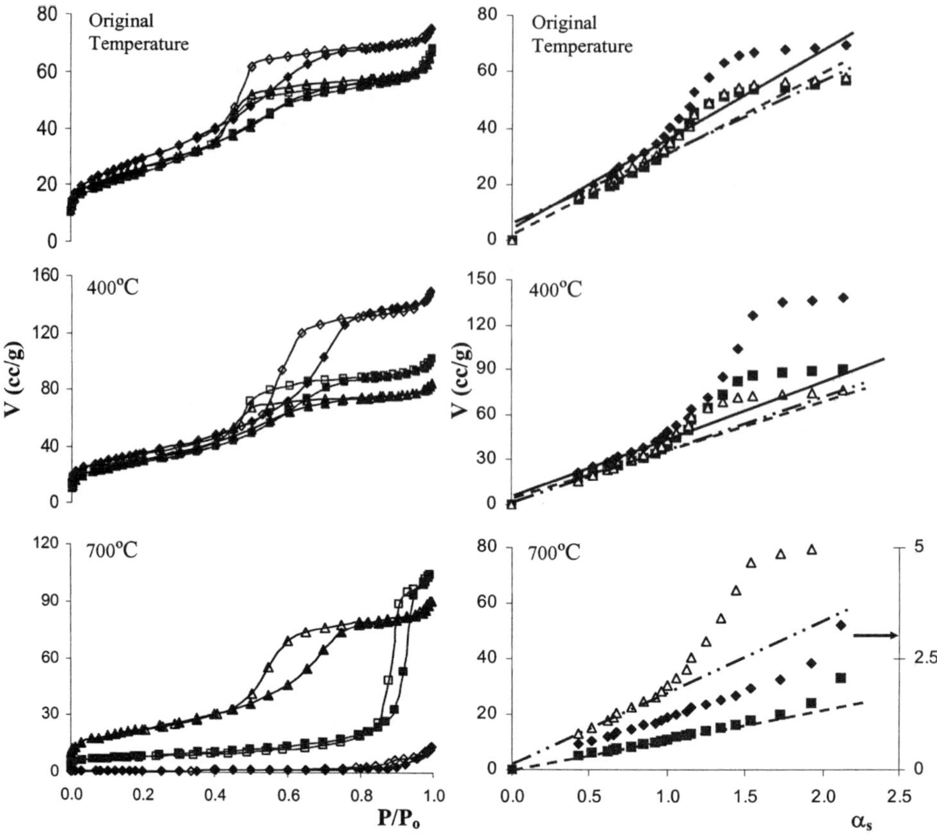

Figure 3 *Adsorption isotherms and α_s plots for the different TiO$_2$ samples: TW ($\blacklozenge$/$\lozenge$); TS ($\blacksquare$/$\square$); TP ($\triangle$/$\blacktriangle$)*

4 CONCLUSIONS

The titanium oxides free of sulfate and phosphate groups exhibit increased surface areas and mixed microporosity and mesoporosity up till 600°C. At higher temperatures, surface area values decrease with a predominance of mesoporosity and some macroporosity. Observed acidity values decrease with heat treatment, as expected with the elimination of bridging hydroxo groups with heating.

For the sulfated oxides, the sulfate groups, present as free sulfate below temperatures of 200°C, develop into predominantly bidentate ligands with heat treatment. This is associated with an enhancement of surface acidity. At temperatures beyond 400°C, the loss of the sulfate groups takes place, with an marked decrease in surface acidity. This is coupled with a reduction of surface area and microporosity.

For the phosphated samples, the phosphate groups, present as free phosphate below temperatures of 300°C, develop into bidentate ligands with heat treatment. Similarly to the sulfated oxides, this is associated with an enhancement of surface acidity. However the observed acidity values are lower for the phosphated samples than the sulfated samples at the corresponding temperatures till 400°C. Beyond this temperature, and contrary to the sulfated oxides, the phosphated oxides still exhibit significant enhancement of surface acidity, up to 800°C where a drastic decrease is observed. Even at these high temperatures, the infrared spectra show that the phosphate bidentate groups are still present in the titanium oxide.

References

1. L.K. Noda, R.M. de Almeida, N.S. Gonçalves, L.F.D. Probst, O. Sala, *Catal. Today*, 2003, **85**, 69.
2. M. Waqif, J. Bachelier, O. Saur, J.C. Lavalley, *J. Mol. Catal.*, 1992, **72**, 127.
3. C. Azer, S. Selim, N. Yacoub, J. Ragai, *Proceedings of COPS IV*, 1997, 306.
4. C. Azer, S. Selim, N. Yacoub, J. Ragai, *Proceedings of the Sixth International Conference on Fundamentals of Adsorption*, 1998, 306.
5. S. K. Samantaray, T. Mishra, K. M. Parida, *J. of Mol. Catal. A: Chem.* 2000, **156**, 267.
6. A. Mantilla, G. Ferrat, A. López-Ortega, E. Romero, F. Tzompantzi, M. Torres, E. Ortíz-Islas, E. Ortiz-Islas, T. López, J. Navarrete, X. Bokhimi, R. Gómez, *J. of Mol. Catal. A: Chem.*, 2005, **228**, 345.
7. L.K. Noda, R.M. de Almeida, L.F.D. Probst, N.S. Gonçalves, *J. of Mol. Catal. A: Chem.*, 2005, **225**, 39.
8. R. Gómez, *J. of Mol. Catal. A: Chem.*, 2005, **228**, 333.
9. V.M. Benítez, C.R. Vera, C.L. Pieck, F.G. Lacamoire, J.C. Yori, J.M. Grau, J.M. Parera, *Catal. Today*, 2005, **107-108**, 651.
10. A.R. Ramadan, N. Yacoub, S. Bahgat, J. Ragai, *Colloids Surf. A*, 2007, **302**, 36.
11. S.K. Samantaray, K. Parida, *J. of Mol. Catal. A: Chem.*, 2001, 176, 151.
12. S.K. Samantaray, K. Parida, *Appl. Catal. A*, 2001, 220, 9.
13. Z.C. Wang, H.F. Shui, *J. of Mol. Catal. A: Chem.*, 2006, 263, 20.
14. L. Korosi, I. Dekany, *Colloids Surf. A*, 2006, 280, 146.
15. E. Ortiz-Islas, R. Gomez, T. Lopez, J. Navarrete, D.H. Aguilar, P. Quintana, *Appl. Surf. Sci.*, 2005, 252, 807.
16. H.P. Boehm, *Discuss. Faraday Soc.*, 1971, **52**, 264.
17. B. E. Yoldas, *J Mater. Sci.*, 1986, **21**, 1087.
18. K. Nakamoto, *Infrared and Raman Spectra of Inorganic and Coordination Compounds*, Wiley, New York, 1986.
19. M.L. Hair, *J. Phys. Chem.*, 1970, **74**, 1290.
20. N.D. Parkyns, *Chemisorption and Catalysis*, Hepple Publication, 1970.
21. S.F. Lincoln, D.R. Stranks, *Aust. J. Chem.*, 1968, **21**, 37.
22. M. Nagao, Y. Suda, *Langmuir*, 1989, **5**, 42.
23. S.J. Gregg, K.S.W. Sing, *Adsorption Surface Area and Porosity*, 2nd edition, Academic Press, 1982.

CORRELATION OF SURFACE TEXTURE, CATALYTIC ACTIVITY AND CRYSTAL MORPHOLOGY FOR TERNARY MIXED METAL OXIDES

Christothea Attipa and Charis R. Theocharis*

Porous Solids Group, Department of Chemistry, University of Cyprus, P.O. 20537, 1678, Nicosia, Cyprus

1 INTRODUCTION

In recent years cerium dioxide (CeO_2) has been increasingly used in catalytic applications, as a promoter, for example, in the so-called three-way catalysts (TWCs) but also as an oxygen ion conductor in solid oxide fuel cells (SOFCs) and oxygen monitors.[1] Also, ceria is widely used for the oxidation of CO by O_2, for the selective catalytic reduction of SO_2 by CO, as an excellent support for precious metals, as UV absorbent and filter, as buffer layer with silicon wafer, as component in solid oxide fuel cells, as catalyst in chemical-mechanical polishing (CMP) processes, etc.[2-4] The factor that makes ceria such a versatile material, is its capacity for oxygen storage and diffusion through lattice defects in CeO_{2-x}.[1, 5, 6] Several metal- doped cerium oxides exhibit high oxygen ionic conductivity which makes them interesting materials for many applications.[7]

Vanadium-oxide-based solids are widely used in oxidation reactions such as partial oxidation of methanol to formaldehyde and methyl formate, selective oxidation of *o*-xylene to phthalic anhydrite, oxidation of butane to acetic acid, oxidation of organic compounds, selective oxidation with rupture of C-C bond and also are effective for the selective catalytic reduction (SCR) of nitric oxide NO, etc.[1, 8, 9] Samples of vanadia on ceria exhibit an interesting performance in oxidative dehydrogenation (ODH) reactions, the oxidative conversion of propane to propylene and ethylene, and the preferential oxidation of CO to CO_2, and others.[10]

Cu-Ce mixed oxides exhibit high activity for CO oxidation, SO_2 and NO reduction, and total oxidation of hydrocarbons. These materials are known as oxygen storage promoters in gas catalytic reactions, as precursors in gas catalytic reactions, for methanol synthesis and catalytic wet oxidation (CWO) of phenol.[2, 11, 12] The oxides CeO_2, V_2O_5 and CuO are used as catalysts in depollution reactions such as the oxidation of diesel soot particles.[1, 6, 13, 14] However, the effect of combining two different heteroatoms in the ceria structure has been scarcely investigated.[14]

The catalytic test of isopropanol (IPA) decomposition is used for the study of surface acidity of solids. In general: (a) the nature of the solid or the type of the alcohol influences the mechanism of catalytic reaction, (b) the type of the atmosphere where the catalytic reaction is held (oxidant or inactive) and (c) the reaction temperature of the IPA decomposition affects directly the selectivity of the products.[15] The main products of reaction are acetone, propene or diethylether.[16, 17] Even if the reaction of IPA decomposition is used widely in the past few years its application in ternary oxides of the group V-Cu-Ce and V-Fe-Ce is not reported anywhere in the bibliography. The aim of this study, which is part of a longer-term investigation of the properties of metal oxides at the University of Cyprus, was to examine the chemistry of the surface in terms of surface acidity (or basicity), as well as on bulk properties such as crystal structure and morphology.[18]

2. EXPERIMENTAL

The mixed cerium oxide samples were prepared by the co-precipitation method from aqueous solutions of the appropriate salts, using 1M ammonia solution as base. The metal precursors used were cerium (IV) ammonium nitrate, ammonium metavanadate (V) and copper (II) nitrate or iron (III) nitrate. The concentrations of all the solutions used were 0.01M throughout. The appropriate solutions were mixed in the respective ratios, to achieve the desired results. The precipitate was left to stand in the mother liquor for 24h, centrifuged and dried at 373K for 24h.

Six ternary mixed cerium oxides were prepared, namely $V_{0.05}Fe_{0.05}Ce_{0.9}O_2$ (A), $V_{0.05}Fe_{0.05}Ce_{0.9}O_2$ (363K) (B), $V_{0.05}Fe_{0.05}Ce_{0.9}O_2$ (ageing 6h) (C), $V_{0.1}Fe_{0.3}Ce_{0.6}O_2$ (D), $V_{0.1}Cu_{0.1}Ce_{0.8}O_2$ (E) and $V_{0.05}Cu_{0.35}Ce_{0.6}O_2$ (F). The mixed oxide $V_{0.05}Fe_{0.05}Ce_{0.9}O_2$ (B) was heated during the precipitation at 363K for 2.5 hours. The mixed oxide $V_{0.05}Fe_{0.05}Ce_{0.9}O_2$ (C) was left to stand in the mother liquor for 6h.

The catalytic decomposition of isopropanol (IPA) was performed in a bench-scale flow reactor. The samples were calcined in flowing air in the 673K for 2 hours before the catalytic study. A quantity of calcined solid sample (m~ 0.25g) was placed in a PFR (plug flow reactor) glass reactor. Before the catalytic experiments, all solids were heated at 473K for 2 h under helium flow. Analyses of the reactants and products were carried out by sampling 1 cm^3 of gas in a Shimadzu GS-15A gas chromatograph equipped with FID detector.[19-21] The main products of reaction were acetone and propene. The partial pressure of IPA was about 43 mbar. The experiments were done in a temperature range of 333 to 673K.

The ternary mixed cerium oxides were characterized both before and after calcination with various techniques. Surface areas (BET method) were obtained from nitrogen adsorption isotherms at T=77K using a Micromeritics ASAP 2010 apparatus. The analysis of pore distribution was performed using the DFT- plus software. Structure characterization was carried out using a Shimadzu XRD-6000 X-ray diffraction unit with Cu Kα radiation. The data were collected in the 2θ range 10-80^0 with scan rate 2^0/min.. The morphology was examined by scanning electron microscopy (SEM) using a Jeol JSM-5600 microscope.

3. RESULTS AND DISCUSSION

Figure 1 presents the diagrams of IPA conversion as function of temperature. The type of the ternary oxide used influences the conversion of IPA. At the lower end of the temperature range used the highest activity was presented by sample F. The extent of conversion for samples A, B, G, D and E were 96, 84, 62, 93 and 92% respectively. The diverging behavior exhibited by sample F could possibly be attributed to the increased copper content, a metal promoting dehydrogenation at low temperatures.

It is noteworthy that sample F presents the highest conversion of IPA, but has the lowest BET surface area, while sample C that presented the lower conversion has the highest BET surface area (Figure 2). Consequence of those observations is that the extent of catalytic decomposition of IPA in ternary oxides does not depend on their BET surface.

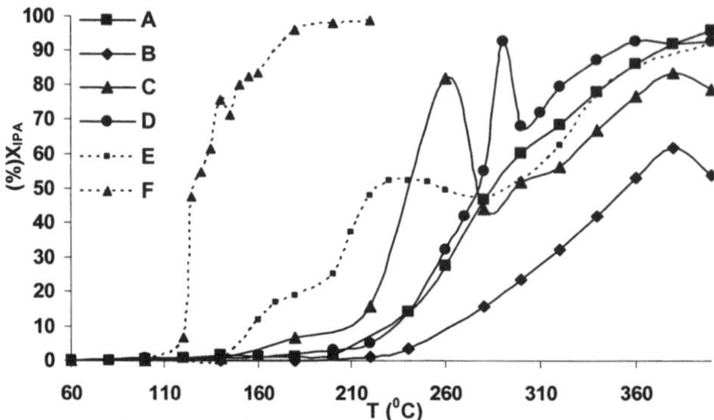

Figure 1 *Comparative diagram of isopropanol conversion ($\%X_{IPA}$) with temperature*

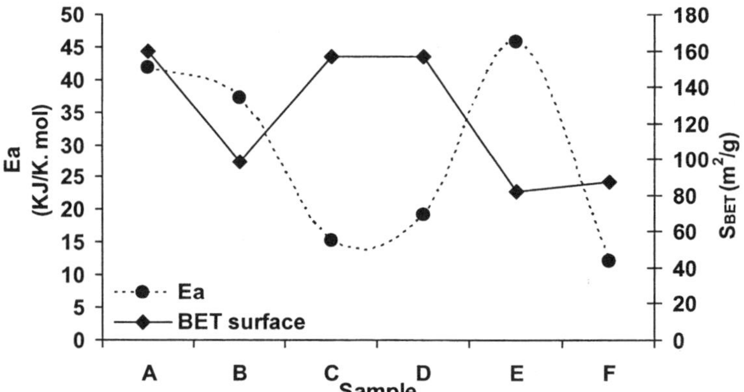

Figure 2 *Comparative diagrams of activation energy (Ea) for the total conversion of IPA and special surface (BET) for samples A, B, C, D, E, F, calcined at 673K*

Depending on the reaction temperature, different products are obtained. For all samples studied, at low temperatures (T<553K) the main product obtained was acetone, while at higher temperatures acetone is obtained together with propene. Simultaneously, in certain instances, trace amounts of diethylether (DIE, <3%, Figure 3 (a, and b)) were also observed.

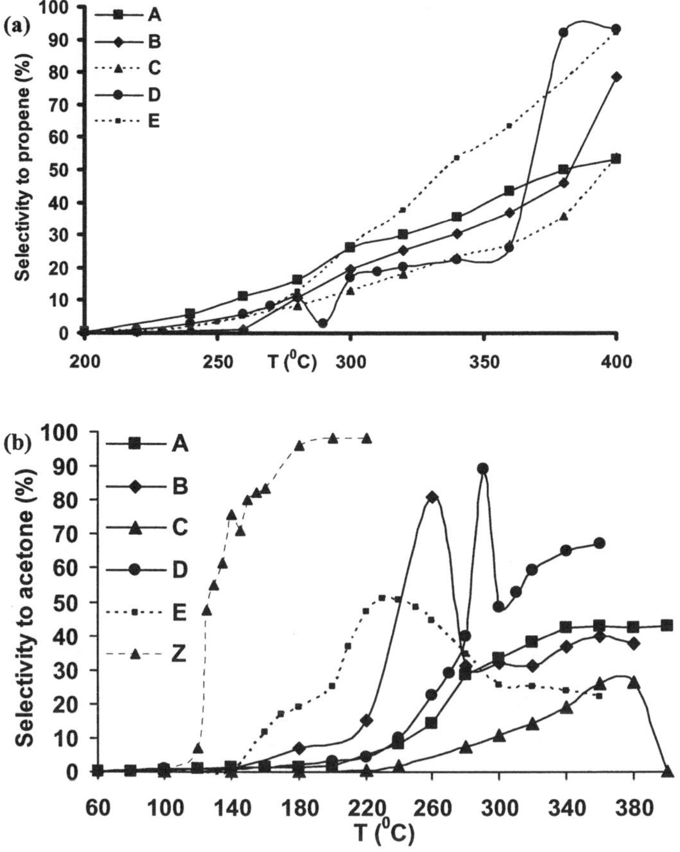

Figure 3: *Comparative diagrams (a) of the selectivity to propene and (b) of the selectivity to acetone. Z denotes pure ceria*

It is remarkable that the highest selectivity in acetone is presented by the copper containing samples. It is also obvious that the highest variance in catalytic activity is presented by the samples prepared by a method different than the standard one. For example, $V_{0.05}Fe_{0.05}Ce_{0.9}O_{2-y}$ which was aged for 6 rather than 24h. Also, increasing the precipitation temperature to 363K, has resulted in a change of catalytic behaviour.

According to Figure 3a, and Figure 3b the decomposition of IPA to acetone and propene (and traces of ether) is independent from the value of the BET surface area. However the highest conversion to acetone (Figure 3a) was exhibited by $V_{0.05}Cu_{0.35}Ce_{0.6}O_{2-y}$ solid that has the lowest surface area.

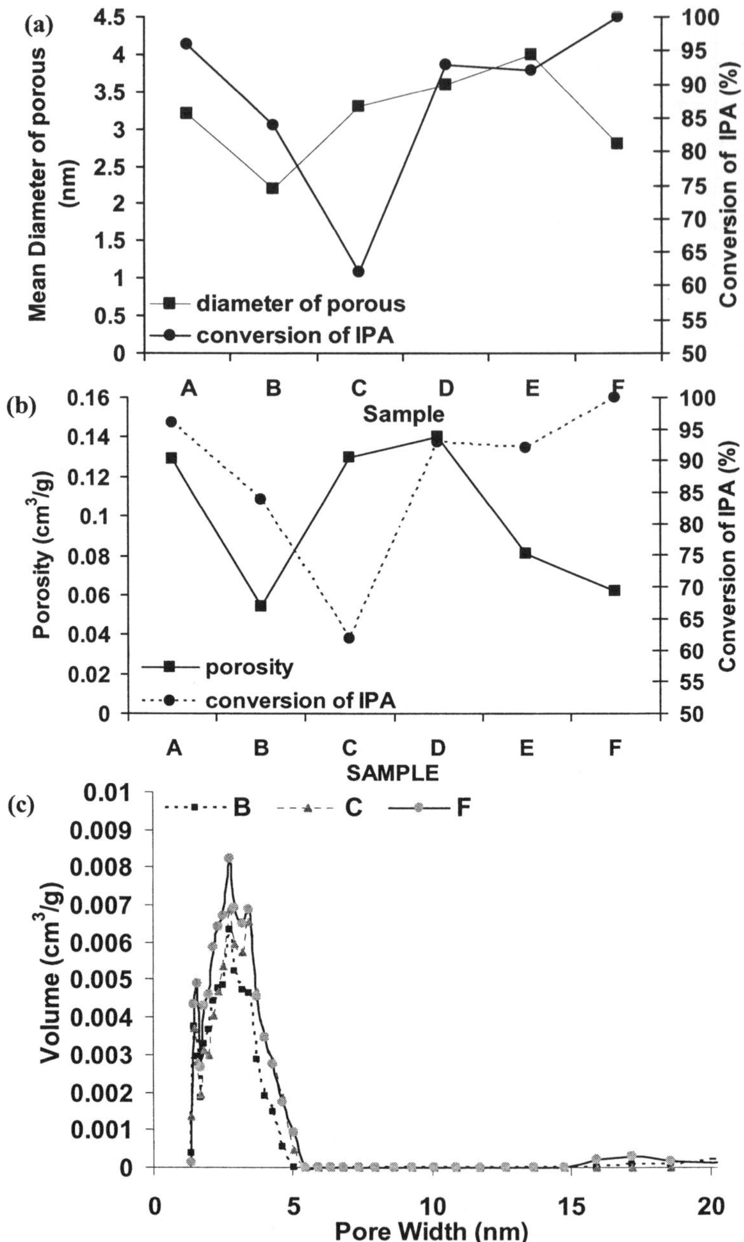

Figure 4: *Comparative diagrams of IPA decomposition versus (a) the mean pore diameter, (b) the pore volume of samples A, B, C, D, E and F. (c) DFT Pore Size Distribution of samples B, C and F that presented the intermediary, lower and higher value of IPA conversion respectively*

In order to understand the diversity in the catalytic behavior exhibited by the samples under investigation, an effort was made to cross-correlation the catalytic results with the surface attributes, the morphology and the crystalline structure of the ternary oxides.

As shown in Figure 4, the extent of conversion for IPA can be well correlated with pore diameter and volume. It is therefore suggested that these two parameters play a role in the catalytic behaviour of this class of solids. It is observed that the sample that presents the highest conversion of IPA also has very small pore volume and diameter (Figure 4a) and narrow pore size distribution (Figure 4c), implying these to be desirable characteristics for catalysts of the specific reaction. However, Sample B with lower pore volume and diameter, exhibits significantly lower values for IPA conversion, whereas sample C, exhibiting the lowest IPA conversion, had pore volume and diameter comparable to sample A which exhibits significantly larger IPA conversion. It is possible, therefore, that in addition to the effect of surface texture on the catalytic behaviour of these solids, the extent of surface acidity is also significant.

Table 1 presents the phases present, as well as the mean particle size for the mixed ceria samples under investigation. It is noted that only one phase is present in all samples, the fluorite-like one of cerium oxide indication that the solids are solid solutions, or in other words that the hetero-atoms are more or less uniformly distributed in the sample. This was confirmed by EDX analysis carried out in the SEM, where spectra obtained for different areas of the samples showed little variation in composition. It is also noted from Table 1, that the mean particle size for all samples was very similar. Therefore, any differences in catalytic activity is more probably correlated with differences in the acidity of the surface of the solids.

Table 1: *Crystalline phases and particle size of calcined samples*

Sample, calcinated 400 ^{0}C	Crystalline phases	Mean particle size (nm)
A	ceria solid solution	4.3
B	ceria solid solution	4.6
C	ceria solid solution	4.1
D	ceria solid solution	3.2
E	ceria solid solution	5.4
F	ceria solid solution	4.1

As reported by numerous researchers, such Pepe et al.,[22] the surface sites where dehydration or dehydrogenation occurs differ in numbers, depending on the crystalline planes present on the surface of the solid, and hence depend on the morphology. For example Grzybowska- Swierkorz et al.,[23] studied the decomposition of IPA on V_2O_5, and observed that the main product obtained was propene and that selectivity towards this product depends on the morphology of vanadium pentoxide. Moreover, the selectivity towards propene decreased with the concentration of levels (001) at the surface of V_2O_5. This was presumably because in the particular level the V=O bonds, which are involved in the process of dehydrogenation are exposed.

In Figure 5 we show the micrographs for samples A, B and C all of which have the composition $V_{0.05}Fe_{0.05}Ce_{0.9}O_2$, but were synthesized using different conditions. There are distinct differences in the morphology obtained for the three synthetic routes. The differences, presumably, lead to different proportions of crystallographic planes being exposed, and thus different numbers of active catalytic sites. These differences may well be reflected in the differences in the distinctly catalytic behaviour of the three samples.

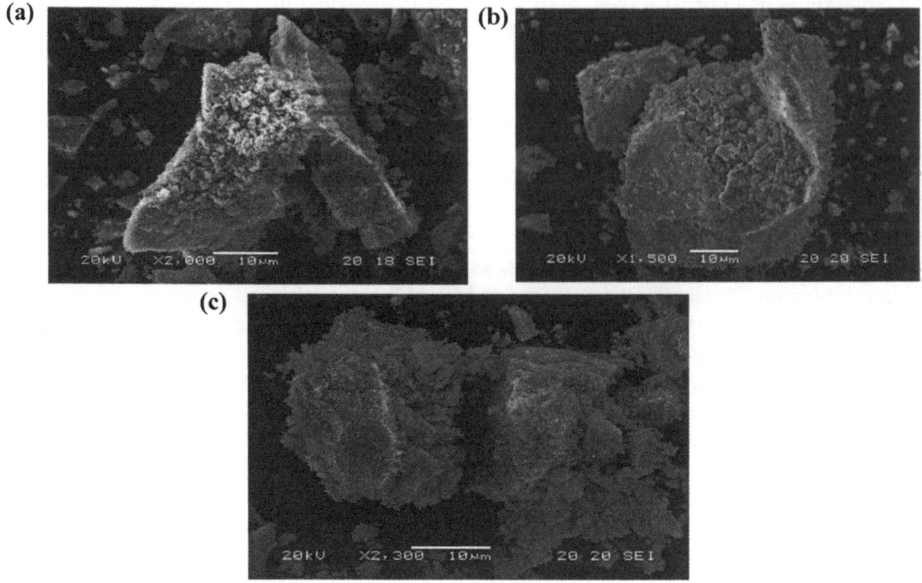

Figure 5 *SEM pictures of $V_{0.05}Fe_{0.05}Ce_{0.6}O_2$ sample: (a) ageing 24 hours at room temperature (A), (b) ageing for 6 hours (B) and (c) precipitation at 363K (C)*

4. CONCLUSION

From the above results it is obvious that the ternary mixed oxides V-Cu-Ce and V-Fe-Ce have acidic (basic) as well as redox active sites because they exhibit higher selectivity towards acetone at lower temperatures in IPA decomposition. The extent of conversion for this reaction does not depend on the surface area of the samples, but on the number and the strength of the active sites. There is also a correlation with pore diameter and volume. Furthermore, a change in synthesis conditions, such as the ageing time or the temperature of precipitation has as a result in a modification of the catalytic behaviour of the solids.

We acknowledge the financial assistance of the Cyprus Foundation for the Promotion of Research (ΙΠΕ) for a partial studentship to CA and the University of Cyprus for financial support. We are grateful to Professor P. Pomonis and his group at the University of Ioannina with assistance with the catalysis and the SEM experiments.

References

1. J. Matta, D. Coursot, E. Abi-Aad and A. Aboukais, *Chemistry Materials*, 2002, **14**, 4118.
2. A. Tschope, M. L. Trudeau and J.Y. Ying, *J. Physical Chemistry B* 1999, **103**, 8858.
3. H. I. Chen and H. Y. Chang, *Ceramics International*, 2005.
4. H. Y. Chang and H. I. Chen, *Journal of Crystal Growth* 2005.
5. T. Radhika and S. Sugunan, *Journal of Molecular Catalysis A: Chemical*, 2006, **250**, 1.
6. R. Cousin, S. Capelle, E. A.-. Aad, D. Coursot and A. Aboukais, *Chem. Mater.*, 2001, **13**, 3862.
7. S. Zhao and R.J. Gorte, *Applied Catalysis A: General*, 2003, **248**, 9.
8. J.-J. Shyue and M.R. De Guire, *Chemistry Materials* 2005, **17**, 5550.
9. X. Gu, J. Ge, H. Zhang, A. Auroux and J. Shen, *Thermochimica Acta* 2006, **451**, 84.
10. B. M. Reddy, P. Lakshmanan and A. Khan, *J. Phys. Chem. B*, 2003, **108**, 16855.
11. S. Hocevar, J. Batista and J. Levec, *Journal of Catalysis* 1999, **184**, 39.
12. B. Sharman, T. Nakayama, D. Grandjean, R. E. Benfield, E. Olsson, K. Niihara and L.R. Wallenberg, *Chemistry Materials*, 2002, **14**, 3686.
13. E. A.-. Aad, J. Matta, D. Coursot and A. Aboukais, *Journal of Materials Science* 2006, **41**, 1827.
14. R. Cousin, S. Capelle, E. A.-. Aad, D. Coursot and A. Aboukais, *Applied Catalysis B: Environmental*, 2006.
15. A. Gervasini, J. Fenyvesi and A.Auroux, *Catalysis Letters* 1997, **43**, 219.
16. A. Gervasini and A.Auroux, *Journal of Catalysis*, 1991, **131**, 190.
17. D. Haffad, A. Chambellan and C. Lavelley, *Journal of Molecular Catalysis A: Chemistry*.
18. C. Attipa and C. R. Theocharis, *Studies in Surface Science and Catalysis*, 2007, **160**, 615.
19. P. Trens, V. Stathoploulos, M. J. Hudson and P. Pomonis, *Applied Catalysis A: General* 2004, **263**, 103.
20. P. N. Trikalitis and P.J. Pomonis, *Applied Catalysis A: General*, 1995, **131**, 309.
21. V. N. Stathopoulos, A. K. Ladavos, K. M. Kolonia, S. P. Skaribas, D. E. Petrakis and P.J. Pomonis, *Microporous and Mesoporous Materials* 1999, **31** 111- 121.
22. F. Pepe, C. Angeletti and S. D. Rossi, *Journal of Catalysis*, 1989, **118**, 1.
23. B. Grzybowska-Swierkosz, G. Coudurier, J. C. Vedrine and I. Gressel, *Catalysis Today*, 1994, **20**, 165.

SYNTHESIS AND CHARACTERIZATION OF Al/Ce INCORPORATED
MESOPOROUS MOLECULAR SIEVES

C. Blanco, C. Pesquera, F. González

Department of Chemical Engineering and Inorganic Chemistry. University of Cantabria.
Avda de los Castros, s/n 39005-Santander. SPAIN. E-mail: pesquerc@unican.es

A series of mesoporous materials with different Si/Al and Si/Ce molar ratios have been
synthesized under reflux. The samples were characterized by XRD, EDXRA, nitrogen
adsorption isotherm and ammonia-TPD. It was found that all the samples increase the total
acidity with respect to the sample synthesized only with silicon, due to the aluminium
introduced in the samples, even when the cerium molar ratio tried was high. The specific
surface area gradually decreases with the increase of the cerium content, independently of
the aluminium tried in the samples. The pore size distribution is similar for all the samples
except when the Si/Ce molar ratio was high.

1 INTRODUCTION

Since the discovery of M41S in 1992[1], mesoporous silicas with very high surface area and
narrow pore distribution in the range from 2 to 50 nm have attracted much research
attention in the fields of adsorption, separation[2-4]. However, pure siliceous M41S materials
possess limited applications in catalytic processes due to its poor acidity and redox
properties. In order to increase the acidity of these materials, isomorphous substitution of
Si by other metal cations such as Al[5], V[6], Ti[7] can be incorporated in these materials.
However, the surface acid sites including Brönsted and Lewis acid in such Al-containing
mesoporous silica materials are very low, which limits their applications on the conversion
of large molecules[8-9]. Moreover, Al-MCM-41 compounds suffer from their weak thermal
stability at high temperatures, which would greatly limit its applications in acid catalysis[10].
In order to study the evolution of the acidity and textural properties of M41S, lanthanides
have been incorporated in these materials[11-12]. La and Ce are the most common elements
used as lanthanides cations.

In this work, we have synthesized a series of mesoporous materials under reflux
conditions with two silicon sources and adding aluminium and cerium with different molar
ratios. The structural and textural characteristics of these mesoporous materials have been
characterized by using different physico-chemical techniques in order to know the
influence of the cerium content in the final properties of the mesoporous materials.

2 EXPERIMENTAL SECTION

2.1 Samples Preparation

Al-Ce-MCM-41 samples with different Al/Ce molar ratio were synthesized. The reagents used in the synthesis were silica fumed and sodium silicate as the silicon source, while sodium aluminate was used as the aluminium source and cerium chloride as the cerium source. Cethyltrimethylammonium bromide (CTMABr) served as template while tetramethylammonium hydroxide (TMAOH) as mineralizer.

The reactants were mixed to obtain reactive gels of the following composition: $1.00SiO_2$: $0.25Na_2SiO_3$ $0.25CTMABr$: $0.10TMAOH$: $100H_2O$. The Si/Al/Ce molar ratio tried in the samples was $100/1/a$ and $100/2/a$ where a is 4, 2, 1.33, 1.

A gel containing silicon, aluminium, cerium and TMAOH in the desired molar ratio was stirred for 1 h and then, an aqueous solution of CTMABr, aged for 1 h, was added gradually under stirring at room temperature and the stirring was continued for 5 min. After that, the gels were heated under reflux for 48 hours. After cooling to room temperature, the solid products were recovered by filtration, washed with water. Finally, the solids were dried in an oven at 60°C overnight. All the samples were calcined in air at 550°C for 6 hours. The samples will be denoted as follows: AlxCeyMS-41 where x (100, 50) indicated the Si/Al molar ratio and y (100, 75, 50 and 25) the Si/Ce molar ratio. In order to study the effects that produce the addition of Al and Ce in these types of materials, a sample with only silicon in the framework was also prepared and it will be denoted as MS-41.

2.2 Samples Characterization

The powder X-ray diffraction (XRD) patterns of the calcined samples were collected at room temperature in a Bruker diffractometer using Cu-Kα (λ= 1.5418Å) radiation, with 6 s for the step time and 0.01 step size from 1.5° (2θ) to 10.0° (2θ). For the unit cell parameter (a_o) the following equation was used: a_o= 2d(100)/$\sqrt{3}$. The chemical composition of the samples with Al and Ce was determined by Energy Dispersive X-ray analysis (EDXRA) in a Jeol electron microscopy (Model JSTM-T 300A) with a Link Analytical AN 10.000 microanalyzer.

The total acidity of the samples was analyzed by temperature programmed desorption (TPD) of ammonia and for this, it was determined in a Micromeritics Autochem 2920 apparatus as follows, the sample was pretreated at 500°C for 2 hours under He flow (50 ml/min). Then, the sample was cooled at 120°C and the adsorption was then performed by contacted with a mixture containing 10% vol. NH_3 in He until saturation was achieved. Afterwards, physisorbed ammonia was removed with He (50ml/min) until ammonia was not detected in the effluent. After that, the sample was submitted to a continuous heating in a flow of He (50ml/min) up to 500°C at a heating rate of 10°/min. Ammonia desorption was analyzed by using a TCD.

Nitrogen adsorption-desorption isotherms at 77K for the samples were performed in a Micromeritics ASAP 2010 apparatus. Before the experiments, the samples were outgassed at 140°C for at least 16 hours under vacuum. Specific surface area was determined by applying the BET equation[13] to the isotherm. The pore size distributions were analyzed by the Barrett-Joyner-Halenda (BJH) method[14] with Halsey equation for multilayer thickness using the adsorption branch of the isotherm. The pore diameter medium (D_{BJH}) was calculated and the cumulative pore volume, V_{BJH}, of the mesopores was obtained from the mesopore size distribution curves.

3   RESULTS

3.1 X-ray Diffraction and EDXR Analysis

The powder X-ray diffraction patterns of all samples exhibit one peak at low angle region. This peak is attributed to the d(001) basal plane, assuming a hexagonal lattice typical of these samples and are collected in Table 1. Incorporation of aluminium and cerium into the materials produces a slight shift to lower diffraction angles when the Si/Ce molar ratio is 25, independently of the amount of aluminium tried. This indicates that the d-spacing and the unit cell parameter compared with the sample with only silicon in the structure (MS-41) increase in these samples. However, when the amount of cerium incorporated in the samples is lower, the values of these parameters are similar, except when Si/Ce molar ratio tried was 100 that this parameter is lower, independently of the amount of aluminium tried.

The chemical composition of the samples obtained by EDRX analysis is summarized in Table 1. From Table 1 it can be observed that when the Si/Al molar ratio tried is 100, the obtained results differ significantly from this value, independently of the cerium content essays. Nevertheless, when the Si/Al molar ratio is 50, this difference is smaller (Si/Al molar ratio lay 33.5-39.5) and similar for all the samples prepared with different Si/Ce molar ratio.

On the other way, the percentage of cerium incorporated in the samples is higher than the aluminium one, independently of the aluminium tried in the synthesis process. This indicated that cerium is more efficient introduced that aluminium in the materials.

3.2 Temperature-programmed Desorption of Ammonia

The total acidity of the synthesized samples has been measured by TPD-NH$_3$ at 120°C. Table 1 gathered the ammonia desorbed in the temperature range of 120-500°C. For the original sample (MS-41), the value is 52 µmol/g of ammonia. This value is one order of magnitude smaller than that obtained with a USY zeolite or an amorphous silica-alumina[15]. This value indicates the poor acidity of this mesoporous material with only silicon in the structure, probably due to the surface Si-OH groups. However, the samples synthesized with aluminium and cerium present an increase of the total acidity. Moreover, for the samples prepared with the same percentage of cerium, the acidity is higher when the Si/Al molar ratio is smaller. This can be due to the effect of more quantity of the aluminium incorporated in the samples that generates higher concentration of acid sites in the structure.The samples with the lower acidity correspond to samples with the higher cerium content (Al100Ce25MS-41 and Al50Ce25MS-41), even that the acidity is more than three times of the original sample (MS-41). These results can be due to the basic nature of the cerium and this is more appreciable in the samples with a higher content of cerium in the synthesized samples.

3.3 Nitrogen Adsorption Isotherms Analysis

Nitrogen adsorption isotherms of the synthesized mesoporous samples are shown in Figure 1. All materials give type IV isotherms with sharp capillarity condensation step in the medium range of the relative pressure (P/Po = 0.32-0.46), characteristic of mesoporous

Table 1 *Chemical composition, structural parameters and total concentration of NH$_3$ of the mesoporous materials*

SAMPLE	Si/Al ratio (EDXRA)	Si/Ce ratio (EDXRA)	d(100) (nm)	a_o (nm)	NH$_3$ (μmol/g)
Al100Ce25MS-41	50.5	27.3	4.0	4.6	178
Al100Ce50MS-41	62.8	52.2	3.5	4.1	226
Al100Ce75MS-41	58.3	60.1	3.4	4.0	192
Al100Ce100MS-41	61.1	88.7	2.9	3.3	217
Al50Ce25MS-41	33.5	20.7	4.3	4.9	199
Al50Ce50MS-41	39.5	63.7	3.6	4.1	235
Al50Ce75MS-41	36.4	73.7	3.5	4.0	301
Al50Ce100MS-41	38.4	78.2	3.0	3.4	217
MS-41	-	-	3.6	4.2	52

materials[16]. The capillarity condensation step of the samples synthesized with the highest amount of cerium (Al100Ce25MS-41 and Al50Ce25MS-41), independently of the Al tried, is smoother than the other samples. After this step, a plateau with a slight inclination at high relative pressure indicative to multilayer adsorption on the external surface of the materials. Finally, a sharp rise in nitrogen uptake filling all other available pores, pores among particles as the relative pressure reaches saturation (P/Po= 1.0). Table 2 collected the textural parameters of the synthesized materials.

As can be observed in Table 2, the lower specific surface areas correspond to samples with the highest amount of cerium tried (Al100Ce25MS-41 and Al50Ce25MS-41), independently of the amount of aluminium tried. This effect is parallel to the one observed with the acidity with these two samples. For the other samples, this parameter is very similar to the sample synthesized with only silicon (MS-41). Even that, it can be observed a small, but gradually, decreases with the increase in the cerium content in the samples. This effect is independently of the aluminium tried.

Table 2 *Textural parameters of the mesoporous materials*

SAMPLE	S_{BET} (m^2/g)	V_{BJH} (cc/g)	D_{BJH} (Å)
Al100Ce25MS-41	614	0.52	30.4
Al100Ce50MS-41	1018	0.90	31.4
Al100Ce75MS-41	1292	1.20	32.0
Al100Ce100MS-41	1155	1.07	30.6
Al50Ce25MS-41	614	0.46	30.2
Al50Ce50MS-41	1091	1.01	32.0
Al50Ce75MS-41	1267	1.19	30.8
Al50Ce100MS-41	1144	1.03	30.8
MS-41	1253	1.14	31.8

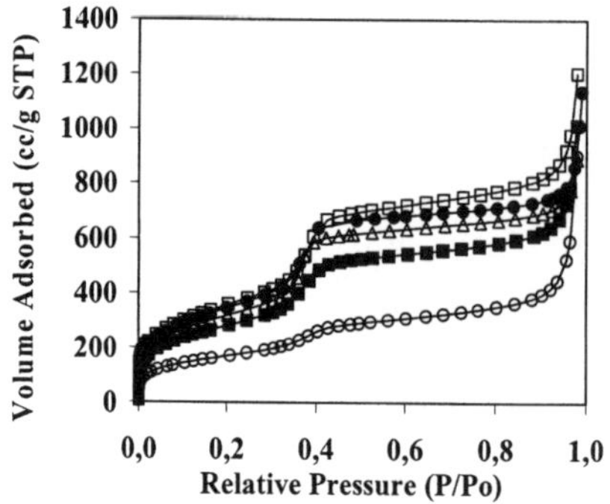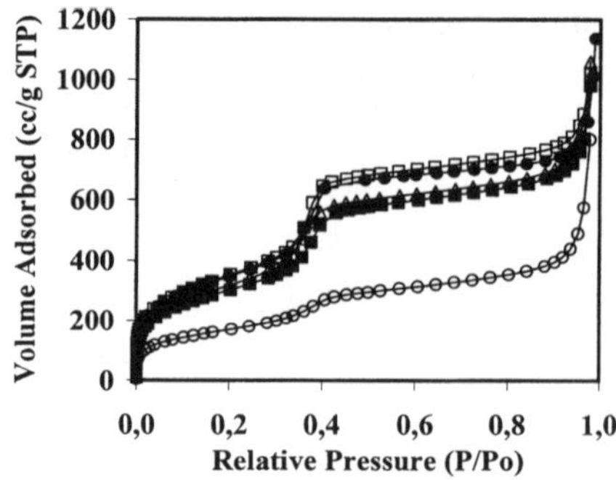

Figure 1. *Nitrogen adsorption isotherm of the samples*: •, MS-41; Δ, AlxCe100MS-41; □, AlxCe75MS-41; ■, AlxCe50MS-41; o, AlxCe25MS-41. Left: Si/Al=100; right: Si/Al=50.

3.4 Pore Size Distribution

Figure 2 shows the pore size distribution curves calculated based on the adsorption isotherms. The step increases of nitrogen gas observed in the isotherms is attributed to the presence of a uniform size of mesopores in the samples. This is confirmed from a very sharp peak at pore diameter (D_{BJH}) around 31Å (Table 2). The pore size distribution is narrow for all de samples except for the samples with the highest amount of cerium tried (Al100Ce25MS-41 and Al50Ce25MS-41), independently of the amount of aluminium tried.

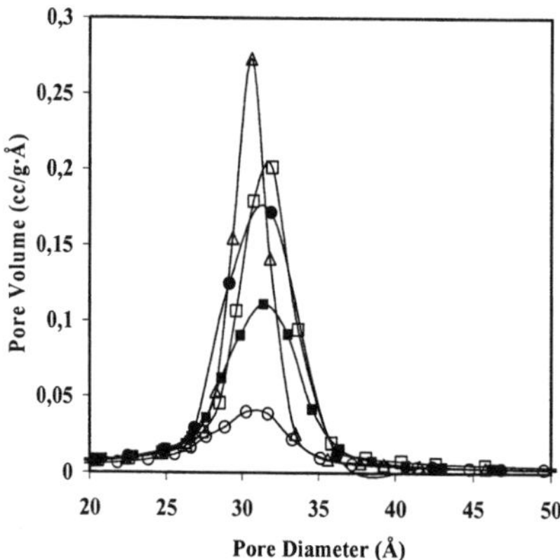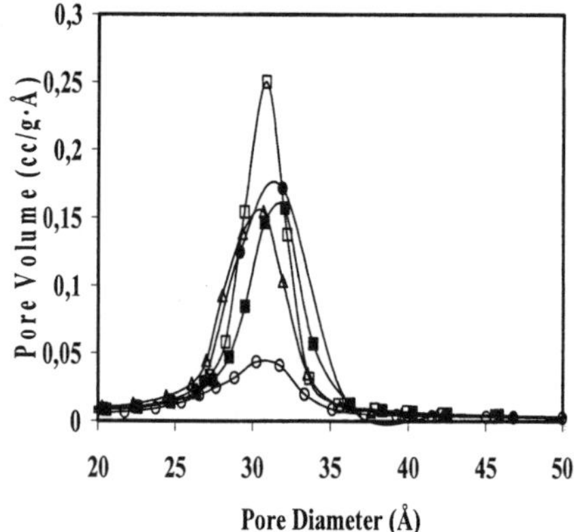

Figure 2. *Pore size distribution of the samples:* •, MS-41; Δ, AlxCe100MS-41; □, AlxCe75MS-41; ■, AlxCe50MS-41; o, AlxCe25MS-41. Left: Si/Al=100; right: Si/Al=50.

For Al100Ce100MS-41 and Al50Ce75MS-41 samples the intensity of the peak due to the pore size distribution is the highest, that correspond with the highest accumulate volume of nitrogen adsorption (Table 2).

4   CONCLUSIONS

A series of mesoporous materials with different Si/Al and Si/Ce molar ratios have been synthesized under reflux. All the samples increase the total acidity with respect to the sample synthesized only with silicon, due to the aluminium introduced in the samples, even when the cerium molar ratio tried was high. The specific surface area gradually decreases with the increase of the cerium content, independently of the aluminium tried in the samples. The pore size distribution is similar for all the samples except when the Si/Ce molar ratio was high.

Acknowledgments

We wish to acknowledge the financial support for this work, provided by the Spanish Government Ministry of Education and Science under Project MAT2006-03683.

References

1 C.T. Kresge, M.E. Leonowicz, W.J. Roth, J.C. Vartuli, J.S. Beck, *Nature*, 1992, **359**, 710.

2 J.M. Kisler, A. Dähler, G.W. Stevens, A.J. O'Connor, *Micropor. Mesopor. Mater.*, 2001, **44-45**, 769.

3 M. Grün, A. Kurganov, S. Scha, F. Schülth, K.K. Unger, *J. Chromatogr. A*, 1996, **740**, 1.

4 S.S. Kim, W. Zhang, T.J. Pinnavaia, *Science*, 1998, **282**, 1302.

5 P.A. Jalil, M.S. Kariapper, M. Faiz, N. Tabet, N.M. Hamdan, J. Diaz, Z. Hussain, *Appl. Catal. A: General*, 2005, **290**, 159.

6 K.M. Reddy, I. Moudrakovki, A. Sayari, *J. Chem. Soc. Chem. Commun.*, 1994, 1059.

7 J. Yu, X. Zhao, Q. Zhao, G. Wang, *Mater. Chem. Phys.*, 2001, **68**, 253.

8 R. Mokaya, W. Jones, Z. Luan, M.D. Alba, J. Klinnowski, *Catal. Lett.*, 1996, **37**, 113.

9 J. Weglarski, J. Datka, H.Y. He, J. Klinnowski, *J. Chem. Soc., Faraday Trans.*, 1995, **91**, 2995.

10 R. Ryoo, S. Jun, *J. Phys. Chem.*, 1997, **101**, 317.

11 A.S. Araujo, M. Jaroniec, *J. Coll. Interf. Sci.*, 1999, **218**, 462.

12 N.H. He, Z. Lu, C. Yuan, J. Hong, C. Yang, S.L. Bao, Q.H. Xu, *Supramol. Sci.*, 1998, **5**, 533.

13 S.J. Gregg, K.S.W. Sing, *Adsorption, Surface Area and Porosity*, Academic Press, New York, 1982.

14 E.P. Barrett, L.G. Joyner, P.P. Halenda, *J. Am. Chem. Soc.*, 1951, **73**, 373.

15 A. Corma, V. Fornés, M.T. Navarro, J. Pérez-Pariente, *J. Catal.*, 1994, **148**, 569.

16 D. Zhao, J. Sun, Q. Li, G.D. Stucky, *Chem. Mater.*, 2000, **12**, 275.

SURFACE NATURE AND STRUCTURE OF LIMESTONE POWDER - EVALUATION OF PROPERTIES USING SORPTION AND STANDARD TECHNIQUES

E. Vogt, B. Buczek

University of Science and Technology, Faculty of Fuels and Energy, al. Mickiewicza 30, 30-059 Cracow, Poland

1 INTRODUCTION

Porous powder materials are used in everyday life and in many industries. Their properties are quite different from the massive solids, in particular there are cohesive interactions between fine particles. This effect causes that the particles stick together and determines the behaviour of powders in many technological processes, e.g. fluidization, pneumatic conveying, mixing or multiphase processes. Value of the cohesion forces that occur in a fluidized bed depends not only on the material, but also on particles size distribution, particles shape, its surface nature and structure. Determination of such characteristics is very important for many processes because it enables us to control the course of such processes, allows modeling and gives us indispensable knowledge for production of new materials.

2 METHOD AND RESULTS

2.1 Material

Raw limestone powder, from the Czatkowice Quarry of Lime, Poland and modified limestone powder, which is used as an anti-explosive agent in the mining industry were examined materials. The modification (hydrophobization) protects it against the loss of volatility,[1] which is caused by humidity, often occurring in the mines of hard coal. Stearic acid is usually studied as the hydrophobizing agent for coal mine applications.[2] A typical method of hydrophobic dust production (milling limestone with stearic acid) is no longer profitable, or sometimes even impossible in modern quarries, and accordingly a new manufacturing method has been developed. This method of limestone powder hydrophobization consists of stearic acid vapour and dust countercurrent flow.[3] The investigated sample contains 0.18 % of stearic acid, being an acceptable level according to the Polish Standard.[2] The specific application of this powder as a substance protecting human life requires that the limestone dust properties must be known very well.

2.2 Measurement methods

2.2.1 Particle size analysis. The results obtained with the laser analyzer (type: Analisette 22 C-Version, made by Fritch GmbH Laborgerätebau) for the material are showed in the Figure 1.

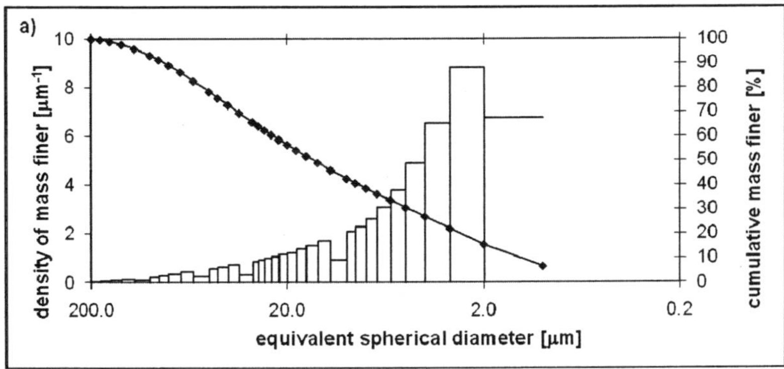

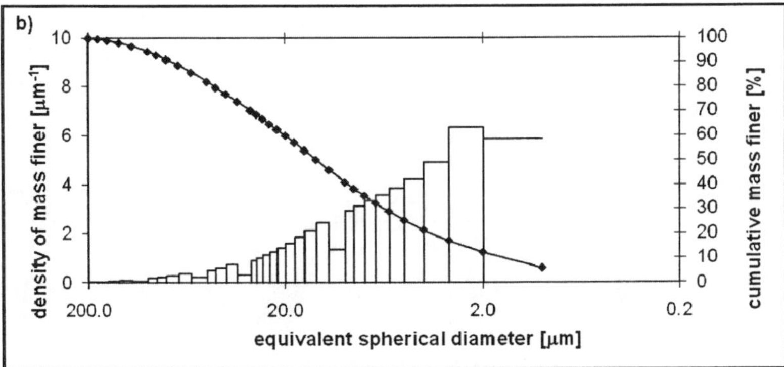

Figure 1 *Particle size analysis: a – raw, b - modified lime dust.*

2.2.2 Powder Characteristics Tester – type PT-E, Ser. No. 901331. Parameters which were measured: bulk density, packed bulk density, compressibility, and angles of: spatula repose, fall, difference and dispersibility. Carr[4] has tried to evaluate powder's flowability and floodability in a numerical manner with the combination of this various physical characteristics. The tables for the conversion of the measured figures into a common index were published. These parameters also allow us to calculate the Carr's[4] or Hausner's[5] ratios, which are applied in contemporary researches of powder materials. Powder Characteristics Tester offers quick and reliable measurements with a single unit. More consistent and accurate data can be obtained with controlled mechanical conditions in this apparatus, then by manual methods.

Table 1 *The characteristic of raw and hydrophobized limestone dust*

Characteristics	Limestone powder	
	Raw	Hydrophobized
Bulk density [kg/m^3]	724	798
Packed bulk density [kg/m^3]	1475	1377
Compressibility [%]	50.9	42
Repose angle [deg]	52	47
Fall angle [deg]	35	33
Difference angle [deg]	17	14
Dispersibility [%]	20	41
Carr's ratio [%]	50.1	42
Hausner's ratio	2.0	1.7

2.2.3 Measurement of the adhesive force. The adhesive force was measured between individual particles and metallic flat surface, characterized by the 50-100 μm roughness (Figure 2a). The technique for measuring of the adhesive force for array of individual particles is described in paper.[6] Measurement was carried out on the coarse limestone (grain fraction: 0.385-0.400 μm) of the same type, modified in the same way. The array composed of about 250 particles placed on a tacky surface is showed in Figure 2b-c.

Figure 2 *The measurement of the adhesive force: a) metallic flat surfaces, b, c) array composed of particles placed on a tacky surface*

The adhesive force was measured with an electronic balance by firstly contacting the array of particles and the metallic surface, separating them, and than recording the greatest negative balance reading. The measurement was carrying out in different humidity conditions. The adhesive force of modified material was neglected apart from the humidity of an environment. The result for raw material is showed in Figure 3.

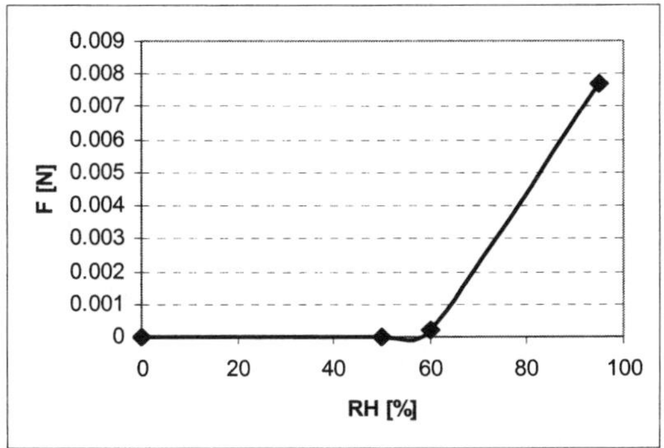

Figure 3 *The adhesive force for raw powder in relation to relative humidity*

2.2.4 Adsorption measurements. Adsorption techniques (low temperature adsorption of nitrogen and adsorption of water vapour) allow us to characterize porous structure of powder and its nature. The adsorption-desorption isotherms of nitrogen are shown in Figure 4. The parameters of the porous structure of the materials: specific surface area using BET (S_{BET}) and Dollimore-Heal's (S_{DH}) methods and volume of pores calculated for $p/p_o = 0.98$ (V_p) are presented in Table 2.

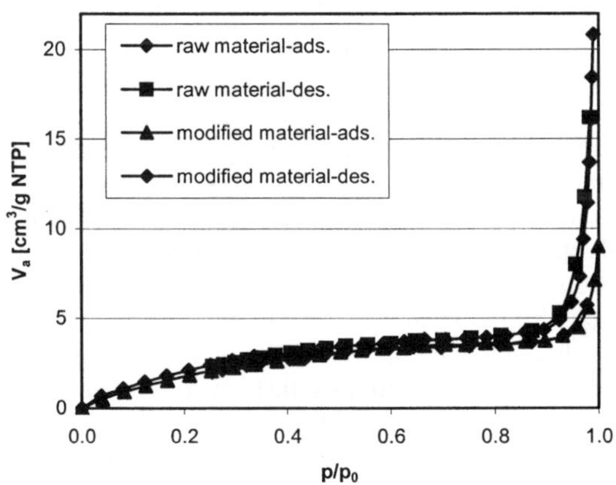

Figure 4 *Adsorption-desorption isotherm of nitrogen at 77 K*

Table 2 *Characterization of the solids by adsorption measurements*

Material	S_{BET} [m^2/g]	S_{DH} [m^2/g]	V_p [cm^3/g]
Raw powder	10.9	6.1	0.019
Hydrophobized powder	9.5	5.5	0.009

Moreover the character of the raw and modified dusts has been studied by determination of the water adsorption isotherms, using the liquid microburette apparatus (Figure 5).[8]

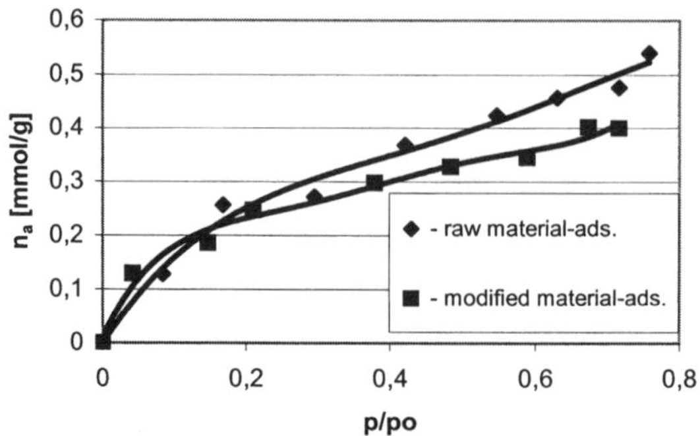

Figure 5 *Water vapour adsorption isotherm on limestone powder*

3 CONCLUSIONS

Obtained results enable us to make a complete characterization of the limestone dust and give us knowledge of the phenomena taking place during the modification.

Limestone dust is very fine; the equivalent spherical diameter of the particles is in the range 1-20 μm. These materials usually have a very high cohesion.

The characteristic of the limestone powder showed in Table 1 gives us the possibility of the determination of properties of this material from the cohesion point of view. Powder Tester is a useful and practical aid for the comparison of the properties of the same assortment of powders. In the case of the raw material, the packed bulk density is about twice as large than the bulk density and this gives the compressibility value that is equal to about 50 %. The compressibility is slightly decreased for the modified sample, but both the raw limestone dust and the hydrophobized powder will settle in elbows, and the outlet will clog because the powders, which have compressibilities above 45 % are characterized by a very bad degree of flowability.[4] This behaviour of dust was observed and special apparatus and techniques are then required. If the angle of fall is smaller than 10 degree, the material has a free flowing characteristic. The degree of floodability is very high.[4] The repose and fall angles for the hydrophobized and raw dust are variant and the difference angles calculated on the basis of these are different too. When we look at the fall angles, we can state that the both dusts have fairly high degrees of floodability, but the values of difference angles point that the hydrophobised dust obtained new properties, i.e. it tends to flush. However the value of the difference angle for both the raw and the hydrophobized dust is low, as in the case of cohesive materials.

A great difference between the dispersibility values was found. It seems that the measurement of compressibility or dispersibility could be used as the criterion of the evaluation of the degree of hydrophobization of dust. In the case of the latter parameter, the difference between the raw and the hydrophobized material is equal, about 50 %. However, these values for both samples are less than 50 % - that is substandard for easy flowing powders.

The Hausner's ratios,[5] equal to 2.0 (raw dust) and 1.7 (hydrophobized dust), are bigger than 1.4, i.e. the two samples of the dust are characterized by all properties of cohesive powders. The Carr's ratios,[4] within suitable ranges, determine the flow rate of powder: very good ($I_C < 18$), medium ($I_C = 18\text{-}25$), weak ($I_C = 25\text{-}30$), very weak ($I_C > 30$). The obtained values 50.1 and 42, which are much bigger than 30, confirm the low flow rate of powder.

Adhesive force measurements show that the modified material lost its adhesive properties, but only when in contact with metallic flat surfaces. It would be better to make the same survey with the use of limestone plate; these investigations are planned.

It seems that the adsorption measurements give us the curious results. It is noticeable that the total adsorption of the nitrogen at $p/p_o = 0.98$ is about twice bigger for raw material than for the modified one. The value of total adsorption in the case of very fine particles is the result of location of adsorbate not only in pores but also in the intergranular spaces. The obtained low values of the specific surface area and the volume of mesopores point that the limestone powder is not a porous material i.e. that whole adsorbate was placed in the intergranular spaces. The porosity of bed for more cohesive raw dust would have to be bigger than for the porosity of modified dust bed. Probably the particles interlock each other making an overhang (vault). The hydrophobized particles were better packed. The bigger value of the bulk density of modified material confirms this fact. During measuring of the packed bulk density adhesive forces probably caused the inverse relation between obtained density values.

The greater water vapour adsorption for the raw material than for the hydrophobized one (Figure 5) proves that the modified product achieved water-resistance, but on the basis of these results, the hydrophobization degree of the limestone dust cannot be determined.

It can be stated that the process of hydrophobization causes changes of the limestone dust features and the proposed investigation cycle enables us to observe these changes in properties, especially the adhesive force for modified material is decreased. It is visible not only when we look at the results obtained with the use of Powder Characteristics Tester but also from adsorption measurements. In the second case we obtained more interesting results than we expected, because usually measurements of adsorption are used to investigate porous materials. Results obtained in this way usually enable us to draw correct conclusions for porous materials. The adhesive force measurements were carried out as pre-investigations in order to prove the possibility of performing such kind of researches for nonspherical, fine particles. The technique of measurement was elaborated too.

The authors are grateful to Ministry of Scientific Research and Information Technology (Project No. N524 036 32/3934) for its financial support of this work

References

1. K. Cybulski, Wiad. Górn., 2003, **10**, 466.
2. Polish Standart PN-G-11020, Mining-Antiexplosive stone dust.
3. B. Buczek and E.Vogt, Twenty - Fourth Annual International Pittsburgh Coal Conference: Coal - Energy, Environment and Sustainable Development: September 10–14, 2007, Johannesburg, South Africa, PCC©2007, pp 1-9.
4. R. Carr, Chem. Eng., 1965, **72**, 163.
5. H.H. Hausner, International Journal of Powder Metallurgy, 1967, **3**, 7.
6. N. Harnby and A.E. Hawkins and I. Opaliński, Trans IChemE, 1996, **74**, Part 2, 616.
7. M. Lasoń and M. Żyła, Chem. Anal., 1963, **8**, 279.

INVESTIGATION OF THE MICROSTRUCTURE OF ULTRA HIGH PERFORMANCE CONCRETE

P. Klobes[1], K. Rübner[1], S. Hempel[2], C. Prinz[1]

[1] BAM Federal Institute for Materials Research and Testing, 12200 Berlin, Germany
[2] Technische Universität Dresden, Institute for Construction Materials, 01062 Dresden, Germany

1 INTRODUCTION

Ordinary concrete is a heterogeneous and porous material, which is produced from cement, water and aggregates in the simplest case. The aggregates are embedded in the products of the hydration reaction of the cement (various calcium silicate hydrates (CSH) and calcium hydroxide ($Ca(OH)_2$). Since the seventies the concrete mixtures have changed from the simple ternary system to more complex compositions with the objective of tailor-made concrete properties. The addition of pozzolanic fillers, such as silica fume or fly ash, in combination with the use of superplasticizers and the reduction of water/cement ratios lead to concretes with high strengths above 60 MPa. Furthermore, the strength of the aggregates as well as the fineness and composition of cement contribute to the increase in concrete strength. Additionally, high strength concrete (HSC), which is also named high performance concrete, show a better durability, such as higher resistance against physical and chemical attacks, lower gas and liquid permeability as well as a denser microstructure.

Ultra high performance concrete (UHPC) is an outstanding development pushing the boundary of classical concrete technology. It is characterised by compressive strengths above 150 MPa and an excellent durability. These properties are achieved by optimisation of the mixture composition, the mixing procedure as well as the curing conditions of the concrete. UHPC is produced using a very low water/cement ratio of 0.25 or smaller in combination with adding superplasticizers based on polycarboxylate ethers. Furthermore, finest cements with contents of 500 kg/m³, defined selections of coarse and fine aggregates with a maximum grain size between 0.5 and 8 mm and fine pozzolanic and inert fillers are used. Heat curing as well as the use of vacuum mixers may contribute to the high strength. The very high brittleness of UHPC can be compensated by the addition of steel or polymer fibres. In doing so, the aim is to obtain a very high packing density of the cement paste matrix and the aggregate/paste interface while a very homogeneous microstructure with a high CSH portion is formed.[1-5]

A particular feature of UHPC is the extremely dense microstructure. In the case of optimal mixing and curing conditions, UHPC contains almost no pores and microcracks.

Therefore, studies of porosity and pore structure are very important to characterise the materials in connection with the mixture optimisation. Essential experimental results for UHPC are presented in comparison with those of high strength concrete and normal strength concrete.

2 EXPERIMENTAL

2.1 Materials

The different concretes are characterised by compressive strengths of 37, 109 and 224 MPa, respectively. The laboratory concretes were made using Portland cements of different strength and fineness as well as siliceous sand and gravel aggregates. To produce a high strength concrete, the water/cement ratio (w/c) was reduced and silica fume was added. For the UHPC, a special mixture was developed, which contains cement, aggregates and several pozzolanic and inert fillers with particle sizes harmonised with regard to a very high packing density. Furthermore, a heat treatment was used. The composition of the concretes and the curing conditions are summarised in Table 1.

Concrete specimens in the age of 90 days were used for the pore structure measurements. Each material was crushed to 4-8 mm granules. Then they were dried at 22 °C and 4 kPa above a cold trap. The samples were more crushed to 0.25-2 mm granules for the sorption measurements. Nonporous siliceous aggregates with grain sizes greater than the maximum particle size of the granules were removed in each case. To determine the density, the samples were ground and dried at 105 °C. The details of sample preparation and sample masses used are summarised in Table 2.

Table 1 *Composition, curing and strength of investigated concretes*

Concrete		Normal strength concrete (NSC)	High strength concrete (HSC)	Ultra high performance concrete (UHPC)
Composition				
Cement type		CEM I 32.5 R	CEM I 42.5 R	CEM I 52.5 R
Cement content	(kg/m³)	310	500	500
Silica fume	(kg/m³)	—	50	116
Fly ash	(kg/m³)	—	—	123
Quartz filler	(kg/m³)	—	—	82
w/c		0.60	0.24	0.28
w/b		0.60	0.22	0.23
Superplasticizer	(%)	—	4.5	2.8
Max. grain size of aggregates	(mm)	32	16	2
Aggregate content	(kg/m³)	1827	1672	1340
Curing		under water (6 d) 20 °C, 65 % r.h.	under water (6 d) 20 °C, 65 % r.h.	under water(3 d) heat treatment 250 °C (2 d)
Compressive strength	(MPa)	37	109	224

Table 2 *Samples for pore structure measurements*

Method	Drying method	Sample size	Sample mass (g)
ESEM		Crushed surface of single 4-8 mm granules	
Mercury porosimetry	vacuum drying at 22 °C, 4 kPa, above cold trap	4-8 mm granules[a]	8
N_2 sorption		0.25-2 mm granules[b]	10
H_2O_{vapour} sorption		0.25-2 mm granules[b]	0.16
Bulk density	oven drying at 105 °C	4-8 mm granules	100
Density		powder	60

[a]aggregates > 8 mm removed   [b]aggregates > 4 mm removed

2.2 Methods

The microscopic images of the microstructure of concrete were taken with an Environmental Scanning Electron Microscope (ESEM). The ESEM technology enables - in contrast to the traditional SEM - the investigation of the specimens in a low vacuum atmosphere without applying a conductive layer on the sample's surface. Using the EDX (Energy Dispersive X-ray) detector installed inside the ESEM it is possible to detect the elements existing in the investigated area of the sample.

The pore structure was measured by mercury intrusion porosimetry applying a PASCAL 140/440 porosimeter system (FISONS) with maximum pressure of 400 MPa according to ISO 15901-1:2005.[6] By assuming a contact angle of 140° and a mercury surface tension of 0.48 N/m a range of pores and pore entrances, respectively, from 60 µm to 1.8 nm pore radius was assessable.

Nitrogen isotherms at 77.3 K were obtained using a Sorptomatic 1990 automated gas adsorption analyser (CE Instruments / FISONS).

Water vapour isotherms of the concretes were measured by a dynamic vapour sorption analyser DVS (POROTEC - SMS London) at 298 K.

The total porosity of the concretes was calculated from the ratio of bulk density and He skeleton density. The bulk density was determined from weighing and volume measurement by water displacement. The skeleton density analysis was performed with a gas pycnometer ACCUPYC 1330 (Micromeritics) using helium according to DIN 66137 part 2.[7]

3 RESULTS AND DISCUSSION

ESEM images of the cement paste matrix of ordinary concrete, high-strength concrete and ultra high performance concrete samples taken for this study are shown in Figure 1. The change of morphology is very pronounced. As expected, microstructure becomes more uniform and denser from the ordinary concrete to high strength concrete and to UHPC. This is associated with notably smaller porosity.

Figure 1 *ESEM images of the cement paste matrix of ordinary concrete, high-strength concrete and ultra high performance concrete (from left to right)*

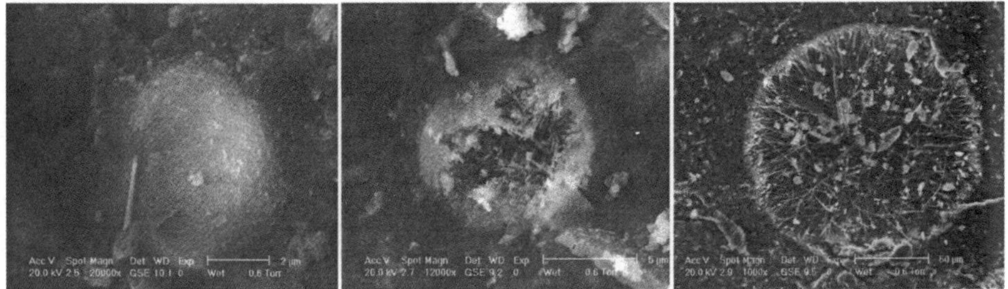

Figure 2 *ESEM images of the cement paste matrix of ultrahigh performance concrete with reacted fly ash particles (left, middle) and a large pore filled with CSH needles (right)*

After heat treatment of UHPC, a considerable amount of fly ash particles are already participating in the hydration reactions in early ages of concrete. The left and middle ESEM image in Figure 2 show such reacted fly ash particles. They are characteristic for the cementitious matrix of heat treated UHPC. The aluminium and potassium ions, which are released due to fly ash reaction, spread in the whole cement paste matrix by migration. As a result, up to 30 μm long needle-like reaction products of Al- und K-rich calcium-silicate-hydrates are formed in larger pores.

The cumulative and log-differential pore volume distributions of the concrete samples as obtained by mercury porosimetry up to 400 MPa are shown in Figures 3 and 4. In comparison with normal strength concrete (NSC) both concretes with higher strength (HSC and UHPC) exhibit a considerable decrease of the total pore volume. It can be concluded from the differential pore size distribution curves in Figure 4 that the pore size distribution of NSC is broader than in HSC and UHPC. This result corresponds with the expected more uniform and denser microstructure of high strength concrete and UHPC compared with ordinary concrete. The distinctly smallest pores are found in UHPC. The numerical results of the pore structure measurements on the different concrete samples by mercury porosimetry are summarised in Table 3.

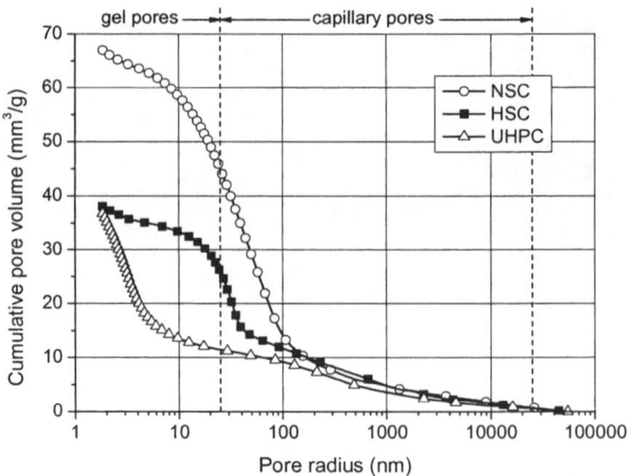

Figure 3   *Cumulative pore volume curves of the concretes obtained by mercury porosimetry*

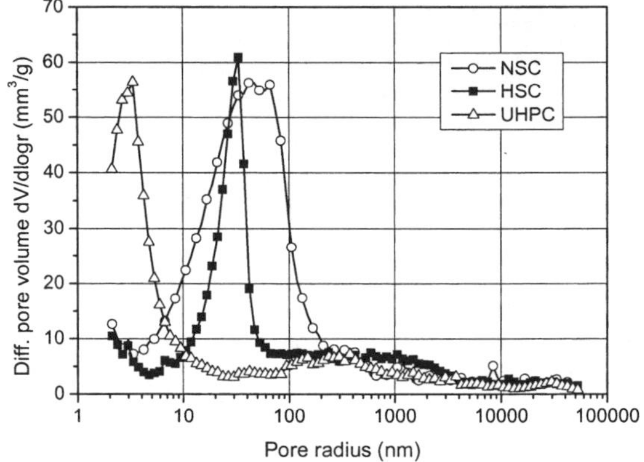

Figure 4   *Log. differential pore volume distributions of the concretes derived from the cumulative pore volume curves in Figure 3*

The BET specific surface areas (SSA) calculated from nitrogen and water vapour isotherms (examples for isotherms are in Figures 5 and 6) indicate for the three concretes investigated here (Table 4) that the BET SSA from the water vapour adsorption exceed BET SSA values of nitrogen adsorption. This seems to be a general problem with nitrogen adsorption measurements of concretes and is in agreement with results reported earlier in the literature.[8-10] One general explanation may be that the lower BET_{N2} surface areas arise from a low rate of thermally activated diffusion through the smallest pore entrances at the low nitrogen adsorption measuring temperature of 77.3 K. Espinosa et. al.[11] discussed the presence of inkbottle shaped pores in hardened cement paste.

Table 3 *Numerical results of pore structure measurements by mercury intrusion porosimetry*

	Total porosity[a] (%)	Total pore volume (mm³/g)	Capillary pore volume (%)	Gel pore volume (%)	Average pore radius (nm)
NSC	16.9	67.0	66.2	32.7	54.1
HSC	11.4	38.1	66.7	31.6	29.8
UHPC	8.8	36.6	29.5	68.9	3.0

[a] determined from helium density and bulk density measurements

Table 4 *Specific surface area determination by N_2 and H_2O multipoint BET analysis in the relative pressure range 0.05 to 0.3*

	N_2 adsorption at 77 K			H_2O adsorption at 298 K		
	BET surface area[a] (m²/g)	BET constant C	coefficient of determination r	BET surface area[b] (m²/g)	BET constant C	coefficient of determination r
NSC	5.0	440.4	0.999833	21.3	15.5	0.999758
HSC	2.9	148.1	0.999896	33.6	6.5	0.999051
UHPC	5.3	408.1	0.999715	29.7	18.4	0.999788

[a] N_2 cross sectional area 0.162 nm²   [b] H_2O cross sectional area 0.125 nm²

Jennings[12] has proposed a more quantitative model for the structure of CSH gel in cement paste postulating two types of CSH, low density (LD) and high density (HD), which are formed by different packing arrangements of the same basic approximately 2-nm-sized units of CSH globules. The globules then cluster more loosely and randomly into the characteristic structure of LD CSH. Both CSH structures are structurally similar but the globules in HD CSH are more tightly packed. As postulated by this model, nitrogen can penetrate the larger pores between the LD CSH structures but, for kinetic reasons, cannot penetrate the intraglobule pores inside the LD CSH structure. Water, on the other hand, can enter most of the intraglobule porosity in both the LD and HD CSH, explaining the higher BET_{H2O} surface areas.

Using the definitions of Setzer[13] who called pores in hardened cement paste with radii < 25 nm gel pores and pores in the pore radius range between 25 nm and 25 μm capillary pores (Figure 3), the total pore volume obtained in our investigations by mercury porosimetry measurements was divided into the gel pore and capillary pore proportions. UHPC decisively exhibits the largest gel pore percentage although its total porosity is the smallest one.

The cumulative pore volume curves calculated with the BJH model from the desorption branch of nitrogen isotherms (Figure 6) are close to those from mercury porosity intrusion measurements (Figure 3). This observation may be interpreted as further evidence for inkbottle-like pore morphology in hardened cement paste because both methods give mainly information about the pore entrances.

The pronounced triangular shaped hysteresis loop of the water vapour isotherms in the cement pastes also indicates inkbottle capillaries with a varying range of narrow short necks but can be indicative of plate-like pore morphologies as well.[14]

Every water isotherm measured here exhibits low pressure hysteresis which may be associated with the deformation of the gel pore structure by swelling effects or with a more specific interaction between the water molecules and surface groups of the adsorbent, such as silanol groups, leading to a stronger uptake of water molecules in the gel pore structure.

                                                    Characterisation of Porous Solids VIII

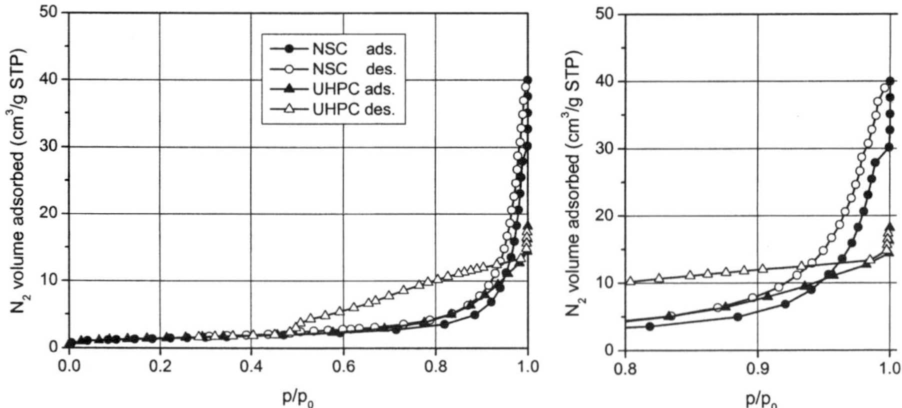

Figure 5   *Nitrogen isotherms of NSC and UHPC (on the right: details for $p/p_0 > 0.8$)*

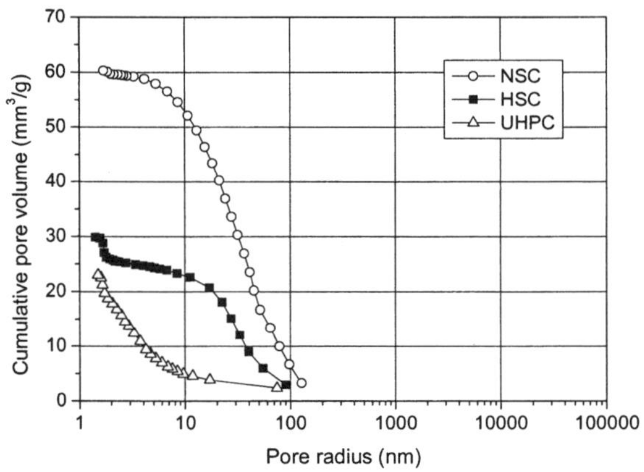

Figure 6   *Cumulative pore volume distribution of NSC, HSC and UHPC from BJH desorption branch of N_2 isotherms shown in Figure 5*

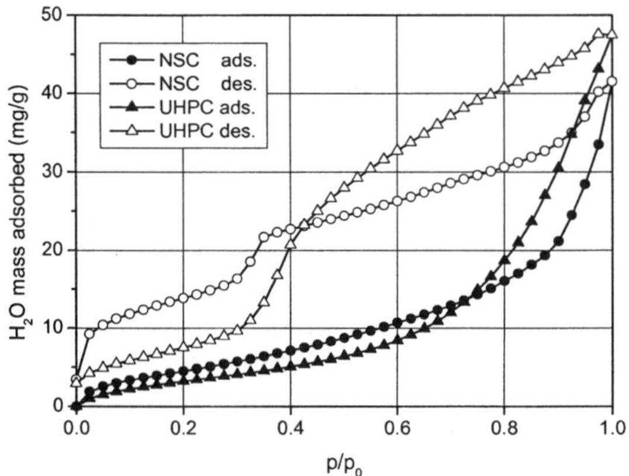

Figure 7   *Water isotherms of NSC and UHPC*

4   SUMMARY AND CONCLUSIONS

In spite of not always completely consistent results, pore size analysis by mercury porosimetry and gas adsorption in combination with complementary methods yields important insight into the structure of hardened cement paste.

In contrast to ordinary concrete and high-strength concrete, UHPC features a more uniform and denser microstructure associated with notably smaller porosity.

In UHPC, the total pore volume and the fraction of capillary pores are reduced, but the fraction of gel pores is increased and the maximum of the pore volume distribution is shifted to lower pore radii.

After heat treatment of UHPC, a considerable amount of fly ash particles are already participating in the hydration reactions in early ages of concrete. The thereby released aluminium and potassium ions spread in the whole cement paste matrix by migration. Up to 30 nm long needle-like reaction products of Al- und K-rich calcium silicate hydrates are formed in large air voids.

The triangular form of the hysteresis loops both of the nitrogen and water isotherms indicate that in addition to inkbottle pores also a plate-like morphology of the CSH phase of UHPC may be present.

The low pressure hysteresis of the water vapour isotherms may be indicative for swelling effects of the non-rigid gel pore structure or for specific interactions between the water molecules and surface groups, such as silanol groups, in the gel pore network.

References

1   M. Schmidt, D. Stephan, R. Krelaus and C. Geisenhansküke, Cement International, 2007, **5**, 86.
2   E. Fehling, M. Schmidt, T. Teichmann, K. Bunje, R. Bornemann, and B. Middendorf, (eds.), Structural Materials and Engeeinring Series, 1, Kassel University Press, 2005
3   C. Geisenhanslüke and M. Schmidt, Proc. 15. Ibausil, Bauhaus-Universität Weimar, 2003.
4   P. Richard and M. Cheyrezy, Cem. Concr. Res., 1995, **25**, 1501.
5   M. Cheyrezy, V. Maret and L. Frouin, Cem. Con. Res., 1995, **25**, 1491.
6   ISO 15901-1:2005, ISO International Organization for Standardization, 2005.
7   DIN 66137-2, Beuth-Verlag, Berlin, 2004.
8   J. J. Thomas, J. Hsieh, and H. M. Jennings, Advn. Cem. Bas. Mat. 1996, **77**, 76.
9   I. Odler, Cem. Concr. Res., 2003, **33**, 2049.
10   M. C. G. Juenger and H. M. Jennings, Cem. Concr. Res., 2001, **31**, 883.
11   R. M. Espinosa and L. Franke, Cem. Concr. Res., 2006, **36**, 1969.
12   H. M. Jennings, Cem. Concr. Res., 2000, **30**, 101.
13   M. J. Setzer, DAfSt, 1977, **280**, 48.
14   J. Adolphs and A. Schreiber, Proc. Int. Symp. On Ultra High Performance Concrete, Structural Materials and Engineering Series, 3, Kassel University Press, 2004.

DETERMINATION OF SURFACE AREA, POROSITY & SURFACE PROPERTIES OF LUNAR REGOLITH

A. Dąbrowski[1], E. Robens[2], E. Mendyk[1], A. Bischoff[3], A. Schreiber[4], W. Gac[1]
M. Dumańska-Słowik[5], K. Skrzypiec[1] and J. Goworek[1]

[1]Maria Curie-Skłodowska University, M. Curie-Skłodowska Sq. 3, 20-031 Lublin, Poland
[2]Institut für Anorganische und Analytische Chemie, Johannes Gutenberg-Universität
Duesbergweg, 10-14, D-55099 Mainz, Germany
[3]Institut für Planetologie Wilhelms-Universität Münster Wilhelm-Klemm-Str. 10,
D-48149 Münster, Germany
[4]POROTEC GmbH Niederhofheimer Str. 55a D-65719 Hofheim, Germany
[5]Depertment of Mineralogy, Petrography and Geochemistry. Faculty of Geology,
Geophysics and Environmental Protection. AGH-University of Science and Technology.
Mickiewicz Av. 30, PL 30-059 Cracow, Poland

1 INTRODUCTION

The lunar soil and rock samples of the Apollo (1969-1972) and the Soviet Union Luna missions (1970-1976) have been examined in detail.[1-4] Planning of new missions and establishing of a manned station at the moon require some more information. The assumption of an inventory of water ice suggests studying the surface properties of regolith in contact with water. We expect that water - if present at the moon - exist exclusively in the form of chemisorbed and vicinal water in some depth in regions which are protected from sun irradiation or fixed in minerals.[5] In particular the inner pore surface of soil should be available to adsorption of water. Aim of the planned investigation is to obtain parameters of the ability of storing water near the lunar surface, both amount and kinetics in relation to environmental conditions.

Looking at the Moon with the naked eye, bright (highlands) and dark (maria) areas can be seen on its surface. Centuries ago, the dark areas were called "maria", assuming water oceans. Telescopic observations showed that the maria are very flat, and are very different from the so-called highlands. We have learned that the maria are relatively young areas on the Moon. During later volcanic episodes, liquid magma came to the surface and filled these basins. When it cooled down and solidified, it formed the large flat areas. As this happened in comparatively recent times, the number of impact craters is far less than in the highland areas.

Impact processes have formed the lunar regolith. A fragmental layer of broken melted and otherwise altered debris caused by innumerable cratering event covers the original bedrock. The regolith layer varies from 3 to 5 meters in the maria to 10 to 20 meters in the highlands. Lunar soil is the subcentimeter fraction of the lunar regolith. Detailed studies have shown that five basic particle types make up the lunar soils: mineral fragments, pristine crystalline rock fragments, breccia fragments, glasses of various kinds, and the unique lunar constructional particles called agglutinates.

2 SAMPLES, METHODS AND INSTRUMENTS

Scanning electron microscopy study was performed at the AGH, Cracow (Poland), density, krypton and vapour sorption measurements at POROTEC, Hofheim (Germany), nitrogen adsorption, TPR studies and Raman spectra at UMCS, Lublin (Poland). Additional investigations and results are published elsewhere.[6-10]

2.1 Samples

We investigated three lunar samples, about 3 g each consisting of weakly coherent fines. Sample 10084.2000 from the Apollo 11 and sample 12001.922 from the Apollo 12 mission are from inside mare basalt regions, and consist mainly of abundant mare-derived basaltic rock clasts and mafic minerals like olivine and pyroxene. Sample 64501.228 is from Apollo 16 mission, which landed in highland regions. It contains abundant highland-derived components like lithic fragments of anorthositic rocks and breccia, and anorthitic feldspar. Sample of the Apollo 11 is dark grey others both appear light grey and with few white particles included, especially inside the Apollo 16 soils.

2.2 Scanning electron microscopy

The fine-grained textures of the three lunar soils were studied by SEM. Backscattered electron (BSE) observations were performed using FEI Quanta 200 FEG scanning electron microscope equipped with EDS detector. The system was operated in high vacuum mode at 15 kV accelerating voltage.

2.3 Specific surface area - adsorption methods

The adsorption isotherms of krypton and nitrogen were measured by standard volumetric/ manometric methods.[12] The Kr isotherm was measured stepwise at 77 K by means of SORPTOMATIC apparatus of Thermo Electron S. p. A., Milano, Italy. The specific surface area was calculated using the 2-parameter of Brunauer, Emmett, Teller (BET) equation in the range of $0.07 < p/p_0 < 0.23$, and using a molecular cross section area of 0.195 nm^2 for krypton molecule. Before the measurements, the samples had been degassed at 80 °C for 4 hours. Adsorption and desorption isotherms of N_2 were collected at 77 K according to the micropore analysis procedure using AUTOSORB-1CMS analyzer (Quantachrome Instruments, US). It follows from our thermogravimetric experiments the higher degassing temperature should be applied than in the previous measurements with Krypton.[7,10] Moreover, a stronger outgassing procedure is needful the micropores to be effectively evacuated. Before measurements the samples (about 1 g) were deeply degassed at 200 °C for 2 hours. For the surface areas calculations likewise the 2-parameter BET equation was applied in the range $0.1 < p/p_0 < 0.3$, and using a molecular cross section area of 0.162 nm^2.

Additionally, the specific surface areas are determined from vapour sorption of n-heptane, n-octane and water measured by gravimetric techniques. Adsorption and desorption isotherms were measured using a DVS 1/Advantage apparatus of Surface Measurement Systems, Ltd., Wembley, Middlesex U.K. The apparatus encompasses a thermostat and Cahn microbalance with maximum load of 1.5 g. Vapour pressure is adjusted and varied by means of a nitrogen carrier gas flow loaded with the adsorbed vapour. The isotherms were measured stepwise at a temperature of 24.9 °C slightly above ambient to avoid condensation at parts of the apparatus.

2.4 Density

In the present investigation, the density was measured by means of the standardised method[11] of helium displacement by the sample within a calibrated vessel at ambient temperature and pressure. We used a PYCNOMATIC ATC instrument of Thermo Electron S. p. A., Milano, Italy. The temperature was controlled at 20 °C.

2.5 TPR studies

TPR studies were conducted on an Altamira AMI-1 system (Zeton Altamira) equipped with thermal conductivity detector TCD. The sample was placed in the quartz reactor with inner diameter 10 mm. High purity Argon and the mixture of 6.2 % H_2 in Ar were used. Samples were heated up to 900 °C with ramp rate of 10 °/min. Calibration of the TCD detector was carried out by measurements of reduction of 20 mg of CuO.

2.6 Raman spectroscopy

Raman spectra were recorded using inVia Reflex Raman microscope (Renishaw plc, UK) coupled to a single-grating spectrograph fitted with CCD camera detector. The two lasers of 514 and 785 nm wavelengths were used for Raman spectra acquiring.

3 RESULTS

3.1 Scanning electron microscopy

SEM micrographs of the lunar samples are shown on Figure 1. The Apollo 16 soil contains

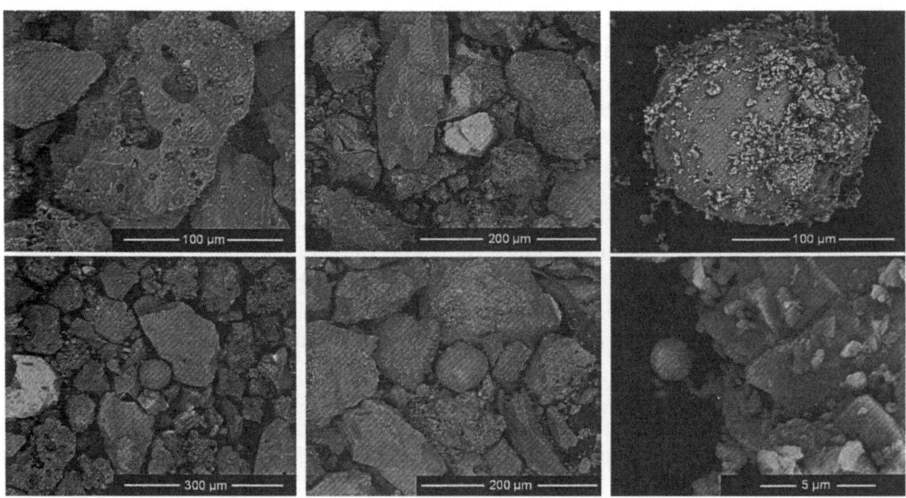

Figure 1 *SEM-BSE pictures of regolith samples: 10084.2000 of the Apollo11 mission (left panel), 12001.922 of the Apollo 12 mission (centre panel), and 64501.228 of the Apollo16 (right panel)*

abundant clastic grains, but also impact molten spherules of various sizes. Fragments with internal pores like those found in the Apollo 11 and 12 soils are either rare or absent. It is reported that in the particle size range of 10-90 µm the Apollo 16 sample is highly feldspathic having more than 50 % plagioclase.[1] In the Apollo 12 sample, more small-grained and more molten and porous particles are found compared with the Apollo 16 soil sample.

The Apollo 12 sample was previously characterised as a well-gardened soil, random mixture of debris fragments that mostly derive from nearby bedrock and whose particles largely consist of mare basalts or degradation products thereof.[2] The Apollo 11 soil has a high abundance of glassy particles or features that indicate melting. The soil is fine-grained with a mean grain size of ~70 µm and poorly sorted. The high concentration of agglutinates (52 %) was determined by Simon et al.[3]

The cleaved and rough particle surfaces of the Apollo 11 and 12 samples pretend a larger surface than those of Apollo 16. However, the structures resolved in the micrometer range do not contribute much to the specific surface area as determined by gas adsorption.

3.2 Krypton and nitrogen adsorption

All N_2 and Kr isotherms at 77 K correspond to type II of the IUPAC classification with nearly linear slopes within $0 < p/p_0 < 0.7$ and practically without hysteresis between adsorption and desorption branch (Figure 2). Resulting values of specific surface area (Table 1) are within the region of values determined from N_2 isotherms by other investigators (0.2-1.5 m^2g^{-1}). However, the values resulting from nitrogen adsorption measurements obtained from an Autosorb analyser differ significantly. We suppose it's due to the stronger outgassing conditions applied.

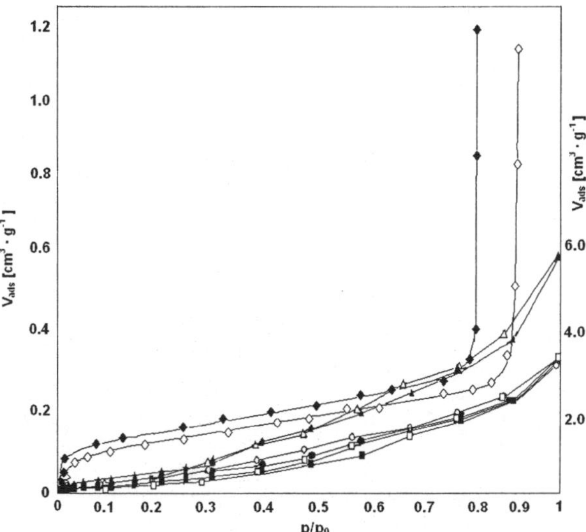

Figure 2 Nitrogen *adsorption (filled points) and desorption (open points) isotherms at 77 K (right V_{ads} scale), triangles - Apollo 16, circles - Apollo 12, squares-Apollo 11. Krypton adsorption at 77 K (left V_{ads} scale), filled diamonds-Apollo16, open diamonds - Apollo 12.*

3.3 Density

In Table 1, mean values of repeated density measurements are summarised. The results are in line with the literature values which range from 2.3 to > 3.2 g cm^{-3} and similar to the commonly accepted value of 3.1 g cm^{-3}.

Table 1 *Helium density and specific surface area (S_{BET}) of the regolith samples*

Sample	He- density g cm^{-3}	S_{Kr} m^2 g^{-1}	S_{N2} m^2 g^{-1}	$S_{n\text{-heptane}}$ m^2 g^{-1}	$S_{n\text{-octane}}$ m^2 g^{-1}	S_{H2O} m^2 g^{-1}
10084.2000. Apollo 11 Mare	-	-	2.39	-	0.48 0.43	0.37
12001.922, Apollo 12 Mare	3.10 ± 0.01	0.58	1.77	0.41	-	0.36
64501.228, Apollo 16 Highland	2.79 ± 0.01	0.65	2.73	-	0.54	0.40

3.4 Water vapour adsorption

Water vapour adsorption and desorption (Figure 3a and 3b) at ambient temperature proceeds fast. A diffusion process through small pores has no effect. The isotherm increases rather slowly from origin which is consistent with literature reports. That is typical for material of a character between hydrophilic and hydrophobic, and it indicates feeble binding. Indeed, investigation reported in the literature show likewise scattering results.[1-3] the specific surface areas that determined with water vapour (diameter of 0.28 nm) is lower than that determined with larger krypton and nitrogen molecules. The reason of this effect is unclear and signalises uncertainty in the measurements. As can be seen in Figure 3 the adsorption of water vapour is almost identical for both investigated materials. Hysteresis loops extending over the whole p/p$_0$ region were observed which may be due to a saturation value of the adsorbed mass was not attained.

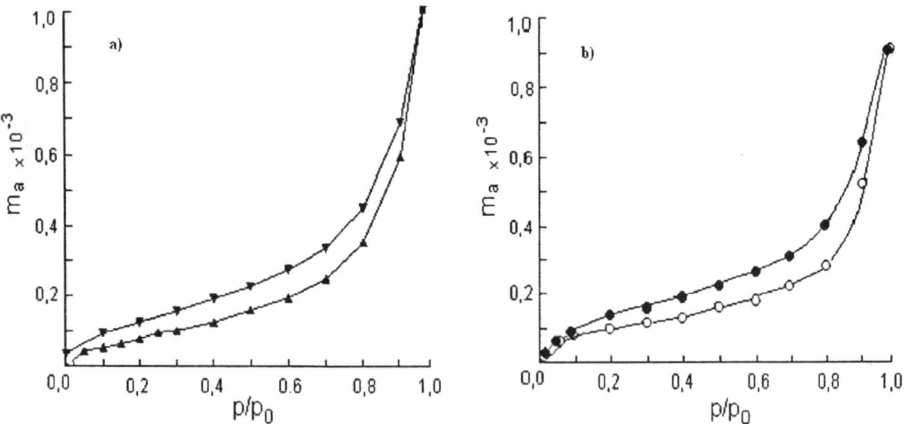

Figure 3 Water adsorption and desorption isotherms, a) Apollo 12 regolith, b) Apollo 16 regolith; m$_a$ - mass adsorbed per gram of sample, p/p$_0$ - relative pressure (humidity).

3.5 TPR studies

Figure 4 shows temperature programmed reduction curves. The minute maximum in the initial TPR run is visible for all samples in the range of 400 - 600 °C. Main reduction process begins above 600 °C. Several overlapping reduction peaks can be distinguished for the Apollo 11 and Apollo 12 samples. It can also be seen that short period of an isothermal reduction at 900°C was insufficient to obtain complete reduction. In Table 2 the calculated amounts of hydrogen consumption from the TPR measurements are shown.

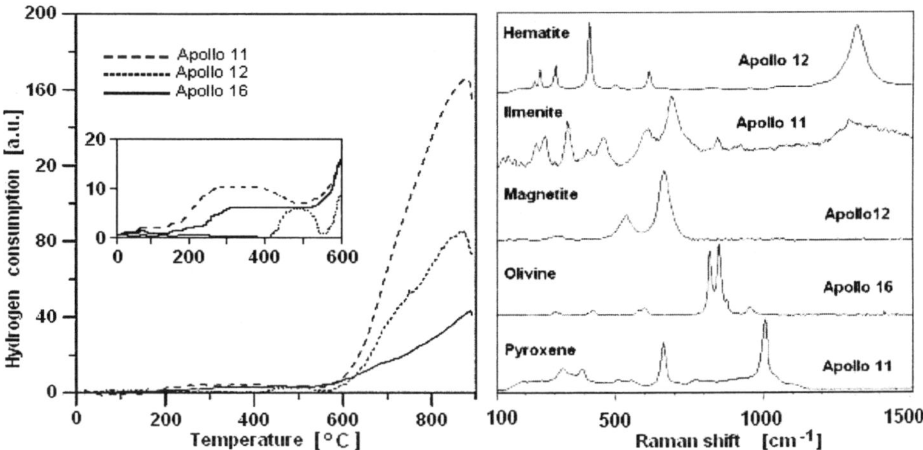

Figure 4 *TPR curves for the regolith samples. First stage of the reduction is shown in enlargement in parent window*

Figure 5 *Raman spectra of iron bearing phases obtained from certain points on the surface of the Apollo 11, Apollo 12 and Apollo 16 samples*

All samples contain large amounts of the species with strong metal-oxygen interactions. The values of hydrogen consumption decrease in the order Apollo 11 > Apollo 12 > Apollo 16. This order correlate quite reasonably with the total iron and titanium abundance (see Table 2, the values in parentheses) obtained by XRF method.[10] Generally, the reduction reaction can be described as follows:

$$\delta H_2 + M\text{-}Fe\text{-}O_x \rightarrow M\text{-}Fe\text{-}O_{x\text{-}\delta} + \delta H_2O \qquad (1)$$

However, it can be noted that other oxide species (i.e. Cr or Mn), especially if they are present on the higher oxidation stage, can be also reduced. Small abundance of these elements in the investigated lunar samples was determined by XRF, ICP-MS and SEM/EDX methods, and reported elsewhere.[10]

Table 2 *TPR experiments. Amount of hydrogen (μmol) consumed during reduction*

Apollo 11 (Fe=10.55%, Ti=5.11%)	Apollo 12 (Fe=10.91%, Ti=1.26%)	Apollo 16 (Fe=2.83%, Ti=0.31%)
56,66	25,32	14,92

The first, small peaks on the TPR curves can be related to the reduction of hematite - iron oxide species to magnetite ($Fe_2O_3 \rightarrow Fe_3O_4$).[14] The second stage of the reduction of these oxides occurs at higher temperatures, and proceeds via successive reduction from magnetite to wustite (FeO) or directly to metallic iron. High temperature maxima of an iron bearing materials have been often ascribed to reduction of an iron with lowest oxidation level (FeII). Similar effect can be observed to chemical compounds, in which Fe-O bonds are stabilised by other elements, such as Al, Mg, Ca, Ti or Si. Our studies of the mineralogical composition of the Apollo 11 sample by Raman spectroscopy indicated the presence of large amount of pyroxene and ilmenite, smaller amount of dispersed species of olivine, and several multiphases of species difficult to identify unambiguously.[9,10] The reduction of ilmenite has been proposed for production of oxygen on the moon. The first stage of that process describes equation 2.

$$FeTiO_3 + H_2 \rightarrow Fe + TiO_2 + H_2O \tag{2}$$

Reduction of the natural ilmenite proceeds through the stages of Fe^{3+} to Fe^{2+} and Fe^{2+} to Fe^0. Moreover, hydrogen can be consumed by Ti-oxides, leading at the beginning to the formation of Ti^{3+} and then, under more severe conditions (especially during reduction at high temperatures by carbon), to the formation of sub-oxides TiO and TiO_{1-x}.[13]

The Apollo 12 and Apollo 16 samples contain small amounts of ilmenite phase, but relatively large amounts of Fe on the low oxidation stage. Hence the peaks on the TPR curves can be regarded as reduction of Fe_3O_4 or FeO and iron silicate minerals. The presence of these iron-bearing phases was clearly confirmed by Raman spectroscopy (see Figure 5).

4 CONCLUSIONS

Commercial density and sorption measuring instruments were proved to be suitable for investigating lunar soil. Sensitivity and resolution are comparable to that of instruments used in investigations reported in the literature. A significant influence on the sorption properties of the samples by heat treatment at temperature about 200 °C is observed. This effect needs to be studied in additional investigations and the results should be verified

No differences in sorption properties have been detected for samples of different origin (mare and highland). Lunar regolith can hardly store water due to low specific surface area and little nano-porosity. On moon never existed conditions that allowed the existence of liquid water. Therefore, gravity induced oozing of liquid water into the depth is impossible. Transport of water under lunar conditions can take place via the gas phase or in a quasi-liquid phase of few molecular layers at the surface of the soil material.

Water-ice is imported to the Moon by comets and cosmic dust particles. Remnants may be stored as thin, incomplete physisorbed or chemisorbed layers at the soil material or fixed in minerals, and bulk stocks are assumed in craters near the poles in which sun radiation cannot enter. Everywhere continuous evaporation takes place. With regard to the diurnal temperature cycle evaporation will be stronger near the outer surface and is less obstructed by diffusion through the regolith layer. Therefore, physisorbed water may be found in some depth below the regolith covering.

TPR studies enriched the knowledge of the nature of the lunar regolith. They show some insight into the redox processes. Ours studies suggest a complex M-O-Fe-O interactions in the lunar minerals (M = Ca, Ti, Al, Mg). These investigations will be continued.

Acknowledgements

The samples had been friendly placed to our disposal by the NASA sample curator Dr. Gary Lofgren, Houston, Texas.

References

1   M..J. Kempa and J.J. Papike, 'The Apollo 16 Regolith: Comparative Petrology of the >20 and <20 μm Soil Fractions' in *Lunar and Planetary Sciences XI*, The Lunar and Planetary Institute, Houston 1980, p. 535.

2   J.A. Wood, U.B Marvin, J.B. Reid, Jr., G.J Taylor, J.F. Bower, B.N. Powell, J.S. Dickey, Jr, *SAO Special Report, SAO*, 1971, #**333**, p. 272

3   S.B. Simon, J.J. Papike and J.C. Laul, in *Proceedings of the Lunar Planet. Sci. Conf.* 12B, 1981, p. 371.

4   D.A. Cadenhead , N.J. Wagner, B.R.Jones, J.R Stetter, *Proceedings of the Third Lunar Sci. Conf. (Suppl. 3, Geochimica et Cosmochimica Acta)*, M.I.T. Press 1972, 2243.

5   M. Chaussidon, *Nature*, 2008, **454**, 170.

6   E. Robens, A. Bischoff, A. Schreiber, A. Dąbrowski, K.K. Unger, *Appl. Surface Sci.,* *2007*, **253**, 5709.

7   M. Iwan, Z. Rzączyńska, E. Mendyk, A. Dąbrowski, E. Robens, 'Thermal Analysis of lunar samples' in *Proceedings of the IXth Seminary on Thermal Analysis,* IChPW, Plock 2007, pp. 164-168

8   E. Robens, A. Bischoff, A. Schreiber, K.K. Unger, *J. Therm. Anal. Cal.* 2008, in press.

9   A. Dąbrowski, E. Mendyk, E. Robens, K. Skrzypiec, J. Goworek, M. Iwan, Z. Rzączyńska, *J. Therm. Anal. Cal.*, 2008, in press

10 E. Robens, A. Dąbrowski, E. Mendyk, J. Goworek, K. Skrzypiec, M. Drewniak, M. Dumańska-Słowik, W. Gac, R. Dobrowolski, S. Pasieczna-Patkowska, M. Huber, M. Iwan, K.J. Kurzydłowski, T. Płociński, J. Ryczkowski, Z. Rzączyńska, J.W.Sobczak, *Annales Universitates Mariae Curie-Skłodowska Sectio AA Chemia* 2008, in press.

11 *DIN 66137-2:* Bestimmung der Dichte fester Stoffe -Teil 2. Gaspycnometrie, Berlin, Beuth 2004.

12 *ISO 9277*: Determination of the specific surface area of solids by gas adsorption using the BET method, Berlin, Beuth 2007.

13 W.K. Jóźwiak, E. Kaczmarek, T.P. Maniecki, W. Ignaczak, W. Maniukiewicz, *Appl. Catalysis A,* 2007, **326,** 17.

14 C.S. Kucukkaragoz, R.H. Eric, *Minerals Engineering*, 2006, **19**, 334.

THE USE OF STARCH-BASED BIOPOLYMER ADDITIVES FOR THE PREPARATION OF POROUS CERAMICS

E. Gregorová,[1] Z. Živcová,[1] W. Pabst,[1] S. Holíková[1] and I. Sedlářová[2]

[1] Department of Glass and Ceramics, Institute of Chemical Technology, Prague, Technická 5, 166 28 Prague, Czech Republic
[2] Department of Inorganic Technology, Institute of Chemical Technology, Prague, Technická 5, 166 28 Prague, Czech Republic

1 INTRODUCTION

During recent years starch has become probably the most frequently used biopolymer in ceramic technology.[1-17] Starch can be used as a simple pore-forming agent (PFA),[1-7] similar to other PFAs of biological origin, such as poppy seed or lycopodium spores,[17-19] or as a combined pore-forming and body-forming (consolidating) agent, as in the process of starch consolidation casting (SCC).[8-17] In both cases the starch type determines the shape and size of pores in the ceramic microstructure. When traditional slip casting (TSC) is used for shaping of porous ceramics (using starch as a PFA) the resulting porosity corresponds very well to the initial PFA-to-ceramic ratio. More precisely, the nominal PFA content (volume fraction of starch related to the ceramic powder in the suspension) is approximately equal to the total porosity in the fired ceramic. The percolation threshold for ceramics prepared by TSC with PFAs is approx. 18 %, i.e. only when this porosity value is exceeded, the pore space can be considered as interconnected.[13] With SCC the situation is different. This method can be sucessfully applied only when the nominal starch content is higher than 5 – 10 vol.%, depending on the starch type, and the maximum content is approx. 60 vol.% or slightly higher, depending on the starch type as well as ceramic powder type and content.[8-17] For low starch contents the porosity achieved with SCC is always higher than the PFA content (nominal starch content) and the pore space is predominantly open.

With respect to the successful use of starch for the preparation of porous ceramics, there is clearly a temptation of using other starch-based biopolymeric materials which are readily available and cheaper than native starch itself. Wheat flour is a mixture of biopolymers, in which the main component is wheat starch. Thus the main chemical elements are C, O, and H. Further components are proteins, which additionally contain small amounts of nitrogen (and sometimes sulphur).[20] Higher proteins (e.g. albumins or globulins) usually exhibit a tendency of forming foams when mechanically agitated or stirred.[20-22] Wheat flour is readily available and commercial grades range from fine flour to semolina. The advantage of using flour or semolina in ceramic technology is their lower price compared to refined products such as native or modified starch. On the other hand it is clear that the presence of non-starch components, in particular proteins, may cause complications or at least require certain modifications in process control. In particular, when used as a pore-and-body-forming agent for the preparation of porous ceramics from suspensions, the resulting process is supposed to be in fact a combination of starch consolidation and protein forming.[21] It is the purpose of this work to examine the possibilities and potential of preparing porous ceramics with fine wheat flour and semolina.

2 METHODS APPLIED AND EXPERIMENTAL RESULTS

2.1. Characterization of pore-forming agents

Two grades of wheat products have been used in this work, a fine flour (Předměřická mouka pšeničná, Mlýny J.Voženílek, s.r.o., Czech Republic, here denoted "F") and semolina (Zátkova jemná krupička, bratři Zátkové a.s., Czech Republic, here denoted "S"). Micrographs of these PFAs were made by optical microscopy (Jenapol, Zeiss, Jena Germany), see Figure 1. It is obvious, that both wheat flour and semolina consist mainly of aggregates of starch granules. The particle size was measured by laser diffraction (Particle Sizer Analysette 22 Nanotec, Fritsch GmbH, Idar-Oberstein, Germany), see Figure 2. Since these measurements were performed in flowing water with ultrasonics, the wheat grain fragments can be partly destroyed and disintegrated into individual starch granules. However, thes measurement conditions are very similar to the conditions in the aqueous ceramic suspensions during and after homogenization. Figure 2 shows the particle size distributions measured for wheat flour (left) and semolina (right) immediately after mixing with water (bottom curves) and after 4 h soaking in (cold) water (top curves). In the case of flour (F), the distribution is monomodal with a mode of approx. 30 μm, while in the case of semolina (S) the distribution is bimodal with a second mode at approx. 500 μm (the fine fraction of semolina has a mode of approx. 25 μm). In both cases the size of the fine fraction corresponds very well to that of refined wheat starch (mode approx. 23 μm [11]). After soaking in cold water for 4 h the size of fine flour is not reduced further, i.e. these small aggregates disintegrate into starch granules more or less immediately).

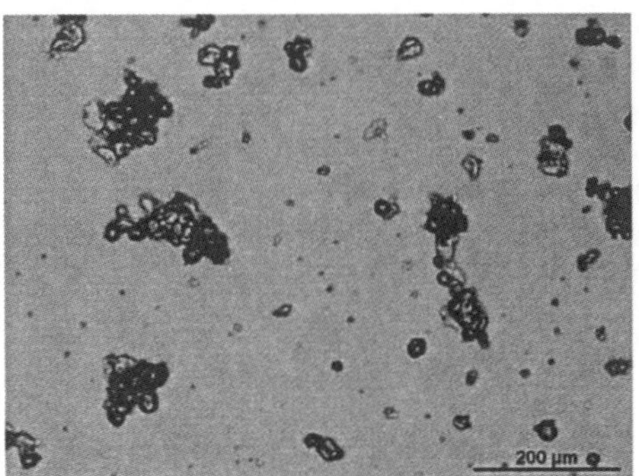

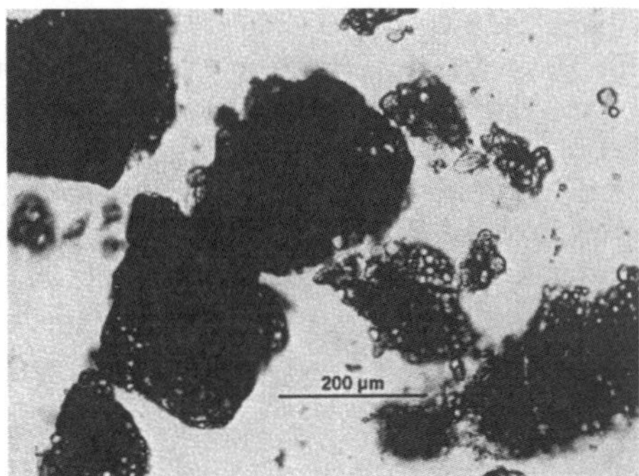

Figure 1 *Optical micrographs of wheat flour (left) and semolina (right); both are wheat grain fragments consisting predominantly of aggregated starch granules*

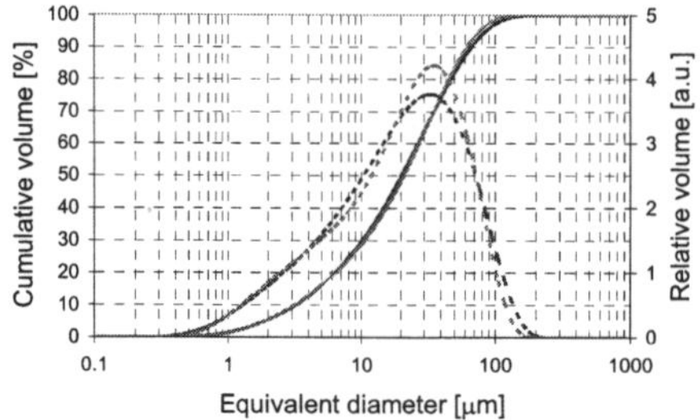
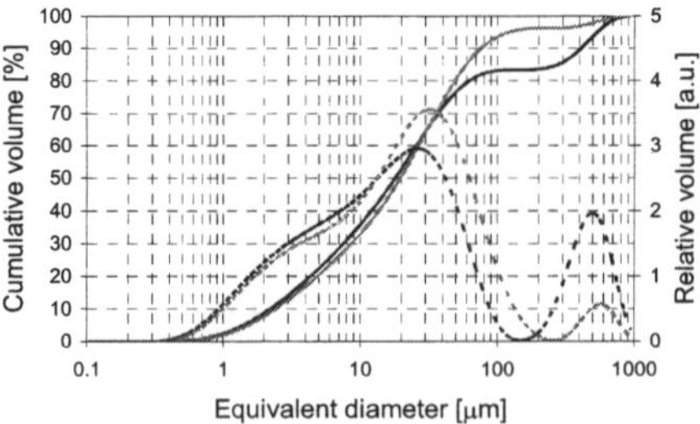

Figure 2 *Particle size distributions of wheat flour (left) and semolina (right) in cold water measured by laser diffraction; bottom curves immediately after mixing, top curves after 4 h*

Table 1 lists the elemental composition of the pore-forming agents (PFAs) wheat flour (F) and semolina (S) as measured by X-ray fluorescence analysis (spectrometer ARL 9400 XP, Switzerland). Note, however, that these trace elements determinable by XFA represent only approx. 0.5 wt.% of the overall sample mass (the rest is C, O, H and N).

Table 1 *Amounts of trace elements [wt.%] in wheat flour (F) and semolina (S) according to X-ray fluorescence analysis*

	Mg	Al	Si	P	S	Cl	K	Ca
F	0.0397	-	-	0.1450	0.2010	0.0670	0.1700	0.0231
S	0.0252	0.1260	0.0087	0.0980	0.1690	0.0620	0.1310	0.0152

2.2. Preparation of porous alumina ceramics

Aqueous alumina suspensions have been prepared with 70 wt.% of submicron alumina powder Al_2O_3 (median size approx. 0.7 µm, CT 3000 SG, Almatis GmbH, Germany) in distilled water, with 1 wt.% (related to Al_2O_3) of deflocculant (Dolapix CE 64, Zschimmer & Schwarz GmbH, Germany). After mixing with the biopolymeric PFAs (F or S) in amounts of 20 or 30 vol.% (related to Al_2O_3, assuming densities of 1.5 g/cm^3 and 4.0 g/cm^3 for the PFAs and alumina, respectively) the suspensions were mechanically agitated in polyethylene bottles with alumina balls on a laboratory shaker at 5 Hz (HS 260, IKA GmbH, Germany) for 120 or 180 min (homogenization step). Suspension rheology was characterized via rotational viscometry (RotoVisco 1, Haake GmbH, Germany). From Figure 3 it is evident that in the case of suspensions with 20 vol.% PFA the homogenization time does not influence the suspension rheology significantly (i.e. differences are within experimental error), whereas for suspensions with 30 vol.% PFA the shear stress (and apparent viscosity) increases with prolonged homogenization time.

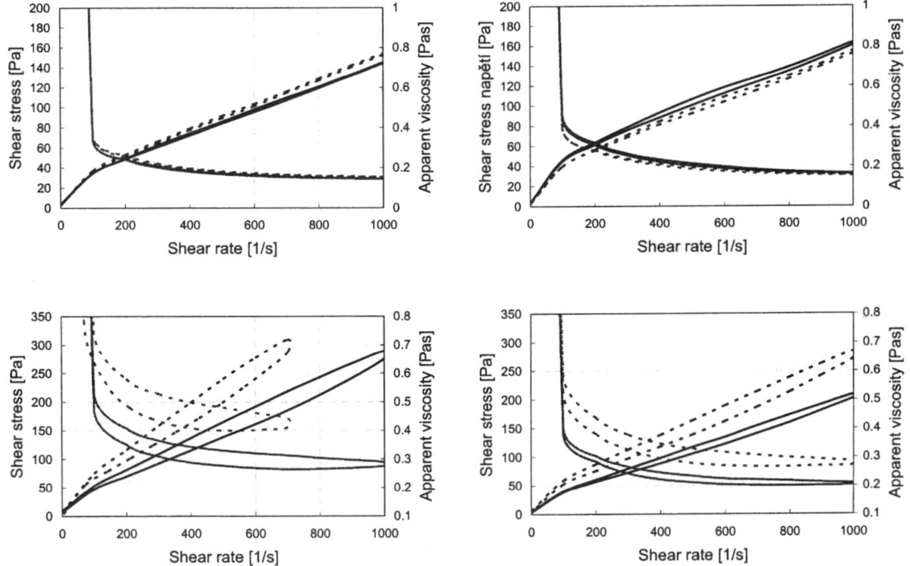

Figure 3 *Flow curves (increasing, up and down) and viscosity curves (decreasing, up and down) for 70 wt.% alumina suspensions with 20 vol.% (top graphs) and 30 vol.% (bottom graphs) PFA after homogenization for 2 h (full curves) and 3 h (dotted curves); wheat flour (left) and semolina (right)*

A further step was the overall size characterization of particles in the as-homogenized suspensions. It could be shown that already after 1 h of homogenization the final PFA particle size (i.e. more or less the starch size) is attained and no further size reduction is possible during this step. It can therefore be assumed that even differently sized wheat grain products (such as fine flour and semolina) lead to the same final PFA size after suspension preparation, and thus the same pore size in the final ceramics, irrespective of the initial state of the PFA. Suspensions mechanically agitated for different times (homogenization for 2, 3, or 8 h) were cast into metal molds. After casting the molds were heated up in a laboratory drier to approx. 80 °C for 2 h. After cooling down to room temperature the samples were demolded, dried to constant mass and fired with a standard schedule (heating rate 2 °C / min) to 1570 °C (hold time 2 h).

2.3. Characterization of porous microstructures

The term "microstructure" is widely used in materials science to denote the structure of heterogeneous materials in general (and of porous materials in particular), and does not imply the presence of "micropores" in the sense of the IUPAC classification. In the present work the microstructure of the as-fired samples was characterized via the Archimedes method (double weighing in water), mercury intrusion porosimetry and microscopic image analysis, see Table 2. It is evident that the porosity increases with increasing homogenization time, because of foaming of the suspension during mechanical agitation. In systems with 20 vol.% wheat flour the porosity increases from 38.2 to 59.1 % for homogenization times 2 and 8 h, respectively. For systems with 30 vol.% the porosity achieved was lower for all homogenization times. This can be explained by the higher suspension viscosity, which makes foaming more difficult (the mechanical agitation being on the same level in both cases). In order to prove that foaming of the suspension is a direct consequence of the PFAs used (more precisely, the protein content in wheat flour and semolina), and not an artifact from suspension preparation, identical alumina suspensions were prepared for comparison purposes with 30 vol.% of pure wheat starch (Pšeničný škrob, Amylon a.s., Czech Republic), which is a refined product with negligible protein content. Porous alumina ceramics have been prepared from these suspensions using exactly the same process steps. The porosity determined for as-fired samples was 31.7 ± 1.3 % after 2 h homogenization and 30.3 ± 1.4 % after 3 h homogenization of the suspension, i.e. within experimental precision the porosity is independent of the homogenization time and the microstructure does not exhibit large spherical pores.

As shown in Table 2, when wheat flour or semolina is used as a PFA in the suspension, the (total) porosity for the ceramics is strongly dependent on the suspension homogenization time (porosity increases with increasing homogenization time) and is usually higher than the nominal PFA content (i.e. the content of F or S in the suspension, related to alumina), except for the 30 vol.% suspension with S, where a longer homogenization time (3h) was required to induce foaming. Most remarkably, porosities of approx. 60 % can be readily achieved in the ceramics even with nominal PFA contents as low as 20 vol.% in the suspension, when the homogenization time is long enough (8h).

Table 2 *Total porosity, mode values from frequency curves, characteristic size values (quantiles) D_{10}, D_{50}, and D_{90}, from cumulative curves measured by mercury intrusion (see Figure 6), and porosity contribution of large pores (> 100 µm), measured by image analysis using a measuring square grid (63 points, area 2100 x 1800 µm) for porous alumina ceramics prepared with wheat flour (F, top) and semolina (S, bottom)*

Homog. time [h]	Nominal PFA content [vol.% F]	Total porosity [%]	Mode [µm]	D_{10} [µm]	D_{50} [µm]	D_{90} [µm]	Porosity due to large pores [%]
2	20 F	38.2	1.7	0.8	2.1	3.1	13.9
8	20 F	59.1	20.8	2.4	19.6	31.6	50.5
2	30 F	30.5	1.3	0.5	1.4	1.7	8.6
3	30 F	41.3	1.7	0.8	1.9	2.5	32.0
2	20 S	34.4	1.3	0.6	1.5	2.4	16.9
8	20 S	62.7	31.3	2.3	29.6	48.8	51.5
2	30 S	30.9	1.3	0.6	1.4	1.7	12.1
3	30 S	50.5	2.6	1.0	2.8	8.2	38.7

Figure 4 shows polished sections of the microstructures corresponding to some of the samples characterized in Table 2, and Figure 5 shows fracture surfaces (after firing) of the samples with highest porosity.

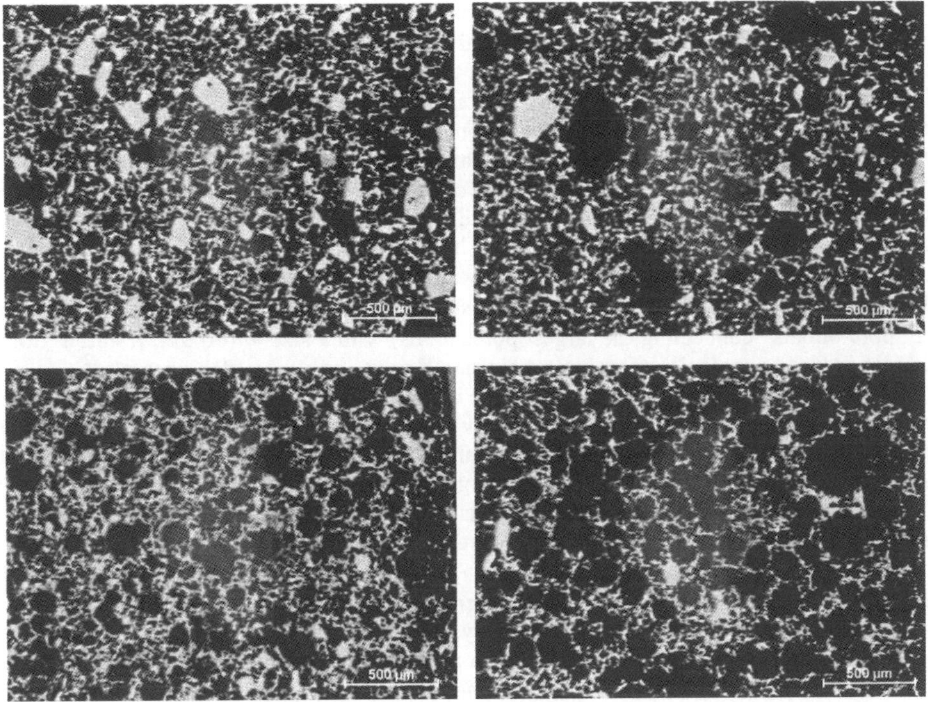

Figure 4 *Microstructures of porous alumina ceramics prepared with 30 vol.% PFA (left: wheat flour F, right: semolina S) after 2 h (top) and 3 h (bottom) suspension homogenization time (optical micrographs of polished sections)*

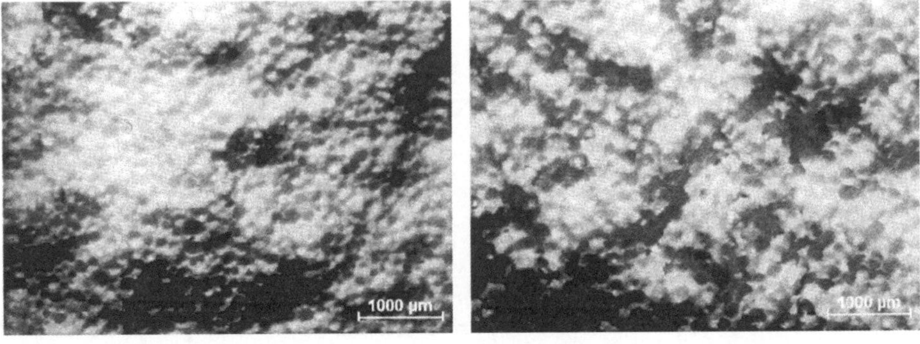

Figure 5 *Fracture surfaces of porous alumina ceramics prepared with 20 vol.% PFA (left: wheat flour F, right: semolina S) after 8 h suspension homogenization time*

From the polished sections it is evident that after 2 h homogenization there are many residual agglomerates of alumina left in the suspension (white regions in the micrographs). Beside the large amount of small many pores (essentially residuals of starch burnout), there is a small amount of large pores of non-uniform size. At this stage the large pores can be considered as singular inclusions in a more or less uniform porous matrix. This is consistent with the finding of mercury porosimetry that after 2 h homogenization the intrusion curves (for F and S) are monomodal with mode and median sizes of 1.3–2.1 µm, corresponding to the pore throat diameter of the interconnections between pores resulting from starch burnout, which must be approx. one order of magnitude smaller than the starch granules, cf. [11-17]. When the homogenization time is increased, the amount of alumina agglomerates is strongly reduced, the amount of large pores increases, and the microstructure as a whole becomes more uniform, finally adopting a cellular type. Figure 5 shows fracture surfaces of the samples with highest porosity (around 60 %), prepared from suspensions with 20 vol.% PFA after 8 h homogenization. At this stage, mercury porosimetry indicates the presence of pores with modes and medians in the size range 20–30 µm. That means, the residual voids from starch burnout now form a second generation of "pore throats" connecting the largest pores (cells) with a size > 100 µm, resulting from protein-induced suspension foaming, where starch serves as a foam stabilizer, cf. [22].

Image analysis results are included in Figure 6 and Table 2. Figure 6 compares the cumulative pore size distribution measured by mercury intrusion (curves on the l.h.s. of the graphs) with the size distribution of the largest pore generation, measured via image analysis (after transformation to volume-weighted curves, r.h.s. curves). It is evident that the medium-size fraction measured as "throats" by mercury porosimetry (in samples where a large amount of large pores is present) corresponds to the size of the pores to due starch burnout, cf. [14]. The large pores have median values of equivalent section area diameters of 216 µm and 164 µm for porous alumina ceramics prepared 20 vol.% suspensions with wheat flour (F) and semolina (S), respectively, after 8 h homogenization. The porosity due to large pores has been evaluated by image analysis using a measuring square grid and counting only those events in which large pores are hit by grid intersection points. The resulting values (see Table 2) clearly show that the contribution of large pores to the total porosity increases significantly with increasing homogenization time. We emphasize that in the sense of the IUPAC classification all pores considered in this paper are macropores.

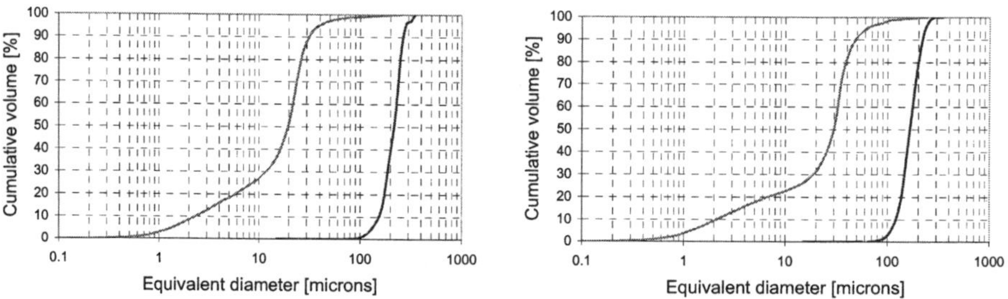

Figure 6 *Mercury intrusion curves (pore throat size distribution curves, l.h.s. curves) and image analysis curves (r.h.s. curves) for porous alumina ceramics prepared with 20 vol.% PFA (left: wheat flour F, right: semolina S) after 8 h suspension homogenization time*

3 CONCLUSIONS

Based on the findings of this paper, the following conclusions can be made: First, due to the protein content of wheat flour and semolina, starch consolidation casting (SCC) using these biopolymer materials as pore-forming agents (PFAs) and body-forming (consolidating) agents can be considered as a combination of starch consolidation and protein casting. The mechanical agitation of the suspension during the homogenization step causes foaming of the proteins, and starch serves as a foam-stabilizing agent. Second, after sufficient homogenization time the initial particle size of wheat flour and semolina has a relatively small influence on the microstructure of the porous ceramics after biopolymer burnout and firing, because the homogenization step causes additional biopolymer milling (desintegration of wheat grain fragments into indivudual starch granules) and protein-induced foaming becomes a dominant factor. Third, the porosity of ceramics increases with homogenization time. The homogenization time also controls the development and amount of large spherical pores. Increasing the homogenization time makes the microstructure more uniform, minimizes the occurrence of alumina agglomerates and reduces the width of the pore size distribution. Fourth, using wheat flour and semolina as pore- and body-forming agents in the SCC process leads to porous ceramics with a hierarchical microstructure consisting of at least three generations of pores: largest spherical ones caused by foaming (D_{50} approx. 200 μm), intermediate pores remaining as residual voids after starch burnout (D_{50} approx. 20 μm), and small pore throats (D_{50} approx. 2 μm). If four generations of pores were desired, residual interstitial pores between the alumina grains with an estimated size of < 0.2 μm could be retained by partial sintering.

Acknowledgement: *This study was part of the project IAA401250703 ("Porous ceramics, ceramic composites and nanoceramics", GAAV ČR), and of the frame research program MSM 6046137302 ("Preparation and Research of Functional Materials and Material Technologies using Micro- and Nanoscopic Methods", MŠMT ČR).*

References

1 S. F. Corbin and P. S. Apte, *J. Am. Ceram. Soc.*, 1999, **82**, 1693.
2 J. G. Kim, J. H. Sim and W. S. Cho, *J. Phys. Chem. Solids*, 2002, **63**, 2079.
3 A. Diaz and S. Hampshire, *J. Eur. Ceram. Soc.*, 2004, **24**, 413.
4 R. Barea, M. I. Osendi, J. M. F. Ferreira and P. Miranzo, *Acta Mater.*, 2005, **53**, 3313.
5 C. Galassi, *J. Eur. Ceram. Soc.*, 2006, **26**, 2951.
6 J. M. LeBeau and Y. Boonyongmaneerat, *Mater. Sci. Eng. A*, 2007, **458**, 17.
7 L. Yang, X. Ning, K. Chen and H. Zhou, *Ceram. Intern.*, 2007, **33**, 483.
8 O. Lyckfeldt and J. M. F. Ferreira, *J. Eur. Ceram. Soc.*, 1998, **18**, 131.
9 H. M Alves, G. Tarí, A. T. Fonseca, J.M.F. Ferreira, *Mater. Res. Bull.*, 1998, **33**, 1439.
10 M. E. Bowden and M. S. Rippey, *Key Eng. Mater.*, 2002, **206-213**, 1957.
11 E. Gregorová, W. Pabst and I. Bohačenko, *J. Eur. Ceram. Soc.*, 2006, **26**, 1301.
12 E. Gregorová, Z. Živcová, W. Pabst, I. Sedlářová, *Ceramika-Ceramics*, 2006, **97**, 219.
13 E. Gregorová, Z. Živcová and W. Pabst, *J. Mater. Sci.*, 2006, **41**, 6119.
14 E. Gregorová and W. Pabst, *J. Eur. Ceram. Soc.*, 2007, **27** (2007), 669.
15 S. Bhattacharjee, L. Besra and B. P. Singh, *J. Eur. Ceram. Soc.*, 2007, **27**, 47.
16 X. Mao, S. Wang and S. Shimai, *Ceram. Intern.*, 2008, **34**, 107.
17 Z. Živcová, E. Gregorová and W. Pabst, *Processes Applications Ceram.*, 2008, **2**, 1.
18 E. Gregorová and W. Pabst, *Ceram. Intern.*, 2007, **33**, 1385.
19 Z. Živcová, E. Gregorová and W. Pabst, *J. Mater. Sci.*, 2007, **42**, 8760.
20 A. M. Stephen, *Food Polysaccharides and Their Applications*, Marcel Dekker, New York, 1995.
21 O. Lyckfeldt, J. Brandt and S. Lesca, *J. Eur. Ceram. Soc.*, 2000, **20**, 2551.
22 U. T. Gonzenbach, A. R. Studart, D. Steinlin, E. Tervoort and L. J. Gauckler, *J. Am. Ceram. Soc.*, 2007, **90**, 3407.

IRON FIXATION TO THE FIBERS OF ACTIVATED CARBON AS AN ALTERNATIVE TO THE FENTON PROCESS

D. Barba[1], L. Giraldo[2], J.C. Moreno-Piraján[1] and V. Sarría[1]

[1] Department of Chemistry, Universidad de los Andes, Carrera 1A No. 18A-10, Bogotá, Colombia.
[2] Department of Chemistry, Universidad Nacional de Colombia, Ciudad Universitaria, Bogotá, Colombia.

1 INTRODUCTION

The Fenton process (Fe^{2+}/H_2O_2) is an alternative technique that is frequently being used for the chemical oxidation of non-biodegradable organic matter since the compounds used for this treatment do not represent risks to the environment or public health. Additionally, this process has been proposed to facilitate the biodegradation of toxic organic compounds and/or biorecalcitrants.[1] For this technique, oxidant agents such as hydrogen peroxide and catalysts like ferrous ion salts are used. With redox reactions at $pH \leq 3$, these agents generate hydroxyl radicals ($^{\bullet}OH$) and organic matter is degraded.

$$Fe^{2+} + HO{-}OH \longrightarrow Fe^{3+} + {}^{\bullet}OH + {}^{-}OH \tag{1}$$

Because the catalyst (Fe^{2+} ion) is in the same phase as the agents, the catalysis involved in this reactions are homogeneous. However, the Fenton reaction has several disadvantages, such as the high acidity that must be used, requiring a posterior neutralization. Additionally, once degradation is achieved, it is needed to remove the iron in compliance with government regulations on compound discharges in bodies of water, leading to an additional separation process that generates a significant quantity of waste. Some investigators have proposed reprocessing this catalyst in order to avoid excess waste. To compensate for these disadvantages, techniques have been tested to affix the catalyst on diverse supports such as fibers, clays, granular carbon and polyethylene membranes.

This research aims to evaluate catalyst fixation, presenting the organic compound adsorption attained through the Fenton process and heterogeneously catalyzed by affixing it to the same adsorbent agent as an alternative. Phenol was used as the model contaminant. This preliminary study shows promising results. Slight iron lixiviation, high levels of removal, the possibility of reusing the material containing the catalyst available for Fenton degradation and working at a nearly neutral pH are some of the advantages of this technique.

2 METHODS AND RESULTS

2.1   Activated Carbon Fibers Analysis

The porous texture of the material was analyzed. Adsorption isotherms of N_2 at 77 K was assessed using the BET method. Additionally, the volume and the porous distribution were assessed using the Dubinin Radushkevich method. Table 1 presents the parameters of the activated carbon fibers, which clearly indicate that the activated carbon fiber used in this investigation is essentially microporous. This ensures that the troublesome phenomenon of diffusion observed on meso and macroporous materials can be avoided.

2.2   Iron Fixation to the Fiber

Iron fixation to the fiber of activated carbon was carried out by immersing approximately 1.0 g of the fiber in a solution of 0.,1% (p/p) heptahydrated ferrous sulfate ($FeSO_4 \cdot 7H_2O$) and 0.5 M sulphuric acid at 20°C during 48 hours to ensure a high saturation level. Later, the material was dried at 40°C for 3 hours.

The iron detection was done using the 1.10-phenantroline method as Fe^{+2} chelant agent. [2-11] When using 0.5 M H_2SO_4 as the solvent of the solution used for the impregnation process, the fiber was subjected simultaneously to an acid treatment with the fixation of the ferrous ion, omitting the previous step done by Yuranova et. al to activate the acid organic groups of the fiber.[2] After the iron fixation process was performed, approximately 8.57mg Fe^{+2}/g fiber were found to have been adsorbed. When the efficiency of the technique was analyzed, it was found that about 92% of the iron presented in the solution was retained by the fiber. In each assay where the adsorbent material had been previously treated, Fe^{+2} lixiviation were measured and just 1% of it was being desorbed, leaving most of the metal available for the heterogeneous catalysis process.

Table 1

Parameters of Zorflex A Activated carbon fiber

	Area BET (m^2/g)	Microporous volume DR (cm^3/g)	total volume P/Po = 0.94 (cm^3/g)	Mean porous diameter (nm)	% Microporous
TZ_A (Zorflex A)	695	0.345	0.392	2.252	88

2.3   Phenol Adsorption Isotherms on the Fiber and Removal and Phenol Adsorption on the Fiber Tests

To obtain the adsorption isotherms on the fiber, diverse phenol concentrations were taken and different quantities of activated carbon were immersed in them for a 24 h-period. Once this time elapsed, the phenol concentration was determined accomplished. Once the Fe^{2+} was affixed to the fiber, the level of phenol removal attained by the adsorption process on the porous material was compared with that of the adsorption process assisted by the hydrogen peroxide. Using this oxidant agent, the organic compound was degraded through Fenton reactions catalyzed by the iron previously affixed to the fiber. For both cases, 100 mg/L phenol solution was used, continuously stirred for 4 hours, period in which aliquots were taken at different time intervals to monitor phenol removal. Phenol detection was done using US-Visible Spectrophotometry (λ= 270 nm). Preliminary assays were carried out to evaluate the pH influence on the process, working with values of 6.7 (neutral pH) and 2.8 (optimal pH according to scientific literature[5]). Assays were also performed to evaluate the fiber re-use potential; five consecutive tests were done with the material employed. These tests were done using an acidic pH attained by adding 1N HCl. The adsorption isotherm obtained after 24 h of contact between phenol solutions and the fabrics is shown on figure 1.

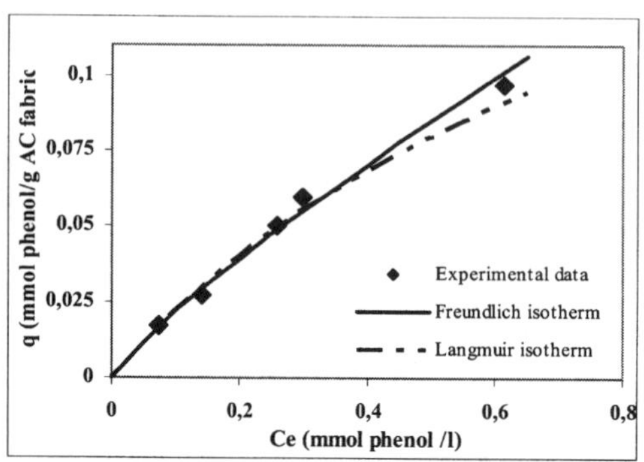

Figure 1. *Phenol adsorption isotherm at 20°C, pH = 6.7 in aqueous solution. The adsorption behavior of phenol into the activated carbon fiber was studied applying Langmuir's and Freundlich's models.*

Table 2

Parameters and correlation coefficients of Langmuir's and Freundlich's isotherm equations for phenol adsorption

	N	K_F ((mmol g^{-1})(Lmmol^{-1})$^{1/n}$)	R^2	P value
Freundlich's parameters	1.178	0.153	0.992	4.806
	Q (mmol g^{-1})	B (L mmol^{-1})	R^2	Pvalue
Langmuir's parameters	0.237	1.021	0.989	5.309

These models have been implemented efficiently in other studies that have used activated carbon fibers to adsorb phenolic compounds and other pollutants.[6-9] Table 2 shows the parameters calculated with a linear regression analysis in which linearized versions of Langmuir's and Freundlich's equations were applied to experimental data, along with their respective correlation coefficients (R^2). Because the calculated R^2 was very similar, it was necessary to determine other statistic parameter in order to know which model is the best for experimental data. This parameter is known as normalized percent deviation or percent relative standard deviation (P) and it has been used in previous works on adsorption isotherms.[6-8,10] The P value is obtained with the following equation:

$$P = (100/N) \sum \left(|q_{e(exp.)} - q_{e(pred.)}| / q_{e(exp.)} \right) \tag{2}$$

Where P is the percent relative deviation; N is the number of experimental data; $q_{e(exp.)}$ is experimental q_e and $q_{e(pred.)}$ are the q_e by Langmuir's and Freundlich's equations. Frequently, a fit is considered excellent when its P value is lower than 5. When the R^2 and P values are observed, it can be concluded that the model that best fits the experimental data is Freundlich's. In this work, the value found for the n parameter was very close to 1. This means that all sites in adsorbent material have equal adsorption energies at all sorbent concentrations. However, this could be due to the fact that the diluted solutions were used in isotherm adsorption tests and, therefore, it could be evidence of Henry's law.

2.4   How pH Influences the Adsorption and Fenton Process

In Figure 2, the influence of the pH on the adsorption process and the adsorption process assisted by the Fenton reaction may be observed. The pH values 2.8 and 6.7 were evaluated. 2.8 is an estimative value in which Fenton reactions are supposed to reach the maximum efficiency. 6.7 is the characteristic pH of the 100

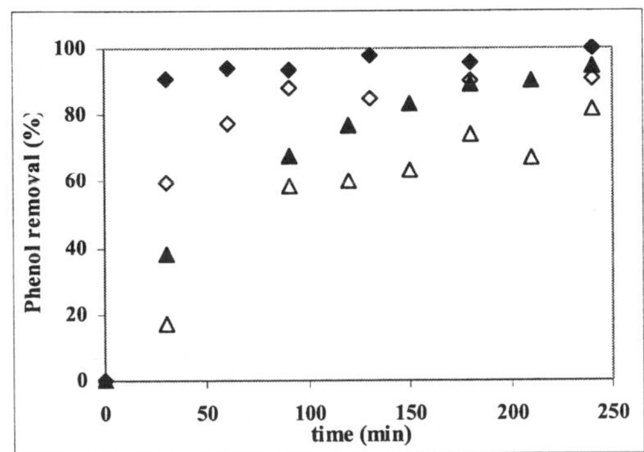

Figure 2. pH efect on phenol removal: ($\triangle$) phenol removal, pH = 6.7 (only adsorption process);($\blacktriangle$)) phenol removal, pH = 6.7 (adsorption process assisted by H_2O_2); ($\Diamond$) phenol removal, pH = 2.8 (only adsorption process); ($\blacklozenge$) phenol removal, pH = 2.8 ((adsorption process assisted by H_2O_2)

mg/l phenol solution, near neutrality. From this figure, it may be seen that during the first 30 minutes, removal takes place most quickly, with a significant difference in the removed phenol percentage of the solution when the tests done at neutral and acidic pH levels are compared. This can also be seen in tests in which only phenol adsorption was monitored, indicating that the adsorption phenomenon is influenced by the pH. At an acidic pH, it may thus be easier for phenolic groups to deprotonate and remain in their ionized state, ready to form a complex with the Fe^{2+} affixed to the fiber.

As time elapsed, this difference became smaller, reaching a minimum at 240 minutes and attaining final removal percentages of 94% when working with acid pH and hydrogen peroxide and 82% with acid pH and hydrogen peroxide. The evolution of the pH was monitored in each test in which the initial pH was close to 7, with the aim of determining whetherto find if treated solutions reached very low pH values. In these experiments, final pH values were close to 4.6. This is because phenol adsorption release H^+ and because phenol degradation caused by Fenton reactions produce by-products such as acetic and oxalic acids.[12]

2.5 Reaction Kinetics

In order to calculate the order in which reactions occur order and the kinetic constants associated to each process, the integral method was used. To do so, data was taken from the first 60 minutes of the experiments because the first

Table 3

Kinetic constants and correlation coefficients obtained from the linearization of three kinetic models

Process	Kinetic Model					
	First order		Pseudo-first order		Second order	
	$-$	R^2	$k2\ (min^{-1})$	R^2	k^3 $(mmol*l*min^{-1})$	R^2
Phenol adsorption on ACF	0.0389	0.7875	-	-	0.1034	0.9842
Phenol removal (Adsorption and Fenton reactions)	0.0433	0.9377	0.002	0.9377	0.1487	0.9916

part of the processes present the fastest reaction rate and, therefore, their associated data allowed the order of reactions to be determined. The kinetics applied on experimental data were the first, pseudo-first and second-order kinetics. Table 3 shows the kinetic constants found. By comparing the correlation coefficients (R^2), it was determined that the kinetic reaction that rules both adsorption and Fenton-assisted adsorption is second-order kinetics. This is uncommon due to the fact that in scientific literature, first-order kinetics are often reported for adsorption reactions, which could be due to the high affinity of commercial activated carbon fiber for phenol compounds. In addition, phenol is removed from the solution faster when the adsorption process and Fenton reactions are combined and therefore, a second-order reaction could be more reasonable in this process.

2.6 Fiber Re-use Assays

In order to evaluate the degree to which fiber is re-used as an adsorbent agent and as catalytic support, five consecutive tests were carried out whereas the fiber was re-used from the preceding experiment. Additionally, the levels of efficiency attained during the monitored processes of this study were compared. Figures 3 and 4 show that in the first two assays, the efficiency of phenol removal does not seem to be significantly affected. However, in the third one, a decrease resulting from the fiber's action as a catalytic support and adsorbent agent was observed. This process promotes the saturation of the fiber and consequently, the loss of its adsorptive properties.

In Figure 4, only minimal changes to efficiency were expected, taking into account that this graph presents the process assisted by Fenton reactions. However, a small decrease is observed because although the predominant phenomenon is adsorption, the Fenton process to degrade phenol is efficient only when it is assisted by ultraviolet light.[12]

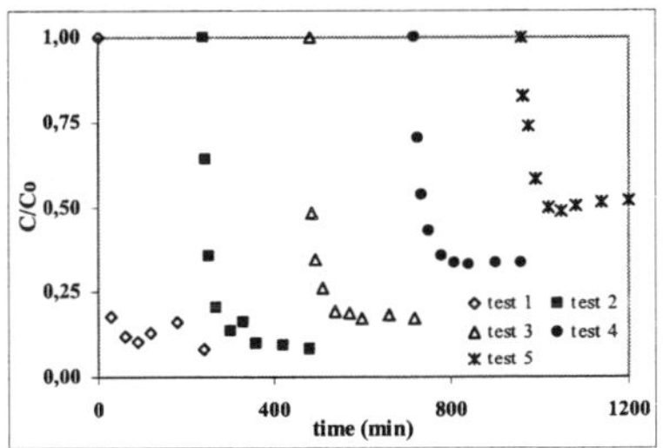

Figure 3. *Consecutive tests of phenol adsorption to evaluate the re-use of fiber*

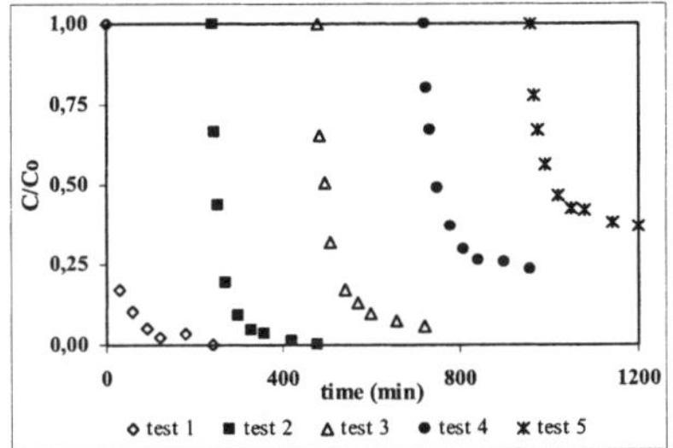

Figure 4. *Consecutive tests of phenol adsorption assisted by Fenton to evaluate the re-use of fiber*

The results reveal the first evidence of how the Fenton process works because in each assay, the removal percentage is higher for the Fenton-assisted adsorption process. This difference becomes more significant as the number of re-use cycles increases. The second evidence that Fenton reactions support adsorption was noted during previous assays carried out to oxidize phenol using hydrogen peroxide without Fe^{+2}. When phenol concentration in the solution was monitored during the experiments, only a minimal change was noted, a finding in line with studies by Kavitha y Palanivelu.[12] This contradicts the notion that an increase in the efficiency of the Fenton-assisted adsorption assays is exclusively owed to phenol oxidation by the peroxide.

3 Conclusions

In this study, efficiencies above 90% were found for the Fe^{+2} fixation processes on fiber, indicating that the chosen method and material worked very well for the purpose of this investigation. It was also found that only ≈1% of the iron became lixiviated from fiber at the end of each experiment. The pH influence was determinant within the first 30 minutes of phenol removal since better results were obtained in the experiments carried out at acidic pH levels. However, it could be observed that after 240 minutes, the difference between the percentages of final removal was not so significant. This allows us to conclude that to work with a nearly neutral pH and reach a high degree of efficiency, the process must be continued for at least 180 minutes. In summary, the model that may be concluded to best describe the adsorption behavior is Freundlich's; in addition, both the adsorption reaction and the Fenton-assisted adsorption reaction follow second-order kinetics, despite the uncertainty generated in the construction of the adsorption reaction model. Regarding the fiber re-use assays, fiber was found to work very well during first three cycles. From the fourth cycle on, fiber became saturated and lost its adsorptive properties. When the removal efficiencies were compared for both processes studied, the highest removal percentages found corresponded in all cases to the Fenton-assisted adsorption tests and as the number of tests increase, the difference becomes more evident, showing that Fe^{+2} was playing its role as a catalytic agent.

Acknowledgements

The authors wish to thank EPFL@Cooperation Unity and its program Seed Money, and the science department at Universidad de los Andes for their financial support.

References
1. M. Rodriguez, V. Sarria, S. Esplugas and C. Pulgarin. J. Photochem. Photobiol. , 2002, **151** , 129.
2. T. Yuranova , O. Enea, E. Mielczarski, J. Mielczarski , P. Albers and J. Kiwi. Appl. Catal. B: Environ. 2004, **49**, 23.
3. D.Gumy, P. Fernández-Ibáñez, S. Malato, C. Pulgarin, O. Enea and J. Kiwi., Catal. Today., 2005, **101**, 234.
4. S. Sabhi and J. Kiwi. Water Res. 2001, **35**, 1994.
5. M. Cheng, W. Ma, J. Li, Y. Huang, J. Zhao, Y. Xiang Wen and Y. Xu. Environ. Sci. Technol., 2004, **38** 287.
6. E. Ayranci, E. Bayram and N. Hoda. J. Hazard. Mater. B 2006, **137**, 344.
7. E. Ayranci and N. Hoda. Chemosphere, 2005, **60** , 1600.
8. E. Ayranci and O. Duman. J. Hazard. Mater. B ., 2005, **124** ,125.
9. K. László. Colloids Surf. A: Physicochem. Eng. Aspects, 2005, **265**, 32.
10. R.-S. Juang, R.-L.Tseng, F.-C. Wu and S.-H. Lee, Separ. Sci. Technol., 1996, **31**, 1915.
11. H.J. Kuhn, S.E. Braslavsky and R. Schmidt Pure Appl. Chem., 2004, **76**, 2105.
12. V. Kavitha, K. Palanivelu. Chemosphere , 2004, **55**, 1235.

A RELATIONSHIP BETWEEN PORE SHAPES AND MERCURY POROSIMETRY CURVES AS REVEALED BY RANDOM PORE NETWORKS

P. Čapek[1] and V. Hejtmánek[2]

[1] Institute of Chemical Technology, Prague, Technická 5, 166 28 Prague, CZ
[2] Institute of Chemical Process Fundamentals of the ASCR, v.v.i., Rozvojová 135, 165 02 Prague, CZ

1 INTRODUCTION

Mercury porosimetry is widely accepted as a standard method to characterise porous solids with respect to their pore volume and pore size distribution. When external pressure is gradually increased, mercury, which does not wet most of solids, displaces a low-pressure gas remaining as the wetting phase in pores after initial evacuation (a drainage process). During the reverse procedure the external pressure is gradually lowered, which results in withdrawal or retraction of mercury threads from pores (an imbibition process).

The modern interpretation of data from mercury porosimetry has been addressed in many articles and books. Due to the increasing performance of computers in the eighties and nineties, pore networks became a powerful tool for simulation of the drainage and imbibition processes[1-10]. Besides numerical calculations, the pore-scale mechanisms of drainage and imbibition were experimentally investigated in two-dimensional chamber-and-throat pore networks etched in glass plates (e.g.[11,12]).

In spite of numerous experiments and theoretical concepts that have been assembled, there is still an ambiguity concerning the simulation of mercury intrusion and retraction in pore networks of realistic geometrical and topological parameters. Therefore, the objective of this paper is to explore the capillary pressure curves of mercury in relation to the shape of throats (constant rectangular cross section or convergent-divergent throats), the throat aspect ratio, chamber-to-throat correlation and the ratio of mean chamber size to mean throat size. Effects of the intrusion and retraction contact angles are also studied. In order to make conclusions of our work more reliable, we derived reference network parameters from a stochastic replica of an α-alumina (macroporous) sample[13]. Pore networks of realistic parameters are particularly significant when convergent-divergent throats are considered or when mercury retraction is simulated. In both cases throat lengths that depend on total porosity and pore size distributions have an effect on mercury intrusion and retraction.

2 PORE NETWORK MODELS

Pore networks were modelled as three-dimensional arrays of chambers connected by throats

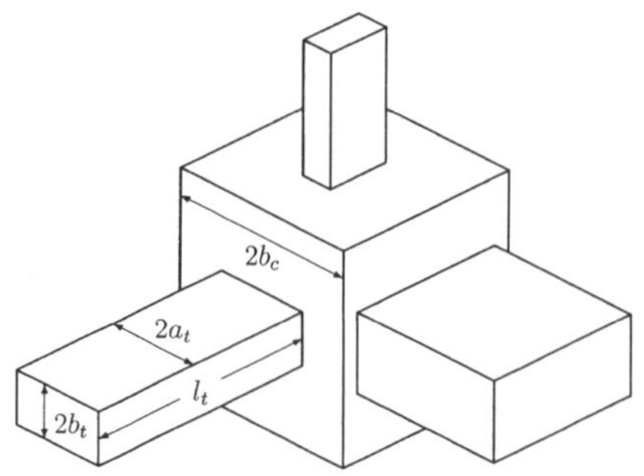
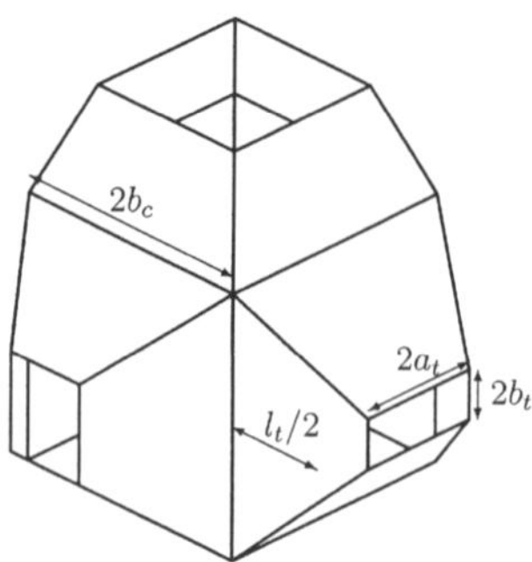

Figure 1 *Schematic representation of a cubic chamber of volume $V_c = 8\,b_c^3$ and the halves of three adjacent throats: rectangular parallelepiped (left) and convergent-divergent (right). Each convergent-divergent throat consists of two obelisks (one of them is only shown). For throat sizes, a_t and b_t, the following inequalities hold: $a_t \leq b_t$ and $b_t < b_c$. The throat aspect ratio, $\lambda = a_t / b_t$, satisfies $\lambda \geq 1$.*

(chamber-and-throat networks). We applied two types of pore networks that differed in the shapes of throats. The first network type (I-type network) consisted of throats with constant rectangular cross section (rectangular parallelepiped throats) and cubic or rectangular parallelepiped chambers. In the second type (X-type network) pairs of cubic or rectangular parallelepiped chambers were connected by a throat consisted of two interconnecting obelisks. For each obelisk, its larger, square base was of the same size as its adjacent chamber. For each throat, the smaller, rectangular bases of the conjoined obelisks were also of the same size, and represented the narrowest cross-section of the throat. A chamber and three throats are depicted in Figure 1 for both network types.

Cubic or rectangular parallelepiped chambers were hierarchically arranged on nodes of two cubic arrays[14]. This specific arrangement of chambers enabled the construction of networks of relatively high total porosity $\phi = 0.4292$ and a broad chamber size distribution (CSD). Our procedure introduced positive chamber-to-chamber (c-c) correlation in the pore network, i.e. large chambers were surrounded by chambers of similar sizes. An algorithm for the efficient assignment of throat sizes to network bonds was implemented in such a way that we were able to introduce positive chamber-to-throat (c-t) correlation of a desired level. In our case of overlapping distributions, a certain level of c-t correlation was unavoidable regardless of input parameters of the algorithm.

When a cubic network is to be of irregular topology, its mean coordination number, $\langle Z \rangle$, must be reduced by removing bonds at random. Then the coordination number, Z, of each chamber is equal to a random integer between 1 and 6. In this work $\langle Z \rangle = 4.0$ was estimated from a skeleton of reconstructed pore space. The chamber size and throat size (TSD) distributions were also derived from replicas of the α-alumina sample that was a subject of stochastic reconstruction[13]. The throat size distributions of the reference networks were modified using a procedure for iterative refinement of network parameters[14]. Having assigned chamber sizes to network nodes and throat sizes to network bonds, and having specified connectivity, the distance between the centres of gravity of two adjacent chambers was adjusted so that the total porosity, ϕ, of the network agreed with that of the prototype. Note that both chambers and throats contributed to the total pore volume. Further details of the network construction can be found elsewhere[14].

3 MERCURY POROSIMETRY SIMULATOR

We derived approximate expressions for the capillary pressure assuming the angular cross-sectional shape of chambers and throats, and assuming the constant intrusion and retraction contact angles, θ_i and θ_r respectively, in the entire network. Values of the contact angles were always measured in the wetting fluid: $0 \leq \theta_i \leq \pi/2$ and $0 \leq \theta_r \leq \pi/2$.

3.1 Mercury intrusion

Following Tsakiroglou and Payatakes's approach[2] we assumed that stable menisci formed at chamber-throat interfaces in both types of our pore network models.

The capillary pressure, P_{ti}, for the mercury intrusion into a throat of constant rectangular cross-section was calculated using a method proposed by Mason and Morrow[15]

$$\frac{P_{ti}b_t}{\gamma} = \begin{cases} f_3 f_2/f_1 + \sqrt{(f_3 f_2/f_1)^2 - f_2/\lambda} \ , & \text{if } \theta < \pi/4 \\ f_3 \ , & \text{otherwise} \end{cases} \tag{1}$$

where $f_1 = 2\theta_i - \pi/2 + 2\sqrt{2}\cos\theta_i \sin(\pi/4 - \theta_i)$, $f_2 = 1/2 \sin(\pi/2 - 2\theta_i) + \sin^2(\pi/4 - \theta_i) - \pi/4 + \theta_i$, and $f_3 = (1 + 1/\lambda)\cos\theta_i$. The symbol γ denotes the interfacial tension between mercury and air. From geometrical considerations the capillary pressure, P_{ci}, for penetration of an interface between a rectangular parallelepiped throat and a cubic/rectangular parallelepiped chamber was derived

$$\frac{P_{ci}}{\gamma} = \frac{\sin\theta_i}{a_t} + \frac{\sin\theta_i}{b_t} \tag{2}$$

If $\theta_i < \pi/4$, a stable meniscus only formed inside a throat because always $P_{ti} \geq P_{ci}$. Otherwise, two positions of menisci were stable. When external pressure reached a value of P_{ti}, mercury penetrated a throat in which the stable meniscus occurred. When external pressure was further increased towards P_{ci}, mercury moved to the interface and created the stable meniscus at the interface. Special care was paid to events when a meniscus did not fit a chamber.

In the case of convergent-divergent throats, we considered the formation of stable menisci in the narrowest and divergent throat parts. The capillary pressure, P_{ti}, in a throat formed by two conjoined obelisks was given by

$$\frac{P_{ti}}{\gamma} = \frac{f_a}{a_t} + \frac{f_b}{b_t} \tag{3}$$

Introducing geometrical simplexes, $p_a = l_t/(b_c - a_t)/2$ and $p_b = l_t/(b_c - b_t)/2$, and their critical value, $p_c = (1/\tan\theta_i - \tan\theta_i)/2$, above which the highest capillary pressure in a throat corresponded to the stable meniscus in its divergent part, we derived the following expressions for f_a and f_b

$$f_a = \begin{cases} (p_a \cos\theta_i + \sin\theta_i)/\sqrt{1 + p_a^2} \ , & \text{if } p_a > p_c \\ \cos\theta_i \ , & \text{otherwise} \end{cases}$$

$$f_b = \begin{cases} \left(p_b \cos \theta_i + \sin \theta_i\right)\big/\sqrt{1+p_b^2}\ , & \text{if } p_b > p_c \\ \cos \theta_i & , \quad \text{otherwise} \end{cases}$$

An algorithm of mercury intrusion applied in our simulator is usually called invasion percolation[16]. The external pressure was increased in a stepwise manner. In each step all menisci were examined one by one to see whether they were stable. If a meniscus got unstable, mercury filled an empty throat/chamber. New menisci were placed at new positions and they were again checked for stability. This procedure was continued until no more unstable menisci were found. Next, the above process was repeated at a slightly higher external pressure. Note that the order in which the network was scanned did not affect the process of intrusion[2].

3.2 Mercury withdrawal

Simulations of withdrawal began at the equilibrium state at which the simulations of intrusion ended. The initial stage of withdrawal implied the existence of a gas compressed in cusps of throats and chambers. We assumed that the wetting phase in pores formed saddle-shaped interfaces with negative and positive principal curvatures. As the external pressure was lowered, the curvature of these interfaces decreased, until some of them got unstable and broke ("snap-off" mechanism). The first menisci formed either in chambers with $Z = 1$ or in throats in which mercury threads snapped off. As soon as unstable menisci occurred in a chamber or a throat, a flow event could follow. However, such an event was only possible if there was at least one continuous thread of mercury connecting an unstable meniscus with a mercury sink. Four types of flow events were considered in our simulator:
Event A. A mercury thread ruptured in a throat connected to two chambers full of mercury. Tsakiroglou and Payatakes[2] derived a functional relationship between the capillary pressure for snap-off in cylindrical throats and the retraction contact angle. Our analysis of their results revealed that the ratio of the capillary pressures for snap-off, P_{ts}, and intrusion, P_{ti}, could be approximated using a formula

$$\frac{P_{ts}}{P_{ti}} = \frac{1.4348 - 0.8609\,\theta_r^2}{2\cos\theta_r} \tag{4}$$

in which P_{ti} was calculated using the retraction contact angle, θ_r. We further assumed that this ratio was also applicable to throats of rectangular cross-sectional shape with the aspect ratio, λ, close to one. If the external pressure is smaller than P_{ts}, mercury was withdrawn from the throat and two menisci were placed at throat-chamber interfaces. Note that a thread was allowed to snap off if a length of the throat was greater or equal to the maximum collar length calculated for the corresponding retraction contact angle[2].
Event B. A throat was filled up with mercury and there was an empty chamber at one of its ends. In this case mercury was withdrawn from the throat when the external pressure was smaller than the capillary pressure for piston-type retraction evaluated from (1) in which θ_r replaced θ_i. After retraction a meniscus was placed at the boundary of the other chamber that was still occupied by mercury.
Event C. Let N_m be a number of menisci in a chamber. Piston-type retraction could occur when $Z = 1 + N_m$ and when the external pressure was smaller than the capillary pressure evaluated from (1) in which the chamber sizes substituted the throat sizes.

Event D. A mercury thread snapped off in a chamber in which $Z = 1 + N_m$. The capillary pressures for snap-off were functions mercury thread shapes that could occur in cubic or rectangular parallelepiped chambers and adjacent throats.

The sequence of flow events during retraction is important in contrast with intrusion[2]. In our simulator mercury left chambers and throats according to descending order of the driving force.

4 RESULTS AND DISCUSSION

In this paper the pore volume, V_{Hg}, occupied by mercury was related to the particle volume, V_p: $V^* = V_{Hg} / V_p$, i.e. $\lim_{p \to \infty} V^* = \phi$. With the only exception of section 4.1, the contact angles θ_i and θ_r were respectively equal to 40° and 60°. Simulations are presented in the form of the capillary pressure curves, $V^*(P)$. In all runs mercury surrounded six sides of a simulated cubic particle. To ensure excellent statistical stability of results, we used, as reference cases, 112×112×112 nodes for the I-type network and 100×100×100 nodes for the X-type network. The networks corresponded to cubic particles (side length of 5 mm), whose sizes were close to a real cylindrical pellet (height × diameter = 4.5 mm × 4.8 mm).

4.1 Effect of contact angles

In this section values of θ_i and θ_r were selected from the interval of [20°, 70°]. The I-type network revealed the smallest resistance for the intrusion contact angle of 45°, cf. Figure 2. For a different value of θ_i, the intrusion curves moved toward a higher pressure range, which was clear from (1) and (2). The effect of θ_i was asymmetrical around the value of 45°. This was caused by exceptional events at interfaces of relatively large throats and small chambers when a fully developed meniscus did not fit a chamber (the meniscus radius was greater than the chamber size, b_c, see section 3.1). The I-type network reduced the overall effect of different values of θ_i even more strongly than the X-type network. Obviously, variations of the obelisk sizes and, consequently, the geometrical simplexes p_a

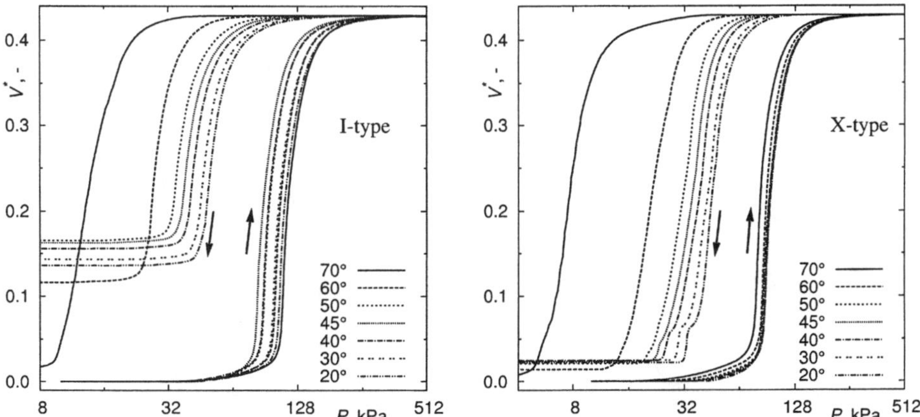

Figure 2 *Capillary pressure curves simulated using the reference I-/X-type networks for various values of θ_i and θ_r.*

and p_b created conditions favourable to compensation of the contact angle effect. There was no plain dependence of $V^*(P)$ on θ_i because some curves intersected. As can be seen from Figure 2, the largest relative shift among the intrusion curves in the pressure ranges in which mercury was filling most of pores was about 1.25 and 1.15 for the I-type network and the X-type network respectively. It contrasted with the constant largest relative shift of $\cos(\pi/9)/\cos(7\pi/18) \approx 2.75$ generated using the bundle of capillary tubes model[1]. It should be noted that the relatively weak dependence of $V^*(P)$ on θ_i simulated using the pore network models solely stemmed from the geometrical consideration (pore-scale models given by (1–3)) and network effects (narrow pores shield wide ones).

The angle θ_r influenced the capillary pressure curves more significantly than θ_i in both cases. In the I-type network the values of θ_r within the interval of $[20°, 50°]$ enabled snap-off in throats and emptying of chambers at external pressures that were significantly lower than the capillary pressures for emptying of chambers. It resulted in large entrapment of mercury whose maximum value was observed for $\theta_r \approx 45°$. In the interval of $\theta_r \in (50°, 70°]$ the beginning of retraction moved to a lower pressure range. The differences between capillary pressures for snap-off in throats and emptying of chambers as well as the residual amount of mercury were smaller than in the first part of the contact angle interval. The retraction curves obtained using the X-type network were well ordered according to θ_r. Within the interval of $[20°, 50°]$ the residual amount of mercury was almost independent of θ_r. When $\theta_r > 50°$, the retraction curves widened and emptying of pores was even more efficient than in the previous case.

Effects of the contact angles were also studied by other researchers, e.g. by Tsakiroglou and Payatakes[2]. In pore networks consisted of spherical chambers and cylindrical throats, they found that the angle θ_r had a stronger effect on capillary pressure curves than θ_i.

4.2 Effect of the throat aspect ratio

The throat aspect ratio, λ, was assigned to each throat of the reference networks at random. Values of λ were sampled from the uniform probability distribution in the interval of $[1, 1.4]$. This choice ensured that there was no correlation between λ and b_t in our networks.

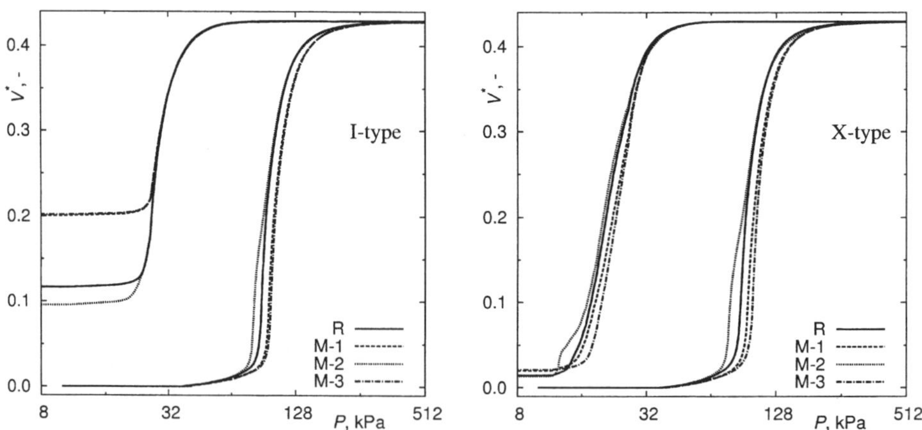

Figure 3 *Effects of the throat aspect ratio, λ, (M-1), the level of c-t correlation (M-2), and the TSD domain width (M-3) on $V^*(P)$. Reference (R) and modified (M) networks*

The capillary pressure curves of the reference networks were compared with similar curves obtained using pore networks in which the same value of $\lambda = 1$ was assigned to each throat, cf. Figure 3.

Symmetric constrictions ($\lambda = 1$ for each throat) caused a mild shift of the intrusion curves to higher pressure ranges, which directly followed from the application of (1–3). Mercury retraction from the I-type network was almost, with the exception of the low pressure range, independent of the constriction shape. It indicated that the capillary pressure for retraction from chambers of small and mild sizes rather than the capillary pressure of snap-off in throats played a significant role in the process. This behaviour contrasted with the retraction curves corresponding to the X-type network. The differences among them were observed in the wide pressure range. However, the residual amount of mercury was affected only slightly.

Ioannidis and Chatzis[5] pointed out that the throat aspect ratio provides an additional degree of freedom, which is important in simultaneous modelling of capillary and conductivity phenomena. However, they only applied the constant value of 6 throughout their networks.

4.3 Effect of chamber-to-throat correlation

Due to the overlapping domains of CSD and TSD and due to the combination of relatively high total porosity and the wide pore size distributions, positive c-c and c-t correlation was inherent in our pore networks, see section 2. We studied the influence of c-t correlation, which we were able to modify easily, on the capillary pressure curve. For this purpose, we strengthened the positive c-t correlation so that the correlation coefficient of the modified networks was about 20% greater than in the reference networks.

The positive c-t correlation caused a decrease of the capillary resistances during intrusion in both cases (Figure 3), see also papers[3,10]. Our algorithm for the introduction of positive c-t correlation was designed to handle the overlapping CSD and TSD. Therefore, large chambers and wide throats were correlated more strongly than small chambers and narrow throats. It also explains why the differences between the capillary pressure curves occurred in the low pressure range.

4.4 Effect of the ratio of mean chamber size to mean throat size, $\langle b_c \rangle / \langle b_t \rangle$

Equations (1–3) imply that the capillary pressures depend on both throat and chamber sizes. Therefore, it was worth analysing the effect of $\langle b_c \rangle / \langle b_t \rangle$ on $V^*(P)$. The modification of TSD was introduced multiplying the upper bounds of both TSD domains by 0.9. The other network parameters matched those of the reference networks.

The intrusion curves moved towards higher pressure ranges (Figure 3). This was expected because the capillary pressures in narrow throats are higher than in wide ones. The effect of the TSD domain reduction on the retraction curves was less obvious. In the I-type network, there was no difference until a critical value of the external pressure was reached. This behaviour suggested that there were different conditions for snap-off in narrow throats but the same conditions for emptying of chambers. Since snap-off in narrow throats was more likely to take place, there were a lesser number of mercury threads connected with the external sink and, consequently, there was the larger residual amount of mercury than in the reference case. In the X-type network there were conditions for simultaneous snap-off in throats and emptying of chambers and, therefore, the residual amounts of mercury were approximately same.

It should be noted that the effect of $\langle b_c \rangle / \langle b_t \rangle$ did not stem only from the presence of narrower throats. Due to our algorithm of the network construction, a higher number of throat sizes were assigned to network bonds at random than in the reference case.

5 CONCLUDING REMARKS

The main conclusions can be formulated as follows:
1. The shape of throats was a significant network parameter that determined the course of the capillary pressure curves. This finding suggests that analysis of experimental capillary pressure curves should not omit pore-shape variability of real media.
2. In contrast with the bundle of capillary tubes model, our pore networks predicted the intrusion curves that were only weakly dependent on the intrusion contact angle.
3. On the other hand, the effect of the retraction contact angle on the retraction curves was strong. There were also great differences in the residual amount (entrapment) of mercury: the X-type network retained less mercury than the I-type network in all cases.
4. A modification of the c-t correlation level had apparently the small effect. This was caused by the inherent c-c and c-t correlation in the pore networks studied.
5. The throat aspect ratio, randomly varied between 1 and 1.4, had a strong effect on the intrusion and retraction curves. It represents a convenient parameter enabling an increase of the pore-shape variability.
6. The effect of a small increase of $\langle b_c \rangle / \langle b_t \rangle$ was two-fold: the capillary resistance of narrower throats was higher and, due to the algorithm of the network construction, the level of c-t correlation decreased in comparison with the reference networks.

The observed effects imply that the quantitative description of immiscible displacement calls for a rather good model of pore space geometry and topology. Similar conclusions are not new and can be found in the literature, e.g. in Dullien's book[1]. However, numerous studies have applied pore networks either formed by chambers and throats of simple geometry or based on artificial parameters. Such models of porous solids can hardly provide a reliable quantitative prediction of immiscible displacement. Since parameters of our pore networks were derived from a stochastic replica of α-alumina, this paper represents an attempt to express functional relationships between pore shapes and mercury porosimetry curves quantitatively.

Acknowledgement

The authors gratefully acknowledge the support by the Grant Agency of the Czech Republic (# 203/05/0347) and by the Czech Ministry of Education (MSM6046137301).

References

1. F.A.L. Dullien, *Porous Media: Fluid Transport and Pore Structure*. Academic Press, San Diego, 2nd edition, 1992.
2. C.D. Tsakiroglou and A.C. Payatakes, *J. Colloid Interface Sci.*, 1990, **137**, 315.
3. C.D. Tsakiroglou and A.C. Payatakes, *J. Colloid Interface Sci.*, 1991, **146**, 479.
4. R.L. Portsmouth and L.F. Gladden, *Chem. Eng. Sci.*, 1991, **46**, 3023.
5. M. A. Ioannidis and I. Chatzis, *Chem. Eng. Sci.*, 1993, **48**, 951.
6. G. P. Matthews, C. J. Ridgway and M. C. Spearing, *J. Colloid Interface Sci.*, 1995, **171**, 8.

7. C.D. Tsakiroglou and A.C. Payatakes, *Adv. Water Resour.*, 2000, **23**, 773.
8. S. Bekri, K. Xu, F. Yousefian, P.M. Adler, J.F. Thovert, J. Muller, K. Iden, A. Psyllos, A.K. Stubos and M.A. Ioannidis, *J. Pet. Sci. Eng.*, 2000, **25**, 107.
9. P. Øren and S. Bakke, *J. Pet. Sci. Eng.*, 2003, **39**, 177.
10. C. Felipe, S. Cordero, I. Kornhauser, G. Zgrablich, R. López and F. Rojas, *Part. Part. Syst. Charact.* 2006, **23**, 48.
11. M.A. Ioannidis, I. Chatzis and A.C. Payatakes, *J. Colloid Interface Sci.*, 1991, **143**, 22.
12. C.D. Tsakiroglou and A.C. Payatakes, *Adv. Colloid Interface Sci.*, 1998, **75**, 215.
13. P. Čapek, V. Hejtmánek, L. Brabec, A. Zikánová and M. Kočiřík, *Transport in Porous Media*, 2008, doi:10.1007/s11242–008–9242–8.
14. P. Čapek, V. Hejtmánek, L. Brabec, A. Zikánová and M. Kočiřík, *Chem. Eng. Res. Des.*, 2008, **86**, 713.
15. G. Mason and N.R. Morrow, *J. Chem. Soc., Faraday Trans. 1*, 1984, **80**, 2375.
16. M. Sahimi, *Flow and Transport in Porous Media and Fractured Rock: From Classical Methods to Modern Approaches*. VCH, Weinheim, 1995.

PHYSICOCHEMICAL CHARACTERIZATION OF NANOSTRUCTURED TUNGSTEN OXIDE: POROUS CLUSTERS AND NANOWIRES

C. Felipe[1], F. Chávez[1], N. Morales[1], E. Lima[2], O. Goiz[3], R. Peña-Sierra[3], M. A. Hernández[4], R. Portillo[5] and M. J. Martínez-Ortiz[6]

[1] Departamento de Fisicoquímica de Materiales, ICUAP, Benemérita Universidad Autónoma de Puebla, 72000, Puebla, México.
[2] Universidad Autónoma Metropolitana, Iztapalapa, Av. San Rafael Atlixco No. 186 Col. Vicentina, 09340, México D. F., México.
[3] Departamento de Ingeniería Eléctrica, CINVESTAV, Instituto Politécnico Nacional, 07000, México D. F., México.
[4] Departamento de Zeolitas ICUAP, Benemérita Universidad Autónoma de Puebla, 72000, Puebla, México.
[5] Facultad de Ciencias Químicas, Benemérita Universidad Autónoma de Puebla, 72000, Puebla, México
[6] Instituto Politécnico Nacional – ESIQIE, Av. IPN UPALM Edif. 7, Zacatenco, 07738 México D.F., México.

1 INTRODUCTION

Tungsten oxides present interesting physicochemical properties which have wide application as sensor of toxic vapours, catalysis, smart windows and thermo-chromic devices.[1] In order to investigate the maximum efficiency of tungsten oxides, the nanostructuration of these materials is claimed as it provides high surface/volume ratio. In this context, the preparation techniques are determinant. For instance, tungsten oxide nanowires have been obtained by direct oxidation of metal tungsten. Liu et al,[2] prepared tungsten oxide nanowires from a hot tungsten filament ($\sim$1400°C) in a vacuum chamber at 1.3×10^{-4} Torr. Other morphologies than nanowires has been also obtained and its application has been reported for the preparation of nanorods, microducts, etc.[3-4]

However, an almost unexplored subject is the array or packing of clusters-shaped aggregation leading to a porous medium useful to guide capillary processes such wetting and non-wetting fluids invasion.[5-6] Of course porous tungsten oxide networks are claimed in applications of heterogeneous catalysis e.g. hydrodesulfurization of fuels, isomerization reactions and photocatalitic activity.[7]

On the other hand, hot filament chemical vapour deposition (HFCVD) technique is an useful *via* to deposit metals on solid surfaces. The main advantage of this technique is that homogeneous films can be achieved. For instance nanospheres of tungsten oxides can be deposited on variations solid surfaces. In addition, the close spaced vapour transport (CSVT) is a potential technique to synthesize tungsten oxide nanostructures, which fulfills the following conditions: a solid source and a reversible chemical reaction with volatile products whose direction is controlled by a temperature gradient.[8-9]

The main advantages of the CSVT technique are: having a higher effectiveness in the mass transport, and forming a subsystem independent of flow and temperature conditions inside the reactor. This indicates that it is independent of the reactor geometric factors. Thus, this work describes the potential application of the HFCVD and CSVT techniques to achieve the evolution of tungsten oxide-containing nanospheres into porous tungsten oxide networks and tungsten oxide nanowires deposited in silicon substrates.

2 EXPERIMENTAL SECTION

Tungsten oxide film is prepared in a tubular quartz reactor (Figure 1), under vacuum (*ca.* 10^{-4} Torr), a W filament was placed 15 mm far from rectangular (24 mm x 40 mm) Pyrex glass plates that is 0.15 mm thick. Glass plates were previously cleaned with xylene, acetone and propanol. Subsequently, a voltage (6.2 V) and a current (3.0 A) were supplied to the filament by a regulated DC power until reach a temperature around 1000 °C. The constant oxidation-sublimation process of the hot filament caused the deposition of a layer of tungsten compound over glass wafers.

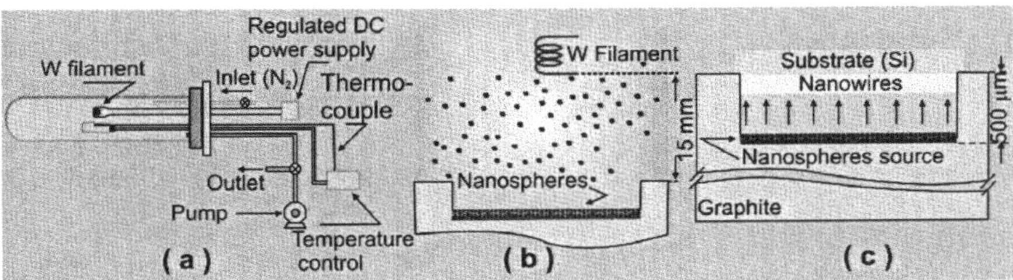

Figure 1 *a) Experimental setup, b) HFCVD nanospheres and c) CSVT nanowires*

The porous tungsten oxide film was deposited on glass wafers, which was thermally treated at temperatures of 120 and 400 °C.

Nanowires were grown for 20 minutes under a N_2 flow and 1 atm of pressure onto Si wafers (substrate) placed on the porous tungsten oxide films (source); these substrate wafers were cleaned previously with xylene, acetone and propanol. The separation between source and substrate was roughly 500 µm. The source was heated at 750 °C, and thermal radiation induced heating at 700 °C temperature.

3 CHARACTERIZATION

The solid materials were characterized by scanning Electron Microscopy (SEM) and its EDS analyses in order to determine qualitative as well as quantitative way using SEM, Philips XL30 equipped with an X-ray energy dispersive spectrometer (EDS).

X-ray diffraction: XRD patterns were recorded on a Bruker axs D8 advance diffractometer coupled to a copper anode X-ray tube, Kα radiation.

Small Angle X-ray Scattering: SAXS experiments were performed using a Kratky camera coupled to a copper anode X-ray tube whose Kα radiation was selected with a nickel filter. The SAXS intensity data, I(h), were collected with a linear proportional counter. Then, they were processed with the ITP program, where the scattering vector, h, is defined as $h = 4\pi\sin\theta/\lambda$; θ and λ represent the scattering angle and the X-ray wavelength, respectively. The powdered samples were introduced into a capillary tube. Measurement time was 9 minutes in order to obtain good quality statistics; indeed linear proportional counter is like a multichannel system therefore each point of the curve was measured for 9 minutes.

N_2 sorption: N_2 adsorption-desorption isotherms at 76 K (at 2250 m altitude of Puebla, Mexico) were measured with an automatic volumetric adsorption instrument (Autosorb1-LC, Quantachrome Instruments). N_2 sorption measurements were determined in the interval of relative pressure, p/p_0, extending from 10^{-6} to 0.995. The saturation pressure, p_0, was continuously registered in the course of the adsorption–desorption measurements. In order to introduce the sample in the appropriate N_2 adsorption cell, the original glass plates where cut in small pieces. Prior to every sorption experiment, the sample was annealing and outgassed at 120 and 400 °C during 20 hours.

4 RESULTS AND DISCUSSION.

4.1 Nanospheres

Figure 2 displays SEM images of tungsten oxide particles, which were taken immediately after deposition on the glass plates. Tungsten oxide particles morphology are seen nearly-spheroid. Nanospheres were agglomerated, apparently in a random way, originating free spaces of different sizes between the particles. An EDS analysis (inset of the Figure 2a) revealed the presence of oxygen and tungsten confirming the oxidation of tungsten. XRD pattern of the porous tungsten oxide (inset of the Figure 2b), recorded at 120 and 400 °C indicated the presence of the WO_3 phase.

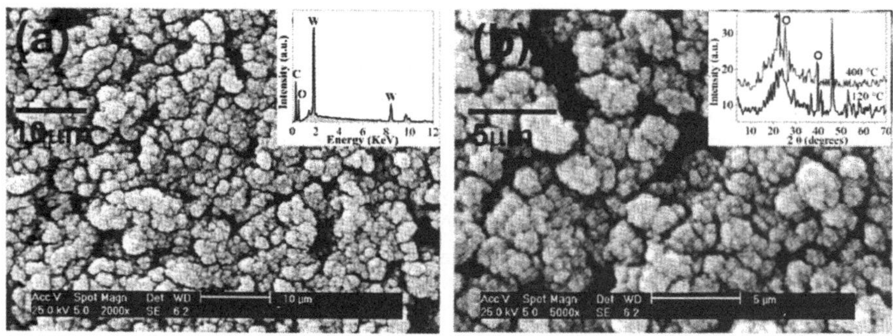

Figure 2 *SEM images of a porous tungsten oxide film deposited by the HFCVD technique at: (a) Low magnification. The inset shows EDS analisis. (b) High magnification. The inset shows XRD patterns of tungsten oxide nanospheres. ° indicates $W_{25}O_{73}$ and $^+$ indicates WO_3, JCPDS files 20-1323 and 71-0070, respectively.*

SAXS results agree with the SEM images. The Kratky plot exhibited the typical profile of the scattered heterogeneities having a globular or spheroid shape.
From SAXS data the size distribution of spheres was determined, Figure 3 a. Note that, the spheres of diameters as small as 20-30 Å dominated the distribution in the sample after treatment at low temperature (120 °C). The size distribution was modified with the heating (400 °C); indeed the amount of big spheres increases with decreasing of the small ones.
Heating of the W filament promoted a spreading of tungsten oxide spheres on the Pyrex glass. The spheres fall, as a rain; they were distributed in a random way on the glass surface. Some spheres fall on the other forming then small clusters. The dynamic of these spheres increased with temperature. Thus, they distributed in the space and connected in a form that they form bigger spheres which were detected by the SAXS technique. The connectivity of spheres caused changes in the fractal dimension and then the diameter of big spheres was not exactly a multiple of the small ones. The increasing of fractal dimension parameter with temperature confirms that spheres increases their connectivity.

In other words, the geometrical figure is maintained but bigger objects are formed which are not perfect spheres but they present pores and defects which is highly desirable in materials with potential properties to be used as catalysts.

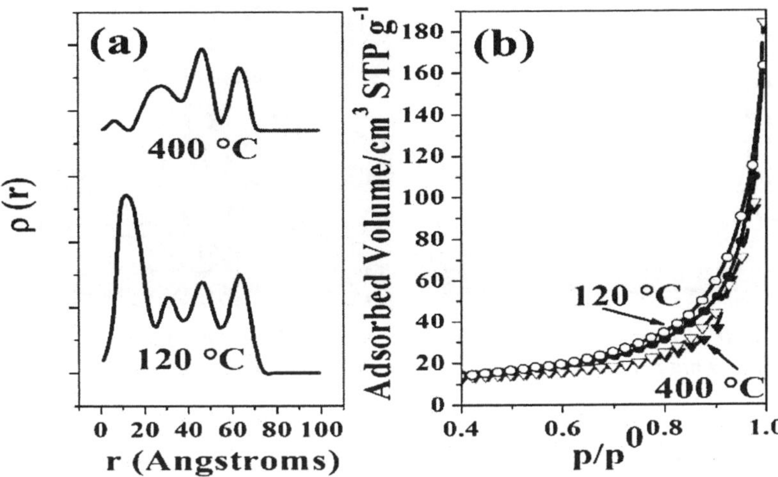

Figure 3. *Particle-size distributions, as determined from SAXS data, of porous tungsten oxide film outgassed at (a) 120 and 400 °C and (b) N_2 sorption isotherms on porous tungsten oxide film outgassed at 120 and 400 °C.*

The N_2 adsorption-desorption isotherms of sample outgassed at 120 and 400 °C correspond to a type II isotherm in the IUPAC classification with a hysteresis loops type H3, Figure 3 b, is a characteristic of capillary condensation in slit-like pores. The sorption properties of this kind of porous structures depend on the spherical shape of the particles. As the temperature of the thermal treatment increases, the BET surface changes from 38.67 to 34.50 m^2g^{-1} because the nanospheres have a tendency to stick together to form clusters; this aggregation process is accelerated at elevated temperature. Thus, the area is reduced by an amount equal to the area lost by formation of nanosphere-to-nanosphere joints. The pore structure of this class of aggregates contains two kinds of holes; i.e. cavities between nanospheres (mesoporous) and between the clusters (macroporous).

4.2 Nanowires

Oxide tungsten nanospheres synthesised previously by the HFCVD technique can be used to made nanowires when these nanospheres on the source are transported and deposited by the CSVT method on silicon wafer substrates placed at 500 μm. The processes involved are: WO_{3-x} nanospheres coming from the hot filament and remaining water of the reactor (oxidant reagent) react at the source temperature (750 °C), to produce a volatile compound $(WO_{2-x}(OH)_2)^g$. After that, $(WO_{2-x}(OH)_2)^g$ is transported through a diffusive mechanism to the substrate surface (700 °C) to gives the following reaction $(WO_{2-x}(OH)_2)^{(g)} \rightarrow WO_{3-x}^{(s)} + H_2O^{(g)}$.

Figure 4 shows SEM images of nanowires grown on the quartz and silicon substrates through the CSVT technique. Nanowires were homogeneously built up. They were sized within diameters about 100 nm and elongated to many tenths of micrometers. EDS analysis (inset of the figure 4 a) supports that they were composed only of tungsten and oxygen. XRD pattern (inset of the figure 4b) revealed that nanowires crystallize as $W_{18}O_{49}$, JCPDS card number 36-101.

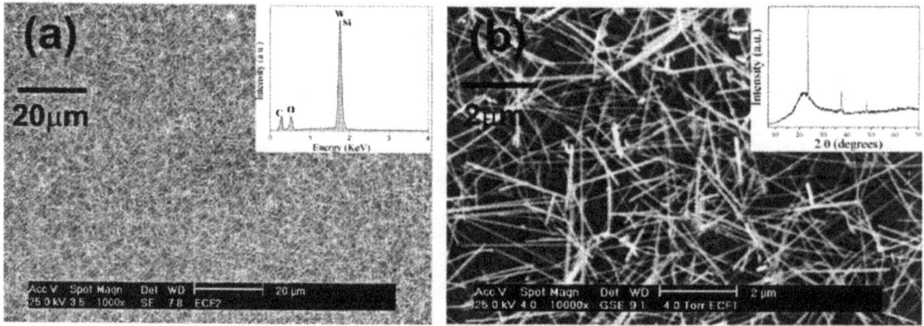

Figure 4. *SEM images of nanowires grown on the silicon and quartz substrates through the CSVT technique at: (a) low magnification and (b) high magnification. .*

5 CONCLUSION

The HFCVD technique was used to prepare nanosphere clusters of WO_{3-X} on glass substrates. The nanospheres are randomly distributed on the substrate originating free spaces of different sizes between the particles as an interconnected 3-D network of porous. As the temperature of the thermal treatment increases, the BET surface changes from 38.67 to 34.50 m^2g^{-1} because the nanospheres have a tendency to stick together to form aggregates. The tungsten oxide nanospheres can be subsequently treated either thermally to leads a porous tungsten oxide network or apply the CSVT technique to promotes the formation of nanowires of $W_{18}O_{49}$ on a substrate at moderates temperatures (750 °C), short times (20 minutes) under a N_2 flow and 1 atm of pressure.

Acknowledgments

Financial support from the Vicerrectoria de Investigación y Posgrado (VIEP-11/ING/07) and from the Programa del Mejoramiento del Profesorado (Promep/103.5/07/2594) is gratefully acknowledged. Thank to Victor Hugo Lara for the use of the Bruker axs D8 advance diffractometer and for his technical assistance.

References

1 D. Lee, S. Han, J. Huh, D. Lee, *Sens. Actuators B,* 1999, **60**, 57.
2 G. Gu, B. Zheng, ; W. Q. Han, S. Roth, J. Liu, *Nano Lett.*, 2002, **2**, 849.
3 K. Hong, M. Xie, H. Wu, *Nanotechnology*, 2006, **17**, 4830.
4 Y. Baek, K. Yong, *J. Phys. Chem. C*, 2007, **111**, 1213.
5 C. Felipe, F. Rojas, I. Kornhauser, M. Thommes, G. Zgrablich, *Ads. Sci. & Tech.,* 2006, **24**, 623.

6 P. Yang, D. Zhao, D. I. Margolese,B. F. Chmelka, G. D. Stucky, *Nature*, 1998, **396**, 152.
7 G. M. Dhar, B. N. Srinivas, M.S. Rana, M. Kumar, S. K. Maity, *Catal. Today*, 2003, **86**, 45.
8 F. Chávez, Y. E. Bravo-García, F. Silva-Andrade, A. Galindo, *Material Letters*, 2000, **43**, 324.
9 C. Felipe, F. Chávez, C. Ángeles-Chávez, E. Lima, O. Goiz, R. Peña-Sierra, *Chemical Physics Letters*, 2007, **439**, 127.

POSITRON POROSIMETRY STUDY OF MECHANICAL STABILITY OF ORDERED MESOPOROUS MATERIALS

R. Zaleski,[1] J. Goworek,[2] A. Borówka,[2] A. Kierys[2] and M. Wiśniewski[1]

[1]Department of Nuclear Methods, Institute of Physics, Maria Curie-Sklodowska University, 20-031 Lublin, Poland. E-mail: radek@zaleski.umcs.pl
[2]Department of Adsorption, Faculty of Chemistry, Maria Curie-Sklodowska University, 20-031 Lublin, Poland

1 INTRODUCTION

In the last years a family of ordered mesoporous materials called M41S[1] was intensively developed. It resulted in invention of variety of M41S class materials differing in shape, size and ordering of the pores. Remarkable properties of these molecular sieves, such as high specific surface area (over 1000 m^2/g), high pore volume (about 1 m^3/g) and a narrow distribution of pore sizes, make it attractive for various industrial applications. In order to successfully use M41S molecular sieves as supports, adsorbents, catalysts etc. their mechanical stability have to be known. So far structural stability was investigated mostly by adsorption/desorption techniques (nitrogen or organics), mercury porosimetry and X-ray diffraction (XRD)[2,3,4]. These methods have some limitations that may distort obtained results in the case of studies of structural stability.

The adsorption methods are sensitive to open porosity only. So their results can be misleading if open pores close, what probably occurs in pressed samples. Besides the same problem, mercury intrusion requires application of very high pressure to fill mesopores. Such pressure can damage the sample and lead to distorted results. Instead of porosity measured by adsorption methods, XRD peaks reproduce periodic structure of samples. For a structure stability experiments usually mechanical presses are used. They are capable of exerting directional pressure only. In such experiment disorder can be introduced in the periodic structure of the sample resulting in the decrease of XRD peaks intensity, while the porosity could remain intact. All specified methods are not suitable for *in situ* measurements, so only the results of the stress applied to the sample can be observed.

The limitations listed above do not apply to Positron Annihilation Lifetime Spectroscopy (PALS). Positrons can freely penetrate solid medium and form positronium (Ps) in both closed and opened pores. PALS porosimetry uses long-lived positronium triplet state (bound state of electron and positron with parallel spins named ortho-positronium, o-Ps) as a probe. The lifetime of o-Ps (142 ns in vacuum) is shortened due to the pick-off process (fast positron annihilation with one of electrons from the pore wall) and depends on pore size and shape[5]. Lifetime measurements can be carried on samples under pressure, what is another advantage over standard porosimetric methods.

In this paper we present the study of mechanical stability of three different ordered mesoporous materials: classic MCM-41 with straight cylindrical pores of diameter about 2 nm, SBA-15 with similar pore structure, but three times larger pore diameter and mesoporous silica fibres (MSF) with pore diameter similar as in MCM-41, but of toroidal pore structure.

2 EXPERIMENTAL

Classical MCM-41 mesoporous sieve was synthesized[6] with the use of tetraethylorthosilicate (TEOS) as the silica source and cetyltrimethylammonium bromide surfactant (CTAB) as a template. Tetrabutoxysilan (TBOS) silica source and CTAB template was used for the synthesis of MSF, that was carried out in a static two-phase acidic system[7] without any auxiliary water-insoluble additive. Both syntheses were done at room temperature. During SBA-15 synthesis[8] amphiphilic triblock copolymer, poly(ethylene glycol)-block-poly(propylene glycol)-block-poly(ethylene glycol) from Aldrich (average molecular weight 5800) was mixed with water and HCl while stirring. TEOS was used as the silica source. The crystallization was carried out at 370K. After synthesis all samples were washed, dried at 360 K and calcined by heating up to 820 K in the air. Next, the material was thermally treated in oxygen at 820 K to remove the carbon deposits remaining on the surface after calcination.

Volumetric adsorption analyzer ASAP 2405 Micromeritics (Norcross, GA) was used for nitrogen adsorption/desorption isotherms measurements, which were performed at 77 K. X-ray diffraction patterns for MCM-41 and MSF were measured by HZG-4 diffractometer (Carl Zeiss Jena) and for SBA-15 by modified DRON-5 diffractometer (USSR). In both cases Cu Kα monochromatic radiation was used and a step in 2θ was 0.02°.

Positron annihilation lifetime spectra were acquired using a conventional fast-slow delayed coincidence spectrometer. Scintillation detectors were equipped with BaF_2 crystals. In order to evaluate properly long-lived o-Ps component intensities, energy windows were wide opened assuring the collection of three gamma annihilation events. The sample in the 'sandwich' configuration (two layers of a sample with the ^{22}Na positron source encapsulated in a Kapton envelope between them) was placed inside the appropriate (vacuum or pressure) sample chamber. During the vacuum measurements the pressure of about 0.3 Pa was kept. Before the measurements, the samples were annealed in vacuum for 1 hour at 420 K to assure that surface of the pores is free of adsorbates. About 1.7×10^7 counts per spectrum were collected in the vacuum measurements. The lifetime spectra were analyzed numerically using Bayesian methods, implemented in MELT routine[9]. The pressure measurement was carried on the samples exposed to air. The samples (with positron source inside) were compressed in a steel die using manual operated hydraulic press. The applied pressure was maintained during the measurements lasting usually 4 hours for each pressure value, resulting in 6×10^6 counts per spectrum. In this case mean lifetime values and intensities of the components were fitted using the LT program[10]. All PALS measurements were done at room temperature.

3 RESULTS AND DISCUSSION

3.1 Materials characterization

Properties of investigated materials were characterized using standard methods: nitrogen adsorption/desorption and XRD. Nitrogen isotherms (Fig.1a) exhibit typical for mesopores steep increase: for MCM-41 at relative pressure $p/p_0 \approx 0.35$ and for SBA-15 at $p/p_0 \approx 0.75$, with distinct capillary hysteresis loop. Another increase and a small hysteresis, probably originated from nitrogen condensation in the intergranular spaces, can be found in adsorption on MCM-41 at $p/p_0 > 0.9$. It is almost indistinguishable in SBA-15 and absent in MSF. Adsorption isotherm of nitrogen on MSF is quite different than on both other materials. The increase of adsorption within mesopores in the range $p/p_0 \approx 0.3$-0.4 is very small. Most likely it is a result of the presence of closed pores inaccessible for nitrogen. Above $p/p_0 = 0.5$ the isotherm is flat.

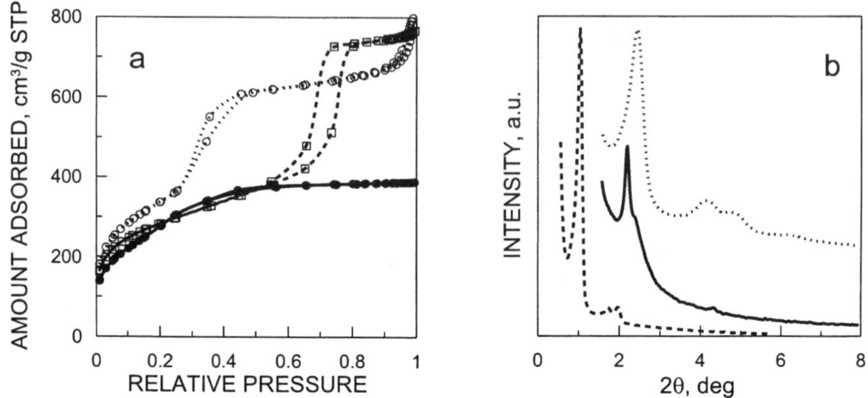

Figure 1   *Nitrogen adsorption/desorption isotherms (a) and X-ray diffraction patterns (b) of MSF (solid line), MCM-41 (dotted line) and SBA-15 (dashed line)*

Pore size distributions were calculated from the N_2 adsorption isotherms using the BJH method[11]. The values between the experimental points were interpolated assuming Gaussian distribution of pore sizes. Mean pore radius was taken at the maximum of the distribution. The specific surface area was estimated from the adsorption data by the standard BET method in a relative pressure range 0.04 - 0.25. The total pore volume was obtained from a single point adsorption at the relative pressure of 0.983. Mesopore radii R_{N_2}, total pore volumes V_{N_2} and specific surface areas S_{BET} were summarized for all the samples in Table 1. Almost the same pore radii in MSF and MCM-41 are over two times smaller than in SBA-15 satisfying authors expectation. Total pore volume of MSF, two times smaller than for other samples, is probably underestimated because of inaccessibility of the pores to N_2 molecules. The specific surface area is almost the same for MSF and SBA-15, what was expected keeping in mind larger size of pores and total pore volume in SBA-15. More surprising is only 20% higher specific surface area of MCM-41.

Table 1   *Parameters characterizing structure of the pores obtained from nitrogen desorption isotherms, XRD and PALS for MSF, MCM-41 and SBA-15 samples*

Sample	R_{N_2} (nm)	V_{N_2} (cm³/g)	S_{BET} (m²/g)	D_{XRD} (nm)	R_{XRD} (nm)	W_{XRD} (nm)	R_{PALS} (nm)	P_{esc} (ns⁻¹)
MSF	1.31	0.60	985	4.03	1.84	1.15	1.23	0.002
MCM-41	1.27	1.17	1209	3.58	1.78	0.75	0.94	0.015
SBA-15	3.08	1.18	1012	8.43	4.31	1.52	0.85	0.036

Clearly discernible reflections in the range $2\theta < 6°$ for all investigated samples (Fig.1b) confirm periodic structure of the pores. Three peaks in MCM-41 and SBA-15 patterns are typical for hexagonal ordering of silica framework. The reflections can be indexed as (100), (110) and (200) assuming hexagonal symmetry. The XRD pattern of MSF sample exhibits only two (100) and (110) reflections and the shape of (100) peak seems to be distorted, probably due to the toroidal arrangement of the pores. Assuming that the pores have cylindrical shape with (100) interplanar spacing D_{XRD}, the radius of mesopores R_{XRD} and the thickness of silica wall W_{XRD} can be calculated from the relations[12,13]

$$R_{XRD} = \frac{1}{2} c D_{XRD} \sqrt{\frac{\rho V_p}{1 + \rho V_p}}, \tag{1}$$

$$W_{XRD} = \frac{2}{\sqrt{3}} D_{XRD} \frac{R_{XRD}}{b}, \tag{2}$$

where ρ = 2.2 g/cm^3 is the density of a sample walls, V_p is a volume of the primary mesopores (it can be estimated from nitrogen isotherms), c = 1.213 and b = 1.050 are constants. The results presented in Table 1 show, that the pore walls are almost two times thicker in MSF and SBA-15 comparing to MCM-41. The radii calculated from the above equations are 40% larger than those from N$_2$ isotherms for MCM-41 and SBA-15, that is typical for MCM-41 class materials[13].

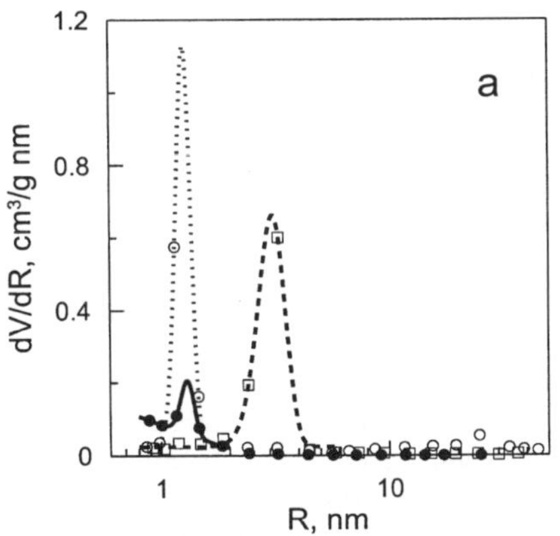
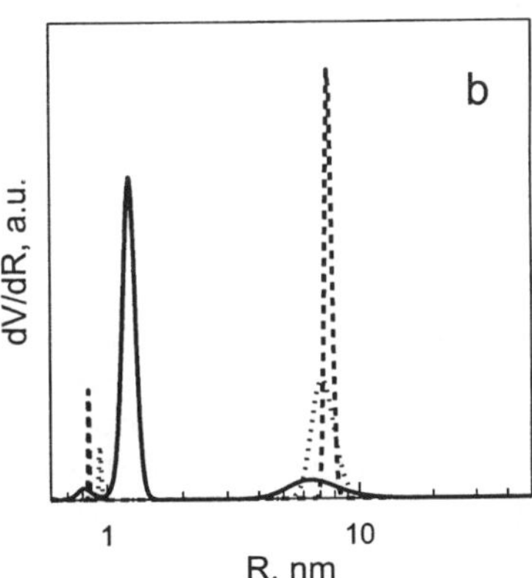

Figure 2   *Pore size distributions calculated using the BJH method from nitrogen desorption isotherms (a) the TEG model from PALS spectra (b): MSF (solid line, full circles), MCM-41 (dotted line, open circles) and SBA-15 (dashed line, open squares)*

Positron annihilation experiment was performed in vacuum in order to eliminate the ortho-para conversion on paramagnetic atmospheric oxygen, resulting in o-Ps lifetimes dependence on probability of pick-off process only. In this case relation between the lifetime and pore radius is given by the Extended Tao-Eldrup (ETE) model[14]. The distribution of the pore radii, similar to the one obtained using BJH method, can be calculated using the relation

$$\frac{dV}{dR} \propto I(\tau) \frac{d\tau}{dR}, \tag{3}$$

where $I(\tau)$ is the intensity distribution from an experiment, $d\tau/dR$ is a derivative of the $\tau(R)$ relation given by the ETE model with assumption of cylindrical shape of the pores. The pore distributions calculated using eq. (3) are shown in Fig.2b, the radii R_{PALS} at maximum of the distribution are compared in Table 1. The distributions calculated from the PALS data show two fractions of free spaces. The first group with radii about 1 nm, comparable to these seen in the BJH distributions, can be ascribed to the primary mesopores. The second fraction with the radii 5-10 nm is a result of Ps annihilation in intergranular spaces. Predominance of a volume of the intergranular spaces over a volume of the mesopores in MCM-41 and SBA-15 reflects positronium preference to migrate to the largest free volume which Ps can reach during its lifetime. Such migration is possible only if the pores are open, thus this effect is neglible in the case of MSF. Another, more disadvantageous, effect caused by the positronium migration is shortening of o-Ps lifetimes related to the open mesopores in MCM-41 and SBA-15. Application of the ETE model to these lifetimes results in underrated values of the pore radius. On the other hand the shortening allows to obtain information about mean pore length. The pore length is proportional to the probability of Ps escape, which can be estimated using equation

$$P_{esc} = \tau_{PALS}^{-1} - \tau_{N_2}^{-1}, \tag{4}$$

where τ_{PALS} is the lifetime value from PALS experiment and τ_{N_2} is the value obtained from the TEG model for radii R_{N_2} from BJH calculation. Results presented in Table 1 suggest that mean pore length is over two times larger in MCM-41 than in SBA-15. The value of P_{esc} obtained for dominating peak in MSF is very close to zero indicating the presence of closed pores. However there is also small peak at 0.8-1.0 nm. Probably it originate from the fraction of open pores near to the surface of the fibres, where the o-Ps lifetimes are shortened.

3.2   Mechanical stability

It was reported earlier that hydrolysis of the pore surface influences the process of mechanical compression[3]. Presence of water is usually unavoidable in practical applications, therefore the study of the structural stability of the samples was not performed in vacuum, but in standard conditions in air, assuring water accessibility to the samples. In this case the relation of o-Ps lifetime versus pore size is more complex. Processes other than pick-off and o-Ps migration influence the annihilation probability. Atmospheric oxygen induces ortho-para conversion[15] resulting in both the o-Ps lifetime and the intensity reduction. Moreover a species adsorbed from air reduce pore size and changes electron density on the surface making the ETG model inadequate[16]. Even though relative, quantitative changes of pore size and porosity ascribed to each group of the pores can be easily observed. The results presented in Fig.3 show mean lifetimes, related to mean pore radius, and intensity, related to pore concentration. The o-Ps components can be ascribed to following groups of empty spaces: short-lived component – small free volumes in the silica walls (SC, Fig.3a), intermediate component – the primary mesopores (MC, Fig.3b) and long-lived component – the intergranular spaces (LC, Fig.3c). The PALS data show that SC is not influenced by pressure up to 1 GPa – the structure of silica is intact. Increase of its intensity is mostly the result of disappearance other free volumes competing in positronium trapping. Contrary to linear rise observed in MCM-41 and SBA-15, there is almost no change of SC intensity below 300MPa, what is in coincidence with the intensity of LC.

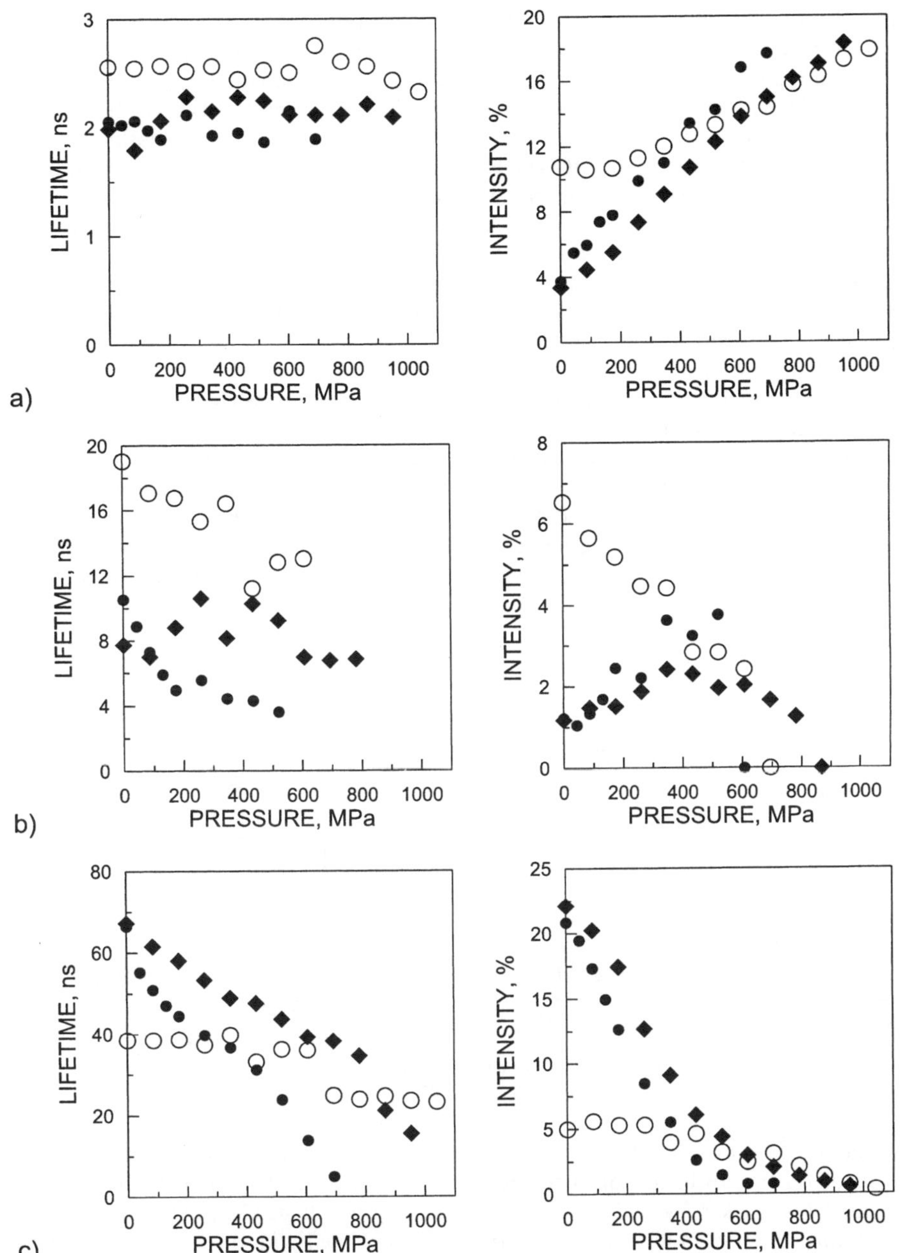

Figure 3 *Mean lifetime and intensity of three o-Ps components (a,b,c) found in PALS spectra of MSF (open circles), MCM-41 (full circles) and SBA-15 (diamonds) as function of pressure*

The lifetime of MC decreases approximately linearly. Only in MCM-41 change of the slope is visible at about 150 MPa. The lifetime decrease is followed by the intensity of MC only in the case of MSF. Initial increase of the intensity in MCM-41 and SBA-15 samples is caused by closing the ends of the primary mesopores. In consequence, the probability of positronium escape decreases, so the intensity of Ps annihilating inside of the pores increases. Above ~600 MPa the MC intensity starts decreasing in SBA-15. It means that reduction of the number of the pores exceeds decrease of the escape probability. In all the samples, before the lifetime of MC reach zero, its intensity suddenly reduces resulting in the component disappearance. It occurs at ~600 MPa for MCM-41, at ~700 MPa for MSF and the most resistant is SBA-15, where MC vanishes at ~850 MPa. At these pressures all primary mesopores finally collapse. This sequence is in agreement with pore walls thickness presented in Table 1. Moreover the suspicion of overestimating W_{XRD} value for MSF seems to be confirmed – the pressure of the pore collapse in MSF is close to that in MCM-41, while calculated W_{XRD} is much larger, almost the same as in SBA-15.

The largest changes with pressure are observed for LC. Its intensity decreases smoothly to zero in all the samples. The samples which consist of small grains (MCM-41 and SBA-15) shows roughly exponential decrease, while the intensity of Ps annihilating in free spaces between many times larger fibres (MSF) is preserved up to 300 MPa and then decreases linearly. Also the pressure dependence of the LC lifetimes differs for the granular and fibrous samples. In MCM-41 and SBA-15 the continuous decrease of the long-lived lifetime is observed, which changes its slope after the collapse of the pores. In MSF the LC lifetime is constant with increasing pressure. There is only stepwise change when the pore structure collapses.

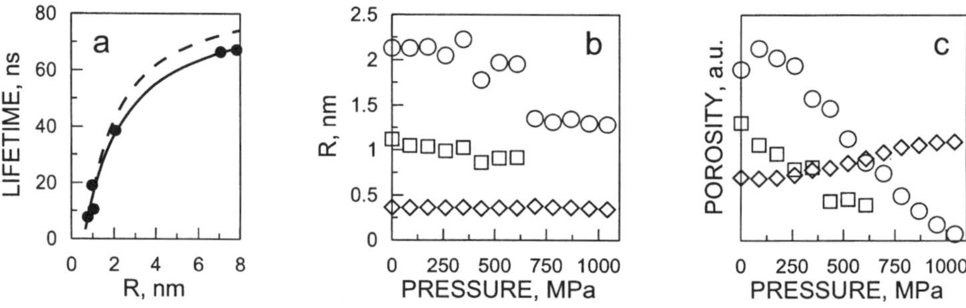

Figure 4   *o-Ps lifetime vs. free volume radius dependence in air (a) – dashed line after Sudarshan et al.[16], solid line is fitted to the experimental data for MSF, MCM-41 and SBA-15. Radius (b) and contribution to porosity (c) of free volumes in silica walls (diamonds), primary mesopores (squares) and intergranular spaces (circles) in MSF as a function of exerted pressure*

A model taking into account processes influencing the lifetime of positronium in air has to be used in order to estimate pore sizes from data shown in Fig.3. An attempt to create such 'air' model was taken by Sudarshan et al.[16] resulting in a dependence shown in Fig.4a as dashed line. In order to test this model, the radii calculated with the ETE model from the lifetimes measured in vacuum were assigned to the lifetimes measured in air for the same samples. The result diverge a little from the original model curve. However the 'air' model can be adjusted to investigated class of materials by changing values of the model parameters to fit present experimental data (solid line in Fig.4a). The obtained new values of the model parameters: reduction of the pore size due to presence of adsorbed layer $D_{ad} =$ 0.58 nm and air quenching rate $\lambda_q = 0.0047$ ns^{-1} are still in good agreement with experimental data[16]. The new relation allowed to calculate values of radius for o-Ps lifetimes observed in air during mechanical stability tests. Because only MSF is free of distortions related to Ps migration, the pore radii were calculated for this sample (Fig.4b). The values of SC lifetime does not seem to be affected by air presence and moreover they lie beyond applicability of the 'air' model. Therefore the radii related to these lifetimes were calculated using the ETE model. The knowledge of pore radii and appropriate component intensity allows to estimate the porosity related to particular group of free spaces[17] (Fig.4c).

4 CONCLUSIONS

The results obtained by PALS allow to observe qualitative changes occurring in investigated ordered mesoporous materials subjected to a mechanical pressure in air. Evolution of various groups of free spaces (sub nanometer voids in the silica walls, primary mesopores and intergranular space) can be monitored. The quantitative results (pore size and porosity) can be calculated for the system where positronium migration between free volumes is neglible. More complicated systems with open pores need better understanding of processes accompanying positronium annihilation (Ps migration mostly) to be described correctly.

References

1   C.T. Kresge, M.E. Leonowicz, W.J. Roth, J.C. Vartuli and J.S. Beck, *Nature*, 1992, **359**, 710.
2   V. Y. Gusev, X. Feng, Z. Bu, G. L. Haller and J. A. O'Brien, *J. Phys Chem.*, 1996, **100**, 1985.
3   T. Tatsumi, K.A. Koyano, Y. Tanaka and S. Nakata, *J. Porous Mat.*, 1999, **6**, 13.
4   M. Hartmann and A. Vinu, *Langmuir*, 2002, **18**, 8010.
5   R. Zaleski , J. Wawryszczuk and T. Goworek, *Radiat. Phys. Chem.*, 2007, **76**, 243
6   M. Grün, K.K. Unger, A. Matsumoto and K. Tsutsumi, in: *Characterization of Porous Solids IV*, eds. B. McEnaney, J.T. Mays, J. Rouquerol, F. Rodriguez-Reinoso, K.S.W. Sing and K.K. Unger, The Royal Society of Chemistry, 1997, p. 81.
7   F. Marlow, M. McGehee, D. Zhao, B. Chmelka and G. Stucky, *Adv. Mater.*, 1999, **11**, 632.
8   Z. Luan, M. Hartmann, D. Zhao, L. Kevan, *Chem. Mater.*, 1999, **11**, 1621.
9   A. Shukla, M. Peter, L. Hoffman, *Nucl. Instr. and Meth. A*, 1993, **335**, 310.
10 J. Kansy, *Nucl. Instr. and Meth. A*, 1996, **374**, 235.
11 E.P. Barrett, L.G. Joyner and P.H. Halenda, *J. Am. Chem. Soc.,* 1951, **73**, 373.
12 T. Dabadie, A. Ayral, Ch. Guizard, L. Cot and P. Lacan, *J. Mater. Chem.*, 1996, **6**, 1789.
13 A. Sayari, P. Liu, M. Kruk and M. Jaroniec, *Chem. Mater.*, 1997, **9**, 2499.
14 T. Goworek, K. Ciesielski, B. Jasińska, J. Wawryszczuk, *Chem. Phys. Lett.*, 1997, **272**, 91.
15 F.F. Heymann, P. Osman, J.J. Veit and W.F. William, *Proc. Phys. Soc. (London)*, 1961, **78**, 1038.
16 K. Sudarshan, D. Dutta, S.K. Sharma, A. Goswami and P.K. Pujari, *J. Phys.: Condens. Matter*, 2007, **19**, 386204.
17 C.L. Wang, M.H. Weber and K.G. Lynn, *Appl. Phys. Lett.*, 2002, **81** 4413

CHARACTERISATION OF MODIFIED POROUS SILICA BY INVERSE GAS CHROMATOGRAPHY

D. Enke[1], M. Rückriem[1], D. Stoltenberg[2] and A. Böhme[3]

[1]Institute of Chemistry, University of Halle-Wittenberg
[2]Max Planck Institute for Dynamics of Complex Technical Systems, Magdeburg
[3]Institute of Chemical Engineering, University of Erlangen-Nuremberg

1 INTRODUCTION

Porous silicas are of great interest for applications in the fields of separation, catalysis or sensors. The performance in the different applications is always correlated with their microstructure and surface chemistry. An improvement of the performance of porous silicas is possible via various surface modifications. Organic modifications allow a tailoring of the surface properties for desirable applications. As an example, hydrophobic silica surfaces can be established by the reaction with hexamethyldisilazane (HMDS). This procedure enhances the resistance to water adsorption significantly. The optimisation of drug release (ibuprofen) from MCM-41 is an actual application of this modification procedure. The design of the carrier system by the chemical modification of the drug-impregnated ordered mesoporous silica material with HMDS is an effective approach for well-controlled drug release.[1]

The measurement of the effect of surface modification is often difficult. Inverse gas chromatography (IGC) is an innovative technique to investigate the relations between surface modification and surface properties by detecting physicochemical parameters correlated to the solids surface. IGC is very sensitive to differences in the surface chemistry. Thus it is a suitable tool to characterise (modified) porous materials.

The principle of this technique is to conduct gaseous or liquid probe molecules with known properties over the surface of the solid of interest. The retention time of the probe molecules is then influenced by the surface of the solid, hence enabling the calculation of physiochemical properties. Such properties are for example the surface energy of solids, adsorption isotherms, differential enthalpy and entropy of adsorption and diffusion coefficients.[2-4] Recently, Thielmann used the IGC for the characterisation of MCM-41, activated carbons and porous alumina by the determination of the dispersive surface energy γ_S^D .[4]

The aim of this work was to investigate the correlations between the dispersive surface energy and the surface chemistry of porous materials. Porous silica glass beads are a suitable model system for the study of these correlations. Porous glasses are characterised by a controlled pore size in the range between 0.3 and 1000 nm.[5] The surface chemistry of porous glasses is well known. The surface properties of porous glasses can be modified by the reaction of the hydroxyl groups.[6] In this study the surface chemistry of two porous glasses with different pore size was modified by (i) thermal treatment, (ii) reaction with HMDS and (iii) a combined procedure to investigate the dependency of the dispersive surface energy on the change of the surface chemistry.

2. EXPERIMENTAL

2.1 Sample preparation

Mesoporous and macroporous glass beads were synthesized by the following procedure. Sodium borosilicate initial glass beads (70 wt.% SiO_2, 23 wt.% B_2O_3, 7 wt.% Na_2O), 0.1-0.2 mm in diameter were heat treated for 24 hours at 630°C and 48 hours at 550°C respectively, for phase separation into an alkali-rich borate phase and an almost pure silica phase. Afterwards a combined acid and alkaline leaching treatment followed. The resulting porous glass beads were washed with deionised water and dried at 80°C after each leaching step. Figure 1 shows the preparation procedure of the porous glass beads.

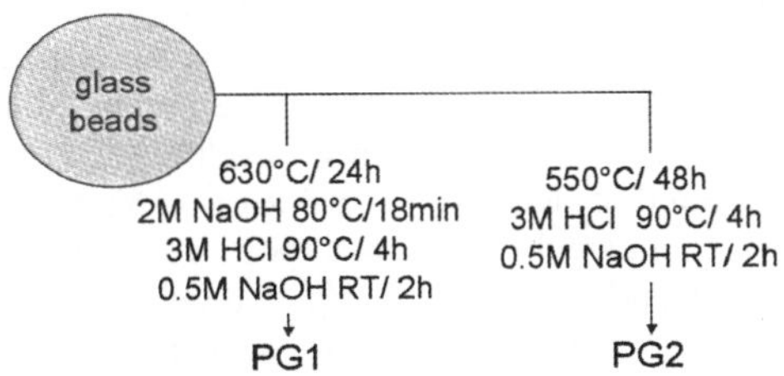

Figure 1 *Preparation procedure of the porous glass samples*

The surface of the macroporous glass PG1 was modified by the reaction of HMDS with the surface hydroxyl groups in *n*-hexane solvent. The reaction mixture was heated to reflux during the reaction period. Trimethylsilyl (TMS) groups were generated on the surface by reaction (1). The degree of silylation was controlled by the reaction period (0 - 24 h).[7] The modified porous glass beads were washed with *n*-hexane several times and dried at 100°C for 24 hours. Figure 2 summarises the chemical modifications and sample names.

$$(CH_3)_3SiNHSi(CH_3)_3 + 2 \equiv Si\text{-}OH \rightarrow 2 \equiv Si\text{-}O\text{-}Si(CH_3)_3 + NH_3 \qquad (1)$$

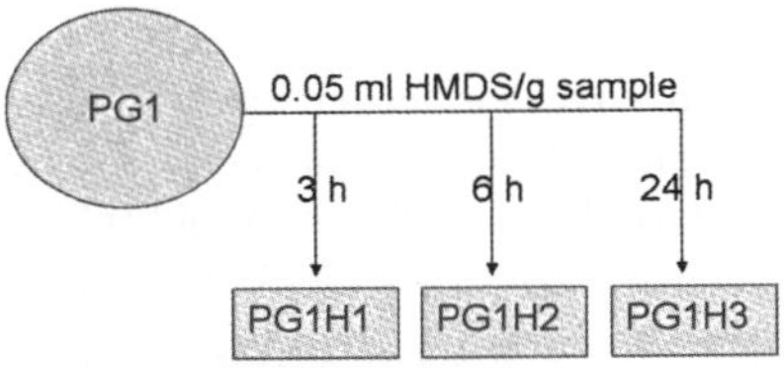

Figure 2 *Chemical modification of PG1*

Furthermore the mesoporous glass beads PG2 were modified by thermal treatment in a temperature range between 300 and 500°C for 24 hours. The aim of the thermal treatment was to create different concentrations of the surface hydroxyl groups by the temperature dependency of the condensation of adjacent hydroxyl groups. As a result, siloxane species were formed (equation 2). The temperatures of the pretreatment were annexed to the sample names.

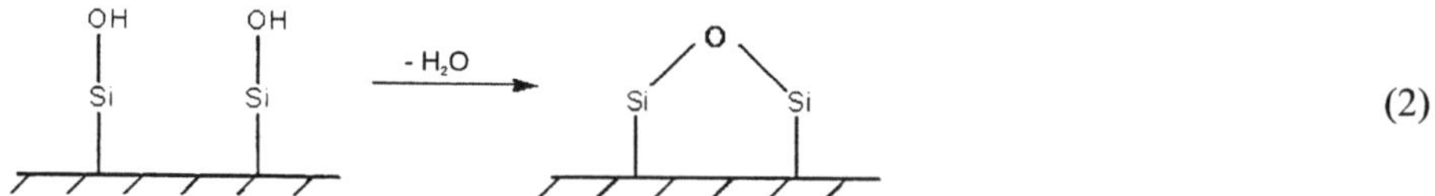

$$\tag{2}$$

In addition a combined modification of PG2 was performed by pretreatment at several temperatures (300, 400, 500°C) for 24 hours followed by a chemical reaction with HMDS (0.5 ml HMDS/g sample, 48 hours) afterwards. In this case the sample name is composed of the temperature of pretreatment and an "H" for the HMDS modification.

2.2 Characterisation

Nitrogen sorption measurements were performed by using a Sorptomatic 1990 apparatus (ThermoFinnigan). All samples were degassed at 120°C before measurement for at least 24 hours at 10^{-5} mbar. Adsorption and desorption isotherms were measured over a range of relative pressures (p/p^0) from 0 to 1.0. Surface areas were determined from the linear part of the Brunauer-Emmett-Teller (BET) equation in a relative pressure range (p/p^0) of the adsorption isotherms between 0.05 and 0.25.[8] A value of 0.162 nm^2 was used for the cross-sectional area per nitrogen molecule. The total pore volume was estimated from the amount of gas adsorbed at the relative pressure $p/p^0 = 0.99$ assuming that pores were filled subsequently with condensed adsorptive in the normal liquid state. The pore size distributions of the mesoporous glass beads were determined from the desorption branch of the nitrogen sorption isotherm according to the BJH method.[9]

For the determination of texture data by mercury intrusion the samples were degassed for 24 hours at 120°C. Afterwards the measurements were performed by using a Pascal 440 porosimeter (ThermoFinnigan). A contact angle of 141.3° for Hg was used. The cumulative pore volume of the macroporous glass beads at a given pressure represents the total volume of mercury taken up by this sample at that pressure. The pore diameter was calculated by applying the Washburn equation and a cylindrical pore model.

Elemental analysis was performed on a CHNS 932 apparatus to calculate the concentration of the TMS groups from the carbon content.

2.3 IGC measurements

The determination of the dispersive surface energy γ_s^D was conducted with an IGC equipment by Surface Measurement Systems. All samples were activated for 24 hours at 120°C prior to the measurements. The samples were put into a column (length = 30 cm, inner diameter = 2 mm) and were fixed with glass wool at the top and bottom of the sample. For the determination of γ_s^D a series of alkanes (hexane, heptane, octane and decane) with a concentration of 4% in a helium carrier gas and a gas flow of 10 ml/min at a column temperature of 93°C were utilized. The retention time of each alkane was measured by a flame ionisation detector (FID) and represents the basic measurement of the

IGC experiment. The retention volume can be calculated from the retention time at peak maximum for symmetrical (Gaussian) peaks by equation 3:

$$V_N = \frac{T}{273.15}(t_r - t_0)FJ/m \qquad (3)$$

where T is the column temperature, t_R is the retention time of the probe molecules and t_0 is the "dead time" of a non-interacting probe molecule like methane. F is defined as the flow rate of the carrier gas, J is the James-Martin correction, which corrects the retention time for the pressure drop in the column.[10]

The relation between the free energy of sorption ΔG and the retention volume is given by:

$$\Delta G = RT \ln V_N + K \qquad (4)$$

where R is the gas constant and K is a constant. ΔG can also be related to the energy of adhesion W_A between the probe molecule and the solid:

$$\Delta G = N_A a W_A \qquad (5)$$

with a as cross sectional area of the adsorbate and N_A as Avogadro constant. The energy of adhesion can be split up into a non-polar dispersive and a specific term caused by polar interactions:

$$W_A = W_A^D + W_A^{SP} \qquad (6)$$

For non-polar interactions W_A^D can be described by:

$$W_A^D = 2\sqrt{\gamma_S^D \gamma_L^D} \qquad (7)$$

with γ_S^D as the dispersive surface energy of the solid adsorbent and γ_L^D as the surface tension of the liquid adsorbate (given in literature).

Combining equations (4), (5) and (7) leads to the linear relationship of equation (8) given by Schultz et al.:[11]

$$RT \ln V_N = 2N_A (\gamma_S^D)^{1/2} a(\gamma_L^D)^{1/2} + const. \qquad (8)$$

The dispersive part of surface energy can now be obtained from the slope $m = 2N_A(\gamma_S^D)^{1/2}$ of this equation. So if non-polar probe molecules are used for the measurement of the retention time, the dispersive term of surface energy can be calculated.

3. RESULTS AND DISCUSSION

3.1 Sample preparation

The textural properties of the porous glasses were measured by low temperature nitrogen adsorption and mercury intrusion. The texture characteristics are shown in Table 1. The samples modified with HMDS show smaller specific pore volumes, what is understandable due to the larger TMS groups in comparison with the hydroxyl species. Additionally, a reduction in surface area appears especially for the chemically modified samples of PG2. The reduction of the surface area of the modified samples could be established by two reasons: on the one hand the TMS groups cover the roughness of the sample surface (very small pores, cracks, and fissures that are not accessible for nitrogen adsorption any more), on the other hand the pore size is slightly narrowed by a TMS group layer on the surface.[7] However, this is not reflected in the mean pore diameter of the mesoporous samples determined by nitrogen adsorption.

Table 1 *Texture properties of the prepared samples*

Sample	Spec. surface area [$m^2 \, g^{-1}$]	Mean pore diameter [nm]	Spec. pore volume [$cm^3 \, g^{-1}$]
PG1	38[a]	52[a]	0,453[a]
PG1H1	37[a]	49[a]	0,446[a]
PG1H2	39[a]	49[a]	0,447[a]
PG1H3	38[a]	50[a]	0,427[a]
PG2	112[b]	11,3[b]	0,484[b]
PG2_300	115[b]	11,3[b]	0,479[b]
PG2_400	109[b]	11,3[b]	0,463[b]
PG2_500	112[b]	11,0[b]	0,472[b]
PG2_H	73[b]	11,3[b]	0,381[b]
PG2_300H	72[b]	11,3[b]	0,410[b]
PG2_400H	70[b]	11,3[b]	0,411[b]
PG2_500H	79[b]	11,0[b]	0,429[b]

[a] textural characterisation by mercury intrusion
[b] textural characterisation by nitrogen adsorption

The results of the elemental analysis and the calculated concentration of TMS groups are shown in Table 2. The concentration of TMS groups was calculated according to equation (9) given by Berendsen et al.[12] and Hemetsberger et al.[13]:

$$[Si] = \frac{\%C \cdot 10^6}{(100 \cdot n \cdot 12 - \%C \cdot M)O_s} \tag{9}$$

where $\%C$ is the weight percentage of bonded carbon, n is the number of carbon atoms of alkylsilane, O_S represents the specific surface area of the unmodified porous glass beads and M is the molecular mass of the bonded organosilane.

Table 2 *Surface concentration of TMS groups of chemically modified samples*

sample	carbon content[a] [%]		TMS concentration [μmol/m^2]
PG1H1	0.35	0.34	2.59
PG1H2	0.45	0.46	3.27
PG1H3	0.70	0.67	5.02
PG2H	1.43	1.45	3.68
PG2_300H	1.33	1.24	3.17
PG2_400H	1.17	1.20	3.09
PG2_500H	1.00	0.97	2.49

[a] two analyses

The surface concentration of TMS groups obtained for the variation of reaction time of the initial macroporous glass beads PG1 are in the range between 2.59 and 5.02 μmol/m^2. There is a correlation between the reaction time of the HMDS treatment and the generated TMS groups on the surface. The amount of TMS groups increases with increasing reaction time.

The initial mesoporous glass beads PG2 and the samples pretreated at different temperatures were modified with HMDS by the same procedure, but different TMS group concentrations are detected by elemental analysis. An increasing temperature of pretreatment causes a decreasing concentration of TMS groups. The reduction of active hydroxyl groups available for the reaction with HMDS induced by the thermal pretreatement and the inaccessibility of residual hydroxyl groups under the "umbrellas" of generated TMS groups are potential reasons for this coherence.

3.2 Dispersive part of surface energy

On the basis of the PG1-series it is noticeable that a correlation between the amount of bonded TMS groups and the dispersive surface energy exists. The dispersive part of surface energy increases with an increasing amount of TMS groups bonded to the surface. This is a result of the modified surface chemistry that becomes more and more hydrophobic with an increasing amount of TMS-groups. In figure 3 some TMS concentrations are plotted versus the dispersive surface energy.

The knowledge of the preparation procedure of the samples is important for the interpretation of the dispersive surface energy in the case of the PG2-series. Two opposite processes take place during the thermal pretreatment. There is a reduction of the hydroxyl groups and the formation of new boron-species on the surface of the porous glass.[14] The effect of both processes depends on the temperature and increases with increasing temperature. Because of this fact, the differences between the dispersive surface energies of the thermally treated samples are minor.

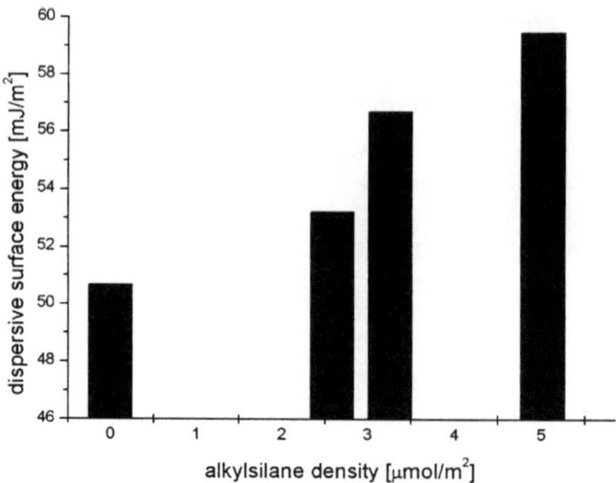

Figure 3 *Correlation between the dispersive surface energy and the TMS-concentration for the PG1-series*

As it is shown in figure 4 the silylated PG2 samples show higher γ_s^d values due to the reduced polarity introduced by the TMS groups. The differences between an exclusive thermal modification and a combined procedure decrease with increasing temperature of pretreatment. This is caused by a reduction of active sites (decreasing concentration of TMS groups) and an increasing amount of boron on the surface of the modified samples.

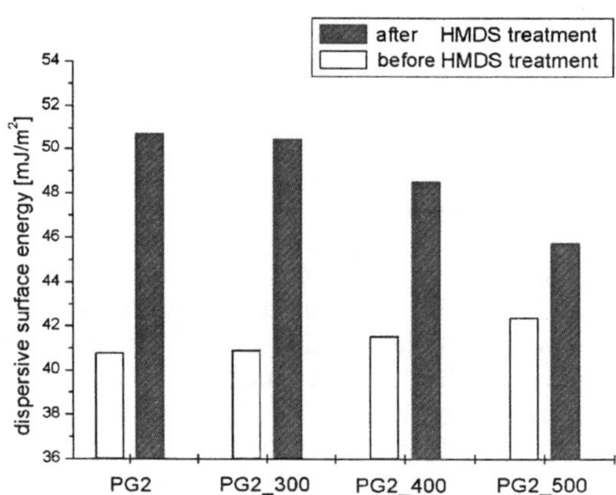

Figure 4 *Comparison of the results of different modifications by means of the IGC- results*

4. CONCLUSIONS

The IGC technique was used to characterise the surface modification of porous glasses. A dependence of the dispersive part of surface energy on the surface chemistry could be demonstrated in this study. Caused by the thermal reduction of hydroxyl groups and the modification by trimethylsilyl groups the solids surface becomes more hydrophobic. The hydrophobic character leads to an increase in the dispersive surface energy. Furthermore there is a correlation between the concentration of the TMS groups and the dispersive part of the surface energy. This allows a quantification of the surface modification.

References

1 Q. Tang, Y Xu, D. Wu, Y. Sun, J. Wang, J. Xu and F. Deng, *J. Controll. Release*, 2006, **114**, 41.

2 V. Kiselev and I. Yashin in *Gas-Adsorption Chromatography*, Plenum Press, New York, 1965, p 120.

3 N. A. Katsanos and G. Karaiskakis in *Time-Resolved Inverse Gas Chromatography and Its Applications*, HNB Publishing, New York, 2004, p 44.

4 F. Thielmann, *J. Chromatogr. A*, 2004, **1037**, 115.

5 D. Enke, F. Janowski, W. Schwieger, *Microporous and Mesoporous Materials*, 2003, **60**, 19.

6 F. Janowski and D. Enke in *Handbook of Porous Solids* ed. F. Schüth, K.S.W. Sing and J. Weitkamp, Wiley-VCH, Weinheim, 2002, p 1432.

7 T. Takei, A. Yamazaki, T. Watanabe, M. Chikazawa, *Journal of Colloid and Interface Science*, 1997, **188,** 409.

8 S. Brunauer, P.H. Emmett, E. Teller, *J. Am. Chem. Soc.*, 1938, **60**, 309.

9 E.P. Barrett, L.G. Joyner, P.H. Halenda, *J. Am. Chem.. Soc.*, 1951, **73**, 373.

10 J. Adolphs, *Materials and Structures*, 2005, **38**, 443.

11 J. Schultz, L. Lavielle and C. Martin, *J. Adhesion*, 1987, **23**, 45.

12 G. E. Berendsen, L. de Galan, *J. Liquid Chromatogr.*, 1978, **1**, 561.

13 H. Hermetsberger, M. Kellermann, H. Ricken, *Chromatographia*, 1977, **10**, 726.

14 F. Janowski and W. Heyer in *Poröse Gläser - Herstellung, Eigenschaften und Anwendung*, Deutscher Verlag für Grundstoffindustrie, Leipzig, 1982.

STRUCTURAL DISTINCTION OF SILICA GELS USING FTIR AND RAMAN SPECTROSCOPY

Istvan Halasz, Mukesh Agarwal, Runbo Li and Neil Miller

PQ Corporation, Research and Development Center, 280 Cedar Grove Road, Conshohocken, PA 19426, USA

1. INTRODUCTION

Micro and mesoporous, crystalline and amorphous silica and silicates are important in adsorption, catalysis, and many other areas[1-4]. Many of these materials are obtained via gel formation from aqueous alkaline silicate solutions which are the least expensive ingredients. Desirable gel characteristics include both physical parameters like porosity, surface area, mechanical strength, or durability and chemical properties like compatibility with and/or adhesion to plastics, metals, adsorbates, reactants, and so on. Various techniques can help in attaining the targeted material properties such as variation of pH, aging, employment of different washing, drying, coating procedures and many more. Na-silicates are most frequently used ingredients but Li- or K-silicate solutions have been found to be preferable for certain applications. The concentration, alkaline/silica ratio, presence of structure directing agents or multivalent metal ions (with Al^{3+} in the front) can also fundamentally affect the outcome of the gelling process.

Despite the broad technical interest and long empirical expertise, very little is known about the molecular structure of silica gels and even less about the transition of dissolved, sub-nano sized silicate molecules into these siloxane polymers. Since ^{29}Si NMR studies suggest that above about Na/Si ~ 2 ratios every Na-silicate solution contains an equally random mix of Si_1-Si_{10} siloxane rings and the composition of dissolved silicates is not a function of alkaline ions present[5, 6], the arbitrary condensation of such siloxane rings into a random 3D network seems to be a plausible description for the formation and molecular structure of silica gels. In agreement with this description, silica gels are generally viewed as polymers with indistinguishably random molecular constitutions. Hence the relation between their molecular structure and the above mentioned physical and chemical properties have barely been investigated.

In a series of recent studies we demonstrated that FTIR and Raman spectroscopy can identify distinct differences between the molecular structures of silicates in aqueous solutions containing different alkaline ions, alkaline/silicon ratios, and silicate concentrations[7-12]. These results triggered some simple but as yet unanswered questions pertaining to the structure of gels made from these solutions. Will the gel retain the fundamental construction of dissolved silicate molecules or it diminishes due to the random coupling of siloxane rings? Will these phenomena depend on the pH, nature of alkaline ion, and/or concentration of the initial solutions? Is there a connection between the molecular structure and the physical/chemical properties of gels? Does the molecular structure of gels alter during aging, washing, or drying? Silicate ions released upon gel re-dissolution, as some assume for example in zeolite synthesis, will have similar or different structure compared to those being present in the solution before gelling? Can structural differences of gels be traced by FTIR and Raman spectroscopy?

We have addressed some of these questions with regard of certain sodium silicates and demonstrated that these molecular spectroscopic methods can differentiate between the molecular constitutions of some silica gels fabricated differently[13-15]. In this paper we compare the FTIR and Raman spectra of fresh wet gels prepared from aqueous solutions of various Li-, K-, and Na-silicates at different pH values.

2. MATERIALS AND METHODS

All starting silicates are commercial products of the PQ Corporation. Some of their properties are shown in Table 1. 3 and 0.2 mol SiO_2/L solutions were made with deionized water (18.2 MΩ*cm) and equilibrated for >24 hours before measurements. Selected properties of these solutions are shown in Table 2. A 6 M, veritas redistilled HCl solution from GFS Chemicals was diluted to 3M for setting the pH for gelling.

A Corning Scholar 425 pH meter equipped with an Accumet electrode was used for measuring the pH. A Nicolet Magna 550 FTIR spectrometer with a single bounce diamond ATR and a 532 nm dispersive laser Raman spectrometer from Kaiser were used for carrying out the spectroscopic measurements. Details of these instruments and techniques have been reported elsewhere[17, 18].

Table 1. *Selected properties of commercial materials used for making alkaline silicate solutions for gelling (further information on www.pqcorp.com)*

Name (A^+ ion)*	Concentration [SiO_2 w%]	SiO_2/A_2O weight ratio	Density [g/cm^3]
LithiSil 25 (Li$^+$)	20.5	8.20	1.20
N$^®$ Silicate (Na$^+$)	28.7	3.22	1.38
STAR$^®$ (Na$^+$)	26.5	2.50	1.40
Metso Beads$^®$ 2048	solid	0.97	2.61
KASIL 1 (K$^+$)	20.8	2.50	1.26
KASIL 1624 (K$^+$)	15.0	1.65	1.22

*A = Li or Na or K

Table 2. *Concentration, pH, dissociation degree (α), and average molweight (AMW) of starting silicate solutions (methods and further data in References 8, 10, 12, 16)*

Commercial name [Nominal composition]	Concentration [mol SiO_2/L]	pH	Dissociation α	AMW [SiO_4] tetrahedron/Silicate ion
LithiSil-25	3.0	11.1	0.30	10.3
[$(Li_2O)_{0.25}SiO_2$]	0.2	11.0	0.61	2.8
N®-silicate	3.0	11.3	0.29	10.5
[$(Na_2O)_{0.30}SiO_2$]	0.3	11.1	0.74	1.2
STAR®	3.0	11.7	0.29	6.1
[$(Na_2O)_{0.38}SiO_2$]	0.2	11.4	0.70	2.7
Metso Beads® 2048	3.0	13.6	0.32	1.0
Na_2SiO_3	0.2	13.0	0.82	1.0
KASIL® 1	3.0	11.6	0.26	13.4
[$(K_2O)_{0.27}SiO_2$]	0.2	11.2	0.59	4.1
KASIL® 1624	3.1	12.0	0.26	4.8
[$(K_2O)_{0.38}SiO_2$]	0.2	11.5	0.58	2.1

3. RESULTS AND DISCUSSION

Figs. 1 and 2 illustrate that the molecular composition of silicate solutions can substantially differ and both FTIR and Raman spectroscopy can indicate such differences. These figures capture the main characteristics of spectra of every starting solution shown in Table 2. Their detailed analysis along with the concept of the structural peak assignments has been reported elsewhere[8, 10, 12]. Briefly, we utilize the traditional association of symmetric and asymmetric Si-O stretch related vibrations (v_s and v_{as}) above about 700 cm^{-1} with the Q^n (n = 0, 1,…4) connectivity of [SiO_4] tetrahedra to 0-4 other tetrahedra in the silicate molecules. These assignments are probably best established in the structural studies of glasses and we found that they also work reliably with the Raman spectra of small, dissolved silicate molecules[10, 12]. We found the FTIR assignments to be somewhat more ambiguous for various reasons. For example the assignment of 1120 cm^{-1} band in Fig. 1 to the usually reported Q^3 connectivity, mostly associated with corner positions in interconnected siloxane rings, seems to contradict to the increase of this band with dilution because this rather breaks up the larger rings into small molecules as the average moleweights (AMW) in Table 2 indicate. One could of course also speculate that the 1120 cm^{-1} IR band might belong to Q^3 type tetrahedra in small branched linear

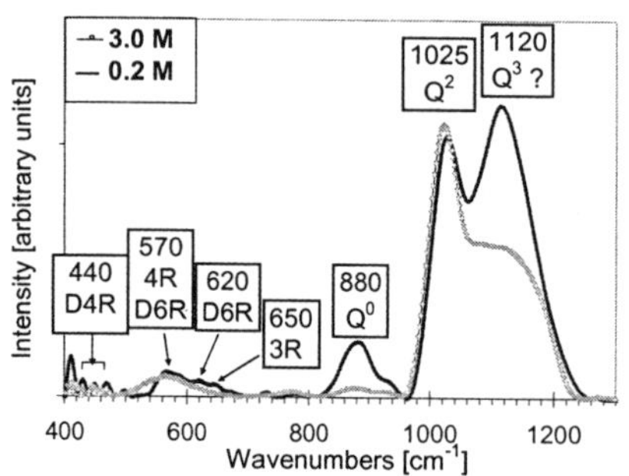

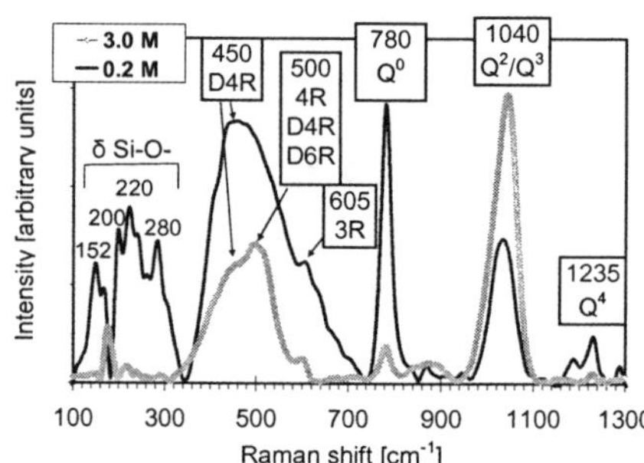

Figure 1. *FTIR of the Li-silicate solutions*   **Figure 2.** *Raman of the Li-silicate solutions*

molecules which can form from the larger siloxane rings Yet this assumption contradicts the Raman results in Fig. 2 which suggest Q^0 monomer formation upon dilution in accordance with the decreasing AMW. Along with some other considerations we haven't been able to fully resolve these contradictions between the FTIR and the Raman data[10, 12].

The assignment of single and double rings (R and DR) to the bending, twisting, and rocking vibrations of Si-O-Si connections at <700 cm^{-1} has also long tradition especially in the spectroscopy of zeolites. Figs. 1 and 2 illustrate that the Raman bands are more intense in this range than the IR bands. Moreover Raman also permits measurement in the <320 cm^{-1} energy range which includes vibrational motions of terminal groups including like non-bridging oxygen ions or Si-O-Na, Si-O-Ag, etc. metal ions in glasses and zeolites[19-21].

Figures 3 and 4 compare the spectra of acid set gels obtained by dosing 3 M alkaline silicate solutions into HCl solution. Every silicate gelled at pH ~ 1 within a few minutes. The higher ratio of Q^3 and Q^4 tetrahedra in the FTIR spectra of gels from Lithisil-25 and Kasil-1 suggests that they form stronger 3D network connection than the other two polymers. It appears that the alkaline/silicon ratio is more important from this point of view than the difference between the alkaline ions. The IR bands near 790 and 960 cm^{-1} have not been associated in the literature with Q^n connectivities or siloxane rings. The high intensity of the 960 cm^{-1} band, which can be assigned to the symmetrical stretching mode of Si-OH bands[22, 23], is unusual. It might suggests a large number of connectivity defects in the polymer network which can be magnified by traces of Ti^{4+}, Fe^{3+} or other metal ions[17, 24].

The Raman spectra in Fig. 4 suggest different relation between the structures of these gels than did their FTIR spectra. Three characteristically different spectra are shown, with which the spectra of other gels substantially overlap. Similar to our conclusion from the IR spectra, it appears that the alkaline/Si ratio is more important in determining structure of gels than the type of alkaline ions. The intense Q^2 and ring vibrations of the high Na/Si ratio Star-originated gel suggest that this material is built from a relatively loose network of siloxane rings. In contrast, the low Li/Si ratio Lithisil-originated gel (also note the larger AMW in Table 2!) seems to contain less from the detected medium sized rings and its strong Q^4 band indicates extensive network connections. The intense and broad ν_{as} band around 1250 cm^{-1} is unusual since it is quite week in the Raman spectra of most solid silicates including for example zeolites which definitely are composed of almost purely

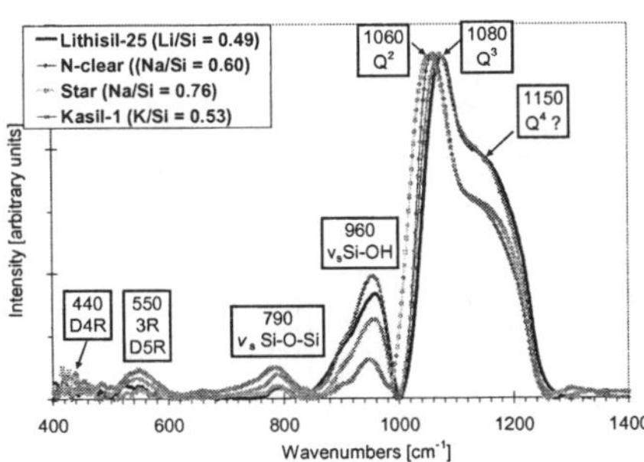

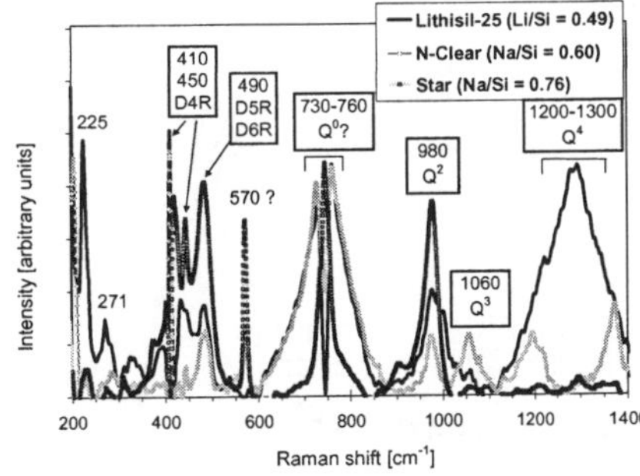

Figure 3. *FTIR of gels made at pH ~ 1 from 3 M solutions of various silicates*   **Figure 4.** *Raman of gels made at pH ~ 1 from 3 M solutions of various silicates*

                                    Characterisation of Porous Solids VIII

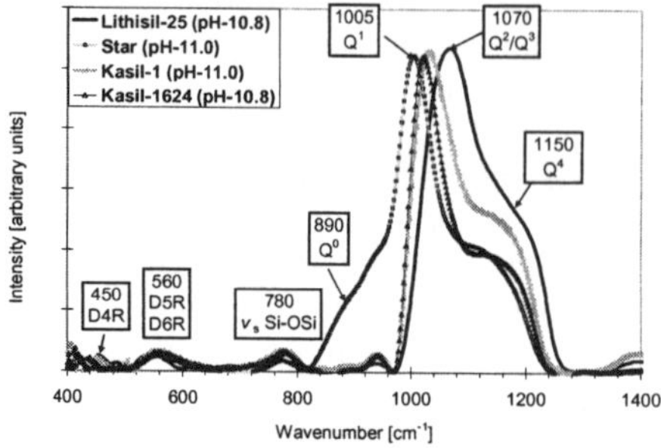

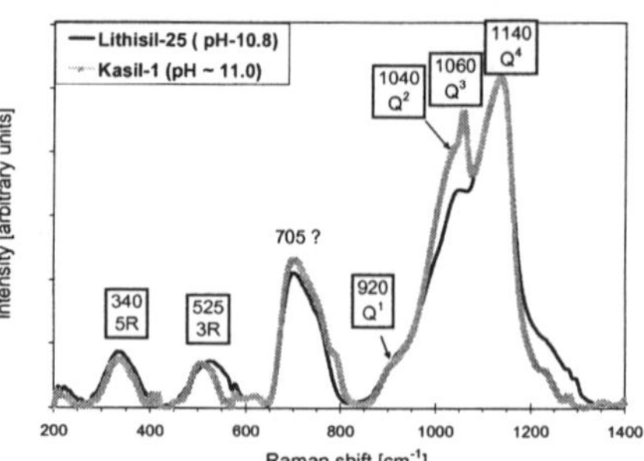

Figure 5. *FTIR of base set gels from 3 M solutions of various silicates*

Figure 6. *Raman of base set gels from 3 M solutions of various silicates*

Q^4 type [SiO$_4$] tetrahedra[17, 18, 24]. The Li-gel presumably contains therefore an unusually high number of asymmetric interconnections. In accordance with this speculation one can see in the Li-silica gel spectrum an intense, connectivity-defect related terminal deformation vibration at 225 cm^{-1}. The assignment of bands in the 730-760 cm^{-1} region to Q^0 monomers is arbitrary based on the fact[8] that monomers in solution give an intense Raman shift near 780 cm^{-1}.

Base set gels were fabricated by dosing HCl solution into the silicate solutions. Figs. 5 and 6 show that depending on their original pH (Table 2) the 3 M solutions gelled at slightly different pH values within a few minutes. The FTIR spectra in Fig. 5 indicate an even stronger shift in the v_{as} stretching Si-O vibrations near 1000 cm^{-1} than those in the FTIR spectra of acid set gels (Fig. 3) but this difference seems still rather depend from the alkaline/Si ratio and not on the nature of alkaline ions present. In contrast to the >Q^2 connectivity of acid set gels, at least the Star originated gel seems to be dominated by Q^1 and Q^0 connections in Fig. 5. The intense 960 cm^{-1} band seen in Fig. 3 is virtually also absent from the FTIR spectra of base set gels.

The Raman spectra in Fig. 6 contradict again some of the FTIR assignments pertinent to the same materials. The Raman spectra of the base set gels were found to be nearly identical with each other and do not indicate substantial amount from Q^0 or Q^1 connected [SiO$_4$] tetrahedra to be present. Fig. 6 shows the two gels which showed the biggest

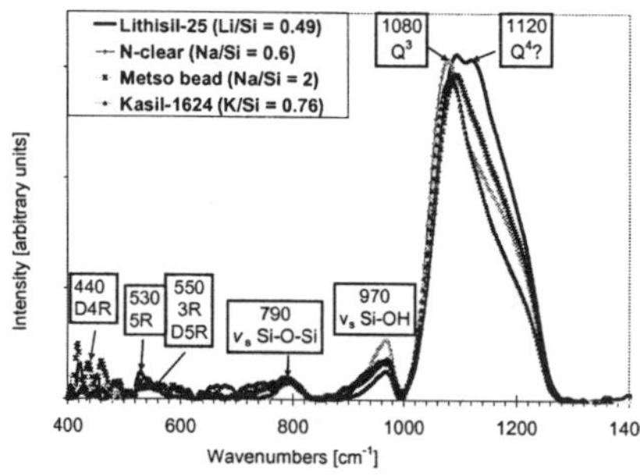

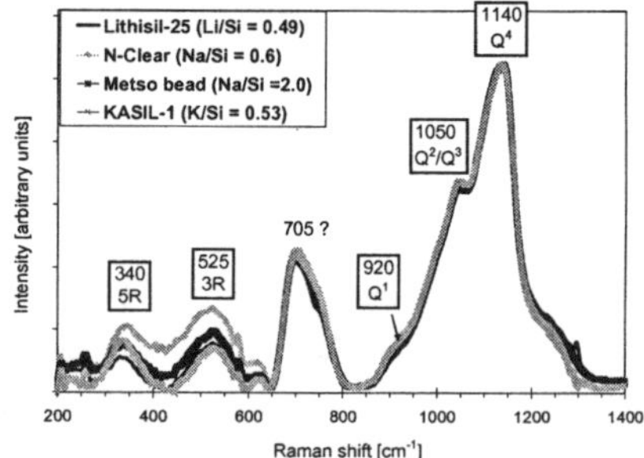

Figure 7. *FTIR of gels made at pH ~ 7.1 from 0.2 M solutions of various silicates*

Figure 8. *Raman of gels made at pH ~ 7.1 from 0.2 M solutions of various silicates*

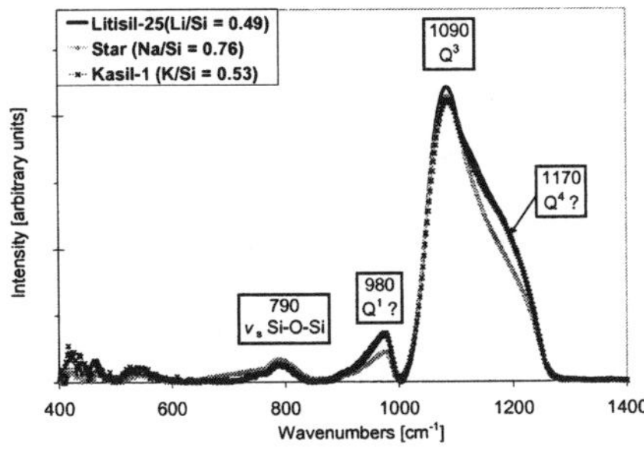

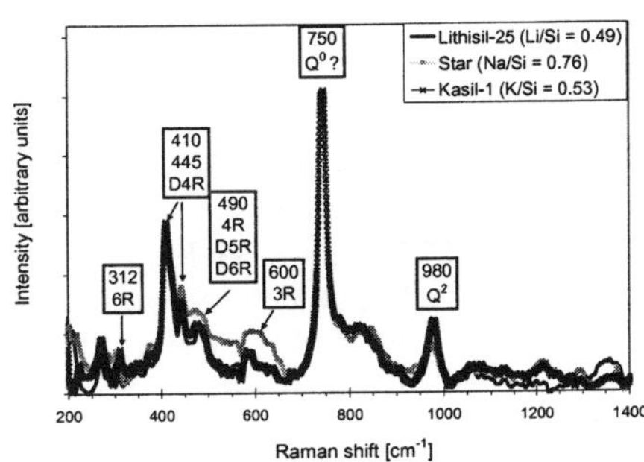

Figure 9. *FTIR of gels made at pH ~ 5 from 0.2 M solutions of various silicates*

Figure 10. *Raman of gels made at pH ~ 5 from 0.2 M solutions of various silicates*

Raman difference. Most other relevant spectra totally cover the spectrum of the Lithisil-25 based siloxane polymer. It is perhaps interesting to note that unlike in the varying structures of acid set gels (Fig. 4) mainly 3R and 5R rings seem to dominate in these base set gels. The formation of these rings agrees both with the COSY ^{29}Si NMR based notion of Knight et al.[6] who argued that most sub-nano sized silicate molecules contain 3R rings which are not appropriate secondary building units for making zeolites and with Houssin at al.[25] who explained the formation of MFI zeolites with appropriate agglomeration of 5R rings in the presence of adequate structure directing agents.

Figs. 7 and 8 indicate remarkable similarity in the molecular structure of the base set gels fabricated from 0.2 M silicate solutions and those which were made from 3 M solutions at basic conditions (Figs. 5 and 6). A slight difference is that the FTIR spectra in Fig. 7 resemble each other more than the spectra in Fig. 5 and do not indicate noticeable Q^0 or Q^1 connectivities. Since both the average molweight and the molecular structure of the dilute silicate solutions were found to be quite different from both each other and the spectra of the more concentrated solutions[8, 10, 12], one can conclude from the largely overlapping spectra of gels in Figs. 5-8 that the initial composition of silicate solutions has practically no effect on the molecular structure of the base set gels. The gels from 0.2 M solutions were also made by dosing acid to the silicate, but one had to go down to pH 7.1 to obtain gel within a 24 hour period (some gels formed faster).

The acid set gels from 0.2 M silicate solutions were also fabricated similar to the acid set gels from the 3 M solutions but in these cases we had to increase the pH to 5 in order to obtain gel within 24 hours except for the 0.2 M N-silicate solution from which we could not get gel by this procedure within 72 hours even by going up to pH ~6.

The overlapping spectra in Figs. 9 and 10 indicate that the molecular constitution of silicates in the initial dilute solutions has basically no effect on the structure of the gels. A comparison with Figs. 3 and 4 suggests however, that the difference between the molecular structures of 3 M and 0.2 M solutions can affect the structure of acid set gels.

The FTIR spectra of the acid and the base set gels in Figs. 7 and 9 do not indicate much structural differences, but their Raman spectra in Figs. 8 and 10 (and also in Figs. 4 and 6 for that matter) are quite different. Consequently one can generally conclude that the molecular structures of the acid and the base set gels are markedly different from each other which conceivably could provide molecular basis for various empirically targeted properties of silica gels.

CONCLUSIONS

1) Molecular spectroscopy is adequate to see differences between the molecular constitutions of silica gels.
2) Data from FTIR and laser Raman spectroscopy complement each other, but further refinement is desirable to resolve contradictions and getting more quantitative pictures about the structures of these siloxane polymers.
3) There are substantial differences between the molecular structures of acid set and base set silica gels.
4) The nature of alkaline ion has only minor effect on the gel structure.
5) The molecular composition of starting solutions plays also only minor role in formation the polymer structure.
6) Dilution of the initial silicate solutions can shift the pH of gel formation significantly, but does not affect substantially the ultimate molecular structure of base set gels.
7) The alkaline/Si ratio affects the structure of gels when made from more concentrated silicate solutions but this effect increasingly diminishes with increasing dilution.

References

1. D. Mravec, J. Hudec, I. Janotka, *Chem. Pap.*, 2005, **59**, 62.
2. R. F. Lobo, *Nature*, 2006, **443**, 757.
3. H. E. Bergna, W. O. Roberts (editors), *Surfactant Sci. Ser.*, 2006, **131**, 1.
4. L. F. Giraldo, B. L. Lopez, L. Perez, S. Urrego, L. Sierra, M. Mesa, *Macromol. Symp.*, 2007, **258**, 129.
5. G. Engelhardt, D. Michel, *High resolution Solid State NMR of Silicates and Zeolites,* J. Wiley & Sons, Chichester, NY, Brisbane, Toronto, Singapore, 1987.
6. C. T. G. Knight, J. Wang, S. D. Kinrade, *Phys. Chem. Chem. Phys.*, 2006, **8**, 3099.
7. I. Halasz, M. Agarwal, R. Li, N. Miller, *Catal. Today*, 2007, **126**, 196.
8. I. Halasz, M. Agarwal, R. Li, N. Miller, *Catal. Lett.*, 2007, **117**, 34.
9. I. Halasz, R. Li, M. Agarwal, N. Miller, *Stud. Surf. Sci. Catal.*, 2007, **170A**, 800.
10. I. Halasz, M. Agarwal, R. Li, N. Miller, *Stud. Surf. Sci. Catal.*, 2008, **174**, in press.
11. I. Halasz, A. Derecskei-Kovacs, *Molecular Simulation*, 2008., in press.
12. I. Halasz, R. Li, M. Agarwal, N. Miller, *Microp. Mesop. Mater.*, 2008, submitted.
13. I. Halasz, R. Li, M. Agarwal, N. Miller, *Proc. 19th. NAM, Philadelphia*, 2005, P-122.
14. I. Halasz, M. Agarwal, R. Li, N. Miller, *Proc. 20th. NAM, Houston*, 2007, O-S2-04.
15. I. Halasz, R. Li, M. Agarwal, N. Miller, *J. Non-Crystalline Solids,* 2008, submitted.
16. R. Li, I. Halasz, M. Agarwal, N. Miller, *Proc. Pittcon, Chicago,* 2007, 500-2P.
17. I. Halasz, M. Agarwal, E. Senderov, B. Marcus, W. Cormier, *Stud. Surf. Sci. Catal.,* 2005, **158**, 647.
18. I. Halasz, M. Agarwal, B. Marcus, W. Cormier, *Microp. Mesop. Mat.*, 2005, **84**, 318.
19. R. Hanna, *J. Phys. Chem.* 1965, **69**, 3846.
20. M. D. Baker, G. A. Ozin, J. Godber, *Catal. Rev. Sci. Eng.,* 1985, **27**, 591.
21. T. Sun, K. Seff, *Chem. Rev.*, 1994, **94**, 857.
22. M. Hino, T. Sato, *Bull. Chem. Soc. Japan*, 1971, **44**, 33.
23. F. Boccuzzi, S. Coluccia, G. Ghiotti, C. Morterra, A. Zecchina, *J. Phys. Chem.*, 1978, **82**, 1298.

24. I. Halasz, M. Agarwal, E. Senderov, B. Marcus, *Appl. Catalysis A: General,* 2003, **241**, 167.
25. C. J. Y. Houssin, C. E. A. Kirschhock, P. C. M. M. Magusin, B. L. Mojet, P. J. Grobet, P. A. Jacobs, J. A. Martens, R. A. van Santen, *PCCP*, 2003, **5**, 3518.

THE POISSON RATIO OF POROUS MATERIALS

W. Pabst and E. Gregorová

Department of Glass and Ceramics, Institute of Chemical Technology, Prague, Technická 5, 166 28 Prague, Czech Republic

1 INTRODUCTION

The Poisson ratio can be defined for solid materials as the negative ratio of the relative diameter change in the transverse direction (negative for contraction) and the relative length change in the direction of tension (positive). When Poisson introduced this ratio in the 1820es, he came to the conclusion that it should be 0.25 for all materials and should therefore not be considered as a material property.[1] This is indeed the case when the solid is composed of particles atoms or molecules governed by spherically symmetric potentials.[2] In this case the so-called Cauchy relations hold and the elastic behavior of isotropic solids is governed by only one elastic constant, e.g. the Young's modulus.[3] For example, quasi-isotropic polycrystalline solids made up of alkali halides (cubic) and certain glasses may be expected to approach to this type of behavior to a certain extent.[4,5] In general, however, isotropic materials are characterized by two elastic constants, concomitant with the fact that (linear) elasticity is a fourth-order tensor property.[6] The Poisson ratio may be chosen as one of these. From the positivity of elastic moduli (assumption of thermodynamic stability) it follows that the Poisson ratio of an isotropic material must lie in the range $-1 < \nu < 0.5$, and although it has been frequently claimed in textbooks that isotropic materials with negative Poisson ratios do not exist, their existence has been definitely proved and the first auxetic materials have been produced in the 1980es.[7-9] It is clear, therefore, that the possibility of auxetic behavior has to be taken into account when the porosity dependence of the Poisson ratio is to be discussed. The Poisson ratio of porous materials is a topic of constant interest,[10-17] since – in combination with one of the elastic moduli (typically the Young's modulus) it determines the linear elastic response of isotropic materials completely and is therefore of practical importance for many problems of wave propagation and acoustics. However, up to now there is still no agreement on the "most realistic" porosity dependence of the Poisson ratio. Nevertheless, based on published literature values and the results of several popular models, most authors seem to recognize only a "slight" porosity dependence, with a converging trend to some asymptotic limit value when the porosity approaches 100 %. In this paper we summarize and discuss the known relations and show that only power-law relations and certain types of exponential relations result, under certain conditions (sufficiently high porosity and appropriate Poisson ratio of the solid phase), in predictions that allow for negative Poisson ratios. According to our findings, auxetic behavior may occur in highly porous materials, e.g. cellular solids (provided their microstructure is appropriately tailored, e.g. to contain "re-entrant" corners), only when the Poisson ratio of the solid phase is higher than 0.2.

2 BASIC RELATIONS OF LINEAR ELASTICITY THEORY

Isotropic materials are characterized by two elastic constants. The most popular choices are the tensile modulus (Young's modulus) E, the bulk modulus (compression modulus) K, the shear modulus (torsion modulus) G and the Poisson ratio v.[6] The elastic moduli are connected to the Poisson ratio via the standard relations

$$v = \frac{3K - 2G}{2(3K + G)} = \frac{E}{2G} - 1 = \frac{3K - E}{6K} . \tag{1}$$

and thus the stiffness ratio K/G can be expressed in terms of v as

$$\frac{K}{G} = \frac{2(1+v)}{3(1-2v)} . \tag{2}$$

It follows from Eq. (1) or (2) that incompressible materials ($K \gg G$) are characterized by a Poisson ratio approaching 0.5 (corresponding to a bulk modulus approaching infinity). Rubber and many liquids, e g. water, are incompressible in a similar sense, but it has to be kept in mind that these materials are not linearly elastic. Cork has a Poisson ratio close to zero, and therefore does not contract significantly in the direction perpendicular to uniaxial tension. Since all three elastic moduli (E, K and G) are positive (corresponding to thermodynamic stability), it follows from the relations $E = 2G(1+v)$ and $E = 3K(1-2v)$ that the Poisson ratio is bounded by the inequality

$$-1 < v < \frac{1}{2} . \tag{3}$$

3 RELATIONS FOR THE EFFECTIVE POISSON RATIO OF POROUS MATERIALS

The elastic moduli of two-phase composite materials are bounded from above and below by micromechanical bounds.[18] However, for porous materials (i.e. composites, for which one phase is the void phase) the lower bounds degenerate to zero.[19] Therefore, the conventional microstructural one-point and two-point bounds (i.e. Voigt-Reuss-Paul bounds and Hashin-Shtrikman-Walpole bounds[20-24]) do not impose additional restrictions on the effective Poisson ratio beyond those given by Eq. (3). Thus, in principle, for all materials – from completely dense (porosity 0 %) to completely porous (porosity approaching 100 %) – microstructures might be found that lead to Poisson ratios ranging from – 1 (auxetic) to 0.5 (incompressible). In practice, of course, this is not observed. Auxetic behavior is usually found only for high porosities (typically in foams[25-28]), and many common materials seem to exhibit a relatively weak porosity dependence of the

Poisson ratio, mostly with a tendency to converge to some positive value quoted variably as 0.2, 0.25 or 0.333, as the porosity approaches 100 %.[10-12, 29-32]

Now, from Eq. (1) it follows that for porous Voigt materials, i.e. materials with microstructures for which the bulk and shear moduli obey the upper Paul bounds,[22] $K = K_0(1-\phi)$ and $G = G_0(1-\phi)$ (where K_0 and G_0 are the solid phase bulk and shear and ϕ the porosity), the effective Poisson ratio of the porous material remains the same as that of the solid phase, independent of porosity. However, when the porous material has a Hashin-Shtrikman microstructure,[23] i.e. the upper bounds are

$$K_{HS} = K_0 \cdot \frac{1-\phi}{1+A_{HS}\phi} \qquad \text{with} \qquad A_{HS} = \frac{1+v_0}{2(1-2v_0)} \tag{4}$$

and

$$G_{HS} = G_0 \cdot \frac{1-\phi}{1+B_{HS}\phi} \qquad \text{with} \qquad B_{HS} = \frac{2(4-5v_0)}{(7-5v_0)} \tag{5}$$

where v_0 is the Poisson ratio of the solid phase (e.g. the dense, i.e. pore-free polycrystalline material), there is a converging trend to the limit Poisson ratio (for $\phi \to 0$)

$$v^* = \frac{1+5v_0}{9+5v_0}. \tag{6}$$

This value (also resulting from the Mori-Tanaka approach[33]) is always between the solid Poisson ratio v_0 and the value 0.2, and implies, that for the special case of a solid Poisson of $v_0 = 0.2$ the effective Poisson ratio v is independent of porosity.

In the low-porosity region the linear approximation (also called dilute or non-interaction approximation),[18,34]

$$v = v_0 + \frac{3(1-v_0^2)(1-5v_0)}{2(7-5v_0)} \cdot \phi, \tag{7}$$

seems to follow a converging trend as well, but the effective Poisson ratio for $v_0 \neq 0.2$ crosses the $v = 0.2$ horizontal line at porosities lower than 100 %, and the resulting v^* values are in the range $0.125 - 0.214$. Remarkably similar are the Poisson ratio predictions following from the Nielsen relations,[35]

$$K_N = K_0 \frac{(1-\phi)^2}{1+A_N\phi} \qquad \text{with} \qquad A_N = \frac{5v_0-1}{2(1-2v_0)} \tag{8}$$

and

$$G_N = G_0 \frac{(1-\phi)^2}{1+B_N\phi} \qquad \text{with} \qquad B_N = \frac{1-5v_0}{7-5v_0} \tag{9}$$

(ν^* values in the range $0.091 - 0.231$), and the Poisson ratio predictions based on the assumption that both the tensile modulus and the shear modulus follow a Spriggs-type exponential relation,[36] i.e. $E_S = E_0 \cdot \exp(-[E]\phi)$ and $G_S = G_0 \cdot \exp(-[G]\phi)$, where $[E]$ and $[G]$ are the ν_0–dependent Einstein coefficients[37,38] of the corresponding linear approximations for E and G,[18,34,38-40]

$$\frac{G}{G_0} = 1 - [G] \cdot \phi \qquad \text{with} \qquad [G] = \frac{15(1 - \nu_0)}{7 - 5\nu_0}, \tag{10}$$

$$\frac{E}{E_0} = 1 - [E] \cdot \phi \qquad \text{with} \qquad [E] = \frac{3(1 - \nu_0)(9 + 5\nu_0)}{2(7 - 5\nu_0)}, \tag{11}$$

so that

$$\frac{\nu + 1}{\nu_0 + 1} = \exp(([G] - [E])\phi). \tag{12}$$

(ν^* values in the range $0.168 - 0.239$). Unfortunately, since the effective elastic moduli of this type of materials violate the upper Hashin-Shtrikman bounds, such materials cannot exist.

An alternative solution, though disputable (due to its incompatibility with the Nielsen solution[35]), has been proposed by Ramakrishnan and Arunachalam.[29,30] It is based on the following relations for the bulk and shear moduli,

$$K_{RA} = K_0 \frac{(1 - \phi)^2}{1 + A_{RA}\phi} \qquad \text{with} \qquad A_{RA} = \frac{1 + \nu_0}{2(1 - 2\nu_0)} \tag{13}$$

and

$$G_{RA} = G_0 \frac{(1 - \phi)^2}{1 + B_{RA}\phi} \qquad \text{with} \qquad B_{RA} = \frac{11 - 19\nu_0}{4(1 + \nu_0)} \tag{14}$$

and predicts the effective Poisson ratio to converge to the value 0.25 as the porosity approaches 100 %, irrespective of the solid Poisson ratio ν_0. On the other hand, the differential approach[41] as well as the self-consistent approach[42,43] both predict an convergence value of 0.2, implying that for porous materials with the "magic" Poisson ratio $\nu_0 = 0.2$ the effective Poisson ratio is independent of porosity. According to Zimmerman[10] the differential solution can be written in terms of the stiffness ratio

$$\frac{G}{K} = \frac{3}{4} + \frac{3(1 - 5\nu_0)}{4(1 + \nu_0)} \cdot \left(\frac{G}{G_0}\right)^{3/5}, \tag{15}$$

where the relative shear modulus can be approximated by the expression

$$\frac{G}{G_0} = (1-\phi)^2 \cdot \left(\frac{2(1+v_0) + (1+5v_0)(1-\phi)^{6/5}}{3(1-v_0)} \right)^{1/3} . \tag{16}$$

The self-consistent solution can be obtained iteratively by invoking the implicit Budiansky relations for porous materials (with finite bulk modulus)[43]

$$K_B = \left(\frac{1}{K_0} + \frac{\phi}{K_B(1-\alpha)} \right)^{-1} \qquad \text{with} \qquad \alpha = \frac{1+v_0}{3(1-v_0)} \tag{17}$$

and

$$G_B = \left(\frac{1}{G_0} + \frac{\phi}{G_B(1-\beta)} \right)^{-1} \qquad \text{with} \qquad \beta = \frac{2(4-5v_0)}{15(1-v_0)} . \tag{18}$$

Also these relations, for which an explicit closed-form expression cannot be given, predict a convergence of the effective Poisson ratio v to the value $v^* = 0.2$ as the porosity approaches 100 % (and porosity independence for $v_0 = 0.2$, as before).

Note that all the aforementioned relations predict a decrease of the effective Poisson ratio with porosity only for high solid Poisson ratios, i.e. when $v_0 > 0.2$ (for the Ramakrishnan-Arunachalam model, Eqs. (13) and (14), $v_0 > 0.25$). Moreover, none of these relations permits negative effective Poisson ratios v (as long as the solid Poisson ratio v_0 is positive). It is, however, an empirical fact that auxetic behavior can be achieved in foams where the solid phase has a positive Poisson ratio.[8,25-28] A simple consideration shows that the Archie-type power-law relation,[44-46]

$$\frac{v+1}{v_0+1} = (1-\phi)^{[E]-[G]} . \tag{19}$$

predicts negative Poisson ratios when $v_0 > 0.2$, but only for very high porosities (> 80 % for $v_0 = 0.5$ and > 94 % for $v_0 = 0.3$). On the other hand, the exponential relation

$$\frac{v+1}{v_0+1} = \exp\left(\frac{([G]-[E])\phi}{1-\phi} \right) . \tag{20}$$

based on the Pabst-Gregorová-type modified exponential relations for tensile and shear moduli,[47]

$$E_{PG} = E_0 \cdot \exp\left(\frac{-[E]\phi}{1-\phi} \right), \tag{21}$$

$$G_{PG} = G_0 \cdot \exp\left(\frac{-[G]\phi}{1-\phi}\right), \tag{22}$$

predicts negative Poisson ratios when $v_0 > 0.2$ for significantly lower porosities (> 62 % for $v_0 = 0.5$ and > 73 % for $v_0 = 0.3$). According to both the Archie-type power-law relation and the Pabst-Gregorová exponential relation, porous materials with a solid Poisson ratio v_0 lower than 0.2 cannot exhibit auxetic behavior. Note that all relations mentioned, including those that predict negative Poisson ratios at high porosities, remain positive and predict a converging trend in the low-porosity region. This is in contrast to the Arnold-Boccaccini-Ondracek relation,[13] which – except for containing an artifact at $\phi = 0.4$ (due to an assumed kink in the porosity dependence of the bulk modulus) – predicts an asymptotic limit value of $v^* = 0.5$ as $\phi \to 1$ (the v-singularity at $\phi = 1$ and the corresponding "boundary layer" in the $v - \phi$–diagram has been analyzed by Zimmerman[11]). It is also in contrast to the empirical approach of Kováčik,[16,17] who derives his Poisson ratio prediction

$$\frac{v+1}{v_0+1} = \left(1 - \frac{\phi}{\phi_c}\right)^N \tag{23}$$

from Phani-Niyogi-type fit relations[48] for E and G, without adopting our theory-based interpretation of the exponent,[38] $N = ([E]-[G]) \cdot \phi_c$ (with ϕ_c denoting a critical porosity, at which the elastic moduli become zero; in Kováčik's approach it is assumed to be the same for all elastic moduli and the Poisson ratio).

Finally, we would like to emphasize that reliable experimental data are still very rare. It is our experience, and seems to be well known,[5,49] that a precision of approx. 1 % in the elastic moduli (a realistic value when resonant frequency or ultrasonic methods are applied) results in an uncertainty of approx. 10 % in the Poisson ratio (apart from external error sources such as sample variability etc.). This allowed e.g. Gibson and Ashby to arrive at the unjustified conclusion that the Poisson ratio of cellular materials would converge to 0.333.[31,32] This erroneous conclusion – based on experimental evidence with limited data – has an analogue more than 150 years ago, when Wertheim claimed this value (0.333), based on his experiments with glass and brass, to be universal.[50] In this sense, one has to agree with Boccaccini's statement from 14 years ago, in that "the experimental trend of v with porosity still should be considered unpredictable".[12] Unpredictable, however, is also the porosity dependence of the elastic moduli, as long as the porosity is the only microstructural feature available. On the other hand, our previous work[34,38,51] has shown, that even qualitative information on the type of microstructure can render the porosity dependence of the tensile modulus to a large degree predictable (thermal conductivity as well[52,53]). The same is true for the Poisson ratio. That means, as soon as a microstructure can be uniquely classified, the porosity dependence of the Poisson ratio becomes predictable to a satisfactory degree. In particular, many materials prepared with pore-forming agents have a microstructure that makes the effective properties follow the Pabst-Gregorová exponential relation.[34,38] Materials with this type of microstructure are therefore hot candidates for negative Poisson ratios at reasonably low porosities (> 62 %).

4 SUMMARY AND CONCLUSIONS

Based on the above theory-based analysis, there are several positive statements that can be made. These are:

1. The conventional micromechanical bounds (Voigt-Reuss-Paul and Hashin-Shtrikman-Walpole) do not impose additional restrictions on the admissible range of values of the effective Poisson ratio $-1 < \nu < 0.5$.
2. Voigt materials have an effective Poisson ratio that is independent of porosity.
3. All aforementioned relations (except for the Ramakrishnan-Arunachalam relation, whose theoretical status is questionable) predict an effective Poisson ratio independent of porosity when the solid Poisson ratio is 0.2.
4. Porous Hashin-Shtrikman materials have an effective Poisson ratio that converges to the asymptotic limit value $\nu^* = (1 + 5\nu_0)/(9 + 5\nu_0)$ as the porosity approaches 100 %.
5. The converging trend is a general feature of all models mentioned in the low-porosity region (except for the Arnold-Boccaccini-Ondracek relation, which contains an artifact making the porosity dependence non-monotonic).
6. Only Archie-type power-law relations and Pabst-Gregorová-type exponential relations allow for negative Poisson ratios (auxetic behavior); in contrast to the Archie relation, which restricts auxetic behavior to extremely high porosities, the Pabst-Gregorová relation allows auxetic behavior for porosities as low as 62 %, in agreement with empirical findings.
7. All aforementioned predictive relations (note that the original Kováčik relation contains an adjustable fit parameter and is therefore per se not predictive) indicate that a solid Poisson ratio higher than 0.2 is necessary to achieve auxetic behavior in a porous material by appropriately tailoring the microstructure; thus, according to the findings of this paper, porous materials with $\nu_0 < 0.2$ cannot have negative Poisson ratios.

Acknowledgment: *This study was part of the project "Porous ceramics, ceramic composites and nanoceramics" (Grant No. IAA401250703), supported by the Grant Agency of the Academy of Sciences of the Czech Republic.*

References

1 S. D. Poisson, *Mémoires de l'Académie (Paris)*, 1829, **8**, 357.
2 D. J. Quesnel, D. S. Rimai and L. P. DeMejo, *Solid State Comm.*, 1993, **85**, 171.
3 A. E. H. Love, *A Treatise on the Mathematical Theory of Elasticity*, fourth edition, Dover Publications, New York, 1944.
4 R. Pampuch, *Ceramic Materials – An Introduction to Their Properties*, Elsevier, Amsterdam, 1976.
5 H. T. Smyth, *J. Am. Ceram. Soc.*, 1959, **42**, 276.
6 W. Pabst and E. Gregorová, *Ceramics-Silikáty*, 2003, **47**, 1.
7 R. F. Almgren, *J. Elasticity*, 1985, **15**, 427.
8 R. Lakes, *Science*, 1987, **235**, 1038.
9 W. Yang, Z.-M. Li, W. Shi, B.-H. Xie and M.-B. Yang, *J. Mater. Sci.*, 2004, **39**, 3269.
10 R. W. Zimmerman, *Mech. Mater.*, 1991, **12**, 17.
11 R. W. Zimmerman, *Appl. Mech. Rev.*, 1994, **47**, S38.

12 A. R. Boccaccini, *J. Am. Ceram. Soc.*, 1994, **77**, 2779.

13 M. Arnold, A. R. Boccaccini and G. Ondracek, *J. Mater. Sci.*, 1996, **31**, 1643.

14 A. P. Roberts and E. J. Garboczi, *J. Mech. Phys. Solids*, 2002, **50**, 33.

15 A. P. Roberts in *Cellular Ceramics – Structure, Manufacturing, Properties and Applications*, ed. M. Scheffler and P. Colombo, Wiley-VCH, Weinheim, 2005, p 267.

16 J. Kováčik, *J. Mater. Sci.*, 2006, **41**, 1247.

17 J. Kováčik, *Adv. Eng. Mater.*, 2008, **10**, 250.

18 S. Torquato, *Random Heterogeneous Materials – Microstructure and Macroscopic Properties*, Springer, New York, 2002.

19 W. Pabst and E. Gregorová, *Ceramics-Silikáty*, 2004, **48**, 14.

20 W. Voigt, *Wied. Ann. (Wiedemann's Annalen)*, 1889, **38**, 573.

21 A. Reuss, *Z. Angew. Math. Mech.*, 1929, **9**, 49.

22 B. Paul, *Trans. Metall. Soc. AIME*, 1960, **218**, 36.

23 Z. Hashin and S. Shtrikman , *J. Mech. Phys. Solids*, 1963, **11**, 127.

24 L. J. Walpole, *J. Mech. Phys. Solids*, 1966, **14**, 151.

25 C. P. Chen and R. S. Lakes, *J. Mater. Sci.*, 1991, **26**, 5397.

26 C. P. Chen and R. S. Lakes, *J. Mater. Sci.*, 1993, **28**, 4288.

27 C. W. Smith, J. N. Grima and K. E. Evans, *Acta Mater.*, 2000, **48**, 4349.

28 J. N. Grima, R. Gatt, N. Ravirala, A. Alderson and K. E. Evans, *Mater. Sci. Eng. A*, 2006, **423**, 214.

29 N. Ramakrishnan and V. S. Arunachalam, *J. Mater. Sci.*, 1990, **25**, 3930.

30 N. Ramakrishnan and V. S. Arunachalam, *J. Am. Ceram. Soc.*, 1993, **76**, 2745.

31 L. J. Gibson and M. F. Ashby, *Proc. R. Soc. Lond. A*, 1982, **382**, 43.

32 L. J. Gibson and M. F. Ashby, *Cellular Solids – Structure and Properties*, second edition, Cambridge University Press, Cambridge, UK, 1997.

33 T. Mori and K. Tanaka, *Acta Metall.*, 1973, **21**, 571.

34 W. Pabst and E. Gregorová, *J. Eur. Ceram. Soc.*, 2006, **26**, 1085.

35 L. F. Nielsen, *J. Am. Ceram. Soc.*, 1984, **67**, 93.

36 R. M. Spriggs, *J. Am. Ceram. Soc.*, 1961, **44**, 628.

37 A. Einstein, *Ann. Phys. (Annalen der Physik)*, 1906, **19**, 289.

38 W. Pabst and E. Gregorová, *Ceramika-Ceramics*, 2006, **97**, 71.

39 R. M. Christensen, *Proc. R. Soc. Lond. A*, 1993, **440**, 461.

40 R. M. Christensen, *Int. J. Solids Struct.*, 2000, **37**, 93.

41 A. N. Norris, *Mech. Mater.*, 1985, **4**, 1.

42 R. Hill, *J. Mech. Phys. Solids*, 1965, **13**, 213.

43 B. Budiansky, *J. Mech. Phys. Solids*, 1965, **13**, 223.

44 G. E. Archie, *Trans. AIME*, 1942, **146**, 54.

45 W. Pabst and E. Gregorová, *J. Mater. Sci. Lett.*, 2003, **22**, 1673.

46 W. Pabst and E. Gregorová, *J. Eur. Ceram. Soc.*, 2007, **27**, 479.

47 W. Pabst and E. Gregorová, *J. Mater. Sci.*, 2004, **39**, 3213.

48 K. K. Phani and S. K. Niyogi, *J. Mater. Sci.*, 1987, **22**, 257.

49 S. Spinner, *J. Am. Ceram. Soc.*, 1954, **37**, 229.

50 W. Wertheim, *Ann. chim. (Annales de chimie)*, 1848, **23**, 52.

51 W. Pabst and E. Gregorová in *Ceramics and Composite Materials – New Research*, ed. B. M. Caruta, Nova Science Publishers, New York, 2006, p 31.

52 W. Pabst and E. Gregorová in *New Developments in Materials Science Research*, ed. B. M. Caruta, Nova Science Publishers, New York, 2007, p 77.

53 W. Pabst and E. Gregorová in *Thermal Conductivity 29 – Thermal Expansion 17 (Proceedings)*, ed. J. R. Koenig and H. Ban, Destech Publications, Lancaster (Pennsylvania), 2008, p 487.

Subject Index

Note: Page numbers refer to the starting page of the article in which the concept occurs.